LIQUID SCINTILLATION COUNTING
and
ORGANIC SCINTILLATORS

Harley Ross
John E. Noakes
Jim D. Spaulding

LEWIS

Library of Congress Cataloging-in-Publication Data

Liquid scintillation counting and organic scintillators/ edited by Harley Ross, John E. Noakes, and Jim D. Spaulding.
    p. cm.
    "Proceedings of the International Conference on New Trends in Liquid Scintillation Counting and Organic Scintillators, 1989"—P.
    Includes bibliographical referencesand index.
    1. Liquid scintillation counting—Congresses. 2. Liquid scintillators—Congresses. 3. Organic scintillators—Congresses. I. Ross Harley. II. Noakes, John E. III Spaulding, Jim D. IV. International Conferenceon New Trends in Liquid Scintillation Counting and Organic Scintillators (19989: University of Georgia)

QC787.S34L57 1991   539.7′75—dc20   90-26005
ISBN 0-87371-246-3

LEWIS PUBLISHERS, INC.
121 South Main Street, Chelsea, Michigan 48118

PRINTED IN THE UNITED STATES OF AMERICA

Preface

It has now been over 30 years since the first conference dedicated exclusively to liquid scintillation counting was held at Northwestern University (Aug. 20–22, 1957). One might expect that over this time period, something as routine as radionuclide assay *via* the liquid scintillation process would have entered the realm of the prosaic or perhaps even stodgy. Fortunately, this is not the case. Exciting, new research initiatives in the areas of organic scintillators, electronic detection systems, and advanced information processing methodology have come together to create completely new measurement concepts that further expand liquid scintillation applications. Today, we are able to make measurements that were unthinkable just a few short years ago. It is probably fair to say that the driving force in this continuing technology development is the growing need for improved sensitivity, selectivity, and accuracy in nuclide assay. Primarily this arises from demands in biomedical and environmental assessment efforts. Only liquid scintillation counting appears to meet the wide range of application that is required in many of these projects.

The broad expanse of problems investigated with liquid scintillation techniques often complicates the reporting, making it difficult for workers in fields outside some specific area of interest. For example, a cursory search of the literature on liquid scintillation in 1988 showed that there were articles of general interest in at least 33 different technical journals! Even the most dedicated researcher would have difficulty keeping up with the latest developments. In response to this diversity of interest, the periodic liquid scintillation conferences, and their proceedings, have served the important role of bringing investigators with different concerns into a common forum. Such conferences have been held continuously every three to five years since 1957. The international conference on New Trends in Liquid Scintillation Counting and Organic Scintillators, held in October, 1989 and sponsored by the Oak Ridge National Laboratory and The University of Georgia, is a continuation of this essential tradition. Our hope is that both the conference and these proceedings have fulfilled the promise of our eminent predecessors.

No conference can take place, nor proceedings published, without the help of a great many people and organizations. Thus, the organizers would like to recognize Beckman Instruments, Inc., Packard Instrument Co., and Pharmacia LKB Nuclear, Inc. for the generous financial support that made the conference possible; B. M. Coursey, C. Ediss, A. Grau Malonda, S. Guan,

C-T. Peng, H. Polach, and F. Schoenhofer, our international Scientific Advisory Committee that served to unify the various research interests into a superb technical program; Carolyn Morrill (UGa) and Linda Plemmons (ORNL) for handling the mountains of paperwork and the massive logistics associated with the conference; Arnold Harrod our conference coordinator for his efforts before and during the conference; the staff of Lewis Publishers, and particularly Vivian Collier, who made the publication of these proceedings as painless as possible. Finally, we would like to acknowledge our conference participants and authors, without whom the above would have meant very little. Our sincerest thanks to all of you!

Harley Ross
John Noakes
Jim Spaulding

Harley H. Ross is a Senior Research Staff Member at the Oak Ridge National Laboratory (ORNL), Oak Ridge, Tennessee and also serves as Professor of Chemistry at the University of Tennessee, Knoxville. His formal training as an analytical-radiochemist has guided his research work into a broad range of theoretical and applied nuclear-related problems, primarily in the area of radionuclide characterization and measurement. His professional interests also include the development of enhanced instrumentation concepts for analytical assay, the role of computer simulation in radiochemistry, and the use of radioisotopes in making a variety of chemical measurements. He has published over 70 papers, has contributed chapters to several books, including two encyclopedias, and has worked as a consultant, both for other national laboratories and private industry. In 1967, he was awarded an IR-100 award, the first at ORNL, for his development of a radioisotope-excited light source spectroscopy system. Since his first publication in liquid scintillation counting almost 30 years ago, he has maintained his interest in the field and continues to explore and publish novel ideas for improved scintillation measurements.

John E. Noakes holds a dual position as Director of the Center for Applied Isotope Studies and Professor of Geology at the University of Georgia in Athens. His formal training has been in the fields of chemistry, geology, and oceanography, in which he has published over 80 scientific papers and holds 9 patents. His scientific interests are directed to applied and basic research with a strong emphasis in the use of stable and radioactive isotopes. He has over 30 years experience in liquid scintillation counting and has consulted for most of the major instrument manufacturers, lectured at the national and international levels, and assisted in the set-up of 28 benzene radiocarbon laboratories world-wide. His current interests in liquid scintillation counting are directed to instrument design and the application of low-level counting to environmental studies, hydrogeology, and radiocarbon dating.

Jim D. Spaulding is a member of the research staff at the Center for Applied Isotope Studies, University of Georgia, Athens, where he is Associate Director and manager of the environmental monitoring program. His formal training in physics and mathematics, and an academic year fellowship at the Special Training Division of the Oak Ridge Associated Universities, Oak Ridge, Tennessee, gave him the background for continued research in low background nuclear instrumentation. He has co-authored in excess of 30 scientific papers, several of which are on the design and use of low-level liquid scintillation counting, especially as applied to environmental monitoring. He has also served as a consultant to a major liquid scintillation counter manufacturer and as an Expert in Indonesia for the Department of Technical Assistance and Co-operation, International Atomic Energy Agency, Vienna, Austria.

In memory of Donald L. Horrocks —1929-1985

A Tribute to Donald L. Horrocks

C.T. Peng

It is fitting for Harley Ross, John Noakes, and Jim Spaulding, the organizers of the International Conference on New Trends in Liquid Scintillation Counting and Organic Scintillators, 1989 to dedicate this Conference to the memory of Donald Leonard Horrocks, who contributed so much to the advancement of liquid scintillation science and technology.

Donald L. Horrocks was born on July 14, 1929 in Dearborn, Michigan. He attended Reed College, Portland, Oregon and graduated in 1951. He continued his study at Iowa State College, Ames, Iowa and graduated in 1955 with a Ph.D. in Physical Chemistry. From there, after two brief research positions, he went to work in Argonne National Laboratory as a research scientist for 15 years. He moved to Beckman Instruments, Inc. in 1971. He passed away on May 22, 1985. He is survived by his wife Margaret and two daughters, Andrea and Cynthia.

Don began his scientific career in Nuclear Chemistry, measuring the tritium yield in the fission of ^{242}Cf and ^{233}U and in fissile elements proposed for future reactor use. He measured tritium, alpha emitters, and fission fragments by liquid scintillation counting, using single phototube instruments and pulse height analysis. He recognized the importance of pulse height analysis and energy resolution in radiation measurement by liquid scintillation techniques, and he was the first to report the use of the relative pulse height of Compton edges in the presence of sample beta continuum for quench correction in liquid scintillation systems.[1]

This was an important observation and was the basis for the "H" (or "Horrocks") number concept for quench correction, which was introduced at a much later date. His interest in improving detection efficiency led him to the study of scintillation processes, especially the pulse height energy relationship of liquid scintillators for alpha particles, fission fragments, and electrons of less than 100 keV in energy. He also studied the effect internal conversion and intersystem crossing of organic scintillator molecules had on their fluorescence and scintillation efficiency. He studied the self- and concentration-quenching of scintillation solutes as a result of molecular coplanarity; he studied Hbonding, proton transfer and excimer emissions, the aromatic solvent transfer efficiencies for different excitation energies, and potential organic scintillator synthesis. He demonstrated the superiority of 1,2,4-trimethylbenzene over p-xylene as a scintillation solvent; it not only increased the transfer efficiency but also decreases solute excimer emission. He also reported the use of pulse-shape discrimination as a particle counting method with high specific ionization. He reported the use of liquid scintillation counting to measure the disintegration rate of beta-emitting nuclides and the use of $^{113}Sn-^{113m}In$ for energy calibration. He measured radioactive noble gases by liquid scintillation counting. His research in organic scintillators and the esteem that his peers had for him lead him to organize and convene a successful international conference on organic scintillators at Argonne, Illinois, in 1966.[2] Its success also afforded the impetus for him to co-organize two additional conferences on liquid scintillation counting held in 1970 and 1979 in San Francisco.[3,4]

Don moved from ANL to Beckman Industries, Inc. shortly after the first San Francisco conference. This change of employment did not interrupt his research career. In fact, it provided him with an opportunity to put many of his theories and observations on liquid scintillation efficiency and quenching into practical use. He completed his book, *Applications of Liquid Scintillation Counting*, which was comprehensive and thorough in both theory and practical applications and is a classic of the scientific literature.[5]

Don introduced the X-ray-X-ray coincidence method for measuring the disintegration rate of ^{125}I by gamma-ray scintillation spectrometry.[6] He introduced the concept of the H number for quench correction in liquid scintillation systems. The H number is a measure of the shift of the Compton edge between the spectra of quenched and unquenched samples.[7] Both of these inventions allowed new counters based on these principles to be designed, manufactured, and successfully marketed.

Don had over 100 scientific publications and held 20 patents related to liquid scintillation science and technology. Many of his contributions, such as quench measure by the H number method, detection of sample-scintillator phase separation by the two phase method, determination of the radionuclide disintegration rate by the extrapolation-H number method, determination of a complete quench curve from a single sample by optical absorption method, development of micro volume counting methods, understanding of system background sources and pulse-height distribution, plastic shift, and detection of measurement of chemiluminescence in samples, are the mainstay of modern liquid scintillation instrumentation.

Don was a quiet, thorough, and persuasive man. He approached problems systemically and logically. He was a trusted source for authoritative and reliable information on every aspect of liquid scintillation science and theory. He will be long remembered.

REFERENCES

1. Horrocks, D.L. "Measurement of Sample Quenching of Liquid Scintillator Solutions with X-ray and Gamma Ray Sources," *Nature* 202(4927):78–9 (1964).
2. D.L. Horrocks, Ed. *Organic Scintillators*, New York: (Gordon and Breach, 1968), p. 413.
3. D.L. Horrocks and C.T. Peng, Eds. *Organic Scintillators and Liquid Scintillation Counting, Proc. Intl. Conf.*, (New York: Academic Press, 1971), p. 1078.
4. C.T. Peng, D.L. Horrocks, and E.L. Alpen, Eds. *Liquid Scintillation Counting. Recent Application and Development. Vol. I and II*, New York: Academic Press, 1980), p.414 and 538.
5. Horrocks, D.L. *Applications of Liquid Scintillation Counting*, New York: Academic Press, 1974). p. 346.
6. Horrocks, D.L. "Measurements of Disintegration Rates of ^{125}I sources with a Single Well-Type Thallium Activated Sodium Iodide Detector," *Nucl. Instrum. Methods* 125(1):105–11 (1975).
7. Horrocks, D.L. "A New Method of Quench Monitoring in Liquid Scintillation Counting: The H Number Concept," *J. Radioanal. Chem.* 43(2):489–521 (1978).

Contents

CHAPTER 1

PMP, a Novel Scintillation Solute with a Large Stokes' Shift

Hans Güsten and Jeffrey Mirsky

ABSTRACT

The excellent fluorescence properties of PMP (1-phenyl-e-mesityl-2-pyrazoline) such as a long wavelength emission of 425 nm, a high fluorescence quantum yield, and a short fluorescence decay time make this compound a promising solute for scintillation counting. Due to a very large Stokes' shift, which is more than twice as large as that of commonly used organic scintillators, PMP is suited as a primary solute requiring no secondary solute as wavelength shifter. The exceptionally large Stokes' shift results in a small overlap between the absorption and emission spectra of the solute, even at high concentrations; hence a much longer light attenuation path length compared to common scintillators. The structural features of the new solute also give rise to a unique self-quenching behavior of PMP. Unlike other scintillators, a shallow maximum of light output vs solute concentration curve is exhibited between 0.01 and 1 M in common solvents or solvent mixtures used in large-volume scintillation chambers. As to other scintillation characteristics such as the scintillation efficiency for ^{14}C and ^{3}H, the performance of PMP is similar to that of good commercial liquid or plastic scintillators. PMP is now used successfully in an 80,000 L large-volume scintillation detector applied in neutrino physics. The unique spectroscopic scintillation properties of PMP also suggest its use in plastic scintillating fibers.

INTRODUCTION

We have shown recently that the class of 1,3-diphenyl-2-pyrazolines (DPs) exhibit exceptionally large Stokes' shifts when properly substituted by bulky substituents in the ortho' positions of the phenyl rings.[1] These sterically hindered DPs have promising potential as novel primary solutes for liquid scintillation counting.[2,3] Among the sterically hindered DPs tested for liquid scintillation counting, 1-phenyl-3-mesityl-2-pyrazoline, henceforth called PMP,[4] is the best compromise in terms of good photophysical and scintillation properties and ease of synthesis. In this chapter the results obtained on the application of PMP in scintillation counting are summarized.

Table 1. Photophysical Properties of PPO, DP, PMP, and POPOP in Benzene at Room Temperature

Scintillator	S^a(g/l)	Q^b_F	$\tau(\mu s)$	λ_A(nm)	λ_F(nm)	ΔSt(cm^{-1})
PPO	414	0.79	1.3	308	365	5,240
DP	49	0.92	3.2	362	444	4,980
PMP	335	0.88	3.0	295	425	10,370
POPOP	$\simeq$1	0.85	1.5	362	418	3,540

[a]In toluene at 20°C.
[b]In degassed solution.

EXPERIMENTAL

A detailed description has ben given[5,6] of the techniques used to measure the photophysical data such as the absolute fluorescence spectra, the fluorescence quantum yields, and fluorescence decay times, as well as the scintillation characteristic.[4] The method used to evaluate and optimize the attenuation length of various scintillators consists in measuring the intensity of light collected by a photomultiplier placed at one end of a 2 m long quartz rod when this is orthogonally crossed by the ionizing radiation of a ^{60}Co or ^{137}Cs source at various distances from the photomultiplier. The light output of the different scintillators has been measured relative to NE 235 H using the techniques described by Kowalski et al.[7] With ^{60}Co, the 70% value of the Compton maximum was used.

RESULTS AND DISCUSSION

The main photophysical properties of PMP, DP, the parent compound of PMP, and the two widely used organic scintillators, PPO and POPOP, are compiled in Table 1. PMP combines very good photophysical properties such as a high fluorescence quantum yield Q_F, a short decay time (τ), and the long wavelength emission of the fluorescence maximum (λ_F), with a high solubility (S). The most striking difference of PMP in comparison to DP and the common scintillators PPO and POPOP is the exceptionally large Stokes' shift (ΔSt) of more than 10,000 cm^{-1}. Since there is no fine structure in the absorption and fluorescence spectra of these large organic molecules at room temperature, the 0–0 transition is difficult to determine, and the energy different of the absorption and emission maxima is therefore taken as a measure of the Stokes' shift. A comparison of the parent compound DP with PMP, where three methyl groups are substituted in the 2,4,6-position of the phenyl ring, reveals that the large Stokes' shift is introduced only by the sterical hindrance of the bulky methyl groups in 2- and 6-positions of the phenyl ring. The result of this sterical hindrance is that in a large hypochromic shift the absorption is shifted from 362 to 295 nm, while the fluorescence spectrum of PMP in comparison to DP remains nearly unaffected energetically.[1,2] Thus, the Stokes' shift of PMP is more than twice as large as that of commonly used organic scintillators. A comparison of the absorption and fluorescence spectra of PMP and POPOP is shown in Figure 1.

The advantage of the large Stokes' shift and the broad absorption spectrum of PMP is that PMP can be used as a single wavelength shifter in scintillation counting. Thus, contrary to the classical combinations of primary and secondary solutes used in scintillation counting for more than 35 years, with the use of PMP, secondary solutes are no longer necessary. In several standard tests in ^{14}C and ^{3}H scintillation counting, the results obtained with PMP and with common organic scintillators were intercompared.[4] The results will not be repeated here. In general, PMP resembles good commercial liquid or plastic scintillators in its scintillation characteristics. Better results are obtained in samples exploiting the large Stokes' shift of PMP, as in the case of samples where color quenching reduces the counting efficiency.[4]

Large-Volume Scintillation

Since all common organic scintillators show some overlap of their absorption and emission spectra (see Figure 1), self-transfer can occur at high concentrations by reabsorption of light emitted from the scintillator molecule. The importance of reducing the reabsorption effects in scintillators by employing solutes with large Stokes' shifts was stressed recently by Renschler and Harrah.[8] Due to the large Stokes' shift of PMP, the self-absorption of its own scintillation light is lower than in conventional scintillators. This is of great importance in large-volume applications such as whole-body counters, neutrino detectors, as well as scintillating fibers of tracking detectors for the new generation of particle accelerators. Since the intensity of the transmitted photons is related to the solute concentration and the length the emitted photons have to travel to reach the photomultiplier, the light attenuation length should be greater with solutes having a larger Stokes' shift. Otherwise, each time reabsorption occurs some quanta of energy are lost due to the deactivation processes inherently associated with the fate of the electronically excited singlet state of the solute molecule. A comparative study of five commercially available large-volume scintillators revealed that the attenuation length of scintillators with PMP is greater by 23–48% while the light output is in the range of 57–62% of an anthracene crystal (= 100%). These light output values are obtained with all scintillators tested. A solvent combination called PPP 2-25 with PMP as the only solute proved to be the best choice for large-volume application. This scintillator had an attenuation length of 320 cm with 57% light output. This scintillator is now operating successfully in the 80,000 L KARMEN detector (Karlsruhe–Rutherford Medium Energy Neutrino experiment) at the Rutherford–Appleton Laboratory.[9] An even greater attenuation length of nearly 4 m has been obtained with PMP in a new solvent combination called PMP ND 380.

There is another benefit when PMP is used as a large-volume scintillator. Unlike the common organic scintillators, PMP shows little tendency to self-

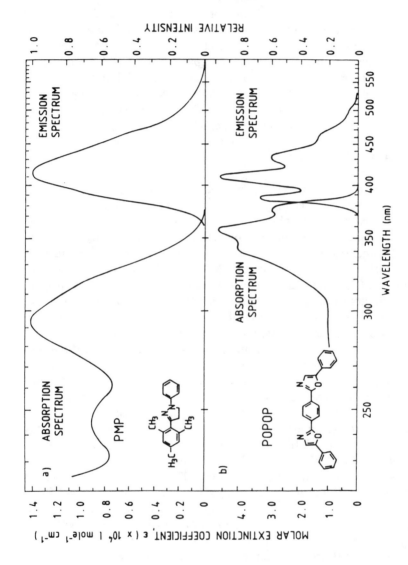

Figure 1.   Absorption and flourescence spectra of PMP and POPOP in cyclohexane at room temperature.

quenching. Most scintillators exhibit a narrow maximum in their light output vs. solute concentration curves. This phenomenon is partly due to the reabsorption processes described, partly to concentration quenching of the fluorescent state of the fluorophore by its own molecule in the ground state.

In Figure 2 the relative pulse heights of PMP and PPO, in a ^{14}C standard toluene solution at 10°C, is plotted vs the solute concentration. While PPO shows a fairly narrow maximum at 0.025 M, PMP displays a shallow maximum between 0.01 and 0.1 M. Even at a concentration of 1 M PMP (264 g/l) the relative pulse height has dropped only by about 10%. Consequently, by use of PMP the doping concentration can be raised to values at which only nonradiative energy transfer is effective. We explain this low tendency to self-quenching of PMP by the different geometries of PMP in the ground and in the electronically excited singlet state.

Scintillating Fibers

The new generation of tracking detectors use parallel bundles of scintillating optical fibers of extremely small diameters.[10] Theoretical reasoning about plastic-based scintillating fibers has pointed to the need for improving the attenuation length of common scintillators.[11] It was shown recently that PMP in polyvinyl toluene (PVT) or polystyrene reach nearly the scintillation efficiency of NE 110 while the transparency of PMP is better by at least one order of magnitude in the concentration range of 0.025 to 0.050 M over the entire wavelength range of its emission spectrum from 400–450 nm.[10,12] Since the

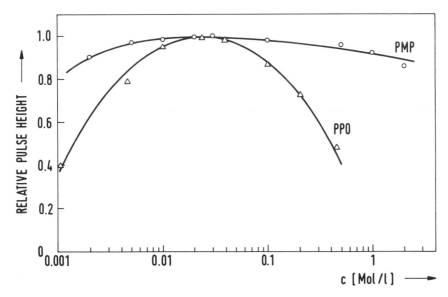

Figure 2.   Dependence on concentration of the relative pulse height of PMP and PPO in a ^{14}C standard toluene solution at 10°.

PVT emission is effectively absorbed at a distance of < 10 μm, plastic scintillators that contain PMP can now be used for the production of efficient small diameter plastic scintillating fibers.[10,12] Furthermore, PMP in polysiloxane showed good radiation resistance in an argon atmosphere.[13] This is another important feature because a radiation hardness of 10^6 rads is required to ensure a reasonable operational lifetime of the scintillating fiber.

ACKNOWLEDGEMENTS

We thank Drs. K. F. Schmidt and R. Maschuw of the Institute of Nuclear Physics of the Kernforschungszentrum Karlsruhe for the measurements of the attenuation lengths.

REFERENCES

1. Strähle, H., W. Seitz, and H. Güsten. "Struktur- und Lösungsmittelabhängigkeit der Fluoreszenz von 1,3-Diphenyl-2-pyrazolinen," *Ber. Bunsenges. Phys. Chem.* 80: 228–294 (1976).

2. Güsten, H., P. Schuster, and W. Seitz. "Organic Scintillators with Unusually Large Stokes' Shifts," *J. Phys. Chem.* 82: 459–463 (1978).

3. Güsten, H. and W. Seitz, "Novel Primary Solutes for Liquid Scintillation Counting," in *Liquid Scintillation Counting: Recent Applications and Development, Vol. 1,* C. T. Peng, D. L. Horrocks, and E. L. Alpen, Eds. (New York, Academic Press, 1980), pp. 51–57.

4. Güsten, H. "PMP, A Novel Solute for Liquid and Plastic Scintillation Counting," in *Advances in Scintillation Counting,* S. A. McQuarrie, C. Ediss and L. I. Wiebe, Eds. (Edmonton, Canada: University of Alberta, 1983), pp. 330–339.

5. Rinke, M., H. Güsten and H.J. Ache. "Photophysical Properties and Laser Performance of Photostable UV Laser Dyes. 1. Substituted p-Quaterphenyls," *J. Phys. Chem.* 90: 2661–2665 (1986).

6. Blume, H. and H. Güsten. In *Ultraviolette Strahlen,* J. Kiefer, Ed., (Berlin: W. de Verlag, 1977), Chapter 6.

7. Kowalski, E., R. Anliker and K. Schmid. "Criteria for the Selection of Solutes in Liquid Scintillation Counting. New Efficient Solutes with High Solubility," *Int. J. Appl. Radiat. Isotopes* 18: 307–323 (1967).

8. Renschler, C.L., and L.A. Harrah. "Reduction of Reabsorption Effects in Scintillators by Employing Solutes with Large Stokes' Shifts," *Nucl. Instr. Meth. Phys. Res.* A 235: 41–45 (1985).

9. Drexlin, G., H. Gemmeke, G. Giorginis, W. Grandegger, J. Kleinfeller, R. Maschuw, P. Plischke, F. Raupp, F.K. Schmidt, J. Wochele, B. Zeitnitz, E. Finckh, W. Kretschmer, D. Vötisch, J.A. Edgington, T. Corringe, N.E. Booth, and A. Dodd. "KARMEN: Neutrino Physics at ISIS," in *Neutrino Physics, Proc. Intern. Workshop, Heidelberg, Oct. 20–22, 1987,* H. V. Klapdor and B. Povh, Eds. (Berlin-Heidelberg: Springer Verlag, 1988), pp. 147–152.

10. D'Ambrosio, C., C. DaVia, J.P. Fabre, T. Gys, J. Kirby, H. Leutz, M. Primout, P. Destruel, M. Taufer, L. Van Hamme, and G. Wilguet, "Supercollider SCIFI Trackers," *Proceedings of the Workshop on Scintillating Fiber Detector Development*

for the SSC, in press. Fermi National Accelerator Laboratory, Batavia, IL, (1988), pp. 14–16.

11. Leutz, H., private communication.

12. Destruel, P., M. Taufer, C. D'Ambrosio, C. DaVia, J. P. Fabre, J. Kirby, and H. Leutz, "A New Plastic Scintillator with Large Stokes' Shift," *Nucl. Inst. Meth. in Phys. Res.* A 276: 69–77 (1989).

13. Bowen, M., S. Majewski, D. Pettey, J. Walker, R. Wojcik, and C. Zorn, "A New Radiation-Hard Plastic Scintillator," *Nucl. Inst. Meth. in Phys. Res.* A 276:391–393 (1989).

CHAPTER 2

Scintillation Counting of Harvested Biological Samples with Low-Energy Beta Emitters, Using Solid Scintillant Filters

C.G. Potter and G.T. Warner

INTRODUCTION

Liquid scintillation counting (LSC) of low–energy beta-emitting isotopes has an important role in the quantitation of biological samples. Although the level of radioactivity is low, the use of liquid scintillant gives rise to disposal problems because of the volume of organic solvents involved.

A significant proportion of LSC work employs samples deposited on filter discs to measure the uptake of labeled compounds into cells and the binding of ligands to receptors. Placing many samples into vials for counting by LSC is a tedious procedure open to identification errors.

Recently methods have been developed whereby a complete filter, bearing many samples, is placed in a thin plastic bag with only a small volume of scintillant (typically 10 mL/96-sample filter) and counted using a flat-bed scintillation counter.[1,2] This counter has a low background count rate and good counting efficiency, and the commercial version has a high sample capacity and rate of throughput. This makes it particularly useful for lymphocyte assays using ^{3}H-thymidine uptake to monitor proliferation, or to measure ^{51}Cr release in cytotoxic assays.[3] The very small volume of scintillant required makes sample disposal easy as well as economical.[4]

Despite these advantageous features, there are other areas of potential application where the flat-bed counter is at some disadvantage. In particular, where samples are soluble in scintillant, diffusion may result in the movement of activity away from the sample area. Movement may also occur if the sample is not firmly bound to the filter and material is released by the addition and spreading of the scintillant while preparing the filter for counting. Another disadvantage not seen in a vial-counting liquid scintillation counter, is that of inter-sample interference, whereby activity in one sample is detected in an adjacent sample area. The degree of this "cross talk" can be reduced to less than 0.01% for low energy isotopes by printing black lines on the filters. These absorb light emitted laterally as well as help identify and index during harvesting and placement of the filter on the plate.[1,2] A further reduction in cross talk, however, is desired when using thick filters, higher energy isotopes, and the most stringent requirements.

The possibility that filters could be composed of scintillant has been previously described.[5] This paper presents preliminary data on a technique using special filters made of fibers coated with a thin layer of solid scintillant. Radioactive samples are deposited on the filter, dried in the usual way and counted without addition of a liquid scintillant. The beta emission excites the solid scintillant and the events are counted using the flat-bed scintillation counter. Cross talk and background are reduced using this method and the samples have little tendency to migrate. Disposal of samples is further simplified because there is no liquid component. The properties of this type of filter are characterized for different solid scintillants as well as describe a heating method to increase the rather low tritium counting efficiency to a level useable for many applications.

MATERIALS AND METHODS

Preparation of Filters

Conventional harvesting uses glass fiber depth filters of a mean porosity suitable for the samples required, e.g., Whatman GF/A or GF/C for cells, the thicker GF/B for membrane preparations, and GF/F for microorganisms. For the flat-bed scintillation counter (Betaplate, Pharmacia-Wallac OY, Finland), special reinforced versions of these filters were available, printed to facilitate orientation of the filter and reduce cross talk. As well as coating these filters with solid scintillant, the counter uses polypropylene depth filters in a variety of thicknesses and porosities. The solid scintillant was prepared from PPO (2,5-diphenyloxazole) dissolved in toluene (Analar grade) at concentrations up to 2.75%, together with a spectral shifter (Bis-MSB, [1,4-di(2-methylsteryl)-benzine]), added at 10% w/w of the PPO. Tetramethylbenzine (durene, from Aldrich, U.K.), which has a low melting point (55°C), was also added at levels of up to 10% w/v of toluene. The filter sheets were immersed in the scintillant solution, drained and hung up to dry in a fume hood having a fast flow of air. After an hour the filters were completely dry and virtually odorless. This method ensured that a thin coating of scintillant covered each fiber of the filter. When durene was used, it formed a solid solvent as a vehicle for the scintillant proper. Even for the finest porosity filters, there was little reduction in filtration capability with the addition of solid scintillant so that suitably cut sheets could be used in a cell harvester or filtration manifold. Similar filter sheets were prepared for tests using polyvinyl toluene/PPO, butyl-PBD, or anthracene in place of PPO as the scintillant.

Cell Harvesting

The cells used for harvesting tests were K562 cells grown in RPMI and 12.5% fetal calf serum. They were labeled overnight in the flask with 5μCi/mL ^{3}H-thymidine or 0.1μCi/mL ^{35}S-methionine. Microtitration plates were plated with 0.2 mL aliquots at cell densities of $1-5 \times 10^5$ cells/mL and the aliquots filtered using a cell harvester (Skaatron, Norway) with a 10-sec water wash, followed by a flow of air for 5–10 sec. The filters were generally dried for 2 or 3 hours in an incubator at 37°, or on a hot plate, or in a microwave oven. A methanol wash could not be used as it washed away some of the solid scintillant.

Heat Treatment and Counting Preparation

After drying, filters were placed in thin plastic bags having a low–melting-point inner layer, laminated to an outer heat resistant plastic. Control filters had liquid scintillant added (Betaplatescint, Pharmacia Wallac OY, Finland) and the bag heat-sealed in the usual way. Samples with solid scintillant were counted directly or after heating to melt the scintillant, which could thus permeate the radioactive samples and be better coupled to the electron emission.[6] Uniformity of heating was achieved in two ways, one by use of a microwave oven where the dielectric losses in the filter and solid scintillant produced heat enough to melt the scintillant after a few minutes exposure. The second method was to pass the dried filter in its bag through the heated rollers of an office laminator (Coated Specialties Ltd., Basildon, U.K.). Thus the filter became compacted and the bag sealed, producing a thin, rugged set of samples ready for counting. The temperature required should not be high enough to produce shrinkage of the composite during lamination, otherwise registration of the samples with the counter support plate will be lost. Too low a temperature will not melt the scintillant. For polypropylene filters the temperature may be made high enough to melt the plastic also producing a transparent set of samples. These may have a pleasing, solid appearance, but the melted plastic can move while under the rollers giving unreliable results.

RESULTS

Glass Fiber Filters

Cells were harvested onto glass fiber filters (Pharmacia-Wallac) with and without solid scintillant. In the first experiment, the amount of PPO in toluene for the coating solution varied between 5.5 and 27.5 g/L also with or without 5% durene. Table 1 shows that the highest amount of PPO provided the highest efficiency, which was enhanced by heating and passing the filter and bag through heated rollers. Addition of durene generally increased efficiency a

Table 1. Relative Counting Efficiencies of Solid Scintillant, with and without Durene, for Different Amounts of PPO + BisMSB (9:1 w/w), Using Samples (n = 12) of ^{3}H-Labeled Cells Filtered onto Scintillant Coated Glass Fiber Sheets

%PPO + BisMSB[a]	Heating[b]	± Durene	Rel. Eff[c] (% ± SD)
0.55	−	−	5.1 ± 0.4
	−	+	4.3 ± 0.5
	+	−	9.4 ± 1.5
	+	+	12.3 ± 1.3
1.10	−	−	5.6 ± 0.7
	−	+	7.2 ± 0.5
	+	−	12.7 ± 2.4
	+	+	17.9 ± 1.2
1.65	−	−	5.9 ± 0.9
	−	+	6.7 ± 0.6
	+	−	18.9 ± 1.5
	+	+	23.1 ± 1.9
2.20	−	−	7.0 ± 0.9
	−	+	8.5 ± 0.6
	+	−	20.5 ± 2.0
	+	+	26.6 ± 2.0
2.75	−	−	10.3 ± 1.7
	−	+	10.7 ± 0.7
	+	−	26.6 ± 3.5
	+	+	26.1 ± 1.8

[a]PPO + BisMSB dissolved in toluene, with or without 5% durene and used to soak filters dried and used in a filtration assay of ^{3}H-thymidine-labeled K562 cells.
[b]Heating by the heated rollers of office laminator.
[c]Relative efficiency, compared with parallel samples counted in the flat-bed scintillation counter using BetaplateScint cocktail.

little and could partly replace PPO. At the higher quantities of PPO or with durene present, the filters became stiffer and filtration became a little impaired. In another experiment (Table 2), butyl-PBD showed a similar efficiency to PPO before heating, but it was reduced after passing the filters through the heated rollers. Addition of durene did not improve efficiency either before or after heating. Table 3 shows the results of an experiment comparing 2% PVT/toluene with 2.75% of either butyl-PBD or PPO, both with and without durene, (increased to 10%) in toluene. Again there was no difference or reduction in counting efficiency for butyl-PBD and PVT, even when using microwave heating, which would not produce the compaction made by heated rollers. Anthracene with or without durene was also tried and heating efficiency improved slightly. One noteworthy aspect was that the energy spectrum for anthracene was shifted to higher levels compared with the others scintillants tested.

^{35}S-labeled Samples

For cells labeled with ^{35}S-methionine and harvested onto plain or printed glass fiber filters with PPO/BisMSB/durene solid scintillant, the efficiency

Table 2. Relative Counting Efficiencies of Solid Scintillant, with and without Durene and for Different Amounts of Butyl-PBD, Using Samples (n = 12) of ^{3}H-Labeled Cells Filtered onto Scintillant-Coated Glass Fiber Sheets, with and without Heating

	Relative Counting Efficiency of BetaplateScint (% ± SD)		
% of Butyl-PBD[a]	Heating[b]	– Durene	+ Durene
0.5	–	3.8 ± 1.3	5.4 ± 1.8
	+	3.4 ± 1.4	4.0 ± 1.2
1.0	–	8.5 ± 2.6	7.8 ± 2.9
	+	6.9 ± 2.1	7.0 ± 2.3
1.5	–	8.1 ± 2.4	9.0 ± 2.5
	+	6.7 ± 2.2	8.1 ± 2.2
2.0	–	13.0 ± 3.5	7.2 ± 1.8
	+	10.7 ± 3.0	7.0 ± 2.0
2.5	–	10.8 ± 2.6	8.3 ± 2.0
	+	8.0 ± 1.8	7.3 ± 1.9

[a]Butyl-PBD dissolved in toluene, with or without 5% durene, and used to soak glass fiber filter, which was dried and used in a filtration assay of ^{3}H-thymidine–labeled cells.
[b]Heating by the heated rollers of an office laminator.

Table 3. Relative Counting Efficiencies of Solid Scintillants with Different Compositions, Coated on Glass Fiber Filters and Used to Harvest ^{3}H-Labeled Cells

Fluor	Heating[a]	± Durene[b]	Rel. Eff.[c] (% ± SD)	N
Liquid scintillant control	None	–	100.0 ± 8.8	20
Polyvinyltoluene plastic scintillant 2% in toluene	None	–	8.4 ± 0.5	20
	μ	–	8.5 ± 0.6	
	μ + roll	–	4.3 ± 0.5	
Butyl-PBC 2.5% in toluene	None	–	8.9 ± 0.7	10
	μ	–	8.9 ± 0.4	
	μ + roll	–	7.6 ± 0.4	
	None	+	9.0 ± 0.4	
	μ	+	9.1 ± 0.7	
	μ + roll	+	7.3 ± 0.6	
PPO + 10% w/w BisMSB; 2.75% in toluene	None	–	7.7 ± 0.7	20
	μ	–	10.8 ± 1.4	
	μ + roll	–	19.2 ± 1.7	
	None	+	7.5 ± 1.3	
	μ	+	10.3 ± 0.8	
	μ + roll	+	22.7 ± 2.5	

[a]μ = microwave oven heating (400 W for 120 sec); roll = microwaved samples passed through office laminator (>90°C).
[b]Durene as 10% of the solution used to make the solid scintillant filters.
[c]Relative efficiency compared with parallel samples counted in the flat-bed scintillation counter using BetaplateScint cocktail.

Table 4. Relative Counting Efficiencies, Cross Talk and Background Count Rates for Glass Fiber Filters Coated with Solid Scintillant (PPO/BisMSB/durene) and Used to Harvest ^{35}S-methionine-Labeled Cells, with and without Subsequent Heating

Treatment	Plain Glass Fiber				Printed Glass Fiber			
	bg	cpm	CV%	xt%	bg	cpm	CV%	xt%[a]
Solid scintillant set 1 (n = 12)	4.3	42181	9.2	0.024 (10.0 cpm)	3.0	50065	6.8	<0.001 (0.3 cpm)
Set 1 + liquid scintillant	11.1	128590	5.0	0.020 (25.7 cpm)	14.8	126234	4.5	0.011 (14.3 cpm)
Relative efficiency			32.8				39.7	
Solid scintillant set 2 (n = 12)	3.9	43746	8.8	0.007 (3.0 cpm)	2.2	53463	4.8	<0.001 (0.4 cpm)
Set 2 + heating by laminator	5.1	53160	8.0	0.002 (1.1 cpm)	2.6	68628	4.9	0.002 (1.6 cpm)
Relative efficiency			40.0				50.1	
Increase in relative efficiency with heat			22				28	

[a]Cross talk—mean of 12 adjacent blanks less backgrounds.

was only slightly increased by heating. Table 4 shows that the efficiency reached up to 50% of glass fiber using liquid scintillant. Background count rates were also low, but in addition, these filters showed good inter-sample cross-talk characteristics even without printing, although with the addition of printed lines the cross talk fell to very low levels which are not easily measured, i.e., <0.001%. Similar experiments with ^{3}H-labeled cells gave no measurable cross talk, with or without printing.

Polypropylene Filters

Samples of several types of polypropylene depth filters were obtained from three suppliers. Each was tested by harvesting cells labeled with ^{3}H-thymidine onto the polypropylene filters with solid scintillant and comparing them with controls harvested on glass fiber or polypropylene filters with liquid scintillant. Any reduction in the effectiveness of filtration, together with any self-absorption or quenching, should produce a reduced count rate. As before, filters bearing solid scintillant were counted directly or after heating by microwave or passing through heated rollers. Some filters were found to have too coarse a mesh to filter out the majority of cells and others were too flimsy to handle satisfactorily. The properties of some suitable filters are illustrated by data on three examples, designated A, B, and C. Table 5 shows that the count rate for the polypropylene filters with liquid scintillant was slightly reduced compared to glass fiber. Counting solid scintillant filters directly gave relative efficiencies between 7.4 and 17.9%, which was increased by microwave radiation. A further increase was produced by hot rolling, to a maximum of 33.5% for sample B. At temperatures high enough to melt the polypropylene, counting efficiency was reduced and variation increased. Background count rates

Table 5. Relative Efficiencies (% ± CV) for ^{3}H-Labeled Cells Filtered onto Solid Scintillant Polypropylene Filters

Treatment	Type of Filter (Porosity)		
	A (2.5μ)	B (10μ)	C (?)
Liquid scintillant[a]	91.8 ± 5.5	78.5 ± 6.8	89.7 ± 7.5
Solid scintillant[b]	13.3 ± 8.1	17.9 ± 4.9	7.4 ± 10.6
+ microwave	16.6 ± 8.4	24.8 ± 15.7	9.0 ± 12.8
+ microwave and heated rollers	23.6 ± 7.8	33.5 ± 8.4	21.3 ± 11.5

[a]BetaplateScint, CV for glass fiber was 5.9%.
[b]2.75% PPO/BisMSB, 10% durene/toluene.

were very low unless the samples were melted, which probably increased optical cross talk between the pairs of photomultipliers.

Linearity with Cell Density

Using glass fiber filters and liquid scintillant, linearity of count rate with cell density is possible over a wide range.[3] In order to test whether this was also true for solid scintillant filters, microtration plates were set up with 6 replicate wells at cell densities ranging from approximately 10^3 to 4×10^5 cells/well and harvested on glass fiber filters with or without solid scintillant. Table 6 shows that linearity was maintained over this range, before and after passing the samples through heated rollers.

Uniformity of Samples

A complete microtitre plate was set up with approximately 2.5×10^5 cells/well and harvested using a glass fiber filter, coated with solid scintillant, giving a matrix of 6×16 samples. Twelve more samples in another plate were also harvested using a standard glass fiber filter with liquid scintillant and gave a mean of 107,800 cpm with a 2.2% coefficient of variation (CV). The solid scintillant filter was counted directly (mean of 10,995 cpm) and after passing through heated rollers (mean of 18,203 cpm). Table 7 shows that the normalized means for each row exhibit a significant positive regression, either when

Table 6. Linearity of Relative Counting Efficiency for Different Cell Densities Harvested onto Glass Fiber Filters Coated with Solid Scintillant.[a] Expressed as Solid Scintillant Filter Count Rate/Liquid Scintillant Control Count Rate (% ± SD)

Cell Density × 10^3/Well	Before Heating	After Heating
380	6.49 ± 0.70	11.19 ± 1.15
114	6.14 ± 0.83	10.97 ± 2.03
38	6.73 ± 1.46	11.82 ± 2.62
11.4	6.18 ± 0.66	10.31 ± 1.71
3.8	5.79 ± 1.15	9.47 ± 2.16
1.1	6.98 ± 1.65	12.25 ± 3.00

[a]2.75% PPO/BisMSB/5% durene in toluene.

Table 7. Regression for Normalized Data (y = a + bx) of Means of Rows (n = 16, 6 in each row) for Cells Filtered onto Glass Fiber, and for Glass Fiber or Polypropylene Filters Coated with Solid Scintillant Spiked with ^{3}H-hexadecane, Before and After Heating.

Type of Filter	± Heating	a ± SD		b ± SD		t	p	r
Glass fiber	−	90.5	± 2.6	1.12	± 0.27	4.1	<0.01	0.74
+ cells	+	84.8	± 2.2	1.78	± 0.23	7.8	<0.001	0.90
Glass fiber +	−	94.4	± 1.1	0.72	± 0.12	6.1	<0.001	0.85
^{3}H-hexadecane	+	100.4	± 1.1	−0.05	± 0.11	0.5	NS	0.13
Polypropylene +	−	86.1	± 2.6	1.64	± 0.27	6.1	<0.001	0.85
^{3}H-hexadecane	+	85.5	± 5.7	1.70	± 0.59	2.9	<0.02	0.61

counting the filters directly, or after heating. The average CV for the rows, which allows for the effects of this regression, was 7.9% for direct counting and 9.5% after heating. These values and the regression when using glass fiber filters, show that the performance falls short of standard methods using liquid scintillant. In contrast, in another test, using polypropylene filters with solid scintillant, a single set of samples from 12 wells was harvested and the sample was dried and passed through the hot rollers. These gave a CV of 4.5%, compared with controls on a standard glass fiber filter with liquid scintillant, which gave a CV of 5.6%. Therefore, these polypropylene filters were capable of as good a consistency in any row as for the standard glass fiber filters using liquid scintillant, although any regression according to the sample row detracts from the technique.

It is possible that these problems of variation could be explained if the deposition of solid scintillant on the filter was not homogeneous. To test this, a standard glass fiber filter was prepared with solid scintillant spiked with ^{3}H-hexadecane. After preparation, the filter was dried and counted, before and after heating. The results, also in Table 7, showed a clear positive regression according to the sample row when counted direct, but the regression became insignificant after heating. The count rate also decreased, indicating that some quenching or self absorption of the activity had occurred which was associated with compaction of the glass fiber filter while melting the solid scintillant. A similar experiment with a thin grade of polypropylene showed no decrease in count rate after heating, and the regression was also unchanged (Table 7). These results are in contrast to the samples derived from cells which showed an increase in count rate after heating. Presumably the increased counting efficiency due to scintillant permeation of the samples is greater than any reduction due to compaction-produced self-absorption. The variation in count rate for harvested cells therefore appears to be associated with different amounts of deposited solid scintillant and is probably due to the solid-scintillant solution pooling at the lower end of the filters during the drying process. It is therefore evident that any production of these filters should minimize this variation.

DISCUSSION

The experiments described, indicate that a useful efficiency can be obtained for solid scintillant filters especially when the scintillant is melted using the hot rollers of an office laminator. Further improvements occur when polypropylene filters are substituted for glass fiber. In general, the higher the efficiency the smaller the coefficient of variation obtained, and satisfactory results were obtained with a PPO/durene/polyropylene combination. There is a rather complex interaction between the increased efficiency obtained by filter compaction, which may bring particles of the sample closer to the solid scintillant, and any possible losses of light output, due to the increased absorption by the filter/scintillant matrix. Reduced optical efficiency produced by compaction was most apparent where the scintillant did not melt at the temperatures achieved during the hot rolling. This reduction was greatest for the PVT plastic coating and less for crystaline butyl-PBD. For anthracene, however, there was an increase in counting efficiency. Perhaps, because the energy output is higher for this scintillant, fewer events are lost through quenching, while with the proximity of the compacted sample to the solid scintillant, more of the sample may be exposed to the scintillant. Any reduction in efficiency produced by compaction appeared to be less deleterious using polypropylene. When polypropylene filters bearing cells were used with liquid scintillant the polypropylene is a slightly less effective filter. Rather than being due to a poor optical performance, preliminary experiments show that these filters, with or without solid scintillant, trap a smaller proportion of material harvested than do glass fiber filters, even though the porosity of 10μ should retain whole cells entirely. It is likely however, that cells are rapidly lysed when washed from the microtitre plate with water. Reduced retention of this partial lysate is due to differences in the surface properties of the polypropylene or the solid scintillant (which are both hydrophobic), as compared with the (hydrophilic) surface of glass fiber. It is possible that the surfaces of other filters and solid scintillants will be developed for other porosities and may retain nearly 100% of the sample. Indeed, other scintillants and solid solvents that increase the efficiency still further may also be developed, although the level described here would be adequate for many applications. The technique is useful in that it reduces background count rate and cross talk as well as eases disposal. Furthermore, the lack of sample movement during preparation or its diffusion afterwards, makes the counting of many filtered particulate samples possible using the flat-bed scintillation counter.

ACKNOWLEDGEMENTS

We wish to thank Mrs. A. C. Potter for technical assistance and Mr. P. Harrison for suggesting durene as a solid solvent. We also thank Pharmacia-Wallac for financial support.

REFERENCES

1. Warner, G.T. and C.G. Potter. "A New Design for a Liquid Scintillation Counter for Microsamples Using a Flat-Bed Geometry." *Int. J. Appl. Isot.* 36:819–821 (1985).
2. Potter, C.G., G.T. Warner, T. Yrjonen, and E. Soini. "A Liquid Scintillation Counter Specifically Designed for Samples Deposited on a Flat Matrix," *Phys. Med. Biol.*, 31:361–369 (1986).
3. Potter, C.G., F. Gotch, G.T. Warner, and J. Oestrup. "Lymphocyte Proliferation and Cytotoxic Assays Using Flat-Bed Scintillation Counting," *J. Immunol. Methods*, 105:171–177 (1987).
4. Warner, G.T. and C.G. Potter. "A New Scintillation Counter Design Eases Vial Disposal Problems," *Health Phys.* 51:385 (1986).
5. Warner, G.T. and C.G. Potter. "Method of, and Apparatus for, Monitoring Radioactivity," U.S. Patent Specification 4,298,796 (1981).
6. Potter, C.G. and G.T. Warner. "Carrier for at Least One Beta-Particle-Emitting Sample to be Measured in a Scintillation Counter," Swedish Patent Specification 8801879-1 (1988).

CHAPTER 3

DI-ISOPROPYLNAPHTHALENE—A New Solvent for Liquid Scintillation Counting

J. Thomson

INTRODUCTION

Although there have recently been a few significant developments in the field of solvents suitable for use in liquid scintillation counting (LSC), the traditional solvents xylene and toluene are still used. A review of the evolution of LSC Solvents (Figure 1) shows that after an initial period of development in the 1950s and '60s very little has happened until quite recently. Toluene and xylene were the initial solvents of choice due to their commercial availability and suitability.[1] Dioxan, on its own and in conjunction with naphthalene, was also widely used at this time due to the ability of this system to accept quantities of aqueous sample, thus expanding the application.

After a brief period of inactivity, the '60s saw the emergence of some alternative solvents of which decalin and aromatic petroleum distillates (C9 to C12) were the most notable. These partly supplanted toluene and xylene because of their lower cost. However, the development of emulsifying cocktails in the '60s further established toluene and xylene as the preferred LSC solvents.

Then, in the 70s changes in market trends saw the introduction of a new type of emulsion cocktail—for Radio-immunoassay—based on pseudocumene.

This change coincided with an increased awareness of safety brought about by the restrictions imposed on the handling and use of certain hazardous solvents. The search was now on for safer alternatives to toluene and xylene that had equivalent or better characteristics with respect to their LSC performance. Another factor which became important at this time was the adverse effect on the environment of waste chemicals. Formulators gradually became aware of these factors and realized that a very different type of solvent would be required to satisfy the changing demands of both the market and the environment. Thus the main driving force behind recent developments in sol-

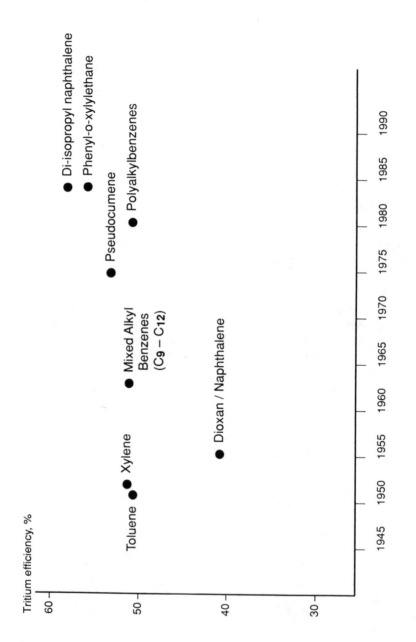

Figure 1. Evolution of LSC solvents.

vents for LSC has not been the search for more efficient solvents, but rather the requirement to substitute less hazardous and more environmentally friendly solvents for those previously used.

In the early '80s the first attempts to resolve these problems involved the use of polyalkylbenzenes which met some but not all of the requirements. The major failings were their low counting efficiency, lack of biodegradability and their toxicity. At Fisons we were active in this field of development. Indeed we used polyalkylbenzene solvents in emulsion cocktails simply because they were the best compromise available. During the course of our investigation about 30 solvents were evaluated for their potential in LSC, but for one or more of the above reasons they all proved to be unsuitable. In due course we became aware of the existence of Di-isopropylnaphthalene (DIN) and quickly realized that here was a solvent that could be ideally suited to LSC.[2] DIN is a mixture of position – isomeric di-isopropylnaphthalenes and is a product of Rutgers–Kureha–Solvents Ltd.

In order to fully appreciate the benefits afforded by DIN consider an ideal LSC solvent.

DEFINING THE IDEAL

In evaluating solvents for use in LSC our start point was to define the characteristics of an ideal LSC solvent. Based on our own experience and comments by other workers in the field we listed the following essential attributes:

1. high flash point
2. low vapor pressure
3. odorlessness
4. low toxicity and irritancy
5. no permeation through plastics
6. biodegradability
7. good fluor solubility
8. low photo- and chemiluminescence
9. high counting efficiency (tritium)
10. good colour and chemical quench resistance

The ideal LSC solvent would obviously satisfy all these requirements but hitherto no solvent has. The evaluation of DIN against these criteria will indicate how it equates to the ideal LSC solvent. Pseudocumene is used in the comparison because it is typical of the solvents currently in wide use in LSC cocktails.

EVALUATION OF DIN VS THE IDEAL

Flash Point

The 148°C (298°F) flash point of DIN makes it safe to use on the laboratory bench since the flammability risk is minimal. DIN is classified as a nondangerous product in accordance with national and international traffic regulations because the flash point exceeds 100°C. By comparison pseudocumene with a flash point of 52°C (125°F) is classified as flammable and should be handled accordingly.

Vapor Pressure

DIN has a very low vapor pressure (1 mm Hg at 25°C) thus ensuring that there is a low vapor concentration at 25°C. This means that there would be very little build up of DIN vapor in the event of a spillage in an enclosed space. By comparison pseudocumene has a vapor pressure twice (2×) that of DIN.

Odor

DIN is virtually odorless and as such makes it pleasant to work with. Pseudocumene, on the other hand, has a highly aromatic penetrating odor, which, if inhaled in appreciable amounts, can cause headaches and narcosis.

Toxicity and Irritancy

Based on the acute data (LD_{50} = 5600 mg/kg oral rat). DIN is not classified as toxic and imposes no acute health hazard to man. DIN is widely used in the manufacture of carbonless copy paper (NCR paper) and consequently its toxicological properties have been extensively studied. A considerable amount of work (to EEC directives) has been completed on DIN and a summary of the toxicological data shows DIN to have the following reactions:

1. acute toxicity (oral—rat) = nontoxic
2. acute toxicity (dermal—rat) = nontoxic
3. skin irritation = nonirritating
4. eye irritation = nonirritating
5. skin sensitization = nonsensitizing
6. toxicokinetics = nonaccumulative
7. mutagenicity = nonmutagenic
8. cancerogenicity = noncarcinogenic
9. teratogenicity = nonteratogenic
10. acute toxicity to fish = nonharmful to fish
11. bioaccumulation (fish) = nonbioaccumulative

A copy of "Toxicological and Physico-Chemical Studies on Di-isopropylnaphthalene" is available from the author upon request. The toxicity of pseudocumene is well known and its principal hazards are that it is a primary skin irritant, is irritating to the eyes, and on inhalation causes respiratory irritation together with central nervous system depression.

Plastic Permeation

In both short- and long-term studies DIN has been shown not to permeate through polyethylene counting vials. Indeed, in a long-term study there was no loss of DIN from a polyethylene vial during a 12 month storage test at room temperature. Pseudocumene, although better than toluene and xylene, is nevertheless steadily lost from polyethylene counting vials. In practice, permeation of the solvent through the vial wall has two effects. Firstly, loss of solvent increases the effect of quench in the vial. Secondly, the solvent and fluors penetrate the vial wall causing the vial to become an additional solid scintillator. This solid scintillator in conjunction with the external standard produces an additional spectrum. This solid scintillator spectrum adds to the "Compton" spectrum produced by the interaction of the external standard with the cocktail solvent. Therefore, there is a gradual change in shape of the overall spectrum. This particularly affects the external standard channels ratio (ESCR) and is manifest as a continual change in the ESCR value. This phenomenon is most evident with ESCR, but the newer parameters (special quench parameter and H-number) are not significantly affected. With ESCR this leads to an underestimate of the efficiency and hence an overestimate of the dpm. The effect is known as the "plastic vial" effect and is shown in Figures 2 and 3.

Biodegradability

By comparison with other aromatic solvents DIN solvent has the remarkable property of being biodegradable in its own right. DIN is described as being greater than 80% biodegraded after 28 d at 4 ppm available oxygen and is therefore classified as biodegradable according to EEC directive 79/831. Annex VII. An independent evaluation by the Severn Trent Water Authority (U.K.) found Optiphase Hi-Safe II (DIN-based emulsion cocktail) to be readily biodegradable by the ISO 7827–1984 (E) method, achieving a degradation level of greater than 80% in 2 d. Greater than 80% degradation in 28 d is the standard necessary for classification as readily biodegradable. A copy of this evaluation is available from the author upon request. However, it is essential that your radiological safety officer and local water authority are consulted to obtain permission for drain disposal before embarking upon any particular course of action.

Fluor Solubility

All the commonly used fluors are soluble in DIN solvent with PPO having particularly good solubility. Figure 4 shows the solubilities of selected fluors in

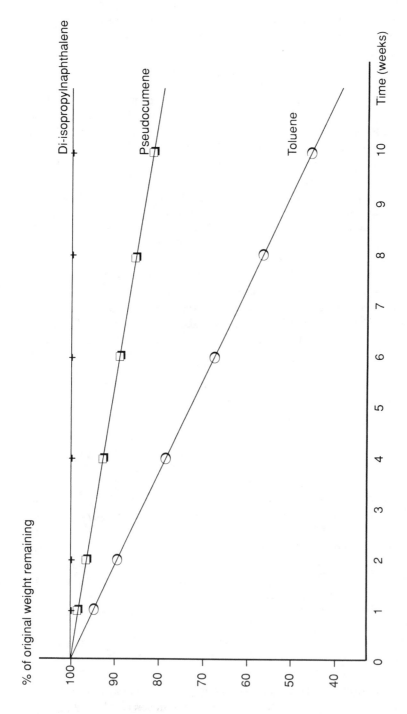

Figure 2. Permeation of various solvents (at 20°C in 20 mL polyethylene vials).

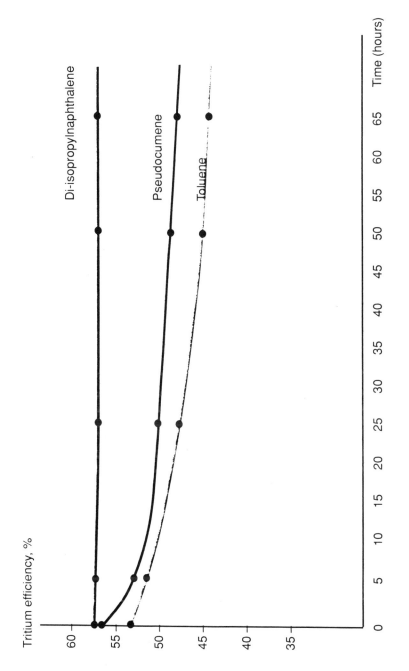

Figure 3. "Plastic vial" effect: variation in tritium efficiency with time for various solvents in 20 mL polyethylene vials.

FLUORS	Max. concentration, g/l
PPO	100
PBD	9.9
Butyl-PBD	40
Bis-MSB	1.8
TP	3.6
POPOP	0.5
Dimethyl-POPOP	1.0

Figure 4. Solubility of various fluors in di-isopropylnapthalene at 15°C.

DIN at 15°C, and easily exceeds the optimum concentration as determined by a simplex optimization. Pseudocumene is also a good solvent for all the common fluors.

Luminescence

DIN solvent is compatible with all the conventional toluene-based alkaline tissue solubilizers and any chemiluminescence should decay within 30 min. Any induced photoluminescence will also decay within 30 min under normal counting conditions. Pseudocumene from some sources will require purification to remove those impurities which can give a color reaction with alkaline tissue solubilizers.

Counting Efficiency (Tritium)

When DIN is used as the solvent in both lipophilic and hydrophilic cocktails then the resulting cocktails exhibit a superior counting efficiency when compared with other types of cocktail. This is especially important when counting to statistical limits since those limits will be reached quicker with DIN based cocktails thus reducing instrument time. The following graphs, Figures 5 through 7, illustrate the superior detection efficiency obtained with tritium-labeled water samples. The DIN based cocktail, Optiphase Hi-Safe II, used in this comparison is a product of Pharmacia–Wallac.

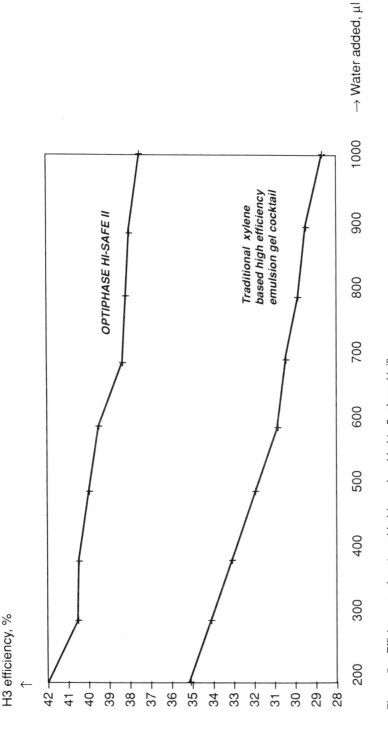

Figure 5.   Efficiency vs μL water added (sample added to 5 mL cocktail).

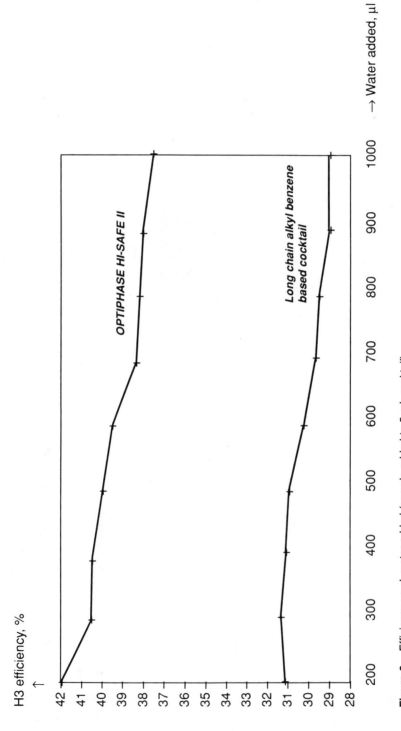

Figure 6. Efficiency vs μL water added (sample added to 5 mL cocktail).

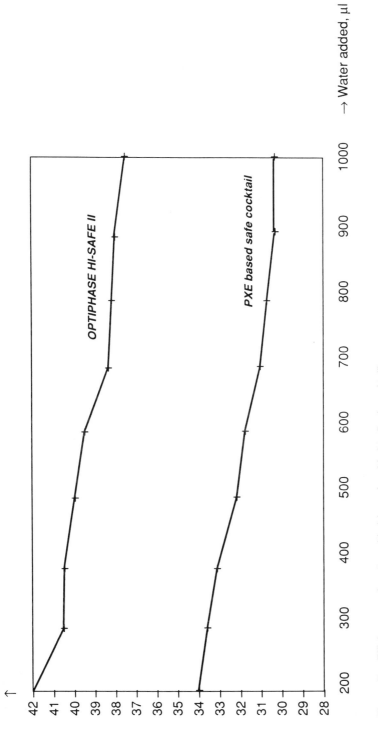

Figure 7. Efficiency vs μL water added (sample added to 5 mL cocktail).

Quench Resistance

DIN exhibits an exceptional resistance to chemical quench as shown in Figure 8. This is important where samples are prepared with acids, solvents etc., which introduce unacceptably high levels of chemical quench with other solvents. DIN also has resistance to color quenching (Figure 9), which is necessary when working with colored samples, e.g., urine and serum. Pseudo-cumene has good resistance to both chemical and color quench, but has been found to be not as good as DIN.

DIN-BASED EMULSION COCKTAILS

DIN can be formulated into emulsion cocktails which achieve a greater degree of safety without compromising performance. This is aptly demonstrated in the Optiphase Hi-Safe range of cocktails which have seen wide acceptance in the U.K. and Europe over the last few years. Taking a closer look at two of these products:

- **Optiphase Hi-Safe II**[©]
  This cocktail possesses all of the safety features previously cited about DIN solvent together with the expected very high counting efficiency and the ability to incorporate a diversity of sample types (Figure 10).
- **Optiphase Hi-Safe III**[©]
  This cocktail combines the capability of accepting large volumes of normal samples with the ability to accept high-ionic strength solutes in moderate to good volume (Figure 11). This makes Optiphase Hi-Safe III almost unique among currently available cocktails. As expected Optiphase Hi-Safe 3 has all the inherent advantages afforded by being a DIN-based cocktail. Optiphase Hi-Safe III has the additional capability of accepting sample types which other cocktails find difficult or impossible to accommodate.

CONCLUSION

Di-isopropylnaphthalene is a significant improvement over existing LSC solvents. Its low vapor pressure, low flammability, low plastic permeability, and low toxicity make it safe and pleasant to handle. It is a more environmentally acceptable product because of its rapid and extensive biodegradability. Lastly and by no means least it is unexcelled in its LSC performance. Therefore DIN comes closest to matching the ideal LSC solvent than any other currently known solvent.

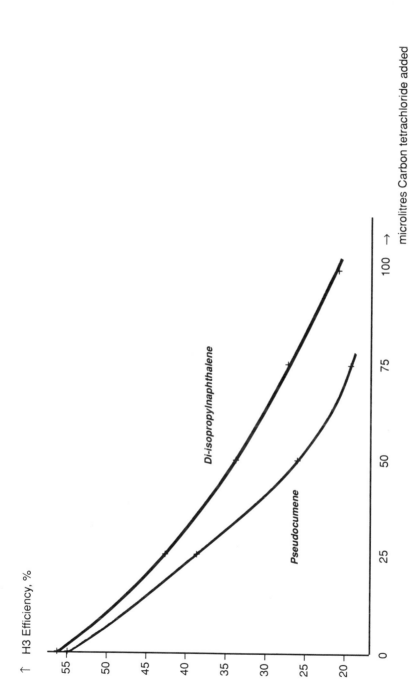

Figure 8. Chemical quench comparison for di-isopropylnapthalene and pseudocumene (μL carbon tetrachloride added to 10 mL cocktail, cocktails = solvent + 4 g/L PPO/0.1 g/L bis-MSB).

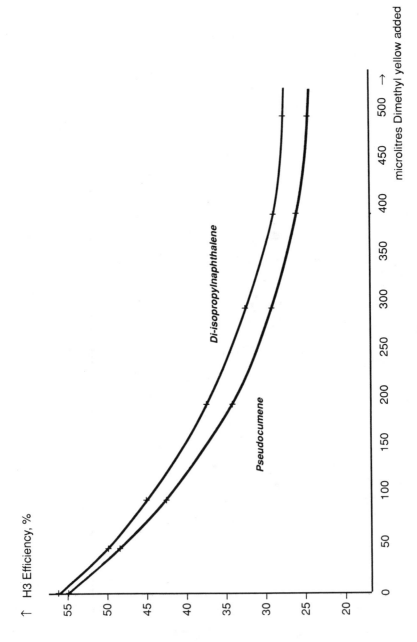

Figure 9. Color quench comparisonf or di-isopropylnapthalene and pseudocumene (μL of 0.001% dimethyl yellow solution added to 10 mL cocktail, cocktails = solvent + g/L PPO/0.1 g/L bis-MSB).

SAMPLE	Temperature, °C			
	10	15	20	25
Water	2.5	2.5	2.6	2.6
0.15M Sodium chloride	3.4	3.4	3.8	3.8
Phosphate saline buffer 0.01M	3.4	3.4	3.4	3.4
0.05M Tris-HCl	3.4	3.4	3.4	3.4
0.1M Hydrochloric acid	9.0	10.0	10.0	10.0
10% Sucrose	3.8	3.8	3.8	3.8
0.2M Sodium hydroxide	4.0	4.6	5.0	5.0
0.1M Sodium hydroxide	4.0	4.0	4.0	4.4
Urine	2.2	2.2	2.2	2.2
Serum - canine	1.0	1.0	1.0	1.0
20% Sucrose	6.6	6.6	5.8	5.8
0.1M Ammonium sulphate	3.4	3.4	3.4	3.6
0.05M Sodium acetate	5.8	5.6	5.0	5.0
0.04 Disodium phosphate	8.0	8.0	8.6	8.6
5mM Hepes	2.8	2.8	2.8	2.8
0.1M Tris-50mM EDTA	5.0	5.2	5.8	5.8
0.2M Ammonium acetate	2.4	2.4	2.6	2.8
1.0M Sodium hydroxide	2.0	2.0	2.2	2.4
10% Trichloroacetic acid	2.2	2.3	2.3	2.3
8.0M Urea	1.1	1.1	1.1	1.1
Optisolve	1.0	1.0	1.0	1.0

Figure 10. Optiphase Hi-Safe II sample acceptance (values [mL] determined by addition of sample to 10 mL of cocktail).

ACKNOWLEDGEMENTS

The author would like to thank the following staff at FSA for their help and support: Mr. H.S. Wagstaff, Mr. B. Illingworth, Mr. R.W.R. Carss, and Mr. G.A. Giblin.

SAMPLE	Temperature, °C			
	10	15	20	25
Water	10.0	10.0	9.0	9.0
Hepes	10.0	10.0	9.5	7.5
Tris HCl 50mM	10.0	10.0	10.0	10.0
Ringers	10.0	10.0	8.5	8.0
0.5M Phosphate saline buffer	9.0	7.0	6.5	5.5
0.1M Phosphate saline buffer	3.0	3.0	8.0	8.0
Sea water	3.0	3.0	7.5	7.5
1.0M Sodium acetate	7.0	6.5	6.0	5.5
4.0M Sodium hydroxide	1.9	1.9	1.9	2.0
40% Caesium chloride	6.5	6.0	5.5	5.0
10% Trichloro acetic acid	3.5	6.5	6.5	6.5
Urine	3.0	3.5	8.0	8.0
4.0M Ammonium formate	2.5	2.5	2.0	1.75
1.0M Ammonium formate	7.5	6.0	5.5	5.0
1.0M Sodium chloride	3.0	7.5	7.5	7.5
2.0M Nitric acid	7.5	6.0	4.0	4.0
0.5M Potassium di-hydrogen orthophosphate	7.5	6.0	5.0	5.0
8.0M Urea	3.25	7.0	7.0	7.5
0.5M Ammonium sulphate	5.0	3.75	3.5	3.25

Figure 11.   Optiphase Hi-Safe III sample acceptance (values [mL] determined by addition of sample to 10 mL of cocktail).

REFERENCES

1. Birks, J.B. *The Theory of Practice of Scintillation Counting* (Pergamon Press, 1964), pp.272–279.
2. Thomson, J. "Scintillation Counting Medium and Counting Method", U.S. Patent 4,657,696, April 14, 1987; European Patent 0176281, September 11, 1985."

CHAPTER 4

Safe Scintillation Chemicals for High Efficiency, High Throughput Counting

Kenneth E. Neumann, Norbert Roessler, Ph.D., and Jan ter Wiel

ABSTRACT

For the last 35 years, the choice of solvent, fluor, and emulsifier for scintillation cocktails has been dictated by the need for efficient energy transfer between the beta electron and the final photon-emitting species. The modern scintillation counting laboratory has several additional requirements which make the design of a scintillation cocktail more critical than ever before: a wide variety of samples, the chemical environment of the radiolabeled sample, and the trend toward microvolume analysis. Furthermore, new environmental regulations are dramatically increasing disposal costs, making the use of high flash point, environmentally safe cocktails quite attractive. These factors combine to create the need for new types of scintillation chemicals capable of safe, high efficiency, low-volume counting. Additionally, the development of many new beta counting assays requires scintillators that are directly suitable for these applications. These requirements have lead to the development of a new generation of liquid cocktail. Data are presented indicating that this formulation provides superior counting performance, while offering a high degree of safety. Additionally, the novel characteristics of this cocktail are discussed with regard to new instrument technologies.

INTRODUCTION

The design of commercial liquid scintillation cocktails has, for many years, been based on the technical requirements of those investigators using the chemicals. These include high radionuclide counting efficiencies, large sample holding capacity, and superior resistance to quench.[1] Issues of chemical safety, storage, and disposability were assigned a relatively minor importance. The solvent used as the cocktail base plays a critical role in determining overall cocktail performance with regard to the above parameters. Acceptable results have classically been achieved with lower order benzene derivatives such as toluene, xylene, and 1,2,4-trimethylbenzene.[2]

In recent years, occupational health and safety and environmental issues have attracted much attention, both in the media and the scientific community. State and federal regulations regarding shipping, use, storage, and disposal of hazardous chemicals are increasingly stringent. Scintillation cocktails, because of the solvents cited above, are recognized as potential environmental and health hazards. As a result, considerable effort has been made to develop scintillation chemicals which offer greater safety.[3]

One of the first significant breakthroughs involved the use of a higher order benzene derivative as the solvent. Optifluor, manufactured by Packard Instrument Company, is based on a long-chain phenylalkane. The use of this solvent permitted the cocktail to be designated as nonflammable, nonhazardous, and drain disposable. However, because of the solvent low scintillation yield, counting efficiencies are less than those of conventional cocktails. The cocktail is also somewhat less quench resistant.

The most recent development is a cocktail possessing all of the characteristics necessary for superior counting performance. Ultima Gold, developed by Packard, is based on a new solvent known as diisopropylnaphthalene.[4] This solvent also has a very high flash point. As a result, Ultima Gold has been classified as nonflammable. The cocktail has been approved for drain disposal, thus reducing many of the costs associated with LS counting. Finally, Ultima Gold presents no unusual, chronic, or severe health hazards to laboratory personnel, and has no noxious or unpleasant odors associated with it.

Of greater significance is that diisopropylnaphthalene, being a derivative of naphthalene, retains the properties of a primary scintillator, while existing in a liquid form. Additional fluors and emulsifiers are readily miscible in the solvent, creating a high performance universal cocktail. The resulting formulation provides high sample capacities, with excellent quench resistance. Due to the scintillating nature of the solvent, excellent counting efficiencies are maintained even with high sample loads or the presence of tissue solubilizers.

The scintillating properties of diisopropylnaphthalene have broad implications for cocktails based on this solvent and, more importantly, liquid scintillation counting equipment. Experimental evidence suggests that the pulse shapes and fluorescent decay characteristics of cocktails based on diisopropylnaphthalene differ markedly from those of conventional cocktails and from thermal noise generated within the PMT. These differences can be used with patented time resolved pulse discrimination circuitry to provide single PMT counting systems capable of high efficiency and low background.[5] One existing application for such a combination is the use of radiochromatography systems, where a radiolabeled HPLC effluent is mixed with scintillation cocktail and presented in front of a single-PMT detector for quantitation.

This chapter summarizes accumulated data on each of the above classes of liquid scintillation cocktails. Experimental results are described which confirm the performance of these new cocktails. Furthermore, data are presented suggesting that Ultima Gold can be effectively used as a liquid cocktail in a single PMT detection system.

Table 1. General Solvent Data

Solvent	Boiling Point°C	Flash-Point°C	TLV ppm	Vapor Pressure 25°C mmHg	CLASSIFICATION International	U.S.A.	Scintillation Yield
Dioxane	101	11	50	40	Flammable liquid	Flamm. class. IB	
Toluene	110	4	100	28	Same	Same	100
Xylene	137–139	25	100	8	Same	Flamm. class. IC	110
Cumene	152	31	50	5	Same	Combustible class. II	100
Pseudo-cumene	168	50	25	2	Same	Same	112
Tri-methyl-benzenes	166–178	50	100	2	Same	Same	100
Phenyl cyclohexane	235–236	100	n.a.		Harmless chemical	Combustible class. IIIB	103
Phenyl alkane (alkylbenzene)	290–310	130–150	n.a.	0.76	Same	Same	94
Phenylxylyl ethane	302–309	150	n.a.	< 1	Same	Same	110
Diisopropyl naphthalene	290–299	132–140	n.a.	1,1	Same	Same	112

EXPERIMENTAL

Physical constants, classifications, and characteristic data of solvents typically used in scintillation cocktails were compiled.[6-9] These are summarized in Table 1.

To evaluate the counting efficiency obtainable from a variety of LS cocktails, 50 μL of ^{3}H-labeled thymidine was spiked into 10 mL samples of each cocktail. Triplicate samples were made. A second set of samples was prepared without any label, to evaluate background countrates. All samples were then assayed for ^{3}H CPM in a Tri-Carb 2250C A liquid scintillation analyzer. From these count data efficiencies and figures of merit (E^2/B) were calculated for each cocktail (Table 2).

Next, to observe the effect of increasing quench on efficiency, a series of samples were prepared in each cocktail. Samples were quenched with increasing volumes of 0.01 M PBS, and spiked with a known and constant amount of

Table 2. LSC Efficiencies and Figures of Merit

Cocktail	Solvent	Classified as safe	Percent tritium efficiency	Bkgrd cpm	FOM E-2/B
Insta-Gel XF	Pseudocumene	No	55.3	14.4	212
Optifluor	Phenylalkane	Yes	48.4	13.4	175
Ultima gold	Diisopropyl-naphthalene	Yes	56.4	13.8	231

[3]H-labeled water. Each was assayed for count rate in the aforementioned LSC. The results of this assay are illustrated in Figure 1.

To explain the performance of the Ultima Gold formulation, electronic signals produced at the anode of the PMT were observed and recorded on a digitizing oscilloscope. Figure 2 displays typical pulses for both Ultima Gold and PMT thermal noise.

The significant differences between these pulses suggests that time-resolved techniques can be used to discriminate against PMT noise, while retaining adequate counting efficiency. Therefore, another set of samples, similar to the ones prepared for the first set of experiments, were prepared. The assay was repeated in an experimental counting system. This system is composed of a single PMT and a pulse discrimination circuit based on time resolution. Table 3 summarizes the results of this experiment.

RESULTS AND CONCLUSIONS

While LS cocktails based on benzene derivatives offer generally excellent performance, most are characterized by some degree of environmental or safety hazard. A comparison of pertinent physical constants and data for these solvents indicates that, in some cases, these hazards and associated costs are significant. This has lead to the development of cocktails based on safe solvents. However, the data presented above suggest that the performance of these cocktails is inferior to xylene or pseudocumene based formulations. Counting efficiencies can be 10–15% lower, leading to a corresponding decrease in figure of merit. Additionally, the poor scintillation energy transfer qualities of the solvents result in cocktails which are easily and severely quenched.

The use of a new class of solvent—diisopropylnaphthalene—results in a cocktail whose counting performance meets or exceeds that of earlier safe cocktails. In fact, the results presented above suggest that this new cocktail (Ultima Gold) performs better than even cocktails based on benzene derivatives. Counting efficiencies can be somewhat higher, resulting in a 5–10% increase in figure of merit. Because diisopropylnaphthalene has scintillation properties of its own, efficient energy transfer is maintained even at extreme sample loading conditions, resulting in a cocktail with excellent quench resistance. Furthermore, this cocktail meets or exceeds all current environmental and safety requirements. Drain disposability also aids in reducing laboratory costs.

Experimental evidence gathered during these studies indicates that Ultima Gold has unique scintillation properties. The width of a typical electronic pulse produced by the interaction of a beta decay event with Ultima Gold is significantly longer than noise pulses produced in the PMT itself. We have seen differences of a factor of four. This is apparently due to the scintillating nature

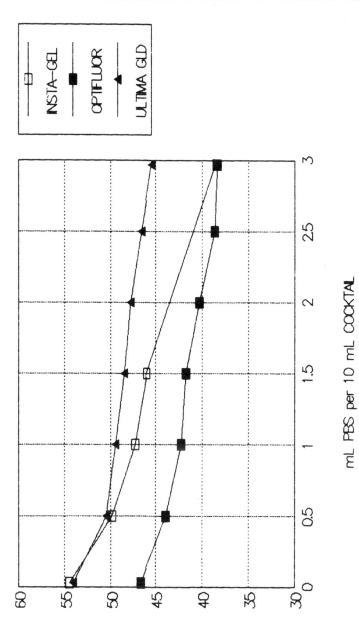

Figure 1.   Cocktail quench resistance (^{3}H, 0.01 M PBS quencher).

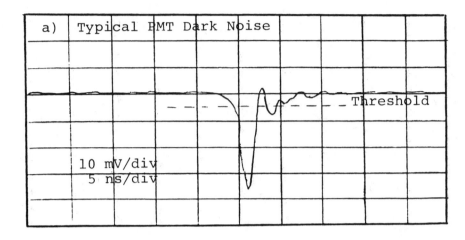

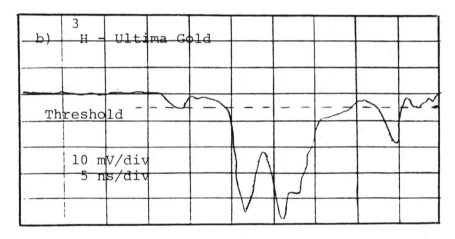

Figure 2. Electric pulses, a) PMT thermal noise, b) Ultima Gold (^{3}H).

of the solvent, which effectively prevents quenching of the slow component of the fluorescent decay.

As has been demonstrated, these differences in pulse width can be exploited in the use of time-resolved noise discrimination techniques. This technology

Table 3. Single PMT Counting Results

Cocktail	Solvent	Classified as safe	Percent tritium efficiency	Bkgrd cpm	FOM E^2/B
Insta-Gel XF	Pseudocumene	No	20.0	114.0	3
Optifluor	Phenylalkane	Yes	14.0	47.0	4
Ultima gold	Diisopropyl-naphthalene	Yes	32.0	54.0	19

offers the possibility of single PMT scintillation counting systems. These systems were not previously feasible with conventional liquid cocktails because of the insignificant differences between decay and noise pulses. Results in an experimental counting system based on these principles show that improvements in efficiency range from 30–60% when Ultima Gold is used. This results in a factor of five improvement in figure of merit. While efficiencies and sensitivities using this technology are somewhat less than in conventional LSC, further developments in both electronics and chemicals are expected to lead to improved performance. Furthermore, the use of time-resolved single PMT technology can lead to the development of high performance multidetector liquid scintillation counting systems.

REFERENCES

1. Bray, G.A. "Determination of Radioactivity in Aqueous Samples," in *The Current Status of Liquid Scintillation Counting*, E.D. Bransome, Ed. (New York: Grune and Stratton, 1970), p. 170.
2. Bray, G.A. "Determination of Radioactivity in Aqueous Samples," in *The Current Status of Liquid Scintillation Counting*, E.D. Bransome (New York: Grune and Stratton, 1970), pp 13–24.
3. Kalbhen, D.A. and V.J. Takkanen. "Review of the Evolution of Safety, Ecological and Economical Aspects of Liquid Scintillation Counting Materials and Techniques," in *Advances in Scintillation Counting*, S.A. McQuarrie, E. Ediss, and L. I. Wiebe, Eds. (Edmonton, Alberta: University of Alberta, 1983), pp. 66–70.
4. U.S. Patent Number 4,657,696, "Scintillation Counting Medium and Counting Method."
5. U.S. Patent Number 4,528,450, "Method and Apparatus for Measuring Radioactive Decay."
6. Weast, R.C., Ed. *CRC Handbook of Chemistry and Physics*, 60th ed. (Boca Raton, FL: CRC Press, Inc., 1979).
7. Data supplied by vendor.
8. Birks, J. B. *The Theory and Practice of Scintillation Counting*, (New York: The Macmillan Company, 1964), pp. 272–278.
9. *Dangerous Goods Regulations*, 30th ed. (Montreal: International Air Transport Association, 1988).

CHAPTER 5

New Organic Scintillators

Stephen W. Wunderly and Joel M. Kauffman

ABSTRACT

Several new organic scintillators have been synthesized based on quaterphenyl and sexiphenyl structures. Physical properties of the new fluors are given. The fluors were compared with commercial scintillators in a variety of applications. The sexiphenyl structure was the most efficient fluor as well as the most resistant to chemical quenching from nitromethane. Both the sexiphenyl and quaterphenyl structures were much more efficient than either PPO or b-PBD based systems for scintillation solutions capable of emulsifying aqueous samples.

INTRODUCTION

Since the development of liquid scintillation counting, those involved in improvements of the technique have sought better and more efficient scintillators. In a paper presented by Birks[1] the focus was minimizing quench rather than compensating for quench. In the same spirit we would like to report on new primary scintillators that lead to improved scintillation efficiency.

In Birks' report, he ranked a variety of scintillator systems with respect to two counting conditions: minimal quench and strong quench due to carbon tetrachloride. He found that while some systems performed well in minimal quench circumstance (the quaterphenyl BBQ or BIBUQ was 2% more efficient than PBD systems), these same systems performed quite poorly in the presence of strong quenchers (BBQ was 49% less efficient than PBD systems with added carbon tetrachloride).

It is obvious from these results that sample conditions causing quench affect the ranking of scintillator systems. Since most liquid scintillation cocktail is used in bioresearch, primarily with aqueous samples, we will rank scintillator systems for aqueous quench as well as minimum quench and strong chemical quench.

Table 1. Fluor Systems Investigated

System[a] Number	1[0] Fluor[b]	2[0] Fluor[c]
1	PPO	—
2	PPO	Bis-MSB
3	b-PBD	—
4	b-PBD	Bis-MSB
5	PPF	—
6	PPF	Bis-MSB
7	d-CH₃O-PPF	—
8	d-amyl-PPF	—
9	(PF)₂	—

[a]Solvent—pseudocumene.
[b]Primary fluor at 2.26×10^{-2} M.
[c]Secondary fluor at 1.6×10^{-3} M.

EXPERIMENTAL

Synthesis

The new primary fluors based on quaterphenyl and sexiphenyl ring structures are pictured in Figure 1. Detailed synthetic procedures will be submitted for publication at a later date. Synthetic details for PPF and d-CH$_3$O-PPF are also found in the Ph.D. thesis of Alem Ghiorghis.[2]

New Fluors: Physical Properties
- PPF 2,7-diphenyl-9,9-dipropylfluorene
    - m.p. 195.5–197.5
    - NMR: 7.8–7.2 (m, 16H, aromatic), 2.0 (m, 4H, CH$_2$CH$_2$CH$_3$) 0.7 (m, 10H, CH$_2$CH$_2$CH$_3$).
    - UV: 330 (4.76)
    - Emission: 357 and 375 (excitation 326)

- *d-amyl*-PPF
    - 2,7-bis(4-t-amylphenyl)-9,9-dipropylfluorene
    - m.p. 202–204.5
    - NMR: 7.8–7.4 (m, 14H, aromatic), 2.05 (m, 4H, CH$_2$CH$_2$CH$_3$) 1.7 (q, 4H, C(CH$_3$)$_2$CH$_2$CH$_3$), 1.35 (s, 12H, C(CH$_3$)$_2$CH$_2$CH$_3$) 0.75 (m, 16H, C(CH$_3$)$_2$CH$_2$-CH$_3$ and CH$_2$-CH$_2$-CH$_3$))
    - UV: 212 (4.785), 328 (4.716)
    - Emission: 365, 383, 401 (excitation 344)

- *d*-CH$_3$O-PPF
    - 2,7-bis(4-methoxyphenyl)-9,9-dipropylfluorene
    - m.p. 197.5–201.5
    - NMR: 7.8–6.9 (m, 14H, aromatic), 3.9 (s, 6H, OCH$_3$), 2.0 (m, 4H, CH$_2$-CH$_2$CH$_3$), 0.65 (m, 10H, CH$_2$CH$_2$CH$_3$)
    - UV: 335 (4.748)
    - Emission: 371, 390 (excitation 337)

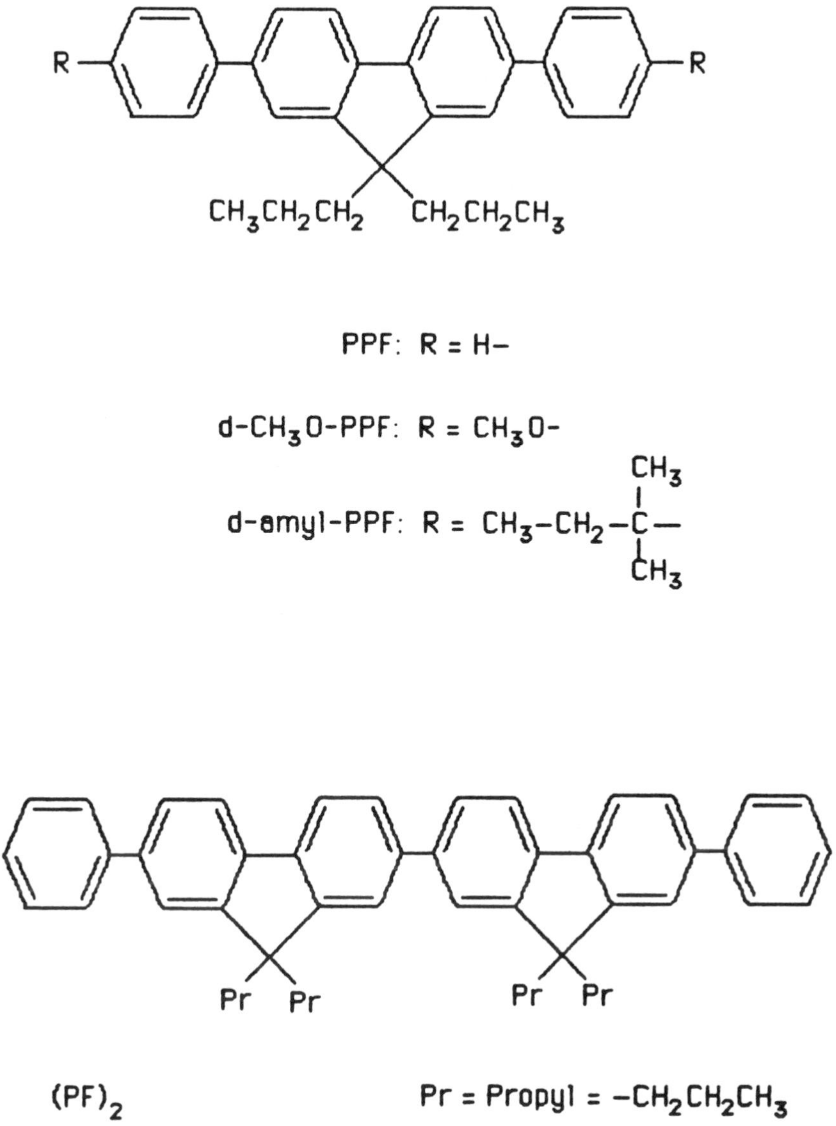

PPF: R = H−

d-CH₃O-PPF: R = CH₃O-

d-amyl-PPF: R = $CH_3-CH_2-\overset{\overset{\textstyle CH_3}{|}}{\underset{\underset{\textstyle CH_3}{|}}{C}}-$

(PF)₂   Pr = Propyl = $-CH_2CH_2CH_3$

Figure 1. Structures of new oligophenylene fluors.

- *(PF)₂*   7,7'-diphenyl-9,9,9',9'-tetrapropyl-2,2'-bifluorene
  m.p. 241.5–242.5
  NMR: 7.9–7.3 (m, 20 H, aromatic), 2.15 (m, 8H, CH_2-CH₂CH₃, 0.8
  (bs, 20 H, CH₂CH_2CH_3)
  UV: 204 (4.98) 347 (4.93)
  Emission: 386, 408, 430 (excitation 347)

- Other Material
  pseudocumene, PPO, b-PBD and bis-MSB purchased from Beckman Instruments, used as received.
  nonylphenolpolyethoxylate purchased from Dorsett and Jackson, Long Beach, CA.
  scintillation measurements made with Model 3801 liquid scintillation counter from Beckman Instruments.

Procedure

1. Primary fluor, 2.26×10^{-4} mol, and 5 mg of secondary fluor (when required) were placed in a maxi glass scintillation vial. Pseudocumene, 10 ml, was added to the vial, capped, and shaken, and an average of 5 H-numbers was determined. This was repeated for each sample. Then each vial was opened and 30 μL of nitromethane were added. The vials were again capped and shaken, and an average of 5 H-numbers was determined for each vial.
2. Primary fluor, 2.26×10^{-4} mol, and 5 mg of secondary fluor (if required) were added to a maxi-glass scintillation vial. Pseudocumene, 6 g, was added to each vial. The vial was capped and shaken until the fluors dissolved. Then 4 g of nonylphenolpolyethoxylate emulsifier and 1 g of water were added to each vial and then capped and shaken. The average of 5 H-numbers was determined for each vial.

CALCULATIONS

Horrocks[3] established a relationship between pulse height (PH) and energy (E) on a Beckman Model 9800 liquid scintillation counter that has been maintained through subsequent models. That relationship is

$$PH = 76 + 280 \log E_i. \tag{1}$$

The pulse height for the Compton edge inflection point for ^{137}Cs, energy 478 KeV, is by definition 826. The amount of quench of a sample is the difference between the inflection point of the Compton edge of that sample compared with the inflection point of the Compton edge of the unquenched standard. This difference is quantified as the H-number and defined as

$$H\# = PH_0 - PH_i \tag{2}$$

where PH_0 = the pulse height of the unquenched standard at the Compton edge inflection point
PH_i = pulse height of the sample in question at the Compton edge inflection point.

By combining Equations 1 and 2 the relationship of Equation 3 may be established

$$H\# = PH_0 - PH_i = (76 + 280 \log 478) - (76 + 280 \log E_i) \tag{3}$$

Table 2. Relative Light Output Air-Quenched Cocktail System

System[a]	H-Number	Relative Light Output
(PF)$_2$	1.5	100
PPF	4.5	98
PPF/bis	7.0	95
d-CH$_3$O-PPF	8.0	95
d-amyl-PPF	8.0	95
b-PBD/bis	9.5	94
b-PBD	11.5	92
PPO/bis	18.8	86
PPO	30.0	78

[a]10 mL of cocktail—air quenched only.

Solving for log E$_i$ we arrive at Equation 4

$$\log E_i = 2.6794 - H\#/280 \qquad (4)$$

E$_i$ is the apparent energy of the quenched sample, i. However, since the decay energy of the ^{137}Cs has not changed, it must be the amount of light from the scintillator per unit of energy input that has changed.

Horrocks[3] has also established a relationship between energy (E$_i$) and number of photoelectrons generated at the photocathode of the PMT, (Equation 5), where #e$_i$ is the number of photoelectrons

$$E_i = 20 + (0.666)\ \#e_i \qquad (5)$$

The number of photoelectrons generated at the photocathode is directly proportional to the scintillator light impinging on the PMT face. Therefore, the ratio of photoelectrons between two scintillators expresses the relative light output of the two scintillators (Equation 6). Using this relationship we may compare and rank the new scintillators with standard commercial scintillators.

$$\text{Relative light output} = \text{RLO}$$
$$\text{RLO} = (e_i \div e_s) \times 100 = [(E_i - 20) \div (E_j - 20)] \times 100 \qquad (6)$$

RESULTS

The traditional fluor systems of PPO or b-PBD, with or without secondary fluor, performed in our tests as would be predicted from Birks' work for minimal quench and heavy chemical quench (Tables 2 and 3). It is suprising to see in Table 4 the equivalence of the two systems in formulations for aqueous samples (one of the most common applications of scintillation cocktails). Therefore, the more expensive fluor, b-PBD, offers no advantage in scintillation performance over the less expensive fluor, PPO, for the measurement of aqueous samples.

Three of the new fluors based on fluorene are structurally related to

Table 3. Relative Light Output Nitromethane Quenched Cocktail System

System[a]	H-Number	Relative Light Output
$(PF)_2$	124	100
b-PBD/bis	150	77.8
b-PBD	152	76.6
d-CH_3O-PPF	176	60.5
PPF/bis	188	53.7
PPO/bis	199	47.8
PPF	208	43.6
d-amyl PPF	212	41.8
PPO	220	38.2

[a]10 mL of cocktail system—quenched with 30 μL of nitromethane.

quaterphenyls. Quaterphenyls are known to be effective fluors. One example, BBQ, is reported by Birks[1]. BBQ was the result of an exhaustive study by Wirth[4] of substitution on oligophenylenes to improve solubility and maximize scintillation pulse height. The latter was achieved only at much higher concentrations than were required with PPO.

Barnett et al.[5] showed that the o,o'-methylene-bridged quaterphenyls, 2,2'-bifluorene, and 2,7-diphenylfluorene (PPF without propyl groups) gave superior pulse heights to quaterphenyl in liquid scintillation counting, a process related to lasing, at least in that the S_1-S_o transition of the fluor or dye occurs mainly by fluorescence.

Pavlopoulos and Hammond[6] suggested that methylene bridged quaterphenyls such as 2,2'-bifluorene and 2,7-diphenylfluorene might prove to be superior laser dyes. All of these quaterphenyls have very low solubility, which severely limits their utility.

Both 2,2' bifluorene and 2,7-diphenylfluorene have recently been reported by Rinke[7] as effective eximer-pumped laser dyes. Both the solubility and photochemical stability of these laser dyes were dramatically improved when nonaromatic hydrogens were replaced by propy groups as reported by Kauffman.[8] PPF and d-CH_3O-PPF were among the quaterphenyls studied as laser dyes.

Recently PPF as well as other fluorenes were examined by Kauffman[9] as

Table 4. Relative Light Output Aqueous Cocktail Systems

System[a]	H-Number	Relative Light Output
$(PF)_2$	75.0	100
PPF	80.0	95.6
d-CH_3O-PPF	81.5	94.3
PPF/bis	82.2	93.7
d-amyl-PPF	82.5	93.4
PPO/bis	100.5	79.4
b-PBD/bis	100.8	79.2
b-PBD	116.8	68.4
PPO	118.5	67.3

[a]6 g of solvent fluor system, 4 g of nonylphenolpolyethoxylate emulsifier and 1 g of D.I. water constitute the aqueous cocktail system.

scintillation fluors in a polysiloxane matrix for detection of gamma rays, mainly in quest of radiation hardness. Despite the low concentrations of fluor obtainable in the polymer, the light output of PPF was surprisingly high, which encouraged us to examine PPF as a fluor for liquid scintillation solutions.

In the current study all three quatraphenyls structures, PPF, di-t-amyl PPF and di-methoxy PPF, have similar scintillation properties. All three have similar performance to b-PBD systems and are much better than PPO based systems for experiments with only air quenching present (see Table 2). All three have poorer performance than b-PBD systems, but better than PPO systems when a strong organic chemical quencher, such as nitromethane, is present (see Table 3). The dimethoxy derivative was more quench resistant to nitromethane than either the parent or di-t-amyl quaterphenyl. Finally, all three diphenyl fluorene systems perform much better than either b-PBD or PPO systems when aqueous samples are a cause of the quench (see Table 4).

The final new fluor, $(PF)_2$, which is named here as a dimer of phenylfluorene, is structurally a sexiphenyl. Its scintillation properties are outstanding compared to any of the other new or traditional fluors. It is the most efficient fluor for all three test systems; air quench, strong chemical quench, and aqueous quench.

The sexiphenyl structure would appear to be a promising direction for research into more efficient and more quench resistant scintillation solutions.

REFERENCES

1. Birks, J.B. "Impurity Quenching of Organic Liquid Scintillators," in *Liquid Scintillation Counting: Proceedings of a Symposium on Liquid Scintillation Counting*; M.A. Cook and P. Johnsons, Eds. (New York: Heyden, 1977), pp. 3–14.
2. Ghiorghis, A. "Synthesis of Oligophenylene Laser Dyes," PhD Thesis, Massachusetts College of Pharmacy and Science, Boston, MA (1988).
3. Horrocks, D.L. "Energy per Photoelectron in a Liquid Scintillation Counter as a Function of Electron Energy," in *Advances in Scintillation Counting*, S.A. McQuarrie, C. Ediss, and L.I. Wiebe, Eds. (Edmonton, Alberta: University of Alberta, 1983), pp. 16–29.
4. Wirth, H.O., F.U. Herrman, G. Herrman, and W. Kern. *Mol. Cryst.* 4:321 (1968).
5. Barnett, M.D., G.H. Daub, F.N. Hayes, and D.G. Ott, *J. Am. Chem. Soc.*, 81:4583 (1959).
6. Pavlopoulos, T.G. and P.R. Hammond. *J. Am. Chem. Soc.*, 96:6568 (1974).
7. Rinke, M., H. Gusten, and J.J. Ache. *J. Phys. Chem.*, 90:2666 (1986).
8. Kauffman, J.M., C.J. Kelley, A. Ghiorghis, E. Neister, L. Armstrong, and P.R. Prause. *Laser Chem.*, 7:343 (1987).
9. Kauffman, J.M. *Proceedings of Workshop on Scintillating Fiber Detector Development for the SSC, Vol. II*, (Batavia, IL: Fermilab, 1988), pp. 677–740.

CHAPTER 6

Advances In Scintillation Cocktails

Jan ter Wiel and Theo Hegge

Liquid scintillation counting techniques have become a wide-spread method to obtain quantitative data for α, β, and γ emitting radionuclides. The detection sensitivity and efficiency for measuring soft β-emitters, tritium, and ^{14}C made liquid scintillation counting the accepted analysis technique.

Ever since the very beginning of liquid scintillation counting, the improvement in instrumentation is clearly demonstrated. The continuous increasing demand for multipurpose scintillation cocktails contributed to the development of a series of investigations leading to a better understanding of the scintillation process and the development of new cocktails.

Early requirements not taken into consideration, liquid scintillation cocktails have been improved in the 70's, and progress was made concerning

- increased sensitivity
- improved compatibility with samples
- volume reduction
- safety of handling and storage
- disposal

The present situation is a result of a development which started in the early 80's. A growing concern of health risk handling scintillation cocktails led to the introduction of new liquid scintillators. Besides the above mentioned, economical factors and continuous development of new surfactant systems contributed considerably to the desires of scientific and routine laboratories using liquid scintillation counting as an analytical tool.

The introduction of new cocktails based on solvents with a high flashpoint with a minimum of safety restrictions was a step forward.

The improvement of surfactants applied in LS cocktails contributed significantly to the simplification of sample preparation procedures. The well known classical sol-gel cocktails have been replaced by cocktails exhibiting a continuous clear liquid phase in applications with aqueous samples. An early example of new types of cocktail construction was discussed by O'Conner and Bransome.[1]

In the newest generation of liquid scintillators, a series of improvements combine a high degree of aqueous sample compatibility with progressive safety characteristics.

COCKTAILS, ASPECTS AND REQUIREMENTS

The major part of samples for radioactivity determination by liquid scintillation counting is of aqueous origin. The requirements to apply the technique in a proper way have been discussed in review articles and conference proceedings (Peng).[2] The general theory is assumed to be known.

Constituents of Cocktails:

The original construction of a cocktail remained the same throughout the years:

- solvent
- surfactant system (emulsifiers)
- scintillators

It is obvious that organic samples or samples soluble in an organic solvent can be quantified in a system without surfactants.

The advances of liquid scintillators are mainly based on progress in application of new solvents and surfactant systems, which will be discussed in general.

Scintillators

The solid primary and secondary scintillators did not change in fact.The most common primary scintillator is still PPO (diphenyl oxazole). Bis-MSB, together with POPOP, are the secondary scintillators in most cocktails.

Solvents

Toluene, dioxane, and xylene were applied initially in liqiud scintillation cocktails, followed by pseudo-cumene (1,2,4-trimethylbenzene). The newest generation of aromatic solvents all have in common the high flashpoint, accompanied by much more preferred safety characteristics. Some influences and requirements have been important in improving the quality of solvents.

Economical:
- availability of aromatic solvents in large quantities
- availability of aromatic solvents economically priced
- improvement of quality
- broadening of range available

Safety/
Economical:
- demand for less volatile products
- application in polyethylene vials
- less restricted storage, handling and transport

Safety: • reduction of volatility of organic solvents resulting in reduced risk to personnel
               • reduction of toxicity
               • reduced fire and hazard risks
               • growing concern of environment

Surfactants

The incorporation of water or aqueous samples in aromatic solvents was initially solved by using, e.g., dioxane and methylglycol. A solution was proposed by Bray.[3] Application of Triton-X-100[4] for example produces results comparable to many commercially available sol-gel scintillation cocktails. Triton-X-200 is a trademark of Rohm and Haas.

As the acceptance of the LSC technique became more important and widespread, its uses demanded improvements. Some criteria of influence were:

• increased detection sensitivity
• compatibility with various types of aqueous samples
• increased sample load in the fluid region
• higher resistance to quench
• economical reasons (counting in small vials)

A combination of investigations on solvents and surfactants created a new generation of cocktails for liquid scintillation counting.

EXPERIMENTAL

Counting efficiencies were determined on Packard Tri-Carb liquid scintillation counters model 2250CA or 1900CA as indicated in the tables. Instrument settings for tritium resulted in 64.1% efficiency for the 1900CA and 67–68% efficiency for the 2250CA for a sealed tritium standard. Instrument efficiency for ^{14}C is 95.8%. All efficiency determinations were performed at room temperature.

Counting efficiency of aqueous samples was obtained by spiking 10 µL tritiated water corresponding to 19,384 dpm in a glass vial containing the sample of interest, all in triplicate. Nonaqueous sample efficiency was obtained by spiking 10 µL tritiated toluene corresponding to 14,000 dpm into the samples, all in triplicate. For determining the counting efficiency of ^{14}C the same procedure as for tritium was followed, except for the amount of ^{14}C, a aliquot of 10 µL ^{14}C corresponding to 3814 dpm was spiked.

All efficiency determinations were obtained by using 20 ml low potassium borosilicate vials.

Efficiency (abs.) was obtained by calculating:

$$\frac{cpm}{dpm} \times 100 = \% \text{ efficiency}$$

Samples were visually and instrumentally checked for homogeneity

$$\% \text{ sample load} = \frac{\text{volume of sample}}{\text{volume of sample} + \text{volume of cocktail}} \times 100$$

Phase diagrams were obtained by adding the aqueous buffer or salt solutions to 10 mL of cocktail with increments of 0.5 mL. Near phase separations, increments of 0.1 mL were taken. Counting of samples was started after equilibration for light and temperature, all samples were counted for 4 min.

Cocktails

Ultima-Gold, Opti-Fluor, Pico-Aqua, Pico-Fluor 40, Emulsifier Safe, and Insta-Gel are commercially available products from Packard Instrument.

The cocktail phenylxylyl ethane was prepared by adding 40% w/w nonionic emulsifier blend to phenylxylyl ethane including 5 g PPO and 0.2 g of bis-MSB (PXE N) per liter cocktail.

SOLVENTS

Physical Constants, Classification:

Some characteristic data of common scintillation solvents are summarized in Table 1. The data show that a considerable improvement on safety has been made:

- Increased boiling point and flashpoint, and decreased vapor pressure indicate a reduced risk in the release of solvent vapors, thus reducing the danger to personnel and increasing laboratory safety.
- Lower flammability not only contributes to less restricted transport and storage regulations, but also lessens the chance of a possible fire hazard.
- Stopping the penetration of solvents through plastic vials is significant to all safety aspects.
- High flashpoint solvents received no TLV values; it is unlikely that vapors will be released in the working atmosphere under normal conditions.

The high flashpoint solvents, e.g., phenylalkane and diisopropylnaphthalene have been well investigated concerning their toxicological properties.[5] Extensive investigations for diisopropylnaphthalene have been performed on bioaccumulation and biodegradability and the outcome is very positive. According to the EEC-directive 79/831 EEC Annex VII, diisopropylnaphthalene is considered a non dangerous substance, having a biodegradability of more than 80% 28 days after it is determined.[6]

Scintillation Characteristics

The question remains whether high flashpoint solvents can compete with established solvents concerning counting efficiency. Tables 2 and 3 summarize

Table 1. Characteristics of Common Scintillation Solvents

Solvent	Boiling Point°C	Flash-Point°C	TLV ppm	Vapor Pressure 25°C mmHg	CLASSIFICATION Interna-tional	U.S.A.	Scintilla-tion Yield
Dioxane	101	11	50	40	Flammable liquid	Flamm. class. IB	
Toluene	110	4	100	28	Same	Same	100
Xylene	137–139	25	100	8	Same	Flamm. class. IC	110
Cumene	152	31	50	5	Same	Combustible class. II	100
Pseudo-cumene	168	50	25	2	Same	Same	112
Tri-methyl-benzenes	166–178	50	100	2	Same	Same	100
Phenyl cyclohexane	235–236	100	n.a.		Harmless chemical	Combustible class. IIIB	103
Phenyl alkane (alkylbenzene)	290–310	130–150	n.a.	0,76	Same	Same	94
Phenylxylyl ethane	302–309	150	n.a.	<1	Same	Same	110
Diisopropyl naphthalene	290–299	132–140	n.a.	1,1	Same	Same	112

TLV = threshold limit value and n.a. = not assigned.

some experimental data for tritium counting efficiency of some commercially available solvents containing PPO and bis-MSB as scintillators.

These data show clearly a very important fact: without suffering in tritium counting efficiency, the safety aspects can be increased significantly.

Table 2. Percentage Tritium Counting Efficiency (TRI CARB 1900 CA)

Pseudo-cumene	Phenylalkane	Phenylxylyl-ethane (PXE)	Diisopropylnaph-thalene (DIPN)
63.6	55.2	62.2	62.8

Table 3. Quench Characteristic with Carbon Tetrachloride; Tritium Counting Efficiency (TRI CARB 1900 CA)

μl Carbon Tetrachloride	Pseudo-Cumene	PXE	Diisopropyl Naphthalene
0	63.3	62.2	62.8
10	50.8	51.8	51.7
20	44.5	43.3	47.5
30	38.5	36.6	41.9
40	33.9	29.5	38.8
50	30.8	23.8	34.4

SURFACTANTS

Surfactant Systems

Surfactants have to be added to organic solvents when, in particular, aqueous samples are the subject of analysis in activity determination. The very well known sol-gel scintillators are most widely used because of their wide applicability. The surfactants, nonionics, applied to obtain these types of cocktails are discussed by Benson[7] and Lieberman and Moghissi.[8] The increasing demand of applications led to the introduction of cocktails based on nonionic and anionic surfactants, an example is given by O'Conner and Bransome.[1] An important advantage is obtained; with aqueous samples give a clear non viscous homogeneous sample, important when small (e.g., 6 mL) counting vials are used.

The diversification of sample types, increase of sample amounts, and whenever possible, minimum of sample preparation procedures influenced the development of new scintillation cocktails, in which in a combination with anionic and nonionics, additional anionic surfactants are applied, e.g., Hegge and Wiel,[9] Sena et al.,[10] Mallik,[11] and references cited therein. The additional surfactants all have improved characteristics of cocktails in common. Some general formulas of surfactants are given in Figure 1.

Alkylphenolethoxylates were the first applied nonionic surfactants, followed by succinates. Some general structures of additional surfactants show that these all have an anionic character.

The characteristics of products differ considerably, as can be seen in Figure 2. Instead of a gel area when a nonionic surfactant is applied, a clear homogeneous fluid results when a combination of non ionics and anionics in an aromatic solvent are the constituents of a cocktail. A general advantage shows that in the application of a nonionic/anionic combination, whatever the solvent is, the relative independence of sample load capacity of temperature. Introduction of combined surfactants results in different sample load capacities for water and 0.01M Pbs (phosphate buffered saline), as shown in Figure 2. This is a general behavior although a wide range of samples is applicable. To each sample type belongs a unique sample load capacity.

Counting Efficiency for Liquid Scintillation Types

The counting efficiency for tritium of cocktails depends on solvent, surfactant system, and amount of surfactants dissolved in the original solvent, assuming the concentration of scintillators is at optimum. Counting efficiency of aforementioned products is given in Table 4.

The counting efficiency with sample of Pico-Aqua is comparable to the sol-gel scintillator, but the surfactant system applied has the advantage that compatible samples form a clear homogeneous mixture.

The other two products contain an anionic/nonionic surfactant system about 25% w/w in different solvents. The lower efficiency of Opti-Fluor is a consequence of the solvent applied.

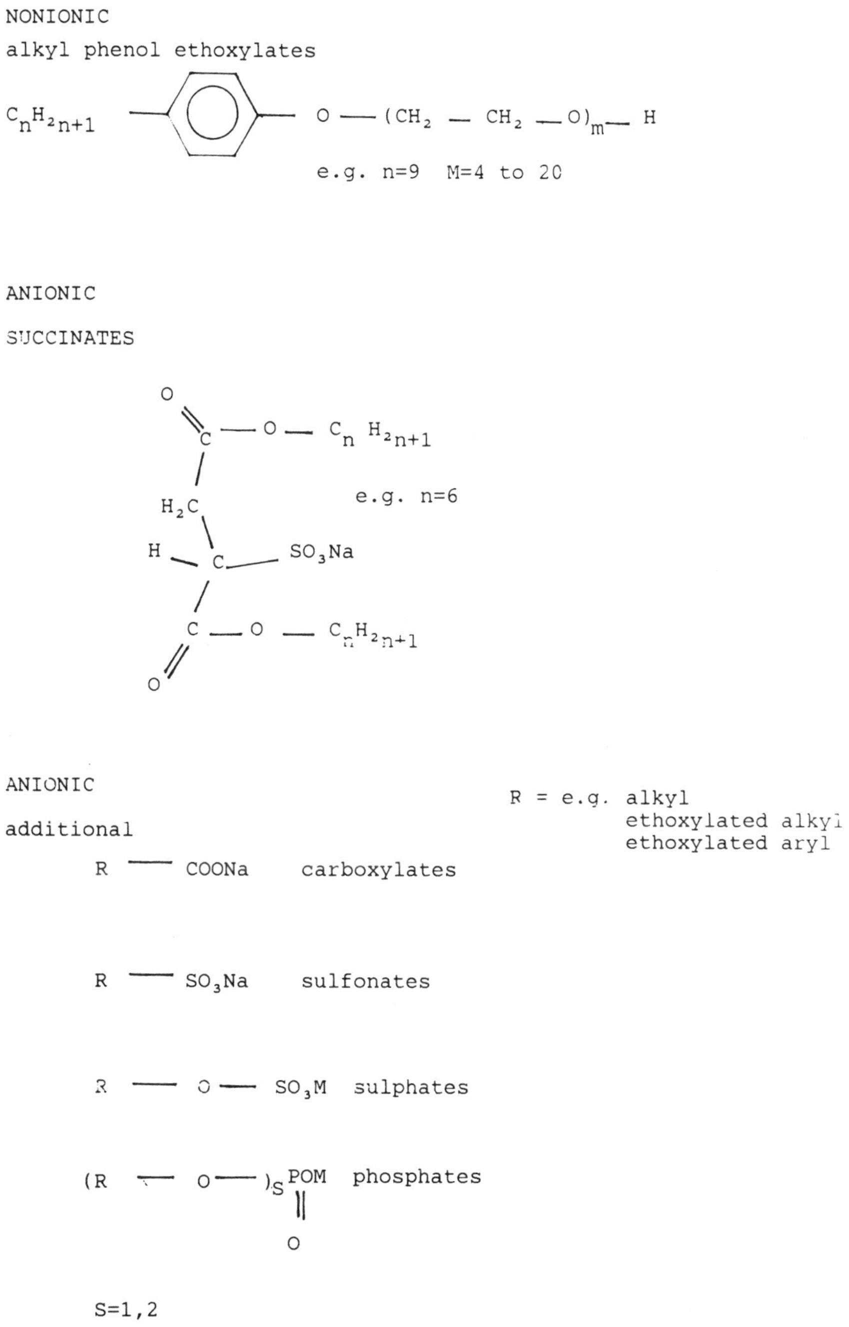

NONIONIC

alkyl phenol ethoxylates

C_nH_{2n+1} —⟨benzene ring⟩— O — (CH$_2$ — CH$_2$ — O)$_m$ — H

e.g. n=9   M=4 to 20

ANIONIC

SUCCINATES

$$\begin{array}{c}
O \\
\parallel \\
C — O — C_n H_{2n+1} \\
| \\
H_2C \\
| \\
H — C — SO_3Na \\
| \\
C — O — C_nH_{2n+1} \\
\parallel \\
O
\end{array}$$

e.g. n=6

ANIONIC

additional

R $= $ e.g. alkyl
            ethoxylated alkyl
            ethoxylated aryl

R —— COONa   carboxylates

R —— SO$_3$Na   sulfonates

R —— O — SO$_3$M   sulphates

(R —— O —)$_s$ POM   phosphates
                        ‖
                        O

S=1,2

Figure 1. General formulas of surfactants.

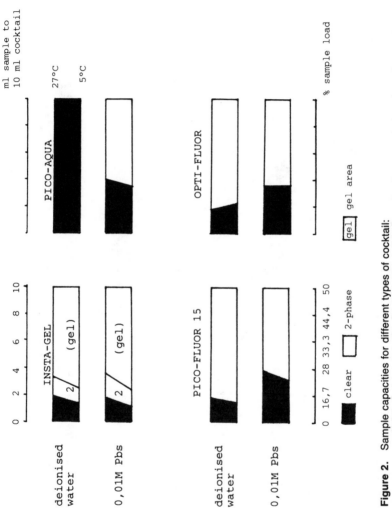

Figure 2. Sample capacities for different types of cocktail:
• Insta-Gel—classical sol-gel liquid scintillator—pseudocumene/xylene based.
• Pico-Aqua—modern liquid scintillator—pseudocumene based.
• Pico-Flour 15—nonionic/anionic scintillator—psuedocumene based.
• Opti-Flour—nonionic/anionic scintillator—phenylalkane based.

Table 4. Counting Efficiency for Tritium for Cocktails from Figure 1 (TRI CARB 2250 CA at 67% Efficiency)

ml of 0.10*M* Pbs to 10 mL of Cocktail	Insta- Gel	Pico- Aqua	Pico- Fluor 15	Opti- Fluor	% Sample Load
0	54.5	48.8	56.9	46.7	0
1	47.3	42.3	52.1	42.2	9.7
3	38.4	36.5	48.2	38.3	23.0
10	27.2	25.3	n.c.	n.c.	50.0

Note: n.c. = no capacity; two-phase sample.

Surfactant Types, Sample Load Capacity

Another illustration showing surfactant influence on sample load capacity of a cocktail for aqueous samples is given in Figure 3. The cocktails differ in composition concerning percentage nonionic surfactants and anionic surfactants as mentioned in, e.g., Benson, Hegge, and O'Conner.

The result is extraordinary. The sample load capacity of a product depends on concentration of buffer (salts) present in the aqueous sample.

COMPARISON OF COCKTAILS

Counting Efficiency General

For a comparison of tritium counting efficiency between some different types of cocktails, data are put together in Figures 4–9, using some typical samples.

The pseudo-cumene based products have the advantage of large sample holding capacities in the fluid region, while a very fast mixing of sample and cocktail is observed. Very high counting efficiency is achievable too, but consequently sample load capacity is smaller.

On the other hand, when applying high flashpoint solvents, reasonable to excellent counting efficiencies are obtained compared to pseudo-cumene based cocktails. A comparison for a common sample type is given in Figures 4 and 5.

It is obvious that the difference observed in tritium counting efficiency is much less pronounced when ^{14}C labeled samples have to be analyzed. In Table 5 some typical examples are given.

Difference in Quenching Agents

A striking difference is observed when tricholoracetic acid solutions are the subject of analysis. The application of aforementioned surfactant systems shows a remarkable difference, as shown in Figures 6 and 7.

The decrease in counting efficiency is much smaller in Pico-Aqua, Pico-Fluor 40, and Ultima-Gold compared to nonionic surfactant based products.

The example given in Figures 8 and 9 of dark yellow colored urine, a big

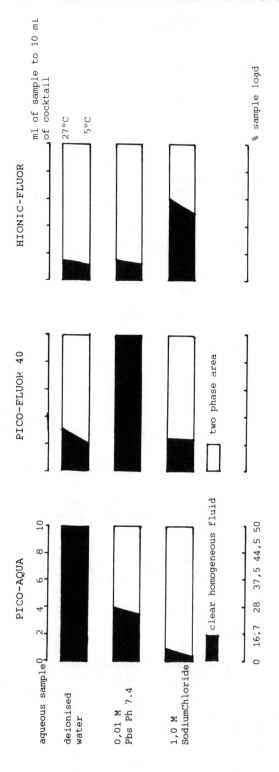

Figure 3. Sample load capacities.

Table 5. Counting Efficiency for ^{14}C (TRI CARB 1900 CA)

Cocktail Type	Capacity	% C.E.
Pseudo-cumene based	Moderate/high	95.6
Pseudo-cumene based	High	92.1
Phenylalkane based	Moderate/high	92.3
Diisopropylnaphthalene based	Moderate/high	92.5

Table 6. Counting Efficiency for Tritium Modern Cocktail (TRI CARB 2250 CA at 68% cH), Example Ultima-Gold

	1 mL of Sample to 10 mL of Cocktail					
Cocktail	None	Water	HC104 1 M	Pbs 0,01 M	TCA 10%	Sucrose 30%
Ultima-Gold	57.6	54.4	54.5	53.8	53.9	53.0

color quencher besides chemical quencher, shows good quench characteristics for modern cocktails compared to established examples.

Background Decay

Another typical aspect of the application of different types of surfactants is background decay when alkaline samples are applied. Sol-gel scintillators, with quaternary ammonium hydroxide solutions, exhibit quite large background levels. Acidification is an answer to this problem, but it needs an extra

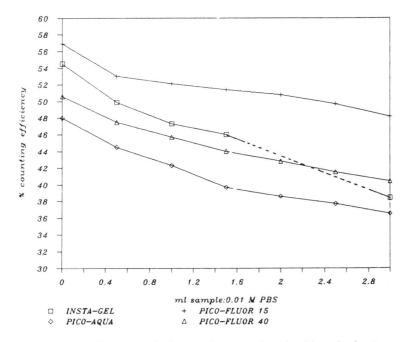

Figure 4. Counting efficiency cocktails, pseudocumene (or xylene) based solvents.

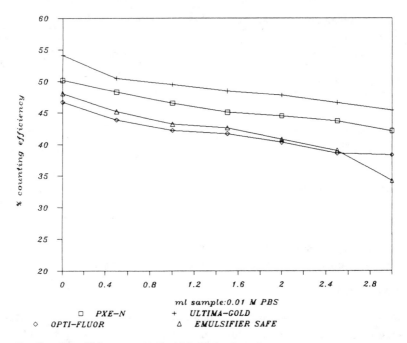

Figure 5. Counting efficiency cocktails, high flashpoint solvents.

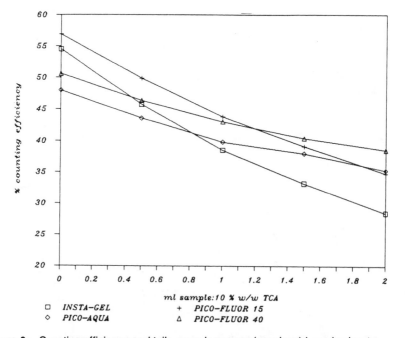

Figure 6. Counting efficiency cocktails, pseudocumene (or xylene) based solvents.

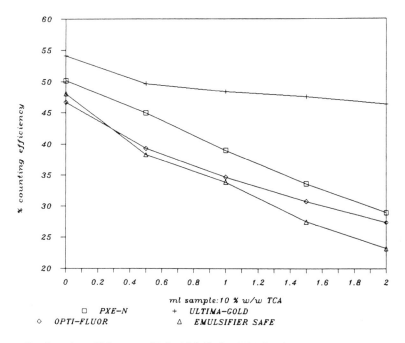

Figure 7. Counting efficiency cocktails, high flashpoint solvents.

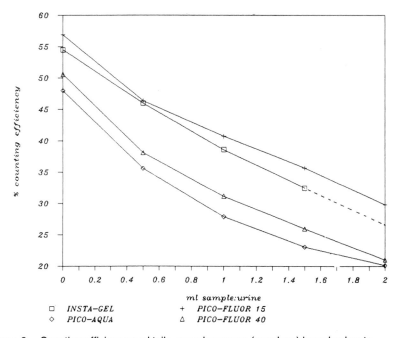

Figure 8. Counting efficiency cocktails, pseudocumene (or xylene) based solvents.

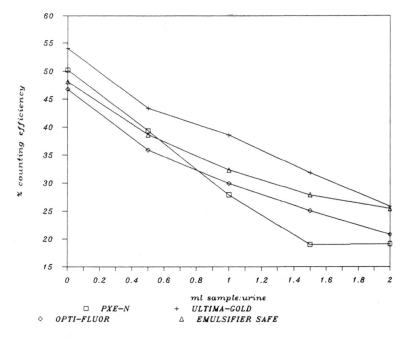

Figure 9. Counting efficiency cocktails, high flashpoint solvents.

processing step. Some modern cocktails, however, make acidification super-fluous. An example is given in Figure 10.

Newest Generation Cocktail

An example of the newest generation cocktails is described below, Ultima-Gold is based on diisopropylnaphthalene:

- high flashpoint solvent, e.g., diisopropylnaphthalene
- safety, no vapor release, very low toxicity
- easy transport and storage
- EPA classified as nonhazardous
- no bioaccumulation
- single phase sample accommodation (some examples are given in Figure 11)
- high quench resistance and counting efficiency for tritium (see Table 6)
- compatibility with alkaline samples

CONCLUSION

Advances in liquid scintillation counting are clearly demonstrated. The application of both new solvents and surfactant systems contributed to the progress. Pseudo-cumene was a first step forward, and it is still important. The introduction of new high flashpoint solvents such as alkylbenzenes and

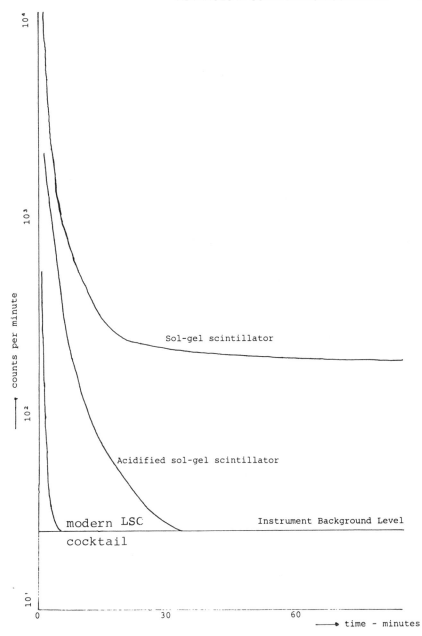

Figure 10. Some modern cocktails make acidification superfluous.

diisopropylnapththalene provided a significant increase in all safety, toxicity, and economic aspects.

The newest generation of liquid scintillation cocktails made improvements in many areas. A high tritium counting efficiency, a fast sample incorporation,

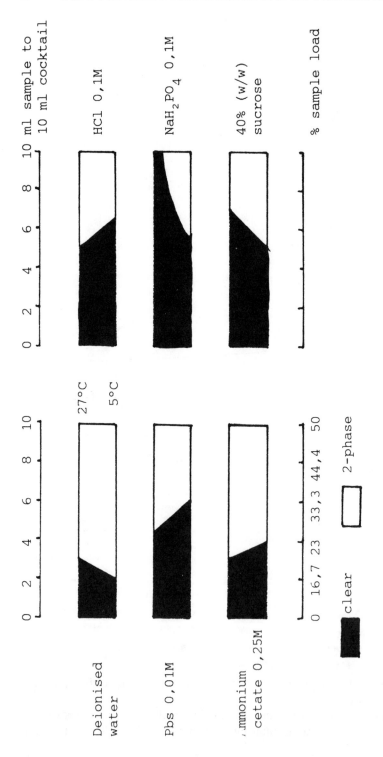

Figure 11. Phase diagram for aqueous samples of a modern high flashpoint liquid scintillation cocktail.

and a high and expanded sample load capacity are achieved. Until now a major drawback to cocktails based on a high flashpoint solvent was the higher viscosity, making sample handling a little more difficult.

These developments all contributed to the safer and easier use of the liquid scintillation technique while enhancing the quality of results.

REFERENCES

1. O'Conner, J.L. and E.D. Bransome. "Difficulties in Counting Emulsions of ^{3}H and ^{14}C Labeled Molecules," in *Liquid Scintillation Counting Recent Application and Developments, Vol. 2*, C.T. Peng, D. Horrocks, and E.L. Alpen, Eds. (New York: Academic Press, Inc., 1980).
2. Peng, C.T. "Sample Preparation in Liquid Scintillation Counting," in *Advances in Scintillation Counting*, S.A. McQuarrie, C. Ediss, and L.I. Wiebe, Eds. (1983), p. 279.
3. Bray, G.A. *Analytical Biochemistry, 1* (4–5):279–285 (1960).
4. Meade, R.C. and R.A. Stiglitz. *Int. J. Appl. Radiat. Isotopes* 13:11 (1962).
5. Thomson, J. *Scintillation Counting Medium and Counting Method*, U.S. Patent 4,657,696.
6. "Toxicological and Physicochemical Studies on KMC," (Rutgers Kureha Solvents GmbH).
7. Benson, R.H. "The Importance of Phase Contact in Sol-Gel Scintillator, Aqueous Sample Systems," in *Liquid Scintillation Counting, Vol. 2*, C.T. Peng, D.L. Horrocks, and E.L. Alpen, Eds. (New York: Academic Press, Inc., 1980), pp. 237–244.
8. Liebermann, R. and A.A. Moghissi. *Int. J. App. Radiat. Isotopes* 21:319–327 (1970).
9. Hegge, Th.C.J.M. and J. ter Wiel. *Mixture for Use in the LSC Analysis Technique* in U.S. Patent 4,624,799, priority 1983, Netherlands 8,303,213.
10. Sena, E.A., B.M. Tolber, and C.L. Sutula. *Liquid Scintillation, Counting and Compositions*, U.S. Patent 3,928,227.
11. Mallik, A., and H. Edelstein. *Liquid Scintillation Composition for Low Volume Specimens*, U.S. Patent 4,443,356.

CHAPTER 7

Solidifying Scintillator for Solid Support Samples

Haruo Fujii, Ph.D. and Norbert Roessler, Ph.D.

ABSTRACT

Although solid support counting suffers from some disadvantages such as self-absorption and relatively poor reproducibility, it is widely employed for screening assays because it makes sample preparation easy. However, as the number of assays increases, the amount of scintillator consumed makes waste disposal costs problematical in the light of environmental restrictions.

A solidifying scintillator formulation will be described which allows for dispensing of the scintillator in liquid form with subsequent solidification at room temperature. Issues of sample handling, counting performance, and configuration of the counting equipment, will be discussed in comparison to liquid cocktails.

INTRODUCTION

Solid support counting of weak beta emitters is an established radioassay technique. It is widely employed because the advantages obtained by easy sample preparation outweigh the disadvantages resulting from poor reproducibility due to self-absorption. In practice, radioactive material is spotted onto chromatography paper or isolated on a filter which is then dried and immersed in a cocktail for counting.[1-3] A common element of many of these assays is that no solubilizer is used and the sample remains on the solid support rather than dispersing into the bulk of the cocktail. This adds the effect of nonreproducible counting geometry to the self-absorption problem. Another drawback of this technique is that excessive amounts of cocktail are used. This leads to a vexing disposal problem when large numbers of samples are processed.

These problems can be overcome by using a paraffin based solidifying scintillator. The solidifying scintillator was formulated to allow the experimenter to reverse the phase of the sample from solid to liquid scintillator. The melting point is chosen to allow for good counting efficiency in the solid phase at room temperature and for easily dispensing the material at moderately elevated temperatures. Since the sample is solid it can be handled easily to obtain good counting geometry. Also, the smaller volume solid samples are more easily disposed of.

Paraffin Scintillator Formulation and Measurement Geometry

Common to all conventional scintillators, whether solid or liquid, is the fact that the phase does not change after sample preparation. To obtain a cocktail which can be dispensed in liquid form and counted as a solid, a conventional organic scintillator is modified by replacing a part of the solvent (p-xylene, diisopropylnaphthalene) with pure paraffin having an appropriate melting point. The melting point can be adjusted within limits by changing the molecular weight distribution of the linear hydrocarbons which make up the paraffin.

The paraffin scintillator consists of PPO, bis-MSB, paraffin, and solvent. The paraffin can be homogeneously mixed with solvent and liquid scintillation fluors without giving rise to quenching. PPO and bis-MSB were used because they are the most popular liquid scintillation fluors.[4] The solvent dissolves the fluors and works effectively as an energy transfer medium to the primary fluor, which is essential for obtaining a high counting efficiency.[5,6]

In use, the scintillator is heated to about 40°C to melt it, and 0.3 mL of the liquid is dropped on a solid support sample. After 10 sec, it solidifies, and a rigid and translucent solid support sample can be obtained. The appearance of the prepared samples is somewhat cloudy, but it is not necessary for them to be fully transparent. Once the solid sample is formed, it is stable and will remain rigid at room temperature.

The solid support samples thus impregnated with the paraffin scintillator can be counted in a liquid scintillation counter with each sample suspended in a polyethylene bag which is vertically supported in a plastic holder as shown in Figure 1. It is well known that the counting efficiency of the solid support sample depends on its orientation in a counter.[7-9] Best results are obtained when the plane of the solid support sample is parallel to the photomultiplier faces, due to improved light dispersion from the sample.[9]

EXPERIMENTAL

Solid Support Samples

Aqueous solutions of ^3H leucine and ^{14}C uridine with known activity were employed; 0.1 mL of these solutions were individually deposited on solid supports, and then dried overnight at room temperature. The solid supports used here are glass fiber filters, membrane filter chromatography paper, and thin layer chromatography plates. The samples were counted on a liquid scintillation counter, Packard TriCarb Model 4000 Series, coupled with a multichannel pulse height analyzer.

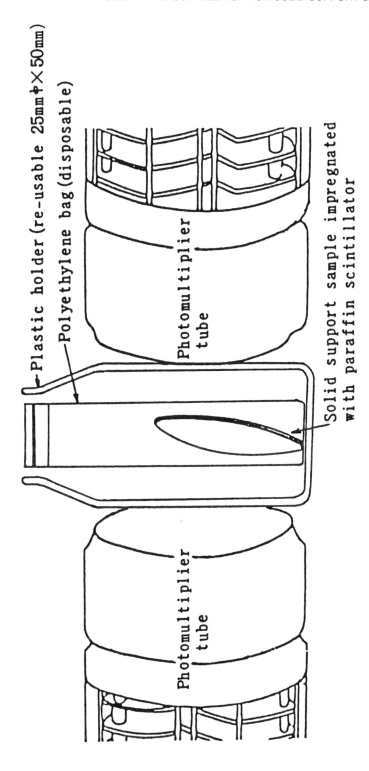

Figure 1. Measurement geometry for solid support sample in LSC.

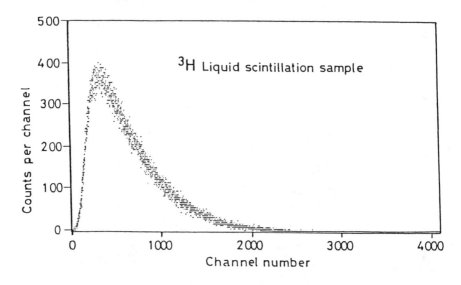

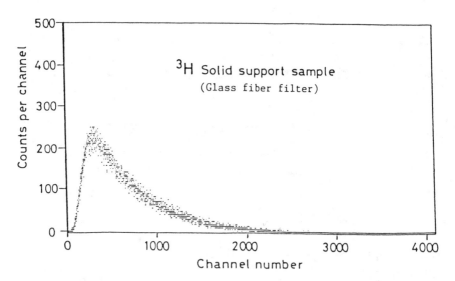

Figure 2. Pulse height spectrum of a ³H solid support sample: in a conventional cocktail (top), treated with the solidifying scintillator.

Homogeneous Samples

Several formulations were also prepared as homogeneous mixtures to test the relative counting performance. In these formulations, 5 g PPO and 0.8 g bis-MSB were dissolved in 250 mL of solvent and mixed with 500 mg of paraffin. After thorough mixing at an elevated temperature, 15 mL aliquots were dispensed into 20 mL scintillation vials. The solvents tested were xylene

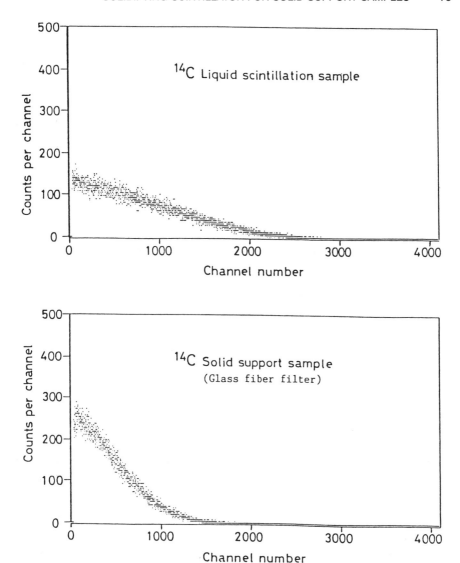

Figure 3. Pulse height spectrum of a ^{14}C solid support sample: in a conventional cocktail (top), tested with the solidiyfying scintillator.

(X), pseudo-cumene (P), and diisopropylnaphthalene (KO, K5). Different batches of paraffin were used for formulations K0 and K5.

The formulations were evaluated by spiking them with ^{3}H or ^{14}C labeled toluene. After thorough mixing, each sample was allowed to solidify. The samples were assayed in a Packard TriCarb Model 2000CA as well as an experimental single tube time-resolved counting apparatus.[10]

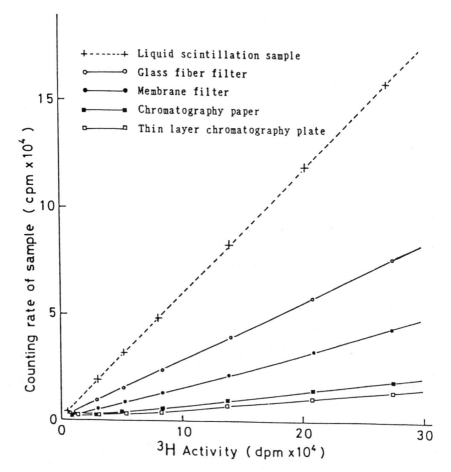

Figure 4. Count rate as a function of ^{3}H activity for various solid support samples.

Results

The scintillation pulse height spectrum was investigated in order to better interpret the counting data. The pulse height distribution of the ^{3}H solid support sample, prepared in the technique described in Figure 1, is very similar to that of the conventional ^{3}H sample, although the counting rate for each channel may be reduced (Figure 2). As for a ^{14}C sample, however, the volume and dispersion area of the paraffin scintillator are not large enough to absorb the β-ray energy of the ^{14}C completely, which seems to cause a shifted pulse height distribution (Figure 3).

The optimum fluor concentration and volume ratio of paraffin to p-xylene were also investigated. The formulation for best counting efficiency was found to be; PPO: 10 g; bis-MSB: 1.0 g; paraffin: 670 mL; and p-xylene: 330 mL.

The relationship between the measured activity and the count rate is shown

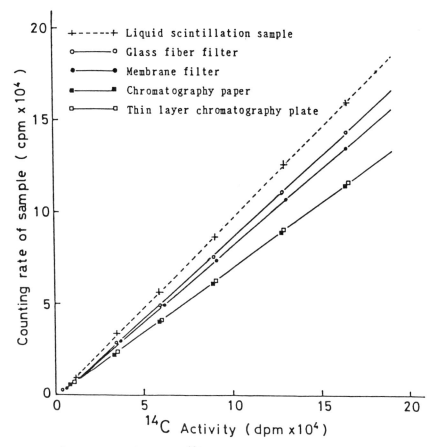

Figure 5.   Count rate as a function of ^{14}C activity for various solid support samples.

in Figures 4 and 5 for various kinds of solid support samples. The relationship is linear over a wide range of activity for each nuclide.

Freshly prepared samples were monitored for several days (Figure 6). The count rate of the sample does not decrease with the time elapsed for more than two weeks. This indicates the absence of any significant vaporization of the paraffin scintillator from the prepared sample.

The counting efficiencies for each support material determined are displayed in Table 1; ^{3}H and ^{14}C can be measured with counting efficiencies of 6–30% and 70–87% respectively. The difference in the counting efficiency for each support material can be attributed to that of the β-ray self-absorption and photon reduction inside the solid support.

Figure 7 displays the ^{3}H and ^{14}C efficiency values of homogeneous paraffin solutions measured on a two tube coincidence counter. The formulations tested used the solvents xylene (X), pseudo-cumene (P), and diisopropylnaph-

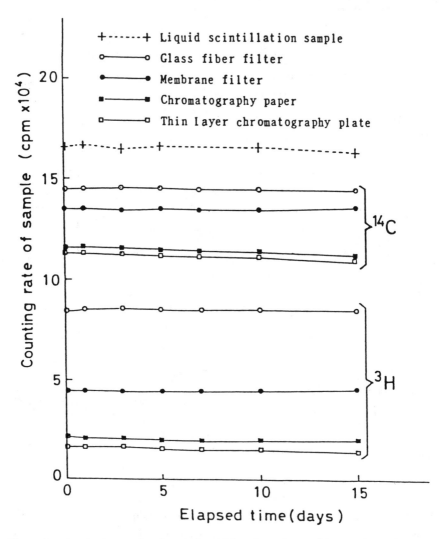

Figure 6.   Count rate as a function of elapsed time for various solid support samples.

thalene (K5, K0). We see that excellent efficiencies are obtained for the solid samples.

In Figure 8, ^3H efficiency values for the formulations are displayed as obtained from the experimental time-resolved single tube counter (Valenta). The apparatus is able to operate as a two pulse coincidence counter — requiring that a second pulse be registered in the tube after the first — or as a three pulse triple coincidence counter. While cocktail formulation had little effect on efficiency in the two-tube counter, we see a dramatic improvement for diiso-propylnaphthalene in the single tube counter. The results for ^{14}C (see Figure 9) are similar.

Table 1. Comparison of Counting Efficiencies Obtained from this Study for some Kinds of Solid Support Samples

Nuclide	Sample No.	Glass Fiber Filter (14 mg/cm²)[a]	Membrane Filter (5.5 mg/cm²)	Chromatography Paper (8.5 mg/cm²)	Thin Layer Chromatography Plate (0.1 mm)[b]
³H	1	29.9	15.5	5.77	7.60
	2	30.9	15.7	5.60	7.02
	3	31.0	15.3	5.56	7.43
	4	30.7	15.3	5.60	7.38
	5	30.0	16.2	5.54	7.58
	Mean ± S.D.[c]	30.5 ± 0.2	15.6 ± 0.2	5.61 ± 0.04	7.40 ± 0.10
¹⁴C	1	85.6	80.1	71.1	71.1
	2	86.5	82.2	71.4	72.8
	3	86.8	82.5	70.2	71.1
	4	86.5	81.9	70.4	69.5
	5	87.1	81.5	70.6	69.6
	Mean ± S.D.	86.5 ± 0.3	81.6 ± 0.4	70.7 ± 0.2	70.8 ± 0.6

[a]Surface density.
[b]Thickness of thin layer.
[c]Standard deviation.

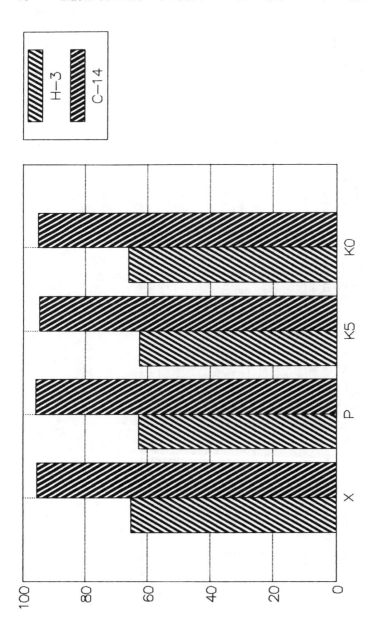

Figure 7. LSC counting efficiencies for paraffin scintillator samples using various solvents: (X) xylene, (P) pseudo-cumene, (K5, KO) diisopropylnaphthalene.

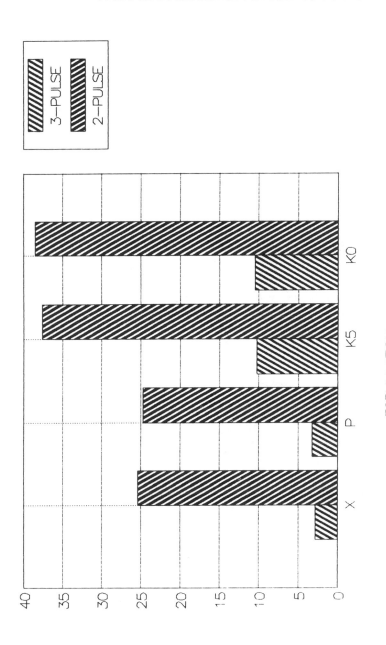

Figure 8. Time resolved single tube counting efficiencies for ^{3}H paraffin scintillator samples using various solvents: (X) xylene, (P) pseudo-cumene, (K5, KO) diisopropylnaphthalene.

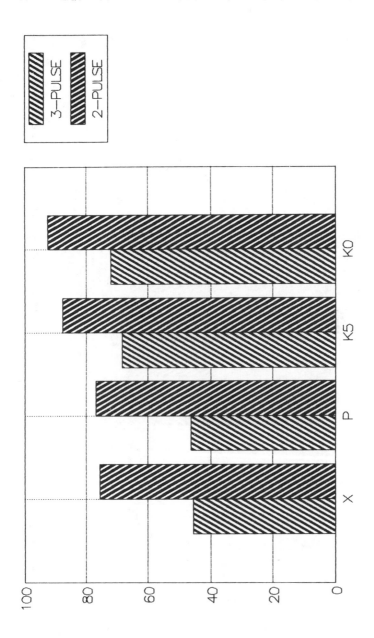

Figure 9. Time resolved single tube counting efficiencies for ^{14}C paraffin scintillator samples using various solvents: (X) xylene, (P) pseudo-cumene, (K5, KO) diisopropylnaphthalene.

Discussion

The results described above indicate that a solidifying scintillator with good counting performance can be formulated using paraffin as a base material. The scintillator is applied to samples isolated on a solid support while melted. This allows it to impregnate the sample while it is in the liquid state. Because it is solid on cooling, an approximate sample geometry for different counting geometries can be assured.

Chemical quenching is not a concern in solid support counting, since the paraffin scintillator becomes rigid before the quenching substance is eluted into the scintillator. Color quenching, on the other hand, may be present when a colored compound is counted. However, considering the thinness of the sample and the possibility of spreading it over a large area, the color quenching effect can be minimized. These considerations mean that samples prepared using a paraffin scintillator will have uniform counting efficiency, although that efficiency will depend, to a large extent, on the kind of solid support.

In summary, the main advantages of this technique are:

1. Decreased cocktail consumption
2. Reduced disposal costs because each sample generates a small volume of solid radioactive waste
3. Good reproducibility due to improved measurement geometry
4. Easy and rapid sample preparation

REFERENCES

1. Peng, C.T. *Sample Preparation in Liquid Scintillation Counting*, RCC Review 17, (The Radiochemical Centre Ltd., England, 1977), p. 64.
2. Horrocks, D.L. *Applications of Liquid Scintillation Counting*, (Academic Press, New York, 1970), p. 156.
3. Furlong, N.B. *The Current Status of Liquid Scintillation Counting*, (Grune & Stratton, New York, 1970), p. 201.
4. L'Annunziata, M.F. *Radionuclide Tracers, Their Detection and Measurement*, (Academic Press, London, 1987), p. 182.
5. Birks, J.B. and G.C. Poullis. *Liquid Scintillation Counting*, Vol. 2, (Heyden, London, 1972), P. 327.
6. Laustriat, G., R. Voltz, and J. Klein. *The Current Status of Liquid Scintillation Counting*, (Grune & Stratton, New York, 1970), p. 13.
7. Winder, F.G. and G.R. Campbell. *Anal. Biochem.*, 57:477 (1974).
8. Vanderheiden, B.S. *Anal. Biochem.* 49:459 (1972).
9. Nakshbandi, M.M. *Int. J. Appl. Radiat. Isot.* 16:157 (1965).
10. Valenta, R.J. *Method and Apparatus for Measuring Radioactive Decay*, U.S. Patent 4,528,450 (1985).

CHAPTER 8

New Red-Emitting Liquid Scintillators with Decay Times Near One Nanosecond*

J.M. Flournoy and C.B. Ashford

ABSTRACT

Several fast liquid scintillators with peak emission wavelengths in the 600 to 750 nm range are described, which are useful for transmitting fast plasma diagnostic information through long fiber optic cables. These fluors all use styryl laser dyes as the final emitters: DCM in the 650 nm fluors, and LDS-722 in the 735 nm fluors. The solvents are either straight benzyl alcohol (BA) or BA mixed with 1-methylnaphthalene. Coumarin-480 is the intermediate wavelength shifter for the DCM fluors and rhodamine-610 perchlorate (Rh-610P) for the LDS-722 fluors. Some systems also contain added tetramethyltin for enhanced X-ray sensitivity. It was found that very high rhodamine concentrations cause a large increase in the solubility of LDS-722 in BA, which leads to a decrease in scintillation decay time. The fastest 735 nm fluor has a WHM of 370 psec and a decay time of about 320 psec. The fastest 600 nm emitter has a FWHM of 1.1 nsec and a decay time of 0.87 nsec.

INTRODUCTION

Several fast, red-emitting liquid scintillator formulations have been described previously.[1,2] They all consist of relatively polar laser dyes dissolved in polar solvents, either benzyl alcohol or benzonitrile. Intermediate wavelength shifters are used in all cases to bridge the gap between the solvent emission spectrum and the absorption spectrum of the final emitter. The new, faster fluors are the result of changes in the solvent and/or the intermediate shifter.

---

*This work was performed under the auspices of the U.S. Department of Energy under Contract No. DE-AC08–88-NV10617.

The efficiencies of all the red-emitting fluors are relatively low, only about 3% or 4% of that of anthracene.[2] However, that is the price one pays for very fast time response, which is usually a result of quenching of the excited states and/or nonradiative internal molecular deactivation processes, both of which reduce the fraction of the excited molecules that actually emit light. It should be noted, though, that the peak efficiencies are affected less than the integral efficiencies when the scintillator is quenched. This comes about because the area of a scintillation pulse can be approximated by the product of the FWHM and the peak height. Thus, when the decay time and the integral light output both become smaller, the peak intensity tends to remain more constant. The extent to which the peak height remains constant depends on the pulse rise time; the faster the rise time, the less quenching affects the peak efficiency. The peak efficiency that is most important when the time response of the fluor is slower than the overall time response of the measurement system. On the other hand, it is actually disadvantageous to employ a fluor whose response time is significantly faster than the overall response of the measurement system, because the peak intensity will be integrated to a lower value by the slower system. It would, therefore, be better to employ a slower but brighter fluor whose time response is appropriate for the bandwidth of the detection system.

The terms "scintillator" and "fluor" are used interchangeably in this paper. For the sake of brevity, abbreviations are used for most of the individual fluor ingredients and some formulations, as shown in the Appendix.

EXPERIMENTAL

All the materials were used as received from the manufacturer, except for 1-methylnaphthalene (AMN), which was purified on a spinning-tape column to remove a distinct yellow color. The fluorescence emission spectra were taken with a SPEX Fluorolog 2 spectrofluorimeter. The exciting wavelength was either 300 or 350 nm. Emission spectra were observed from the same face as the incident excitation light, as well as through a 1 cm thick solution. The latter measurement gives an indication of the extent of self-absorption of the emitted light. Absorption spectra were taken with a Beckman Model UV 5270 absorption spectrophotometer. The dye concentrations are too high to easily determine the complete absorption spectra of the actual fluors, although spectra of the individual ingredients could be taken in dilute solution. As a rough criterion for estimating solutions transparency to their own emission, we have chosen to tabulate the wavelength beyond which the transmittance of a 1-cm thick fluor is greater than 50%.

The time-response data were obtained with a time-correlated single photon counting system that has been described elsewhere,[3] but some of the data were obtained with excitation by a ^{90}Sr beta source, which gave a much faster data rate than the ^{60}Co source mentioned in Reference 3. The time response FWHM of the system was determined by the beta particles response to flashes of

Cerenkov light generated in a piece of high-purity quartz or by Compton electrons produced by the ^{60}Co gamma rays. The system FWHM was less than 200 psec; therefore, the scintillation pulses were not significantly broadened by the instrument response. Start-timing signals were obtained from a disc of the bright, commercial (Bicron) scintillator, BC-422.

Relative scintillation efficiencies of the various fluors were determined from the relative single-photon count rates for ^{60}Co gamma ray excitation. The comparisons were made through a 570-nm long-pass filter, because the emission peaks were not the same. The observed count rates were corrected for background dark counts and for Cerenkov light produced by 1 cm of pure solvent, observed through the same filter. The relative peak efficiencies were estimated by dividing the integral efficiencies by the respective FWHM values.

RESULTS

735-nm Fluors

A fast, 735 nm liquid scintillator, consisting of 0.02 M LDS-722 and 0.10 M C-540A in benzonitrile (BN), has been described previously.[2] This fluor has an impulse response FWHM of about 1.7 nsec, and has been named L-735. When 10% by volume of tetramethyltin (TMSN) was added to improve the absorption cross section for soft X-rays, the solubility of the LDS-722 decreased to just over 0.01 M. A fluor with 10% TMSN, 0.01 M LDS-722, and 0.10 M C-540A is called L-735A-10T, and its pulse parameters appear at the bottom of Table 1. The methyl groups in the TMSN effectively shield the tin atoms from quenching the excited singlet states,[4] as can be seen from the relative efficiencies and decay times of L-735A (without TMSN) and L-735A-10T.

A number of experiments with changes in solvent and intermediate wavelength shifters have now been carried out, partly in an effort to reduce the slight photochemical degradation that was observed with L-735 under ultraviolet illumination. One such combination, benzyl alcohol (BA) as the solvent and rhodamine 610 perchlorate (Rh-610P) as the shifter,* actually decreased the photochemical stability somewhat but resulted in a series of fluors with exceptionally fast time response.

In the course of these studies, it was found that Rh-610P is much more soluble in BA than had been realized: to approximately 0.45 M/L. This amounts to about 25 wt% of the rhodamine salt, so the medium can no longer be expected to have the same solvent properties as benzyl alcohol. One unexpected consequence of this is that the solubility of LDS-722 in BA increases

---

*It might be mentioned that rhodamine 610 (rhodamine "B") has an exceptionally small Stokes' shift, and therefore would not be expected to be useful as an intermediate shifter. We attribute its efficiency in this application to excitation into S2 or S3 by energy transfer from the solvent, since these bands occur near 300 nm.

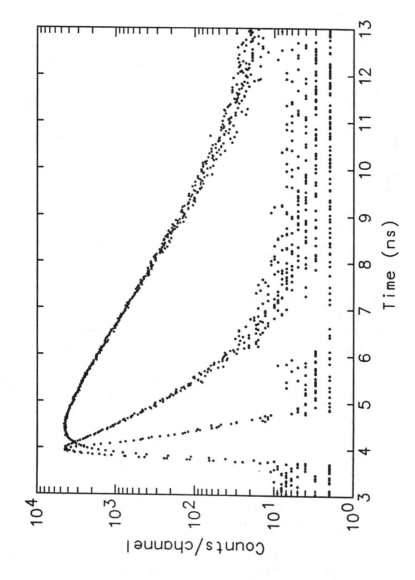

Figure 1.   Time-correlated photon counting data for L-735 and L-735XF, along with the system response function. The slowest decay time shown, 1.27 nsec, is for L-735.

dramatically as the rhodamine concentration is increased, from just over 0.005 M LDS-722 in pure BA to at least 0.02 M in BA containing 0.40 M Rh-610P. The combined effects of increasing the concentration of the LDS-722 and the Rh-610P result in the fastest red-emitting scintillators we have observed to date. Figure 1 contains semilogarithmic plots of data for the fastest of these, L-735XF, as well as for the original L-735 and the pertinent system response function.

Table 1 contains pulse parameters, relative brightness with gamma excitation, and fluorescence emission maxima for a series of these BA-based fluors. The fluorescence emission maxima were determined with an excitation wavelength of 300 nm, and the fluorescence was observed from the opposite face of the 1 cm spectrophotometer cells. The 50% absorption edges for the 1 cm optical path appear in the last column of the table.

The results in Table 1 are presented in order of increasing rhodamine concentration. The fluors described in the last two lines have FWHM values near 500 psec and peak emission wavelengths near 740 nm. The 50% absorption edges for these scintillators increase rather smoothly from 675 nm for the 0.01 M to 0.03 M solutions to 698 nm for 0.20 M Rh-610P. The solution with no rhodamine is the only one out of line. The $1/e$ self-absorption lengths are estimated to be in the range of 30 to 50 cm at 740 nm, which is of considerable interest for possible use in systems involving fluor-filled capillaries.

Incidentally, the decay time of Rh-610P itself in BA was found to decrease from almost 9.0 nsec in very dilute solution ($5.0 \times 10^{-6}\ M$) to 2.1 nsec at 0.10 M and about 1.35 nsec at 0.20 M. This may be either an energy-transfer effect or possible evidence of concentration quenching.

600 to 650 nm Fluors

A fast, 650 nm liquid scintillator consisting of 0.03 M DCM and 0.10 M C-480 in BA has also been described previously.[2] Its impulse response FWHM is 1.8 nsec, and it has been named L-650A. The original motivation for trying AMN as a solvent for DCM-based fluors was twofold: to enhance the efficiency, since BA is not a particularly efficient scintillator solvent, and to take advantage of the high refractive index of AMN, 1.60, for possible use in scintillating capillaries. A number of DCM-containing fluor formulations in BA, AMN, and mixtures of the two have been examined, and the principal results are shown in Table 2, in order of increasing proportion of AMN. It can be seen that the addition of AMN generally causes the integral efficiency to increase, as had been hoped. What was not anticipated was that the AMN-rich fluors were also faster. As a result, the peak efficiencies increase even more than the integral efficiencies as the solvent is changed from BA to AMN.

The effect of coumarin-480 as an intermediate wavelength shifter can be seen in the first two lines and the last two lines of Table 2. When the solvent is BA, there is only fair spectral overlap between the solvent emission and the DCM absorption, resulting in the fairly slow pulse rise time of about 1.05 nsec.

Table 1. Pulse Parameters for ^{60}Co Gamma-excited LDS-722 in BA with and without 10% Tetramethyltin (TMSN), using Rhodamine 610 Perchlorate (Rh-610P) as the Intermediate Wavelength Shifter. A 670-nm Long-pass Filter was used in all the Runs. The Peak and Integral Values are Referenced to L-735-A-10T. RT is the 10% to 90% Pulse Rise Times, TAU is the 1/e Decay Time (taken from 0.7 to 0.7/e of the peak height), and IRT is the 10% to 90% Rise Time of the Integral of the Pulse

Rh-610P (M/l)	LDS-722 (M/l)	RT (nsec)	FWHM (nsec)	TAU (nsec)	IRT (nsec)	Peak (est.)	Integral	Wavelength (nm) Emission maximum	Absorption edge (50%)
0.00	0.010[a,b]	(1.0)	6.44	6.11	16.1	0.2	0.69	720	685
0.00	0.020[a]	—	(6.6)	5.25	12.1	—	0.60	—	—
0.02	0.010[b]	1.18	3.26	2.03	4.6	0.55	0.98	720	675
0.05	0.010[a,b]	0.73	2.08	1.37	3.3	0.75	0.81	722	678
0.10	0.010[b]	0.47	1.39	0.99	2.4	0.6	0.45	730	690
0.20	0.005	0.33	1.52	1.25	2.85	—	—	—	—
0.20	0.010[b]	0.37	0.97	0.63	1.75	0.5	0.25	736	698
0.30	0.005[b]	0.25	0.55	0.58	2.1	0.45	0.14	—	—
0.40	0.020	0.15	0.37	0.32	0.91	0.3	0.06	740	—
L-735	0.020[c]	0.35	1.67	1.27	2.25	1.1	1.0	735	711
L-735A	0.010[c]	0.54	2.12	1.48	3.4	1.2	1.1	735	—
L-735A-10T	0.010[b,c]	0.60	1.89	1.42	3.4	[1.00]	[1.00]	732	698
System response[d]		0.134	0.193	0.085	0.36	—	—	—	—

[a]LDS-722 was supersaturated; some crystallized out over several days.
[b]Contains 10 volume percent tetramethyltin.
[c]Shifter is 0.10 M C-540A; solvent is benzonitrile. See Appendix for composition.
[d]Response to Cerenkov light flashed from Compton electrons generated by the ^{60}Co gamma rays (see Reference 4).

Table 2. Pulse Parameters for DCM Fluors in BA, AMN, and Mixtures of the Two. A 650-nm Band-pass Filter was used in all the Runs. The Peak and Integral Values are Referenced to L-650A. RT is the 10% to 90% Pulse Rise Times, TAU is the 1/e Decay Time (taken from 0.7 to 0.7/e of the peak height), and IRT is the 10% to 90% Rise Time of the Integral of the Pulse.

C-480 (M/l)	DCM (M/l)	AMN %	RT (nsec)	FWHM (nsec)	TAU (nsec)	IRT (nsec)	Peak (est.)	Integral	Wavelength (nm)	
									Emission maximum	Absorption edge (50%)
0.0	0.03	0.0	1.06	3.58	2.93	6.03	0.9	1.77	—	—
0.10	0.03	0.0	0.69	1.79	1.58	4.05	[1.00]	[1.00]	648	629
(L-650-A reference)										
0.10	0.05	50%	0.59	1.62	1.49	3.90	1.4	1.26	640	624
0.20	0.05	67%	0.49	1.37	1.13	3.22	1.25	0.96	632	622
0.10	0.05	100%	0.49	1.11	0.87	2.85	1.7	1.06	600	605
0.0	0.05	100%	0.57	1.61	1.34	4.11	1.55	1.38	600	600

When 0.10 M C-480 is added, the pulse rise time decreases to about 0.7 nsec, due to improved energy transfer. On the other hand, the pulse rise time of DCM in AMN, without C-480, is nearly as short as that of L-650A, because there is much better spectral overlap between the broad emission spectrum of AMN and the absorption band of DCM than there is between BA and DCM. This is why the addition of C-480 to DCM in AMN has less effect on the pulse rise time than it does in BA.

In addition, there is quenching of DCM by C-480 in either solvent, as shown by the decreases in decay time and integral brightness when C-480 is added. Quenching by C-480 accounts for the lower integral efficiency of the fluor with 67% AMN than the fluor with only 50% AMN, since there is twice as much C-480 in the 67% fluor. The fact that there is quenching by C-480 was confirmed by laser-excitation experiments with DCM in BA, in which a similar effect was observed.

Unfortunately, the fluors with AMN show such strong self-absorption that their use in capillaries seems unlikely. This is due to the fact that the emission spectra shift faster towards the blue than do the absorption spectra as the proportion of AMN is increased. When the solvent is changed from pure BA to pure AMN, the emission peak wavelength for DCM moves from 650 to 600 nm, but the absorption peak moves only from 490 to 470 nm. The 50% absorption edge through a 1 cm path of 0.05 M DCM moves from about 630 to about 600 nm, while the peak emission moves from 648 to about 600 nm.

CONCLUSIONS

The time response of scintillator formulations can be strongly dependent on the nature of the solvent, as well as any intermediate wavelength shifter. In each of the systems described above, there is quenching of the final emitter by the intermediate shifter, which was completely unanticipated, and which is largely responsible for the faster time response of the modified formulations. These results also demonstrate that a considerable fraction of the peak efficiency can be retained in the quenched scintillator, provided the pulse rise time is fast enough.

ACKNOWLEDGEMENTS

The authors gratefully acknowledge the contributions of Mr. Steven Lutz, who developed the original L-735 formulation, Dr. Isadore Berlman, who originated the idea of tetramethyltin as a nonquenching, heavy-atom additive, and Prof. Bruce Rickborn for purification of the AMN.

REFERENCES

1. Lutz, S.S., L.A. Franks, J.M. Flournoy, and P.B. Lyons. "New Liquid Scintillators for Fiber Optic Applications," in *Proceedings of Los Alamos Conference on Optics*, SPIE 288, pp. 322–328 (1981).
2. Ogle, J.W., D. Thayer, L. Looney, F. Cverna, G. Yates, C.E. Iverson, S.S. Lutz, M.A. Nelson, and B. Whitcomb. "Radiation-Induced Imaging System Over Long Fiber-Optic Bundles," SPIE 720, pp 24–30 (1986).
3. Flournoy, J.M. "Measurement of Subnanosecond Scintillationn Decay Times by Time-Correlated Single Photon Counting," *Radiat. Phys. Chem.*, 32, pp. 265–268 (1988).
4. Ashford, C.B., I.B. Berlman, J.M. Flournoy, L.A. Franks, S.G. Iversen, and S.S. Lutz. "High-Z Liquid Scintillators Containing Tin," *Nucl. Instr. Meth. Res. Sect. A (Netherlands)*, A243, pp. 131–136 (1986).

APPENDIX

AMN	alpha-methylnaphthalene, or 1-methylnaphthalene
BA	benzyl alcohol (alpha-hydroxy toluene)
BN	benzonitrile (phenyl cyanide, cyanobenzene)
C-480	Coumarin 480 (or Coumarin 102)
C-540A	Coumarin 540A (or Coumarin 153)
DCM	red-emitting styryl laser dye, also Kodak dye No. 14567
L-650A	BA with 0.1 M C-480 and 0.03 M DCM
L-735	BN with 0.10 M C-540A and 0.02 M LDS-722
L-735A	BN with 0.10 M C-540A and 0.01M LDS-722
L-735A-10T	BN with 0.10 M C-540A, 0.01 M LDS-722, and 10% TMSN
L-735XF	BA with 0.40 M Rh-610P and 0.02 M LDS-722
LDS-722	far-red-emitting styryl laser dye (Exciton)
Rh-610P	Rhodamine 610 (or Rhodamine B) perchlorate salt
TMSN	tetramethyltin

CHAPTER 9

New Developments in X-ray Sensitive Liquid Scintillators at EGG/EM*

C.B. Ashford, J.M. Flournoy, S.S. Lutz, and I.B. Berlman

ABSTRACT

A summary of the liquid scintillators developed at EGG/EM SBO is presented. Among the characteristics presented are the emission spectra and efficiency values. All of the scintillation solutions mentioned herein have the potential for sensitivity enhancement to X-rays. One such successful red-emitting solution is L-735A-10T which shows a severalfold enhancement for a sample thickness of 6 mm, excited by 17-keV X-rays.

Other compounds containing heavy atoms such as tetramethylgermanium, tetraethyllead, and tetramethyllead were tested for possible heavy-atom quenching effects using high-energy electrons from a linac. The germanium compound showed no quenching effects, but both lead compounds exhibited considerable quenching of the fluorescence.

INTRODUCTION

At EG&G/Energy Measurements Inc., Santa Barbara Operations, work is in progress in the development of fast and efficient scintillators whose peak emissions are in the range from blue (350 nm) to red (840 nm). Some of the more successful formulations[1] are summarized in Table 1, along with the two commercially available plastic scintillators, BC-422 and BC-400.[2] Their spectral efficiencies are illustrated in Figure 1. The data in both table and figure were obtained by exciting the sample with a pulse of 16-MeV electrons whose FWHM is 50 psec.

In our work incorporating compounds with heavy atoms into scintillation solutions in order to increase their sensitivity to X-rays, we found that tetrabu-

---

*This work was performed under the auspices of the U.S. Department of Energy under Contract No. DE-AC08–88-NV10617.

Table 1. Properties of Liquid Scintillators Developed at EGG/EM Santa Barbara Operations. Excitation by 50 ps FWHM Pulses of 16-MeV Electrons

Scintillator	Final Emitter	Conc. (M/l)	Shifter	Conc. (M/l)	Solvent	PRT (nsec)	FWHM (nsec)	Decay (nsec)	IRT (nsec)	% Eff.[a]	Peak Lambda (nm)
BC-400	—	—	—	—	—	0.9	3.1	2.3	9.4	2.31	420
BC-422	—	—	—	—	—	0.24	1.58	1.72	4.44	1.91	390
L-360	SG-180	0.035	—	—	PC	0.45	2.21	1.56	4.17	1.56	360
L-370	BHTP	0.14	—	—	PC	0.18	0.45	0.40	1.26	0.24	370
Liquid-A	TPB	0.022	PBD	0.027	PC	0.23	1.20	0.90	3.23	0.84	460
L-650A	DCM	0.03	C-480	0.10	BA	0.21	1.65	1.52	3.53	0.13	650
L-660	SR-640	0.002	C-540A	0.02	BA	2.66	14.45	12.09	26.30	0.27	640
L-735	LDS-722	0.02	C-540A	0.10	BN	0.21	1.54	1.27	2.77	0.10	735
L-841	LDS-821	0.005	Rh-610P	0.10	BA	0.52	1.88	1.37	3.99	0.04	850

[a]Refers to the ratio of the optical energy out vs the energy absorbed.

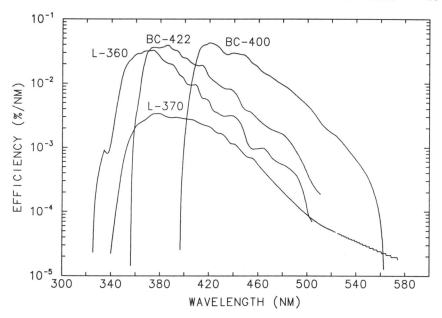

Figure 1a. Absolute scintillation efficiencies vs wavelength for liquid scintillators developed at EGG/EM Santa Barbara Operations.

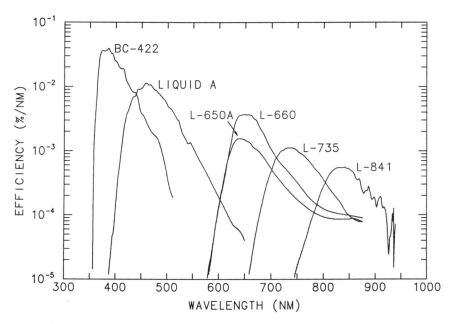

Figure 1b.

tyltin (TBSN) and tetramethyltin (TMSN) would not affect the scintillation properties except for a dilution factor.[3,4] The butyl and methyl groups are effective in isolating the heavy atoms from quenching the excited scintillator molecules. Thus TMSN, because of its higher percentage of tin, is deemed the more useful of the two compounds. Tin-loaded modifications of all the liquid scintillators listed in Table 1 have been successfully prepared.

Tin additives are particularly effective for X-rays in the energy range from about 10 to 100 keV; the K absorption edge for tin is at 29 keV. For lower energies, a lighter additive such as germanium, with a K edge at 11 keV, should be a more efficient absorber; for higher energies something like lead with a K edge of 88 keV would be preferable. Three compounds, tetramethylgermanium (TMGE), tetraethyllead (TEPB), and tetramethyllead (TMPB) were investigated for possible fluorescence quenching based on emission intensity and decay time measurements at the linac.

A new, tin-loaded, 735 nm liquid, L-735A-10T, has been developed, and its enhanced sensitivity to 17 keV X-rays has been measured. This new solution has a measured index of refraction of 1.53 at 740 nm and an optical attenuation of about 50 cm at 740 nm which make it attractive for use in capillaries.

EXPERIMENTAL

The scintillation parameters, including light output, were measured by exciting the solutions with pulses from a linac. The experimental arrangement is shown in Figure 2. These parameters include: 10 to 90% pulse rise time, FWHM, decay time from 0.7 of maximum to 0.7/e, 10 to 90% integral rise time (IRT), and the relative peak and integral of the fluorescence pulse to a reference scintillator.

The solutions were tested in 1 cm-thick Suprasil spectrophotometer cells and were bubbled with argon for two minutes before use. The samples were excited by 6 MeV electron pulses whose FWHM was 50 psec. The emitted light was viewed at an angle of 135° to the direction of the linac beam, as seen in Figure 2. This was done to reduce the Cerenkov contribution to the signal. Light from the samples was focused onto a Varian VPM 221D microchannel plate (MCP) photomultiplier tube (PMT), whose gain was 10^4. The light passed through a 335 to 427 nm FWHM band-pass filter. These signals were then sampled by a Hewlett-Packard sampling system Model 141A, with a remote sampling head. The signals were digitized with a CAMAC ADC and averaged and recorded with a DEC PDP-11/34.

A modification was made on the red-emitting liquid scintillator, L-735, listed in Table 1. The new scintillation solution, L-735A-10T consists of 0.01 M LDS-722 as final emitter and 0.10 M C-540A as intermediate wavelength shifter in a solvent consisting of 90 vol% benzonitrile (BN) and 10 vol% TMSN. The LDS-722 concentration was lowered from 0.02 M in L-735 to 0.01 M in L-735A-10T, because the solubility of LDS-722 decreased when the

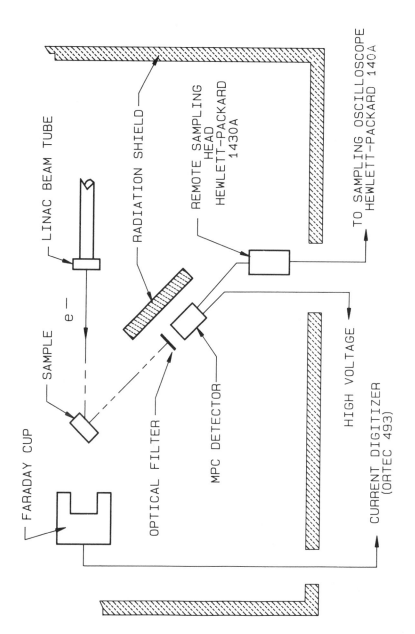

Figure 2. Schematic diagram of test configuration for the linac measurements.

TMSN was added. L-735A has the same dye concentrations as L-735A-10T but no TMSN.

The relative sensitivity of L-735A-10T to L-735A was measured by exposing samples to pulsed X-rays. These pulses were generated by 50 keV electrons, with a FWHM of 850 psec, incident on a molybdenum target as shown in Figure 3. Characteristic X-rays are superimposed on a bremsstrahlung spectrum that extends out to the 50 keV endpoint. The X-rays pass from the vacuum through a 5 m-thick beryllium window and are then filtered by a 3 m molybdenum foil which selectively passes the characteristic X-rays. The X-ray spectrum was determined by placing a 2 atm Xe proportional counter at the same position as the cell. A sample of the resulting spectrum, with the system response unfolded, is shown in Figure 4.

The samples were contained in a cell 12.7 cm in diameter by 0.6 cm in thickness. The window facing the X-ray source was 0.2 mm-thick beryllium, and the one facing the detector was 3.2 mm-thick Spectrasil-B. During measurements the liquids were heated to 43°C, a temperature typical of that encountered under field conditions. The light was filtered with a 630 nm long-pass filter and was detected with an ITT F4129 MCP PMT. The photocathode has an S-20 extended red response. The tube was as close as possible to the Spectrasil window for efficient light collection, and the output was sampled and averaged using a Tektronix 7854 digitizing oscilloscope with an S-12 sampling plug-in and an S-4 sampling head. The sampled pulses were then integrated, and the relative sensitivities were determined from the ratio of the integrals.

All chemicals were used as received. The BN, PC, and TMSN were from Aldrich Chemical Co.. The chemical TMGE was supplied by Alfa Products. The laser dyes LDS-722 and Coumarin 540A were from Exciton Chemical Company, Inc.. The compounds TMPB and TEPB were contributed by Ethyl Corp..

RESULTS

Tetramethylgermanium

As can be seen from the data in Table 2, the addition of TMGE had no measurable effect on the decay time of the 4,4″-di-(5-tridecyl)-p-terphenyl, SG-180,[5] in pseudocumene (PC). The relative pulse integral values actually decreased somewhat less than would have been expected from simple dilution of the PC by addition of the TMGE. Thus there is no evidence of any quenching by the TMGE. It was found previously that when TMSN or TBSN was added, the emission intensity also decreased in proportion only to dilution of the PC.[3,4]

One problem with TMGE is its high volatility: it boils at 43°C. Tetraethylgermanium has a higher boiling point, 163°C, but it contains a lower percent-

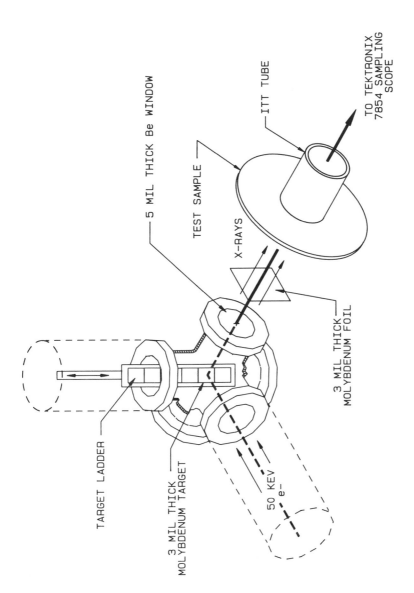

Figure 3. Experimental setup for the relative X-ray sensitivity measurements.

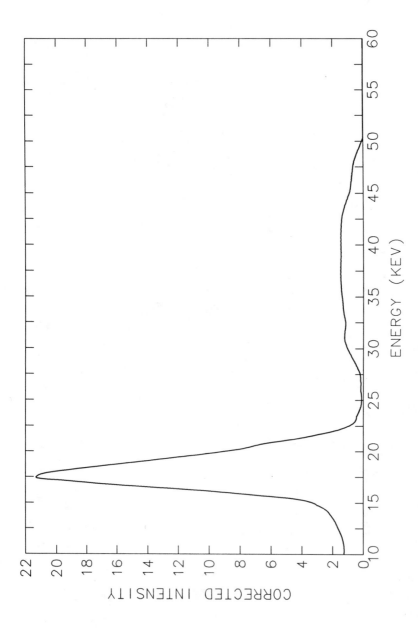

Figure 4. Measured X-ray spectrum for a 3-mil-thick Molybdenum target through a 3-mil-thick Molybdenum filter. The response of the system has been unfolded.

Table 2. Results of Linac Studies of 0.035 *M* SG-180 in PC (L-360) with Added Organometallic Compounds

Scintillator	RT (nsec)	FWHM (nsec)	DECAY (nsec)	IRT (nsec)	Peak[a]	Int.[a]
System Response (Suprasil)	0.15	0.22	0.08	0.54	—	—
100% TMGE	0.16	0.25	0.11	1.25	—	—
0.035 *M* SG-180 (100% PC)	0.55	2.56	1.80	5.26	1.00	1.00
0.035 *M* SG-180 10% TMGE	0.55	2.56	1.75	5.34	1.03	1.04
0.035 *M* SG-180 25% TMGE	0.60	2.61	1.80	5.45	0.90	0.91
0.035 *M* SG-180 35% TMGE	0.55	2.61	1.80	5.69	0.76	0.78
0.035 *M* SG-180 70% TMGE	0.55	2.71	1.80	5.49	0.65	0.68
0.035 *M* SG-180 10% TMPB	0.26	1.92	1.38	3.86	0.42	0.33
0.035 *M* SG-180 20% TMPB	0.18	1.74	1.31	3.28	—	0.22
0.035 *M* SG-180 10% TEPB	0.20	0.89	0.99	2.71	—	—
0.035 *M* SG-180 20% TEPB	0.17	0.41	0.46	1.72	—	—

[a]Peak and Int. values referenced to L-360.

age of germanium by weight, 38.5% vs 55% for the tetramethyl compound. There is also a question of long term chemical stability of solutions containing TMGE. Further studies are contemplated.

Tetramethyllead and Tetraethyllead

The lead compounds were disappointing in that they quenched the emission seriously. Addition of 10 vol% of TMPB reduced the light output to less than 50%. When 20% TMPB was added the signal was dominated by the Cerenkov contribution, precluding any conclusive information. It was hoped that the somewhat larger alkyl groups in TEPB would reduce the extent of quenching, but it appeared to be somewhat worse. In addition, there was some decomposition, possibly photochemical, of the TEPB as the solutions stood in the hood before testing. A small amount of a clear, apparently crystalline, precipitate accumulated on the bottom of the sample vials.

Table 3 shows the relative sensitivities of L-735A and L-735A-10T to 17-keV X-rays. Included in the table are predicted relative sensitivities when 10 vol%

Table 3. Effects of the Addition of 10 Volume Percent TMGE, TMSN, or TMPB or BN Containing 0.01 *M* LDS-722 and 0.10 *M* C-540A. Sensitivities are Relative to the Scintillator with no Organometallics. Excitation was by Molybdenum-filtered Molybdenum X-rays.

Fluor Candidate	Relative Energy Absorption	Predicated Relative Brightness	Observed Relative Brightness
0.01 *M* LDS-722; 0.10 *M* C-540A in:			
100% BN	1.00	1.00	1.0
10% TMGE / 90% BN	3.04	2.74	—
10% TMSN / 90% BN	3.15	2.84	2.8
10% TMPB / 90% BN	4.02	1.33[a]	—

[a]Includes quenching by the TMPB.

TMGE, TMSN, or TMPB is added to BN. Energy absorption coefficients, as tabulated in Reference 6, were used to generate curves for the BN-based scintillators. These curves were folded with the X-ray spectrum shown in Figure 4. The relative energy absorbed, the predicted brightness, and the observed brightness are shown in Table 3. The predicted relative sensitivities include a factor for the dilution of the BN, and in the case of the lead compound, consideration for the quenching by the lead compound. The agreement between the predicted and observed values for TMSN is excellent.

CONCLUSIONS

The compound TMSN still appears to be the best compound that we have tested to enhance the sensitivity of an organic scintillation solution to low energy X-rays. Although the addition of lead would greatly increase the energy deposited, the severity of the quenching more than offsets the gain made, and no further work is anticipated with these lead compounds. BBTMGE shows promise as an additive for lower energy X-rays. There may be more work in this direction once the question of long-term chemical stability is resolved.

ACKNOWLEDGEMENTS

The authors wish to thank our colleagues, Mr. K. Moy and Dr. S. Iverson, for many helpful discussions concerning X-ray dosimetry.

REFERENCES

1. Ogle, J.W., D. Thayer, L.D. Looney, F. Cverna, G. Yates, C. Iverson, S.S. Lutz, M.A. Nelson, and B. Whitcomb. "Radiation-Induced Imaging System Over Long Fiber-Optic Bundles," in *High Bandwidth Analog Applications of Photonics*, SPIE 720, pp. 24–30 (1986).
2. Bicron Corporation, 12345 Kinsman Road, Newbury, OH 44065.
3. Berlman, I.B., L.A. Franks, S.S. Lutz, J.M. Flournoy, C.B. Ashford, and P.B. Lyons. "A High-Z Organic Scintillation Solution," in *Advances in Scintillation Counting,* (Banff, Alberta, Canada: Faculty of Pharmacy and Pharmaceutical Sciences, University of Alberta, 1983), pp. 365–375.
4. Ashford, C.B., I.B. Berlman, J.M. Flournoy, L.A. Franks, S.G. Iversen, and S.S. Lutz. "High-Z Liquid Scintillators Containing Tin," *Nucl. Instruments and Methods in Phys. Res.*, A243:131–136 (1986).
5. Gershuni, S., M. Rabinovitz, I. Agranat, and I.B. Berlman. "Effect of Substituents on The Melting Points and Spectroscopic Characteristics of Some Popular Scintillators," *J. Phys. Chem.*, 84(5):517–520 (1980).
6. Storm, E. and H.I. Israel, "Photon Cross Sections From 1 keV to 100 MeV for Elements Z = 1 to Z = 100," *Nucl. Data Tables*, A7(6):565–681 (1970).

APPENDIX

BA	benzyl alcohol (alpha-hydroxy toluene)
BN	benzonitrile (phenyl cyanide, cyanobenzene)
BHTP (4-BHTP)	4-bromo-4″-(5-hexadecyl)-p-terphenyl
C-480	Coumarin 480 (or Coumarin 102)
C-540A	Coumarin 540A (or Coumarin 153)
DCM	red-emitting styryl laser dye, also Kodak dye No. 14567
LDS-722	far-red-emitting styryl laser dye (Exciton)
LDS-821	near-infrared-emitting styryl laser dye (Exciton)
PC	pseudocumene (1, 2, 4-trimethylbenzene)
PBD	2-phenyl-5-(4-biphenylyl)-1,3,4-oxadiazole
Rh-610P	Rhodamine 610 perchlorate
SG-180	4,4″-di-(5-tridecyl)-p-terphenyl
SR-640	Sulforhodamine 640
TMSN	tetramethyltin
TPB	1,1,4,4-tetraphenylbutadiene

CHAPTER 10

Liquid Scintillation Alpha Spectrometry: A Method for Today and Tomorrow

W. Jack McDowell and Betty L. McDowell

ABSTRACT

Alpha spectrometry using liquid scintillation methods has matured into a technology showing great potential. The nonpenetrating properties shared by beta and alpha radiation make them both candidates for liquid scintillation counting/spectrometry. However, applying liquid scintillation to alpha spectrometry has been difficult, because inefficient light production by alpha particles led to poor alpha energy resolution and alpha-produced scintillation interference of beta- and gamma-produced scintillations. Recent developments in alpha liquid scintillation spectrometers and pulse-shape discrimination have removed the beta-gamma interference problem and greatly improved alpha energy resolution. In addition, selective solvent (liquid-liquid) extraction methods (phase-transfers), setting the nuclide of interest into the scintillator, simplifies sample preparation and provides important additional information for nuclide identification. Such rapid, selective procedures have been developed for radium, uranium, thorium, plutonium, polonium, and the trivalent transplutonium elements. Presently, available energy resolution (230 keV FWHM) allows identification of many of the isotopes of these nuclides. Chemical separations methods and typical alpha-spectral results are presented in this chapter. Promising methods of improving both energy resolution and chemical separations are suggested.

INTRODUCTION

The maturing of alpha liquid scintillation spectrometry into a useful radiometric tool is a fascinating story. It has been known since 1950[1,2] that alpha particles could be counted by liquid scintillation methods, but little practical use has been made of this knowledge. The problems of poor alpha energy resolution, quenching and variable scintillator response, and beta and gamma radiation interference all combined to make useful applications of liquid scintillation to alpha assay extremely limited.

In 1964, D.L. Horrocks and co-workers[3,4] demonstrated that it was possible to obtain useful alpha energy resolution in liquid scintillation systems. They found it necessary to use a detector quite different from that used for beta liquid scintillation. It consisted of a single phototube facing a reflector, with the sample between the phototube and reflector, and efficient optical coupling between the sample, reflector, and phototube. Both Horrocks and Hanschke[5] working independently, demonstrated that a highly-reflective, diffuse-white, hemispheric reflector cavity, oil-coupled to a single phototube provided the optimum alpha energy resolution in a liquid scintillation detector. Hanschke produced both an experimental and a mathematical demonstration of this. A cross section of such a detector may be seen in Figure 1.

Even though the detector described above was able to achieve useful alpha energy resolution in a liquid scintillation system, the problems of variable energy response/quenching and beta-gamma interference with the alpha spectra still prevented useful application of liquid scintillation to alpha counting and spectrometry.

The development of methods to overcome the latter two problems is the subject of this chapter.

EXPERIMENTAL NARRATIVE

The experimental work reported here extended over 20 years. Some of it has never been reported, and none of it has been reported in this context, i.e., an overview of the development of Photon Electron Rejecting Alpha Liquid Scintillation (PERALS) spectrometry.

This development began in 1967 with a need to determine aqueous/organic phase distribution of the alpha-emitting trivalent actinides to an accuracy of ±1% or better. It was found impossible to do this with plate-counting methods. Horrocks[6] suggested that liquid scintillation might be a possible answer to our problem. With well-characterized samples of trivalent actinides, we achieved the needed analytical accuracy using a commercial beta liquid scintillation counter. However, problems with this approach soon appeared. The two-phase distribution data for some nuclides did not fit the pattern we expected.

In order to determine that what we were counting was indeed the alpha from the trivalent actinide under investigation, we needed to see an energy spectrum of the sample. The 1960 model Packard Tricarb liquid scintillation counter was modified to allow the energy signal to be displayed on a multichannel analyzer (MCA). The pulses carrying the energy information did not exit through the window produced by the upper and lower discriminators but instead were converted to logic pulses that were sent to the scaler. In order to see a spectrum and to determine what portion was being counted, the scaler pulses were picked off, modified, and sent to the gate input of the MCA while the energy analog pulses were sent to the energy input. This allowed the discriminators effect on the energy spectrum to be viewed on the MCA. With this arrangement, we were able to observe that some nuclides had interfering beta/gamma-emitting daughters and/or beta- , gamma- , or alpha-emitting impurities, and in all cases the energy resolution was poor, on the order of 800

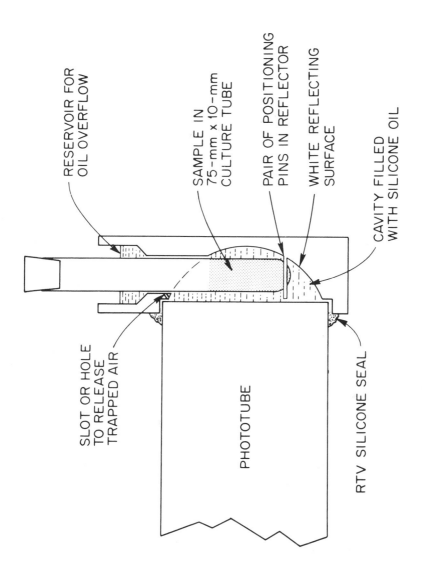

Figure 1. A cross section of a PERALS detector assembly.

to 900 keV full peak width at half maximum height (FWHM). In some cases the additional spectral information allowed a reasonably accurate determination of the test nuclide concentration, while in others it still did not yield the required analytical accuracy due to variations in the energy response of the scintillator, among other problems.

The next step taken was to build a detector that would give improved energy resolution following the work of Horrocks[3,4] and Hanschke.[5] An experimental enclosure (Figure 2) that would accomodate a variety of phototubes and reflectors was built for testing various detector optical arrangements. In true scientific manner, we tested somewhere in the neighborhood of 25 different optical arrangements of our own before we conceded that Horrocks and Hanschke were essentially correct. The contributions to detector construction that we have made are of the nature of refinements and compromises designed to enhance the practical usefulness of the detector. We added an improved reflecting surface, a light-coupling oil reservoir, and an insertion method for a small sample container (a 10×75 mm culture tube). The accurate positioning was an improvement suggested by Steve Musolino of Brookhaven National Laboratory.[7]

One of the crucial requirements uncovered by work on the detector was that the reflective surface must be highly reflective but nonspecular, i.e., diffusing. Magnesium oxide laid down in several layers and bonded by sodium silicate solution (water glass) was effective, but barium sulfate with a very small amount of binder was equally good and was commercially available (Eastman high reflectance coating). The need for a diffusing reflector appears to arise from the nonuniform response of the photocathodes in multiplier phototubes. Figure 3 illustrates typical sensitivity profiles of 2 in phototubes.

Spectra collected with the new detector gave much better energy resolution than those obtainable with the beta liquid scintillation spectrometer (see Figure 4), but we still had the problem of sample nonreproducibility and beta/gamma interference, either one of which alone could make alpha liquid scintillation of very limited practical use. Progress at this point in the development was reported in *Organic Scintillators and Liquid Scintillation Counting.*[8]

Variable energy response and variable quenching are problems associated with the scintillation cocktail. Most of those commonly used for beta assay were, and are, aqueous-phase accepting. These cocktails contain detergents and aqueous/organic coupling solvents. Usually an aqueous sample is added to the scintillation cocktail, where it is incorporated by the constituents of the cocktail into a reasonably clear and homogeneous solution. Variations in the matrix cause variable quenching, but because beta-emitting nuclides produce a continuous spectrum from zero to maximum, this quenching variation can usually be corrected by measuring the radiation produced in that sample by a known external or internal source. Quench corrections of this type cannot be used with alpha spectra. Initial small amounts of quenching do not reduce the counts under an alpha peak but simply shift the alpha peak to a lower voltage/energy scale position. More severe quenching can push the alpha peak out of

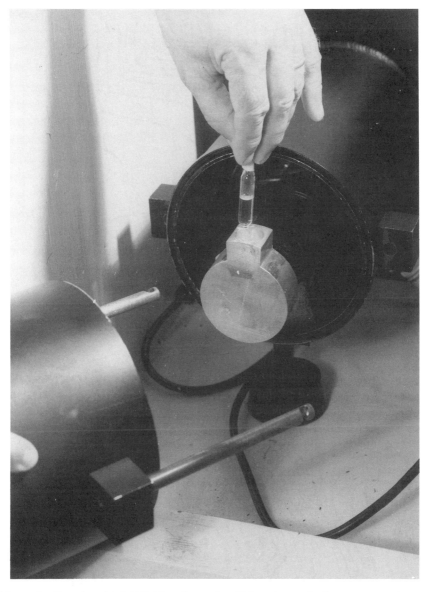

Figure 2. Experimental light-tight enclosure in which various reflector arrangements were tested.

the detectable region, but the reduction in count is not a simple function of the amount of quencher added. First the count reduction is small, then, with more quencher, large, and then, with still more quencher, small again, as the left edge, median and right edge of the bell-shaped curve pass below the detection threshold.

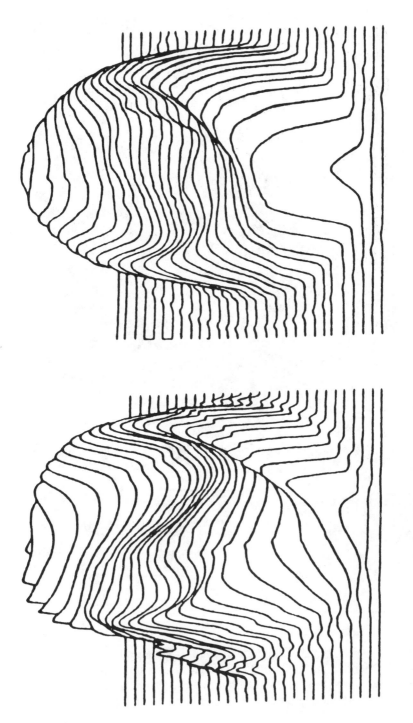

Figure 3. Typical sensitivity profiles of 2-in multiplier phototubes.

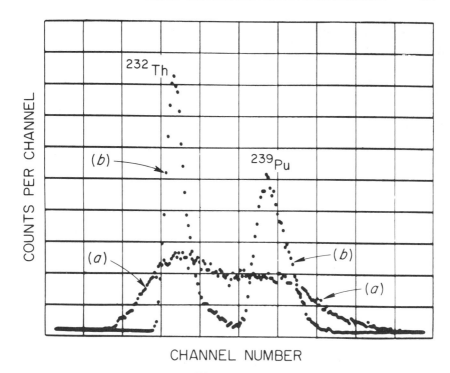

Figure 4. Comparison of the spectra of ^{232}Th(4.01 MeV) and ^{239}Pu(5.19 MeV) obtained on a beta liquid scintillation spectrometer (a) and on a PERALS spectrometer.

Thus, it became clear that a method of incorporating the alpha-emitting nuclide into the scintillation cocktail in some reproducible manner was necessary so that one could expect the peak for a given alpha energy to appear at the same position on the amplifier voltage output; thus at the same position on a spectrum as that displayed on a multichannel analyzer. Application of the extensive knowledge of liquid-liquid extraction existing in the Chemical Separations Research group at Oak Ridge National Laboratory[9] provided a solution to the cocktail reproducibility problem. Progress at this point was reported at the 1974 International Solvent Extraction Conference.[10]

Developing a method of placing the alpha-emitting nuclide into the scintillator so that each time the scintillator and its response to alpha energies would be the same required the production of an identical organic-phase-soluble compound for the nuclide in each sample. Variable quenching, either color quenching or chemical quenching, could not be tolerated. The first approach in many cases has been to transfer the nuclide into an organic phase by using liquid-liquid extraction or "solvent" extraction techniques and then add a portion of the organic phase to a scintillator. Our solution to this problem was to incorporate an extractant into the scintillator thereby producing an "extractive scintillator". There are several extractants one may choose for this purpose,

Table 1. Properties of Selected Solvent Extraction Reagents Useful in Extractive Scintillators

Type	Example	Metals Extracted
Cation Exchange	Bis (2-ethylhexyl) Phosphoric acid	Alkali and alkaline earths weakly extracted; pH 4-14. Lanthanides/actinides (III) strongly extracted; pH 3–5. Actinides (IV–VI) strongly extracted to pH 0.5.
	Neocarboxylic acids	Weak extractant for most ions pH 4–7. Strong Ext. for some ions in combination with appropriate crown ethers.
Ion Pair Coordinators From sulfate systems[a]	Primary Alkyl Amines. MW > 250.	From Sulfate; Fe (III), Y, Zr, Nb(V), Tc(VII), Pd(II), In(III) Sn(II), Eu, Hf, Ta(V), Re(VII) Os(IV), Mo(VI), Pu(IV), U(IV) Th.
	Secondary alkyl amine, MW > 250 (extraction varies with structure.)	From Sulfate: Zr, Nb(V), Mo(VI) Tc(VII), Pd(II), Ta(V), Re(VII), Os(IV).
	Tertiary alkyl amine, MW > 250.	From Sulfate; U(VI).
	Quat. ammonium MW > 300.	From Sulfate; Mo(VI), Tc(VII) Ta(V), Re(VII), Os(VI), Pd(II).
Neutral Coordinators	Trioctyl phosphine oxide.	From Nitrate; U(VI), Zr, Tc(VII) Au(III), Th, Np(IV)(VI), Pu(IV)(VI), Pa, Hf.
		From Chloride; Au(III), Zn, Zr, Sn(IV), Sb(III), Cr(IV), Mo(IV), Fe(III), Th, U(IV)(VI) Pu(IV)(VI), Ga(III), Nb, Bi.

[a]Extraction from nitrate and chloride systems are not listed because the amine salts of these ions are highly quenching and therefore not useful in scintillators.

Note: Amine perchlorates are not extractants. Perchloric acid can be used to strip metals from any of the ion-pair-coordinator extractants.

each having different metal-ion selectivities. Table 1 lists some extractants that are currently used and shows their selectivities from various aqueous systems. Figure 5 shows the quenching (peak shifting) effects of some organic phase compounds. The use of an extractive scintillator involves a simple two-phase equilibration of the scintillator with an appropriate aqueous phase. Then the scintillator is placed in a small culture tube, the tube placed in the PERALS detector, and the spectrum collected.

Thus, we found that in most cases the resolution of the sample nonreproducibility problem was to prepare a water-immiscible scintillator containing a liquid-liquid extraction reagent (in addition to the fluor and an energy transfer agent) and to extract (phase transfer) the nuclide-of-interest into such a scintil-

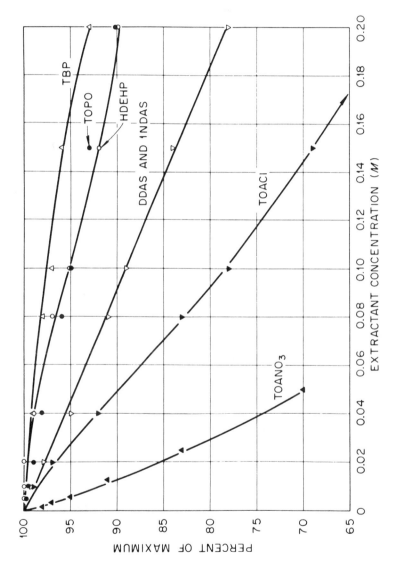

Figure 5. The quenching effects of various reagents.

lation cocktail. The material extracted into the scintillator solution had little or no effect on its energy response characteristics. This development resulted in a reproducible, calibratible system where the peak for 4.0 MeV ^{232}Th or 5.15 MeV ^{239}Pu, for example, would always appear at the same place on an energy scale. A number of applications of alpha liquid scintillation were made at this point and some were reported at the 1979 International Conference on Liquid Scintillation Counting.[11,12]

One additional, serious problem that impeded the use of liquid scintillation as a useful, practical method of alpha spectrometry, was the interference from beta and gamma radiation. Beta and gamma radiation are about 10 times more effective, keV for keV, in producing light in a liquid scintillator than alpha particles.[13] Thus, if a sample contains beta- and/or gamma-emitting nuclides, any alpha peak will be underlain with a beta/gamma continuum. This makes the alpha count background high, variable, and difficult to determine, and it sometimes even makes the alpha peak difficult to locate. In the following we describe how this problem was dramatically and elegantly solved.

The work we were doing with alpha liquid scintillation attracted numerous visitors to our laboratory. Among them was an electrical engineer named John Thorngate. We were explaining to him the terrible problem that the beta and gamma pulses presented when he said, "I can get rid of those pulses." The next day John returned with an armload of NIM modules and within an hour we were seeing liquid scintillation alpha spectra free of beta and gamma interference. By afternoon we had collected sufficient information to publish a paper.[14]

There is about a 40 nanosecond (nsec) difference in the length of beta- or gamma-produced and alpha-produced pulses in a liquid scintillation system.[13] It is possible to sort these pulses electronically and send only the longer, alpha-produced pulses to the multichannel analyzer. This technique is extensively used for separating gamma- and neutron-produced pulses. Similar electronic circuitry was adapted and improved for beta/gamma−alpha separation. In currently available alpha liquid scintillation detectors the rejection of beta/gamma pulses is $>99.95\%$ efficient,[15,16] thus it is possible to collect an alpha spectrum that is essentially free of beta/gamma interference.

At this point, we had the means of removing or minimizing the poor alpha energy resolution, the quenching and variable scintillator response, and the beta and gamma radiation interference. This provided the technology for the development of a practical method of producing liquid scintillation alpha spectra. We called this new development **Photon Electron Rejecting Alpha Liquid Scintillation (PERALS)** spectrometry. The development won an IR-100 award in 1981 and the PERALS spectrometer is now in commercial production.[17] A number of laboratories are having excellent success with the method. The special chemicals and extractive scintillators needed for the PERALS system are also commercially available.[18] Time will tell if the method is to fulfill its promise of a new and useful addition to the existing alpha assay methods.

RESULTS

As in virtually all cases the advance in alpha liquid scintillation methods described above resulted from a search for a suitable radiometric method for our own work, which was primarily a study of the chemistry of solvent extraction separations of metal ions. There was very little separate funding for this work. However, the development of a useful radiometric method for alpha-emitting nuclides seems to fill a gap in the methods already available. Alpha energy resolution was not as good as surface-barrier or Frische-grid methods, but nuclide quantification was better and sample preparation was easier. Sample preparation steps for PERALS alpha spectrometry usually were selective for one ion or a group of ions, and this selectivity supplemented the identification of nuclides by peak position.

Typical PERALS Analysis

Describing a typical analytical procedure may summarize the descriptive material above and bring it to focus on its end use: let us begin with a solid sample presented for thorium and uranium analysis that appears to be soil mixed with leaves and other organic material. One- to two-gram portions of the sample are weighed into crucibles and ignited at 600°C until all the organic material is destroyed. Then the material is placed in a solution of nitric, hydrochloric, and hydrofluoric acids; pressure vessels of organic fluorpolymers are best for this step.[19-22] Sulfuric acid is added and the other acids are boiled away (fusion methods can also be used).[23] Then the acidity of the sulfate system is adjusted to pH 1 to 2 in a 10 mL volume, and a two-phase equilibration extraction is made into a measured 1.2 to 1.5 mL water-immiscible extractive scintillator containing 0.3 M high-molecular-weight tertiary amine sulfate. This uranium in the sample now will have been quantitatively transferred to the scintillator phase. A 1 mL portion is pipetted into a small culture tube, and the spectrum is collected on a PERALS spectrometer. The remaining aqueous phase is contacted with a 1.2 to 1.5 mL portion of an extractive scintillator containing 0.3 M high-molecular-weight primary amine. Again 1 mL of the extract is counted. Multiplying the integral of the peak counts obtained in the first case by 1.2 or 1.5 and dividing by 0.9968 gives the disintegrations per minute (dpm) of uranium, and so calculating the integral of the peak obtained in the second instance gives the dpm of thorium. The accuracy of such a procedure depends on two factors: (1) how carefully the sample is prepared and (2) how many counts are accumulated. Standard mineral samples have been run giving consistent results within 0.1% of the given value.

Solvent Extraction

This operation plays an important part in PERALS spectrometry in that it allows the radionuclide to be selectively and quantitatively transferred to the (extractive) scintillator with little change in the scintillator composition. This makes it possible to obtain a reproducible calibration of the energy scale for a given extractive scintillator/aqueous phase composition. A solvent extraction system is composed of an organic extractant phase (one or more extraction reagents dissolved in an organic diluent) and an aqueous phase. These two phases are immiscible, but under proper conditions phase transfer of one or more species can occur across the organic/aqueous interface. Expansion of the interface by gentle agitation of the two phases allows phase equilibrium to be attained quickly, usually in one to two minutes.

In an aqueous/organic solvent extraction system, the distribution of a metal, M, between the two phases at equilibrium is defined by a distribution coefficient D_M.

$$D_M = (Conc.\ M)_{org}/(Conc.\ M)_{aq}$$

This is true at any phase ratio, V_{org}/V_{aq}. However, it should be remembered that D_M is a concentration ratio, and the amount of metal recovered depends also on the phase ratio. If we call the recovery factor, F_R, the relationship

$$F_R = D_M(V_{org}/V_{aq})$$

describes the total org/aq metal ratio. Thus if D_M is 1000 (not an unusual value) and V_{org}/V_{aq} is 1 then F_R is also 1000, but if $V_{org}/V_{aq} = 1/100$ or $1/1000$ the recovery is much less. The percent recovery in the first case (1:1 phase ratio) is $100 \times 1000/(1000 + 1) = 99.9\%$; while in the second case (phase ratio 1:100), the percent recovery is $100 \times 10/(10 + 1) = 90.9\%$, and in a case where the phase ratio is 1:1000 the recovery will be only 50%. These simple calculations make obvious the importance of phase ratio in analytical applications of solvent extraction. It should also be emphasized that a distribution coefficient depends on the composition of both the organic and aqueous phases. If the composition of either phase is changed, the distribution coefficient can change. It is obvious that one should be aware of distribution coefficients and phase ratios in following, and more particularly in modifying, procedures for PERALS.

OUTLINES OF PROCEDURES

Some procedures that have been tested in our laboratory and shown to work well are briefly described below. Certain requirements that apply to all samples should be noted. Samples to be counted should be optically clear and as colorless as possible. The objective is to get as much light as possible per alpha event to the phototube. All samples should be sparged with a dry, toluene-saturated oxygen-free gas. Dissolved oxygen in the sample impairs both energy and pulse-shape resolution. Any aqueous/organic system that transfers a

Table 2. Quenching and Nonquenching Ions and Molecules

Species in Organic Phase	Relative Degree of Quenching
Any yellow colored material.	Color quenching of varying severity depending on the intensity of the color.
Chloride salts of amines and most chlorine substituted organics and dissolved HCl.	Severe chemical quenching.
Nitrate salts of amines, nitrate substituted organics and dissolved HNO_3.	Severe chemical quenching plus color quenching.
Alcohols and ketones.	Moderate quenching.
Ethers.	Slight quenching
Sulfate salts of amines.	Minimal quenching.
Phosphates, phosphate esters and phosphine oxides.	Minimal quenching.

quenching ion or molecule to the organic phase must be avoided for best results (see Figure 5 and Table 2).

Counting on Glass-Fiber Filters[24]

For these two procedures to succeed the filter must be relatively clean, i.e., free of material of a dark color.

Place the filter (up to 2 in diam) in a 10×75 mm culture tube and add 1 mL of a PERALS aqueous-immiscible scintillator. Use either extractant-free or HDEHP [bis-(2-ethylhexyl)phosphoric acid] — containing scintillator. Momentarily evacuate the tube to water-aspirator pressures and release to argon. Do this twice to remove entrapped air and then count on a PERALS spectrometer. This allows separate determination of alpha and beta events. The spectrum of alpha energies is usually sufficiently resolved to allow identification of the alpha-active material.

Uranium Activity on Cellulose Filters[25]

Cellulose air filters, clean or dirty, can be placed easily in solution and assayed as follows: the filter paper (up to 2 in diam) is placed in a 2 dram, screw-cap borosilicate glass vial and heated, open, in an oven or furnace at 500°C for 2 hr. After removing and cooling, add 2 to 5 drops of concentrated nitric acid, 3 drops of 30% hydrogen peroxide, and 1 drop of a saturated aluminum sulfate solution to the vial. The vial is then heated to 200°C to remove the nitric acid. The solids remaining are redissolved in a solution 1 M in Na_2SO_4 and 0.01 M in H_2SO_4. A measured quantity of an extractive scintillator containing a high molecular weight tertiary amine sulfate is then added to the vial, and the two phases are equilibrated for 2 to 3 min. After the phases have separated, 1 mL of the organic phase is counted on a PERALS spectrometer. Alpha spectra thus obtained usually allow identification of the uranium isotopes present. The results are uniformly more accurate than direct counting of the air filter.

Gross Alpha in Environmental Materials[26]

With the sample in solution in nitric or hydrochloric acid, add 0.5 g of $LiClO_4$ and 1 to 1.5 mL of 0.1 M $HClO_4$. Evaporate under heat lamps or in an aluminum block at 160°C until boiling of the sample ceases and the first perchloric acid fumes appear. Cool the beaker and add 5 to 7 mL of water to dissolve the viscous residue. Measure the pH of the solution; it must be between 2 and 3.5. Transfer the solution to a small equilibration vessel and add a measured quantity, 1.2 to 1.5 mL, of an extractive scintillator that is ~0.2 M in HDEHP. Equilibration will transfer most of the alpha-emitting nuclides, with the exception of radium and radon, to the scintillator. Lanthanides are also extracted. Counting on a PERALS spectrometer will usually allow identification of the nuclides as well as an accurate quantification of them.

Uranium and Thorium[27]

With the sample in solution as above, add sulfuric acid and sodium sulfate and convert to a sulfate system at pH 1 to 2. An initial extraction into a scintillator containing 0.3 M high molecular weight tertiary amine sulfate such as trioctylamine sulfate will quantitatively remove uranium. A second extraction from the same aqueous using a scintillator containing a high molecular weight primary amine sulfate such as 1-nonyldecylamine sulfate will remove thorium. Each organic extractive scintillator solution is then sampled and counted on the PERALS spectrometer. Normal uranium in secular equilibrium will show the double peaks due to ^{238}U and ^{234}U. However, natural processes can concentrate ^{234}U in water, and sediments and artificial processes have concentrated $^{234-235}U$; thus, not all uranium now in the environment is "normal" uranium. If the uranium is highly enriched in ^{235}U, the spectrum will show primarily ^{234}U. Spectra of tailings from the enrichment processes will show a predominance of ^{238}U.

Uranium and Thorium in Phosphates[28]

These elements can be separated from a variety of phosphate-containing materials, e.g., fertilizers, bones, teeth, animal tissues, and wastes. The sample is dissolved and placed in a nitrate or nitrate perchlorate solution. Sufficient aluminum nitrate is added to this solution to complex the phosphate. The solution is then contacted with a toluene solution of trioctyl phosphine oxide

(TOPO). Both uranium and thorium are transferred to the organic phase. The organic phase is then stripped with an equal volume of 0.5 M ammonium carbonate solution, the ammonium carbonate is evaporated, the sample is converted to nitrate or nitrate/perchlorate, and any entrained organics are destroyed. The clear solution is then converted to a sulfate system and treated as above. In many nonphosphate samples uranium can be coprecipitated with magnesium hydroxide or otherwise concentrated and extracted from a sulfate system into a scintillator containing tertiary amine sulfate.

Polonium[29]

The radioisotopes of polonium (usually ^{210}Po) have been difficult to analyze with accuracy using the conventional methods. Using PERALS, however, the procedure is simple, rapid, and accurate. With the sample in solution, add 3 to 5 mL of concentrated phosphoric acid and evaporate to remove other acids. Transfer this phosphoric acid solution to a small equilibration vessel using 3 to 5 mL of water. Add 1 mL of 0.1 M HCl. Add a measured volume, 1.2 to 1.5 mL, of an extractive scintillator that contains 0.1 M TOPO. Equilibrate and count 1 mL on a PERALS spectrometer. Because of the minimal chemical manipulations required, the accuracy of this determination easily can be better than $\pm 1\%$.

Plutonium[30]

Plutonium can be chemically separated from all other elements except neptunium and counted quantitatively by this procedure, and the ^{237}Np peak can be resolved by its energy difference from $^{239-240}$Pu. The initial extractant in this procedure is 0.3 M high-molecular-weight tertiary amine nitrate (TANO$_3$) in toluene. The sample should be in solution in 3 to 4 M total NO$_3^-$ and 0.5 to 1 M HNO$_3$. The plutonium is reduced with ferrous sulfate and reoxidized to Pu(IV) with sodium nitrite. This solution is contacted with not less than $^{1}/_{4}$ its volume of TANO$_3$ solution. Equilibrate and separate the aqueous phase. Wash the organic phase with two $^{1}/_{4}$-vol portions of 0.7 M HNO$_3$. The aqueous from the first equilibration and the washes can be combined and analyzed for uranium, if desired. Plutonium can be stripped from the organic phase either with perchloric acid or, after diluting the organic phase with 2-ethylhexanol, with 1 N H$_2$SO$_4$. The plutonium is reextracted into a scintillator containing HDEHP in the first instance and into a scintillator containing 1-nonyldecylamine sulfate in the second instance.

Radium[31]

A new extraction reagent allows radium to be separated from other alkaline earth elements and the following procedure allows separation from many, if not all, other cations. The sample solution is spiked first with ^{133}Ba, and 10 mg of barium carrier is added. Then radium is precipitated from the sample solution as barium/radium sulfate. Next the precipitate is converted to the carbonate by heating it with a saturated potassium carbonate solution, and the separated barium/radium carbonate is dissolved in dilute acid. The pH of this solution is then adjusted to between 9 and 10, and the radium extracted selectively into an extractive scintillator containing a synergistic reagent mixture of a high-molecular-weight-branched carboxylic acid such as 2-methyl-2-heptylnonanoic acid (HMHN), 0.1 M, and dicyclohexano-21-crown-7 (DC21C7), 0.05 M. The extraction is quantitative, and any chemical losses in the coprecipitation and metathesis steps can be corrected by gamma counting the ^{133}Ba. Radium is counted on a PERALS spectrometer, initially showing a single alpha peak each for both ^{226}Ra and ^{224}Ra. All the alpha peaks from both radium daughters can be seen if one waits for their ingrowth. The percent relative error in this determination, with appropriate counting statistics, is usually less than ±5% and can be much less.

CONCLUSIONS

The basic steps leading to the development of the PERALS system of sample assaying for alpha-emitting nuclides have been presented. The many blind alleys and dead ends explored were not. This is not the end of the story, however. More selective and perhaps simpler procedures are certainly possible. Although pulse-shape resolution (the separation of beta-gamma pulses) hardly needs improvement, energy resolution does. We believe that this is possible. There seems to be no fundamental reason why resolution should be limited in such a system. We hope to be able to pursue this quest in the future.

ACKNOWLEDGEMENTS

The Department of Energy sponsored this work under contract No. DE-AC05-84OR21400 with Union Carbide Nuclear Corp. and with Martin Marietta Energy Systems.

G.N. Case is responsible for most of the laboratory development work described above.

C.F. Coleman, A.P. Malinauskas, and R.G. Wymer provided much-needed encouragement and shielding from many negative comments by higher supervision and our sponsors.

C.F. Coleman invented the name and acronym PERALS.

REFERENCES

1. Broser, I. and H.P. Kallmann. "Uber den Elementarprozess der Lichtanregung in Leuchtstoffen durch alpha-Tielschen, Schnelle Elektronen und Gamma -Quanten II," *Z. Naturforschg.*, 2A:642–650 (1947).
2. Horrocks, D.L. *Applications of Liquid Scintillation Counting*, (New York: Academic Press, Inc., 1974), p. 2.
3. Horrocks, D.L. "Alpha Particle Energy Resolution in a Liquid Scintillator," *Rev. Sci. Instrum.*, 35:334–340 (1964).
4. Horrocks, D.L. and M.H. Studier. "Low Level Plutonium -241 Analysis by Liquid Scintillation Techniques," *Analytical Chem.*, 30:1747–1750 (1958).
5. Hanschke, T. "High Resolution Alpha Spectroscopy by Liquid Scintillation Through Optimization of Geometry," PhD Dissertation, Hannover Technical University, Hannover, FRG (1972).
6. Horrocks, D.L., Personal Communication 1967.
7. Musolino, S., Personal Communication, Ca. 1985.
8. McDowell, W.J. "Liquid Scintillation Counting Techniques for the Higher Actinides," in *Organic Scintillators and Liquid Scintillation Counting*, D.L. Horrocks and C.T. Peng, Eds. (New York: Academic Press, 1971), pp. 937–950.
9. Coleman, C.F., C.A. Blake, and K.B. Brown. "Analytical Potential of Separations by Liquid Ion Exchange," *Talanta*, 9:297 (1962).
10. McDowell, W.J. and C.F. Coleman. "Combined Solvent Extraction-Liquid Scintillation Method for Radioassay of Alpha Emitters," *Proceedings of International Solvent Extraction Conference 1974*, (London: Society of Chemical Industry, 1974), pp. 2123–2135.
11. McDowell, W.J., E.J. Bouwer, and J.W. McKlveen. "Application of the Combined Solvent Extraction-High Resolution Liquid Scintillation Method to the Determination of ^{230}Th and $^{234-238}$U in Phosphate Materials," in *Liquid Scintillation Counting: Recent Applications and Developments, Vol. 1*, C.T. Peng, D.L. Horrocks and E.L. Alpen, Eds. (New York: Academic Press, 1980), pp. 333–346.
12. McDowell, W.J. "Alpha Liquid Scintillation Counting; Past, Present and Future," in *Liquid Scintillation Counting: Recent Applications and Developments, Vol. 1,* C.T. Peng, D.L. Horrocks and E.L. Alpen, Eds. (New York: Academic Press, 1980), pp. 315–332.
13. Horrocks, D.L. *Applications of Liquid Scintillation Counting,* (New York: Academic Press, Inc. 1974), pp. 29–33.
14. Thorngate, J.H., W.J. McDowell, and D.J. Christian. *Health Physics*, 27:123 (1974).
15. McDowell, W.J. "Alpha Counting and Spectrometry Using Liquid Scintillation Methods," NAS-NS -3116, (Technical Information Center, U.S. Dept. of Energy P.O. Box 62, Oak Ridge, TN 37831, 1986), pp. 53–58, 66.
16. Kopp, M.K., Oak Ridge Detector Laboratory, Personal communication 1988.
17. Manufactured by Oak Ridge Detector Laboratory, Inc., 139 Valley Court, Oak Ridge, TN 37830.
18. ETRAC (East Tennessee Radiometric/Analytical Chemicals) Inc., 10903 Melton View Lane, Knoxville, TN 37931.
19. Farrell, R.F., S.A. Matthes, and A.J. Mackie. "A Simple Low-Cost Method for the Dissolution of Metal and Mineral Samples in Plastic Pressure Vessels," Bureau of Mines Report of investigations; No. 8480, (1980).

20. Matthes, S.A., R.F. Farrell, and A.J. Mackie. "A Microwave System for the Acid Dissolution of Metal and Mineral Samples," Bureau of Mines Technical Progress Report-120, (1983).

21. For heating in a furnace or oven, Parr Inst. Co., Teflon-lined pressure vessels numbered 243AC, T303, and 012880 have been used. For use in microwave ovens Parr lists No's 4781 and 4782 as suitable.

22. CEM Corporation in Matthews, NC supplies a microwave dissolution apparatus complete with oven and vessels of their own design.

23. Chiu, N.W., J.R. Dean, and C.W. Sill. "Techniques of Sample Attack Used in Soil Analysis," A Research Report Prepared for the Atomic Energy Control Board, Ottowa, Canada, INFO-0128-1 (1984).

24. McDowell, W.J. and G.N. Case. Unpublished Data.

25. McDowell, W.J. and G.N. Case. "A Procedure for the Determination of Uranium on Cellulose Air-Sampling Filters by Photon-Electron-Rejecting Alpha Liquid Scintillation Spectrometry," ORNL/TM 10175 (Aug 1986).

26. McDowell, W.J. "Alpha Counting and Spectrometry Using Liquid Scintillation Methods," NAS-NS-3116, Technical Information Center, U.S. Dept. of Energy P.O. Box 62, Oak Ridge, TN 37831 (1986), pp 88–89.

27. Bouwer, E.J., J.W. McKlveen, and W.J. McDowell, "Uranium Assay of Phosphate Fertilizers and other Phosphatic Materials," *Health Phys.*, 34:345–352 (1978).

28. Bouwer, E.J., J.W. McKlveen, and W.J. McDowell. "A Solvent Extraction Liquid Scintillation Method for Assay of Uranium and Thorium in Phosphate-Containing Material," *Nucl. Tech.*, 42:102–110 (1979).

29. Case, G.N. and W.J. McDowell. "An Improved Sensitive Assay for Polonium-210 by Use of a Background-Rejecting Extractive Liquid Scintillation Method," *Talanta*, 29:845–848 (1982).

30. McDowell, W.J., D.T. Farrar, and M.R. Billings. "Plutonium and Uranium Determination in Environmental Samples: Combined Solvent Extraction-Liquid Scintillation Method," *Talanta*, 21:1231–1245 (1974).

31. Case, G.N. and W.J. McDowell. "Separation of Radium and Its Determination by Photon-Electron-Rejecting Alpha Liquid Scintillation (PERALS) Spectrometry," in *Proceedings of the 33rd Annual Conference on Bioassay, Analytical & Environmental Radiochemistry, 6–8 Oct. 1987,* Berkeley, California. (No editors, no page numbers given. Proceedings are Xerox copies of papers in ring binder.)

CHAPTER 11

Application of High Purity Synthetic Quartz Vials to Liquid Scintillation Low-Level ^{14}C Counting of Benzene

A. Hogg, H. Polach, S. Robertson, and J. Noakes

ABSTRACT

High purity synthetic quartz is evaluated for low-level ^{14}C detection through liquid scintillation (LS) counting of benzene. A simple cylinder-cell vial design is presented, which incorporates a Teflon® stopper and Delrin shield. The counting characteristics (counting efficiency and background) of the quartz vials are compared to the Wallac Teflon and low-K glass vials, in new technology LS spectrometers (Pharmacia-Wallac, Quantulus, at ANU and Waikato, and the Packard 1050 CA/LL, modified at UGA). The effect of the vial counting characteristics upon maximum and minimum detectable ^{14}C age and the magnitude of the counting error, for both low and high count rate samples, is examined. Synthetic quartz is shown to have counting characteristics superior to low-K glass and is equal to Teflon vials for most applications. Further, quartz does not require the extensive cleaning procedures necessary for Teflon.*

INTRODUCTION

The application of commercially available liquid scintillation (LS) counters for radiocarbon (^{14}C) dating has evolved over the last decade from the utilization of general purpose instruments, to new technology low-level LS spectrometers. The old technology counters, even with extensive in-house modifications, yield at best, between 60 to 75% ^{14}C counting efficiency, characterized by a high background, which at best is >10% of the ^{14}C Modern reference signal (Polach et al., 1983). Two modern low-level LS spectrometers, such as the modified Packard 1050 CA/LL and Pharmacia-Wallac, Quantulus, use electronic optimization (e.g., pulse shape, duration and ratio analyses), and Quantulus uses active and enhanced passive shield, to generally further reduce the background. The performance of these LS spectrometers in relation to ^{14}C dating has been described.[1,2] The best performance is at >80% ^{14}C efficiency characterised by an ultralow background at 0.8% of the ^{14}C reference signal.

---

*Teflon is a Registered Trademark of E. I. DuPont de Nemours.

Such improved performance gives higher precision, extended detectable age limit, or significantly smaller sample ^{14}C LS radiometry.[3] The critical application to low-level ^{14}C spectrometry not only requires evaluation of LS counter performances, but it necessitates consideration of errors and assurances[4] and counter unrelated parameters such as counting vial design, materials, benzene purity, and scintillant performance.

This paper deals with different counting vial materials and their significance to low-level ^{14}C detection characteristics and performance.

COUNTING VIALS

A great variety of counting vial materials are available and were applied to ^{14}C isotope detection in benzene. Glass, quartz, Teflon, and Delrin were found suitable, albeit different in performance characteristics. Polyethylene (PE) and high density PE were not suitable due to their permeability of benzene. Shapes and volumes of vials include special purpose large volume cylinders (50 to 100 mL), square or cylindrical mini-vials within special holders (0.3 to 3 mL), and those based on the standard 20 mL LS counting vial design. High precision ^{14}C low count rate determinations require calibration of each vial independently for efficiency and background; therefore, it is general practice to reuse the same set of vials over many years.

The glass and quartz vials have excellent physical counting properties. Their ^{14}C signal and background detection efficiency remains constant over many years, and memory effects are nil with only minimal washing between samples. Quartz vials give inherently lower backgrounds than K-free glass counting vials.

Teflon and Delrin have excellent counting properties (high efficiency and inherently low backgrounds) but many researchers experience difficulties in their usage due to deformation over a period of time. Memory effects require rigorous cleaning procedures between samples, and alter efficiency over time (Figure 1).

EXPERIMENTAL PROCEDURES AND VIAL DESIGN

Using available LS spectrometers, the authors tested vials of various designs and materials in their laboratories. The University of Georgia, Packard counter, used low-K glass vials in all their measurements. The Australian National University and the University of Waikato used Wallac counters with Wallac Teflon and quartz vials of various origins and sample sizes.

The Teflon vials were used as supplied by their manufacturer (Wallac Oy).[5] One has a wall thickness of 0.9 mm the other of 1.1 mm. This affects their performance (cf. Table 1).

The synthetic quartz vials were manufactured using material from three differ-

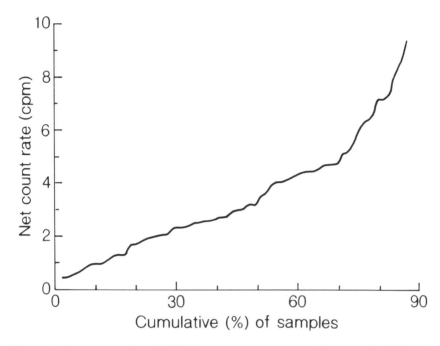

Figure 1. Time series plot of SQPE (external standard endpoint) values for Teflon vials compared to low-K glass vials. Changes in endpoint in Teflon with time correspond to a related change in count rate of the modern reference standard, from 24.2 to 23.7 ± 0.1 CPM, while the background remained the same (within statistical limits). Such changes in performance must be allowed for in high resolution ^{14}C counting.

ent sources: (1) Thermal Syndicate, UK (TS silica), (2) Mikro-Glasstechnik, Germany (MG silica), and (3) GM Associates, USA (GM silica).

The silica vial consists of a flat bottomed cylindrical cell, 34 mm high and 25 mm in diam, sealed by a Teflon stopper containing a Viton 'O'ring (Figure 2). The Teflon stopper contains a partially threaded central opening which is sealed, after the stopper is inserted, by a close fitting tapered stainless steel pin.

Table 1. Counting Characteristics of Teflon, Silica and Low-K vials of 3 mL and 5 mL Benzene Samples at ANU and Waikato, Wallac Quantulus Counters

VIAL	VOL[a]	B[b]	N_0[c]	E[d]	fM[e]	FM[f]	tMAX[g]	tMIN[h]
0.9mm Teflon	3	0.29	27.6	83.7	51.3	24,200	55,400	39
1.1mm Teflon	3	0.24	24.3	73.5	49.5	22,500	55,100	42
MG silica	3	0.32	27.5	83.3	48.6	21,700	55,000	40
TS silica	3	0.43	28.0	84.8	42.7	16,700	54,000	39
GM silica	3	0.36	27.5	83.3	45.8	19,300	54,500	40
Low-K	3	0.50	22.8	69.2	32.2	9,600	51,700	44
0.9mm Teflon	5	0.45	47.5	84.4	70.8	15,800	58,000	30
MG silica	5	0.58	47.7	84.7	62.7	12,400	57,000	30

Note: [a–i]see Table 2.

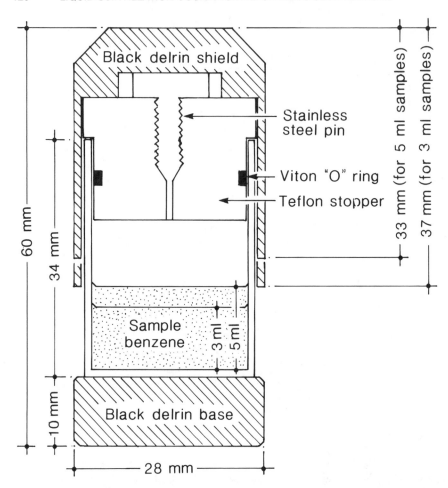

Figure 2.   Cross section of the experimental silica counting vial. A Teflon stopper with a single O-ring is shown. Later experiments gave a significant reduction in vapor loss when a double O-ring was used. The Delrin masks are made in two lengths, one to reach the 5 mL and the other the 3 mL sample meniscus.

The opening prevents the sample from becoming pressurized when the stopper is inserted. A black Delrin base, 10 mm high and 28 mm diam centers the vial along the optical axis of the photomultiplier tubes. A black Delrin shield is fitted over the vial, up to the meniscus of the sample benzene, to reduce optical cross talk. The practical counting volume range is 3 mL to 5 mL of benzene.

The ANU and Waikato experiments with quartz and Teflon used 15 g/L concentrations of Butyl PBD as scintillant, while UGA used 6 g/L PPO with 0.2 g/L secondary pulse length shifter POPOP in benzene.

In all cases the counters were set up to give optimal ^{14}C detection performance, e.g., highest obtainable ^{14}C reference sample signal to background ratio.

Table 2. Counting Characteristics of Low-K vial of 3 mL and 5 mL Benzene Samples at UGA Packard Counter

VIAL	VOL[a]	B[b]	N_o[c]	E[d]	fM[e]	FM[f]	tMAX[g]	tMIN[h]
Low-K[i]								
(1050)	3	0.46	21.08	63.09	31.0	8653	51,400	45
	5	0.75	35.17	64.74	41.0	5588	52,400	35

[a]Volume of benzene (3 mL = 2.637g; 5 mL = 4.5g).
[b]B = background, cpm.
[c]N_o = 95% Oxalic Acid Modern reference standard, cpm.
[d]E = % counting efficiency.
[e]fM = factor of Merit, $N_o/\sqrt{B}$.
[f]FM = Figure of Merit, E^2/B.
[g]Age limit, tMAX—3000 min for B, N_o, and S (2 SD detection criterion, reference 7, p 96), years.
[h]Minimum age, tMIN—3000 min for B, N_o, and S (1 SD detection criterion, Reference 7, p 97), years.
[i]7 mL low ^{40}K borosilicate glass vial with Teflon cap liner. Counted using Packard 1050 modified with scintillating plastic detector guard.

As two important sources of sample ^{14}C count rate variations are known, their effect on performance was tested:

1. Handling the counting vials, exposing them to light, stirring the counting cocktail, or allowing an electrostatic discharge can cause spurious counts at the beginning of the counting cycle.[6] Both Teflon and MG silica were tested for this effect in the Wallac counters. Vials were: (1) rubbed by a Nylon cloth to induce an electrostatic charge, (2) agitated for 30 minutes, (3) exposed for 30 minutes to fluorescent light 15 cm distance, and (4) irradiated by gamma and beta particles from an external high energy and countrate source. A sample of ca., 150% Modern (ANU-Sucrose ^{14}C standard) was counted for 5 min intervals for 30 minutes (6 repeats). All tests were negative. No induced count rate variation could be detected.

2. Loss of sample benzene during prolonged low-level counting times (1 to 3 days) will also cause variation in the observed count rate with time. The evaporative loss in the Teflon vials were nil per day at room temperature. The evaporative loss of the experimental silica vials (Figure 1) was 0.7 mg/d. To minimize this a double 'O'ring stopper was later tested and achieved 0.1 mg/d losses.

Results

The ^{14}C low-level counting characteristics of Teflon, silica, and low-K glass in the Wallac counters are presented in Table 1. Similar data for the low-K glass vial from UGA using the Packard counter is presented in Table 2. The performance of the vials with 3 mL and 5 mL of sample benzene in the various counters is characterized, by their background (B) and net ^{14}C countrate (N_o), at 95% of the Oxalic Acid International Reference Standard. To assist evaluation of data, the radiocarbon dating "factor of Merit" (fM = $N^o/\sqrt{B}$), the conventional "Figure of Merit" (FM = E^2/B), and the oldest (tMAX)[7] and youngest (tMIN)[7] detectable ^{14}C age were included. In the Wallac counters the best performance was obtained with Teflon at 5 mL followed by the MG silica.

DISCUSSION AND CONCLUSIONS

The background and modern reference sample count rates determine the ^{14}C dating performance of both the counter and counting vial: the magnitude of the counting error limits the minimum and maximum determinable ages as well as the minimum countable sample size. Sample size is another variable which affects ^{14}C age resolution.

While counted benzene volumes are usually fixed (in our study to 3 mL or 5 mL), the sample component within the benzene varies in practice. Indeed, all radiometric laboratories dilute the sample, CO_2 or C_6H_6, with gas or benzene containing no ^{14}C. A case study of sample sizes dated at the University of Waikato is given in Figure 3 (count rates are plotted against age). The full size (undiluted) samples must lie on the solid line. The vast majority are diluted and lie below the line. In Figure 4, the samples shown in Figure 3 are expressed as a cumulative plot of sample % for a given count rate (Figure 4). The plot indicates that the count rate rather than sample age is therefore, in practical terms, the limiting variable in ^{14}C age determinations.

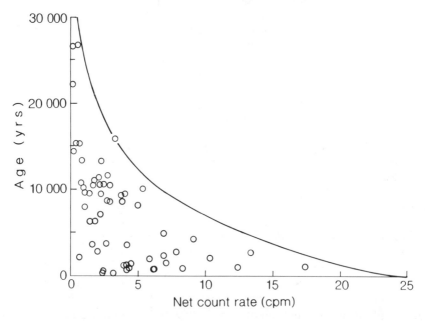

Figure 3. Scatter plot of routine ^{14}C age determination samples counted in the Wallac Quantulus at Waikato over a 12 months period. The solid line gives age and count rate limits for undiluted 3 mL samples. Most samples were diluted with ^{14}C free material to make up the required counting volume.

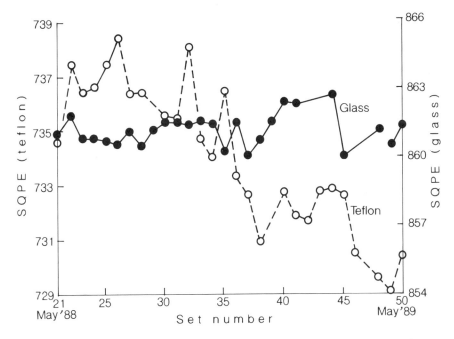

Figure 4.   Cumulative % of samples shown on Figure 3, plotted against the net sample count rate. The plot demonstrates that in routine ^{14}C dating the majority of samples have significantly lower count rates than the modern reference standard, primarily due to the high degree of sample dilution rather than age. High resolution, close to the limit of detection is therefore an important factor in routine ^{14}C dating applications.

To assess the merit of vials, we have taken the 3 mL data (Table 1) for the Quantulus and related each type of counting vial tested to the best vial (0.9 mm thick Teflon). The difference in calculated counting error (1 Standard Deviation, SD), expressed in years, with respect to the net sample count rate is the merit of vials (Figures 5A and 5B).

We can conclude the following:

- The majority of samples ^{14}C dated at ANU and Waikato give low count rates, predominantly due to sample size, rather than the age of material submitted for dating. Whenever this is the case, high resolution ^{14}C counting at the lowest attainable background and highest efficiency is very desirable.
- When dealing with low count rates samples (old or highly diluted, where ^{14}C signal approaches noise), a reduction in counting efficiency (E) is tolerable only if accompanied by a compensating reduction in background (B). For example, a reduction of ~84 to 74% E and almost proportional reduction in B only marginally affects tMAX (e.g., 0.9 and 1.1 Teflon, 3 mL, Table 1) and slightly increases the counting error by 15 years at 30 K years (Figure 5A). A reduction of ~84 to 69% E without compensating reduction in B has a significant effect on tMAX (e.g., 0.9 Teflon and low-K, 3 mL, Table 1) and significant increase in the counting error by 90 years at 30 K years (Figure 5A).
- When dealing with high count rate samples (signal approaches reference stan-

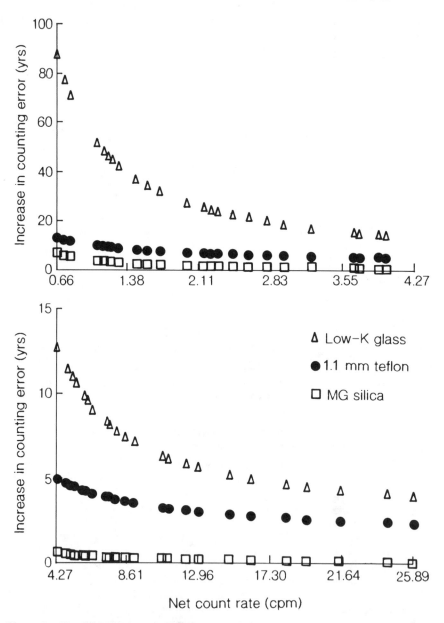

Figure 5.   The difference in the standard deviation (SD) between the best performing vial (0.9 mm wall Teflon) and 3 other vials is given as "increase in counting error years." It is plotted against net count rate (CPM). Figure 5A covers the age range equivalent to 30 to 15 K years for undiluted sample. Figure 5B covers the age range equivalent to 15 K years to 500 years for undiluted samples. All counting times were fixed at 3000 minutes.

dard count rate) an increase in B is tolerable if % E becomes higher (e.g., 0.9 Teflon, B = 0.29 cpm, and TS silica, B = 0.43 cpm, both give the same tMIN, Table 1). However, should the E be significantly decreased and or B significantly increased (e.g., low-K, Table 1 and Figure 5) the tMIN, tMAX, and error deteriorate significantly.

- The performance of the MG silica is closely related to that of the 0.9 Teflon in the Wallac counters. The tMAX and tMIN are not significantly different nor is there a significant difference in the error of determination at the low and high end of count rates (Figure 5).

- Silica varies in performance both in terms of efficiency and background. When suitable material selection is made, and the high resolution and ultralow background characteristics of the Quantalus are used to the utmost, silica will be an effective substitute for Teflon.

- The low-K glass vials in both the Quantulus and Packard counters have similar counting performances (Table 1 and 2). It is hoped the improved silica vial counting characteristics exhibited by the Quantulus measurements can also improve the counting capability for the Packard counters.

- Since the silica vial configuration, as shown in Figure 2, could not be accommodated in the Packard 1050 under optimum 7 mL vial configuration, measurements will be delayed until suitable vials can be fabricated.

REFERENCES

1. Polach, H., G. Calf, D. Harkness, A. Hogg, L. Kaihola, and S. Robertson. "Performance of New Technology Liquid Scintillation Counters for ^{14}C Dating," *Nucl. Geophys.* 2(2):75–79 (1988).

2. Noakes, J. and R. Valenta. "Low Background in Liquid Scintillation Counting Using an Active Vial Holder and Pulse Discrimination Electrons," paper presented at the 13th International Conference on Radiocarbon Dating, Dubrovnik, Yugoslavia, June 20–25, 1988.

3. Otlet, R.L. and H.A. Polach. "Improvements in the Precision of Radiocarbon Dating through the Recent Developments in Liquid Scintillation Counters," paper presented at the 2nd International Symposium in Archaeology and ^{14}C, Groningen, Holland, Sept 18, 1987.

4. Polach, H. "Liquid Scintillation ^{14}C Spectrometry: Errors and Assurances," paper presented at the 13th International Conference on Radiocarbon Dating, Dubrovnik, Yugoslavia, June 20–25, 1988.

5. Polach, H.A., J. Gower, H. Kojola, and A. Heinonen. "An Ideal Vial and Cocktail for Low-Level Scintillation Counting," in *Proceedings of the International Conference on Advances in Scintillation Counting* (Banff, Canada: University of Alberta Press, 1983), pp. 508–525.

6. Haas, H. "Specific Problems with Liquid Scintillation Counting of Small Benzene Volumes and Background Count Rates Estimation," in *Proceedings of the 9th International Conference on Radiocarbon Dating* (Los Angeles: University of California Press, 1979), pp 246–255.

7. Gupta, S.K. and H.A. Polach. *Radiocarbon Dating Practices at ANU* (Canberra, Australia: Radiocarbon Dating Research, 1985), pp 175.

CHAPTER 12

An Introduction to Flat-Bed LSC: The Betaplate Counter

G.T. Warner, C.G. Potter and T. Yrjonen

INTRODUCTION

The use of liquid scintillators was first described in 1950.[1,2] Neither paper refers to the use of liquid scintillators for internal sample counting, but describe their use for alpha, gamma, and neutron counting. Reynolds shows a block diagram of a two PM tube coincidence circuit, and thus was the first to suggest this arrangement for liquid scintillation counting reduction of what he calls, "the well known inconvenience of PM tube background." Both papers refer to the use of benzene, xylene, and toluene, and to the addition of anthracene. Kallman reports the effect of napthalene and the use of terphenyl. The paper usually cited as the first reference to internal sample counting was published in 1953.[3] However a paper in 1952,[4] described a method of counting $^{14}CO_2$ in toluene-PPO solution by condensing the gas at $-80°C$, and thus may have a prior claim.

The first commercial liquid scintillation counter (LSC) is quoted as being marketed in 1954,[5] by Packard, and the earliest reference found to the evaluation of the standard vial was 1957,[6] although this may not have been the first. In this paper, the Wheaton 5 dram vial was compared for background and CPM against 85 mL weighing bottles containing between 10 to 50 mL of scintillant. The Wheaton vial, having a lower background, was shown to be the container of choice. By 1957, Packard seemed to have adopted the Wheaton vial as standard, and their first model in 1954 may also have used a small sized vial. In the U.K. as late as 1968, equipments using the large weighing bottles were still being marketed on the basis of a larger sample volume incorporation.

To reduce thermal noise from the PM tubes, the Packard LSC of 1957, in addition to counting coincidence, used a domestic deep-freeze to house the counting head. Even so, this particular counter had a background for ^3H of 32 cpm and an efficency of 20%, giving a figure of merit of 12.5. However, these figures were very good as a counter marketed in 1968 by a British company, 9 years after the Packard, it had a figure of merit of less than one (0.67). While the Packard counter represents an early version of a liquid scintillation counter, modern equipment is of course based on the same use of vials to contain the scintillant and sample. Up until recently, cooling of the PM tube was a standard feature. The only significant change has been in the instrumentation.

THE FLAT-BED GEOMETRY COUNTER (THE BETAPLATE)

The Betaplate concept was designed in 1976,[7] when it was recognized that the procedure for counting the ^{3}H labeled DNA of cells filtered onto glass fiber discs could be improved. The principle use of this technique was in the mixed lymphocyte assay where the sample preparation proceedure is based on a cell harvester. Suspensions of ^{3}H labeled cells are aspirated from a microtitre plate, filtered onto glass fiber sheets and washed using the harvester. The glass fiber filter is 10 cm wide by 25.5 cm long, and with one particular harvester (Skaatron Inc.), the samples are aspirated 12 at a time, each sample occupying an area 1 cm in diam, a whole filter holding 96 samples in a 6 × 16 matrix.

Conventionally, after the sheet is dried, the discs bearing the samples are individually removed from the sheet with forceps, placed in a standard or minivial and combined with a few milliliters of scintillant. It takes about 20 minutes to prepare 96 samples and load them into a standard LSC, and it is a very tedious operation. Recently, some degree of automation has been available from some manufacturers.

The proposed design (Figure 1) did away with the vial, replacing it with a flat plastic container in which the whole filter sheet could be placed with the 96 samples intact. Scintillant would be added to the whole container and the PM tubes brought up close to each sample in turn; a flat-bed geometry.

In the original drawing, the plate was to be scanned in a raster fashion. The problem of cross talk (that is adjacent sample interference caused by light conduction) was considered and some black material in the form of lines or circles printed on the glass fiber filter between the samples was thought might be an effective method of reducing this problem. It has now been shown [8,9] that the printing reduces cross talk by two orders of magnitude for Tritium, and this forms an important part of the Betaplate concept.

RESULTS AND DISCUSSION

Some basic experiments designed to test the concept were performed using a metal jig made of two sandwiched together aluminium plates, each with a 1 cm diam hole at one end. This device is fitted into a standard vial while a small piece of glass fiber filter bearing ^{3}H labeled cells could be placed between the

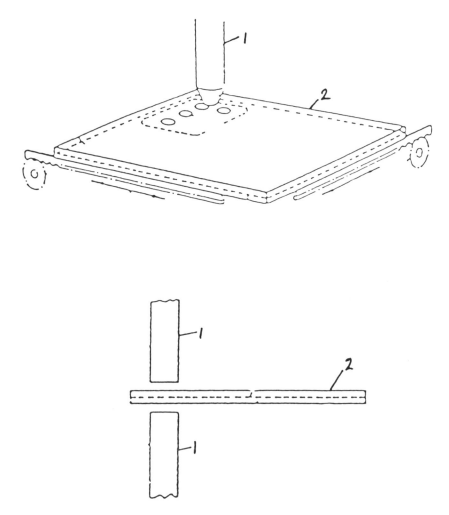

Figure 1.   Layout drawing of the flat-bed counter. The photomultiplier tubes (1), are placed either side of the sample container (2) which could be moved in a raster fashion.

metal plates, either opposite the holes to measure sensitivity, or not opposite, and thus masked, to test for cross talk.

With just a wetting of scintillant, the background countrate in the ^{3}H channel was <10 cpm, and the efficiency was about 50% of that obtained in a standard vial with 2mL of scintillant. After some experimentation, the cross talk was reduced to <0.1%; thus, this crude representation of the principle, showed that cross talk and sensitivity were certainly good enough for work with ^{3}H labeled cells, although at that time cross talk was considered to be too high with ^{14}C.

A simple, manually operated prototype constructed by Wallac Oy of Finland gave efficiency values of 28% for ^{3}H and 78% for ^{14}C, showing that reasonable efficiences could be obtained. Their design staff suggested a sealed plastic bag, rather than a flat plastic box, as the sample container and a metal cassette as a carrier and a means of further cross talk reduction.

A second fully automatic prototype (Figure 2) having a 2016 sample capacity was constructed. In our laboratory, the background is low, but not untypical of sites away from high radiation sources, or high enviromental background; however, the prototype was fitted with quartz tubes so the background figures are lower than the standard production Betaplate, although that is also available with quartz tubes if required. The production 1205 Betaplate is shown in Figure 3.

With the Betaplate system, harvesting is the same save the glass fiber filters. They are especially formulated to be strong enough to prevent the sample areas from easily breaking away, and they are printed with a grid to reduce cross talk. Having harvested the samples and dried the filter, it is then slid into the special plastic bag, 10 mL of scintillant added which, for standard glass fiber sheets is suffient for all 96 samples. The scintillant is gently encouraged to permeate the filter by a hand roller, and the remaining edge is heat sealed. It is then placed in a metal cassette for counting. Time taken about 2 minutes. Glass fiber filters and the special heat-sealable plastic bags are shown in Figure 4.

In addition to glass fiber filters, nylon membranes are available for DNA work and the figures for background, efficiency, and cross talk, include data on nylon membranes (Tables 1–3).

The low values of cross talk (Table 3), show that filter printing not only

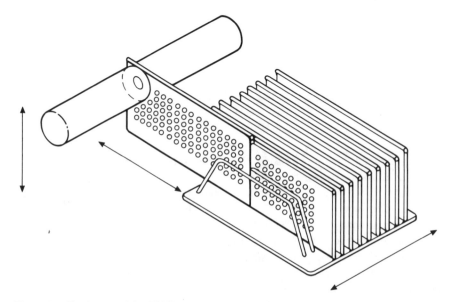

Figure 2.   The layout of the MK II prototype.

Figure 3. Production model Betaplate (Model 1205 Pharamacia/Wallac). The counter has six counting heads and a loading capacity of 1920 samples.

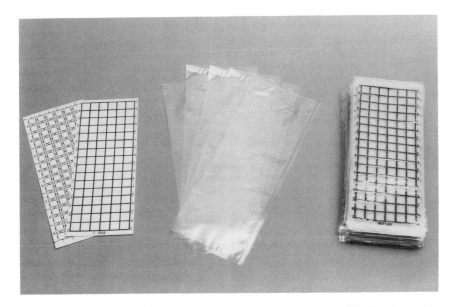

Figure 4. The filter mats with a grid pattern are the standard type used for filtration and the "tile pattern" are the "Spot on" mats taking volumes up to 30 μL.

Table 1. Background Countrates (CPM)

Material	Scintillant Vol.	^{3}H	^{14}C	^{32}P
Glassfiber	10mL	<2	<6	<7
Nylon	4mL	<1	<5	<6

Note: These values were obtained using quartz photomultiplier tubes.

Table 2. Efficiency and Figure of Merit (F)

	Conventional LSC			Flatbed Counter	
Material	Isotope	Eff.	F	Eff.	F
Glassfiber	^{3}H	47%	110	54%	>2000
Glassfiber	^{14}C	96%	180	97%	>1750
Nylon	^{32}P	90%	324	84%	>1000

Note: The samples used were labeled cells and therefore show "working" efficiencies rather than theoretical maximums.

reduces cross talk for low-energy isotopes, but also works well with higher energy emmitters such as ^{14}C and ^{35}S. It has also been found that good results are obtainable with gamma emmitters that have low-energy electron emmission, such as ^{51}Cr and ^{125}I. Specific cytotoxicity and receptor binding assays, therefore, can be performed using the Betaplate.

While the Betaplate was developed for use with filtered cells, plastic backed filters have been developed for spotting small volumes of up to 30 μL, which then can be dried onto the filter. This technique is used for the cytotoxic assay ("Spot-on" mats, Pharmacia-Wallac).

A more recent development is a 400 μL liquid-holding tray (Figure 5), containing 96 wells in the standard Betaplate format ('T' tray Pharmacia-Wallac). It was designed for use with scintillation proximity assay (SPA) kits requiring a 400 μL sample container (Amersham International). The background count and efficiency with a liquid sample are similar to a standard vial geometry LSC. Correlations between standard LSC and Betaplate counting methods are shown in Figures 6 & 7.

The advantages and disadvantages of the Betaplate system are summarised in Table 4.

Another aspect of the Betaplate is the low bulk of disposables (Figure 8). The bags containing filtered samples save 95% of the plastic bulk and a similar reduction in the volume of scintillant used.[10] The 'T' trays, while not having

Table 3. Cross Talk (Nearest Sample)

Material	Isotope	No lines	Printed lines
Glassfiber	^{3}H	0.600%	0.006%
Glassfiber	^{14}C	3.25%	0.080%
Nylon	^{32}P	0.16%	0.015%
Glassfiber	^{51}Cr	0.36%	0.015%
Glassfiber	^{125}I	2.0%	0.3%

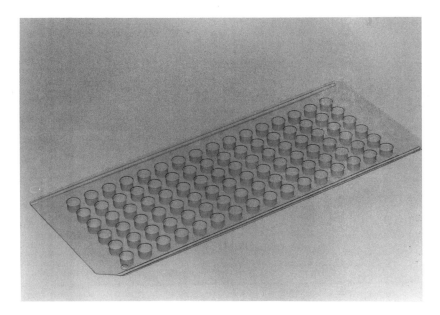

Figure 5. The 'T' tray for liquid samples. (Well capacity 400 μL.)

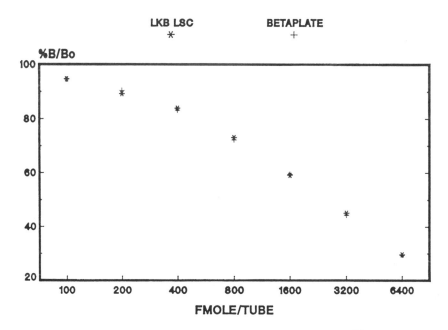

Figure 6. Correlation between a standard LSC and a Betaplate counter: [125]I cAMP Scintillation Proximity Assay. (Data kindly supplied by Dr. Nigel Bosworth, Amersham International.)

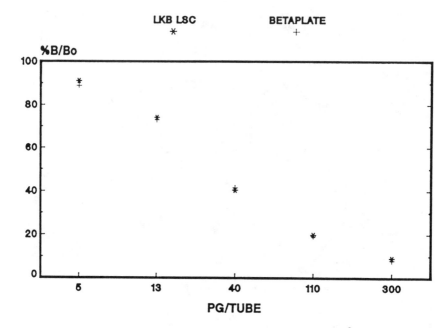

Figure 7. Correlation between a standard LSC and a Betaplate counter: ^{3}H Thromboxane B2 Scintillation Proximity Assay. (Data kindly supplied by Dr. Nigel Bosworth, Amersham International.

quite the same reduction in volume, still show a considerable saving over conventional vials.

In summary, the flat-bed geometry Betaplate has many advantages over the vial based geometry counter when filtered cells or volumes up to 400 μL are being measured; the concept represents the first radical change in LSC design for over 35 years.

Table 4. The Advantages and Disadvantages of the Betaplate System

Disadvantages
1. CPM only, no automatic quench correction possible, no external standard
2. Some cross talk between samples, but much reduced by printing and sensible sample layout
3. Sample size limited to 400 μL

Advantages
1. Ease and speed of sample preparation
2. Economy of disposables
3. Sealed samples reduce biohazards
4. Low background count rate together with good efficiency gives high Figures of Merit
5. Lightweight, compact machine
6. High sample loading capacity
7. Six counting heads gives high sample throughout
8. Low volume of waste for disposal

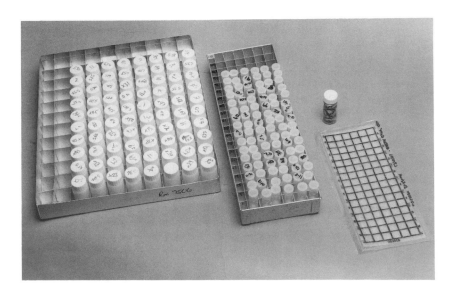

Figure 8.   There is a 95% reduction in the volume of disposables when using filter mats. A 96 sample filter can be contained in a single standard vial.

REFERENCES

1. Reynolds, G.T., F.B Harrison, and G. Salvini. "Liquid Scintillation Counters." *Phys. Rev.* 78:488.(1950).
2. Kallman, H. "Scintillation Counting with Solutions." *Phys. Rev.* 78:621(1950).
3. Haynes, F.N and R.G Gould. "Liquid Scintillation Counting of Tritium Labeled Water and Organic Compounds," *Science*,117:480–482(1953).
4. Williams, D.L. "U.S.A.E.C Document" LA-1484. (1952).
5. Rapkin, E. in *The Current State of Liquid Scintillation Counting*, (New York: Gene & Straton New York. 1970), p. 45
6. Davidson, J.D. in *Liquid Scintillation Counting*, (New York: Pergamon Press. 1958), p 88.
7. Warner, G.T. and C.G. Potter. "A Method of, and Apparatus for, the Monitoring of a Plurality of Samples Incorporating Low Energy Beta-emitting Raioisotopes," Brit. Patent 1586966 (London: HM Patent Office 1980).
8. Warner, G.T. and C.G. Potter. "Sorption Sheet for Sorbing a Plurality of Discrete Samples and a Method of Producing Such a Sheet," US Patent 4,728,792 (Washington: US Patent Office 1988).
9. Potter, C.G., G.T. Warner, T. Yrjonen, and E. Soini. "A Liquid Scintillation Counter Specfically Designed for Samples Deposited on a Flat Matrix," *Phys. Med. Biology*, 31(4):361–369(1986).
10. Warner. G.T., and C.G. Potter. "New Liquid Scintillation Counter Eases Vial Disposal Problems," *Health Phys.* 51(3):385–386(1986).

CHAPTER 13

A Plastic Scintillation Detector with Internal Sample Counting, and Its Applications to Measuring ³H-Labeled Cultured Cells

Shou-li Yang, Ming Hu, Jia-chang Yue, Xiu-ming Wang, Xu Yue, Jie Li, Zi-ang Pan, and Zhong-hou Liu

INTRODUCTION

Liquid scintillation counting has the advantage of homogenous 4π counting geometry and resultant high counting efficiency, but it frequently requires complex sample preparation and produces excessive amounts of radioactive organic waste.[1] As a result, Cerenkov counting and other counting methods have been used.[2] The plastic scintillation detector for determining soft beta nuclides, which is used without cocktail, has about a 30 year history.[3-7] Internal sample counting with gas detectors has been used to measure soft beta emitting nuclides, but internal sample counting with plastic scintillation detectors has been less frequently reported in the literature. This paper describes a simple and convenient plastic scintillation method for routine internal sample counting.[8-9] The application of this method, in measuring the amount of ³H incorporated into cultured cells, was tested for its homogeneity, counting efficiency, reproducibility, and spectrum analysis.

MATERIALS AND INSTRUMENTS

The radioactive isotopes used as reference samples in this work were ³H and ¹⁴C labeled hexadecane. Also used were ³H labeled lysine and thymidine and ³⁵S as Na_2SO_4 solution diluted by an emulsion cocktail. All radio isotopes were made by the Chinese Academy of Atomic Energy.

The plastic scintillator sheet (PSS) was made from styrene monomers purified by vacuum distillation. Fluors of PPO and POPOP at concentration levels of 10 g/L and 0.8 g/L, respectively were added to this monomer. The polymerization procedure was conducted in an oil bath at 110°C for 5 days. The PSS was pressure shaped into rectangular blocks 32 mm long, 15 mm wide, and 0.25 mm thick. Two different detectors were made from the initial plastic sheet. One PSS detector was rectangular with base area dimensions of 24 mm × 8 mm and a volume capacity of 0.996 mL. A second PSS detector was circular with a diameter of 12 mm. A glass filter paper disc used for cell harvesting measured 12 mm in diameter. Cocktails were made using a fluor concentration of 6 g/L of butyl-PBD in a solvent of pure toluene or 2:1 ratio of toluene/Triton X-100.

Radioactivity measurements were made using a Beckman Model LS-9800 liquid scintillation counter. The General Program and the Special Spectrum Analysis Program of the instrument were used in the counting procedures. A Hitachi Model S-520 scanning electron microscope was used for microscopic analyses.

METHODOLOGY AND RESULTS

The plastic detector used in this study sandwiches the radioactive sample between two PSS, and seals it with a gluing optical coupling solvent (GOCS). The radioactive sample may be in solution or on solid support.

The GOCS must have good qualities for both optical coupling and gluing the sample to the PSS. In general, aromatic solvents are superior for gluing. Solvents such as toluene, xylene, anisole, dioxane, ethyl acetate, chloroform, or dichloromethane can be used to soften or dissolve the PSS. Additional emulsifiers such as Triton X-100 in GOCS are necessary for aqueous sample counting. In fact, the toluene cocktail was found to be an excellent GOCS.

To prepare samples on filter paper support, they were dipped in cocktail. A radioactive sample, either dissolved or on support, is then fixed between two PSS by GOCS. It was found the GOCS should not be used in excess, otherwise the plastic sheets would become deformed. Finally, after gluing the two PSS together, they were placed in a 20 mL standard glass vial to be counted.

The sample prepared for counting is a composite of radioactive material, plastic scintillator, and GOCS. Figures 1 and 2 are scanning electron microscope photographs, shown amplified at 5000 and 2500 times. Figure 1 shows a cut section of fiberglass paper, and Figure 2 shows a torn section of the glass filter paper support sandwiched between the plastic scintillator. Figure 2 also shows some broken granules of wrinkled and cracked plastic scintillator folded around and among the glass fibers. Harvested cells in a ruptured broken state due to the treatment by the organic solvents contained in the GOCS are also shown. These photographs reveal that the distance between the glass fiber and plastic scintillator is 0.3 to 2.4 μm.

COUNTING EFFICIENCY AND REPRODUCIBILITY

As known radioactive reference standards, 10 μL of [3]H n-hexadecane and [14]C n-hexadecane were sealed between 2 plastic sheets and placed in an empty,

Figure 1. A cut section of filter glass paper.

standard 20 mL glass liquid scintillation vial. The counting efficiencies of ^{3}H and ^{14}C were 30.6 $\pm$ 1.5% (mean + S.D.) and 85.2 $\pm$ 0.5% using 100 μL of GOCS with a toluene b-PBD-6 g/L cocktail. The counting efficiencies, without the use of GOCS, were reduced to 5.9 $\pm$ 1.3% and 43.1 $\pm$ 6.2%, respectively. These data show that GOCS not only increases ^{3}H and ^{14}C counting efficiencies but also decreases the relative standard deviation. For ^{35}S, as a Na$_2$SO$_4$ aqueous sample, a 2:1 toluene/Triton X-100 b-PBD-6 g/L cocktail was used as the GOCS. The relative counting efficiency of ^{35}S by PSS method is 82% compared to the emulsive cocktail counting method.

Figure 2. A torn section of the glass filter paper sandwiched between the plastic scintillator (broken granules of wrinkled and cracked plastic scintillator are folded among the glass fibers).

The shape of the PSS can be rectangular or circular. In the liquid scintillation counting procedure the rectangular PSS was leaned obliquely against the inner wall of the counting vial so that the radioactive sample was located in an elevated position relative to the counter phototubes. As a result, higher counting efficiencies could be obtained with the rectangular PSS than with the circular PSS, which was laid flat on the bottom of the vial during counting (Table 1).

The counting reproducibilities (coefficient of variation) of three measure-

Table 1. Comparison of the Spectral Parameters of the Plastic Scintillator Method with Homogeneous Counting

		H-3 Hexadecane			C-14 Hexadecane		
		Toluene Cocktail 10 mL	Plastic Scin. + GOCS	Plastic Scin.	Toluene Cocktail 10 mL	Plastic Scin. + GOCS	Plastic Scin.
Compton Electron Spetron	Main Peak Channel	790	640	550	780	640	550
	End Channel	850	760	750	840	760	750
Sample Spectrum	Main Peak Channel	270	240	180	510	510	510
	End Channel						
	1%	395	347	339	650	581	572
	5%	363	312	289	623	554	543
	10%	343	295	262	604	536	524

ments for the rectangular and circular PSS are $1.9 \pm 1.3\%$ and $2.3 \pm 1.2\%$, respectively. The ratio of cpm of circular to rectangular PSS is 0.862 ± 0.056, which shows that the variation of height of the rectangular PSS position may cause larger counting error (coefficient of variation 6.7%). Because of less variation in the position of the circular PSS at the bottom of the vial, a smaller counting error (SD/mean) of $1.3 \pm 1.2\%$ was measured (Table 1). If GOCS is used in excess, the count of the rectangular PSS decreases progressively as its shape changes (Figure 3). The angle in the vial of the rectangular PSS to the PM tubes has little or no effect on counting (Figure 4). From these data, we can see that the rectangular PSS has higher counting efficiency than the circular PSS, but the circular PSS has better reproducibility than the rectangular PSS.

ENERGY SPECTRUM ANALYSIS

Table 2 shows the spectrum analysis from the external standard Compton electron spectrum of an empty 20 mL vial, a circular and rectangular PSS in a vial, and a 10 mL toluene cocktail in a vial. The data reveal that the main peak and end channel for the rectangular PSS, with and without GOCS, are significantly lower than the toluene cocktail sample. They also show the main peak channel for the plastic scintillator with GOCS is significantly higher than without GOCS, though the end channel is approximately the same.

Sample spectrum analyses of the main peak channels for ^{14}C homogeneous counting and rectangular PSS, with and without GOCS, had no obvious change. The ^{3}H main peak channels did decrease in the order mentioned previously. The end channels of the sample spectrum, which were eliminated at

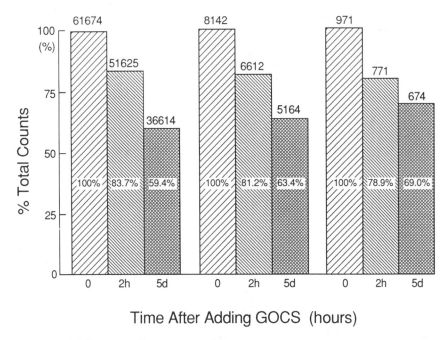

Figure 3. GOCS mount effect on counting efficiency.

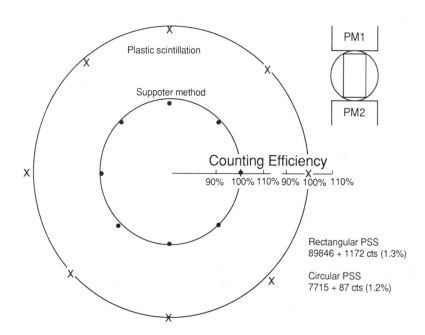

Figure 4. Angle effect on rectangular PSS with PM tube vs counting efficiency.

Table 2. Comparison of Background Spectral Parameters

		20 mL Empty Glass Vial[b]	Two Circle PSS[g] in a Vial[b]	Two Rec-tangle PSS[a] in a Vial[b]	10 mL Toluene Cocktail in a Vial[b]
	0	650	720	890	1000
End Channel	1%	588	611	803	983
Eliminated	5%	488	515	675	963
High energy	10%	451	472	618	905
	Integral	28.5 ± 0.6	25.7 ± 2.1	28.8 ± 0.6	58.8 ± 2.8
Back-Ground (CPM)	C-14 Channel (50—670)	28.5 ± 0.6	25.6 ± 0.5	28.7 ± 0.6	43.3 ± 0.7

[a]PSS: Plastic Scintillator Sheet.
[b]Vial Without Cap.

the 1%, 5%, and 10% high energy counts of the plastic scintillator with GOCS, increased slightly in comparison to that without GOCS. The end channels of homogeneous counting also increased compared to plastic scintillator with and without GOCS. The results for ^3H are similar to that for ^{14}C.

BACKGROUND SPECTRUM

The background spectrum was measured with a 20 mL glass standard vial (no cap), two rectangular PSS in a vial, two circular PSS in a vial, and 10 mL of a toluene-6 g/L of b-PBD cocktail in a vial. The main parameters of the background spectrum are listed in Table 3. The counting time was 90 minutes.

When high energy counts were eliminated at the 10% level, the end channel for the 20 mL empty glass vial decreased from 650 to 451. The rectangular and circular PSS in the glass vial decreased from 720 to 472 and from 890 to 618, respectively. The 10 mL cocktail in the vial decreased from 1000 to 905. These data indicate that the range of the background spectrum decreased about one

Table 3. Comparison of the Relative Counting Efficiency and Reproducibility of the Rectangular PSS with the Circular PSS

No. of Sample	Rectangle PSS		Circle PSS			Effect of Circle
	Mean ± SD (cpm)[a]	SD/Mean (%)	Mean ± SD (cpm)	SD/Mean (%)	Error (%)[b]	Position (%)
1	10495 ± 205	2.0	8335 ± 101	1.2	79.4	+ 2.9
2	11209 ± 26	0.02	10274 ± 134	1.3	91.7	+ 0.03
3	15736 ± 186	1.2	13060 ± 557	4.3	83.0	+ 0.3
4	17114 ± 395	2.3	15066 ± 356	2.4	88.0	− 1.8
5	18033 ± 651	3.6	14187 ± 365	2.6	78.7	− 1.5
Mean + SD	1.9 ± 1.3		2.3 ± 1.2		0.842 ± 1.056	1.3 ± 1.2

[a]n = 3.
[b]Relative Counting Efficiency of a Circle to Rectangle PSS.

third for the two kinds of PSS in their high energy background content. This decrease is greater than the homogeneous counting. This effect is clearly shown in Figure 5 and Table 3.

STABILITY

The radioactivities and backgrounds of PSS have been counted repeatedly over a period exceeding one year. The data show that the counts have not significantly changed. It is from these results that PSS are said to be stable and

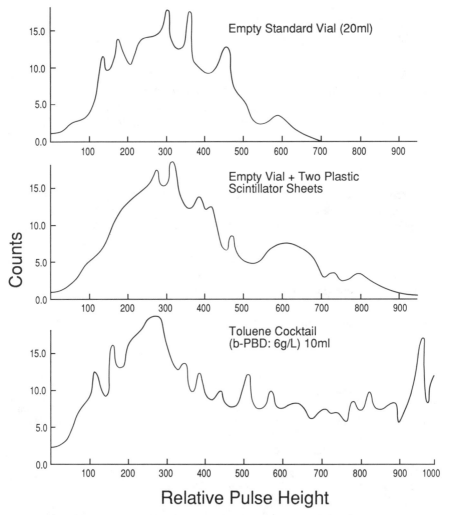

Figure 5. Background spectrum comparison of an empty 20 mL standard glass vial, two rectangular PSS in a vial, and 10 mL of toluene cocktail in a vial.

capable of being stored for long periods of time without undergoing any change.

APPLICATION TO MEASURING OF ³H LABELED CULTURED CELLS

Screening of Chinese medicinal herbs for immunopotentiators was made with the aid of mitogen-induced lymphocyte transformation. Human peripheral blood lymphocytes were cultured in a flat-bottomed 96-well microliter plate at a concentration of 1×10^6 cells/well. A final 0.2 mL contained PRMI 1640 medium, supplemented with 10% fetal calf serum, and 100 units/mL of penicillin G, consisting of 100 μg/mL of streptomycin sulphates. Mitogen and/or various concentrations of herbal extracts were added, and the cultures were kept under 37°C at a 5% CO_2 atmosphere. After 48 hours of incubation 0.5 or 1.0 μCi of ³H-TdR was added to each well and incubation was continued for 6 to 12 hr. Cells were harvested with a Titertek multiharvester and collected on glass filter discs, after which the radioactivities were counted by the methods of supporter and rectangular PSS.

The stimulation index (S.I.) was calculated as follows:

$$S.I. = \frac{\text{cpm of test well}}{\text{cpm of control well}}$$

An S.I. value larger than 1.2 indicates a significant enhancement of lymphocyte transformation induced by the tested herbs.

The results shown on Table 4 indicated that herbs No. 162 and No. 520 exhibited significant enhancement of lymphocyte transformation. The S.I. values obtained in the supporter method were quite similar to that in PSS. Also the relative figure of merit of the PSS (290.0 ± 41.5) was higher than that in the supporter method (151.9 ± 8.1). The ratio of the PSS to supporter method was calculated at 1.91.

The macrophage active factor (MAF) was evaluated by means of macrophage tumor cell cytostasis (MTC). The self-prepared MAF was diluted by 1:25, 1:50, 1:100, and 1:200 and incubated with macrophages (Ms) at a 1×10^5/well concentration for 12 hr at 37°C with a 5% CO_2 humidified air atmosphere. After two vigorous washings with RPMI 1640, p815 mouse macrophage-tumor cells were introduced at a concentration of 1×10^4/well and incubated together with Ms under the same conditions for 30 hr. During the last 6 hr of incubation an alliquot of ³H-TdR, at a 0.5 μCi concentration was added to each well. The degree of incorporation of ³H-TdR by p815 cells was measured by both solid support and PSS counting methods.

The rate of cytostasis at different dilutions of MAF activated Ms were determined by the two counting methods listed in Table 5. The rate of cytostasis was calculated as follows:

Table 4. Plastic Scintillator Method Compared with LS Counting in Screening of Chinese Medicinal Herbs for Immunopotentiators

| Group | Supporter Method | | | | |
	cpm (net) ± SD	S.I.		B (cpm)	E^2/B
Cell ref.	1831 ± 346			65	153.8
PHA: 2r	63213 ± 7465	34.5		66	151.5
PHA: 1r	50032 ± 19042	27.3		68	147.1
PHA: 0.05r	1954 ± 341			63	158.7
No. 162A	4662 ± 222	2.54		66	151.5
No. 520[a]	6821 ± 1269	3.49		60	166.7
No. 251[a]	1671 ± 52	0.85		69	144.9
No. 162A[a]	3527 ± 852	1.80		71	140.8
Mean ± S.D.					151.9 ± 8.1

| Group | Plastic Scintillation Counting | | | | |
	cpm(net) ± SD	E (%)[b]	S.I.	B(cpm)	E^2/B
Cell ref.	1360 ± 79	74.3		19.5	282.8
PHA: 2r	48400 ± 2289	76.6	35.5	19.5	300.9
PHA: 1r	37285 ± 14607	74.5	27.4	19.5	284.8
PHA:0.05r	1406 ± 192	71.9		19.5	265.3
No. 162A	3300 ± 308	70.8	2.42	19.5	257.0
No. 520[a]	5094 ± 535	74.7	3.62	19.5	286.0
No. 251[a]	1185 ± 83	70.9	0.87	19.5	257.9
No. 162A[a]	3059 ± 937	86.7	2.17	19.5	385.6
Mean ± S.D.		75.1 ± 5.1			290 ± 41.4

[a]Plus 0.05r PHA.
[b]Relative Counting Efficiency of Plastic Scintillation to Supporter Method.

$$\text{MTC rate} = \left(1 - \frac{(p815 + M)\ \text{cpm}}{p815\ \text{cpm}} \right) \times 100\%$$

The result shows that even under a relatively high diluting situation (1:200), the MTC rate of MAF activated Ms was still as high as 72.1%, indicating the high activity of the self-prepared MAF. Also, the ratio of the relative figure of merit of the rectangular PSS to that of the supporter method was found to be 1.94.

Monitoring of self-immunity ability in humans was carried out through the assay for interleukin 2 (IL-2) activity. The peripheral blood lymphocytes of a renal transplant patient were stimulated by phytohemagglutinin (PHA) for 24

Table 5. Comparison of the Two Counting Methods in Evaluation of MAF by Means of Anti-Cancer Cell Proliferation

MAF Dilution	1:25	1:50	1:100	1:200	0	B ± S.D.	E^2/B
Supporter Method	1196	3037	10131	19601	70238	59 ± 17	169.9
Plastic Scint.	977	2392	8142	15322	57072	19.5 ± 7.8	329.0
Relative Eff. (%)	81.7	78.8	80.4	78.2	81.3	80.1 ± 1.4	

Table 6. Comparison of the Two Counting Methods in Monitoring of Self-immunity Ability in Renal Transplanting Patient

Titer %	50	25	12.5	6.25	3.125	0	B(cpm)	E^2/B
Supporter	219360	202110	18927	5609	2132	730	65 + 19	153.8
Method	225700	192170	17542	735	3345	855		
Plastic	161010	150460	14450	4237	1661	521	19 + 1.4	286.7
Scint.	158500	134370	12679	520	2436	667		
Relative Eff. (%)	81.7	72.2	74.3	73.1	76.9	74.5	73.8 + 3.0	

hr. The supernant, which was diluted by 1:2, 1:4, 1:8, 1:16, and 1:32, etc., was incubated with C57B1 mouse thymus cells which had been stimulated for 12 hr by Con A. Sixteen hours prior to harvesting ^{3}H-TdR was added. The radioactivities in the cells collected on glass filter paper were compared by PSS counting and the supporter method.

Under proper dilution of the supernant, the thymus cells grew well where the activity of IL-2 was high, which demonstrates the renal transplant patientability for cell immunity. By this method, it is possible to predict the body rejection capability of a renal transplant patient. This procedure is at best a quasi-quantitative assay method, since it may be influenced by many factors. Its primary usefulness is its ability to compare the relative activity of different measurements for one person. As shown in Table 6, the two counting methods obtained comparable results. The ratio of E^2/B of PSS (286.7) to the supporter counting method (153.8) is 1.86.

The PSS method applied in the above three experiments gives the same biological conclusions as the supporter method of counting. The PSS method has the advantage of a higher figure of merit, long storage, and no liquid waste. It is because of these attributes, we strongly recommend the PSS method, especially where small sample size or volume is to be prepared and counted and liquid scintillation waste disposal represents a problem.

Note: This manuscript has undergone extensive revision and condensation in order to meet the style requirements and space limitations of the proceedings. We sincerely hope that we have not changed the substance or connotation of any information intended to be conveyed by the authors. Eds.

REFERENCES

1. Kalbhen, D.A. and V.J. Tarkkanen. "Review of the Evolution of Safety, Ecological and Economical Aspects of Liquid Scintillation Counting Materials and Techniques," in *Advance in Scintillation Counting*, S.A. McQuarrie, C. Ediss, and L.I. Wiele, Eds., (Edmonton, Alberta: University of Alberta Press, 1984), p.66.
2. Yang Shou-li. "The Comment on Literature (1976–1986) for Liquid Scintillation Counting and Prospecting its Future," *Nucl. Electronics & Detection Tech.*, 7(4):222(1978).

3. Schorr, M.G. and F.L. Torney. "Solid Non-crystalline Scintillation Phophors," *Phys. Rev.*, 80:474(1950).
4. Schram, E. *Organic Scintillation Detector*, (Elsevier Pub. Co. Amsterdam,1963), p.66.
5. Basile, L.J. "Progress in Plastic Scintillators," U.S. Government Printing Office , Washington, 1961.
6. Carlsson, S. "The Use of a Plastic Scintillator to Determine the Activity of Solid Samples in a Liquid Scintillation Counter," *Int. J. Appl. Radiat. Isotopes*, 28(7):670–671(1979).
7. Simonnet, F., J. Combe, and G. Simonnet. "Detection of P-32 Scintillation Plastic Vials," *Appl. Radiat. Isotopes*, 38(4):311–312(1987).
8. Yue Jia-chang and Xio Jing-cheng, "A Method in Using Plastic Scintillator Film instead of the Liquid Scintillator Solution for the Low Energy Beta Nuclides," *Annual Report of Chinese Acad. of Agr. Sci.*, (1983), p.141.
9. Yang Shou-li, H. B. Ma, J. H. Sun and Z. H. Liu. "The Plastic Scintillation Detector with Internal Sample of Soft Beta Nuclides," *Nucl. Electronics & Detection Tech.*, 7(6),321–327(1987).

CHAPTER 14

A New, Rapid Analysis Technique for Quantitation of Radioactive Samples Isolated on a Solid Support

Michael J. Kessler, Ph.D.

INTRODUCTION

The quantitation of radioactivity on solid supports has grown extremely rapidly in the past five years due to the increased use of microplate assays. These assays use a special microplate capable of holding 96 samples of volumes up to 300 μL in the 8 × 12 format. A typical microplate is 3 in × 5 in. The incubation of cells, DNA, tissues, or other substrate takes place directly in the microplate. Once the incubation has been completed in the presence of a radioactive labeled substrate, the unreacted/unincorporated components must be separated from the incorporated/bound substrate. This is normally done with a cell harvester, which deposits the cell particulate bound material on a filter media, usually glass fiber filter. Alternatively, the DNA can be spotted with a dot blot apparatus in the 8 × 12 format and hybridization can be performed. A wash solution (buffer or water) removes any unincorporated/ unbound radioactive material on the filter. Each of the filters is punched out of the filter mat into individual scintillation vials, 1 to 12 samples at a time. Five mL of scintillation cocktail is added to each of the individual scintillation vials containing the filters. The 96 vials are capped, shaken, and placed in the counting cassettes of a liquid scintillation counter. Each of the 96 samples is counted for 5 to 10 minutes. The data are printed out and transferred to a computer for final data reduction and graphic presentation. As can be clearly seen from this description of the steps involved in harvesting and quantitation, this procedure is not only time consuming but expensive.

Packard has developed an alternative to this procedure that reduces both the time and cost of harvesting and quantifying each sample by a factor of over ten. This procedure uses two new instruments. The first is a special 96 sample cell harvester (Micromate 196). It can harvest all 96 samples from the microplate simultaneously and is as efficient as the manual harvesters. The second is a

155

special 96 sample radioactivity reader. This Packard Matrix 96 can analyze 96 samples simultaneous using 96 individual detectors using no vial, no cocktail, no bags, and no special filters and causing no destruction to the samples. A description of this harvester and reader will be presented in this article along with applications for which this equipment can be used.

HARVESTING/BLOTTING

Three different methods exist for preparing samples on a solid support. The first is the direct method. This involves spotting the samples in the 8×12 format. The best example of this application is dot blots. For this application the DNA is spotted on a membrane in the microplate format, and ^{32}P or ^{35}S DNA probes are used. A typical DNA hybridization device is shown in Figure 1. The radioisotope, which becomes hybridized to the DNA, is then quantitated using radioautography/densitometry or liquid scintillation counting. This DNA probe hybridization assay has become increasingly popular because of the strong interest in DNA probes and DNA sequencing.

The second method is the use of a cell harvester. This method involves using a special device which is able to aspirate the sample (cell/particulate material) from the wells of the microplate onto a filter media, and wash the wells and the filter to remove any unincorporated/unbound radioactive material. These filter bound samples are punched into liquid scintillation vials, processed, and quantitated in a liquid scintillation counter. With the present manual harvesters, only 6 samples can be harvested, washed or punched out at a time. The new Packard Micromate 196 cell harvester is capable of harvesting and washing all 96 samples simultaneously from a microplate. This is done directly, without any excess tubing in the cell harvester that could become contaminated. This harvester is shown in Figure 2. In order to assess the Micromate 196 cell harvester performance compared to the manual harvester, two separate microplates were prepared using the ^3H-Thymidine cell proliferation assay. These separate plates were analyzed by the manual (6 samples) and the Micromate 196 sample cell harvester. The correlation of the data found an R^2 of 0.949 (Matrix Application Note). From this data it is clear that the new Micromate 196 cell harvester performs as well as the manual harvester but 16 times faster. These samples from the Micromate harvester can either be manually punched out and analyzed by liquid scintillation counter or they can be quantitated directly by the Matrix 96, direct beta counter.

The third method is the use of special filter bottom plates. These filter bottom plates offer the advantage of being able to incubate, wash, and harvest all in a single microplate. Two different types of microplates are available, those with and those without removable filters. Once the cell harvesting and washing is complete each of the 96 filters can either be punched out into liquid scintillation counting or gamma counting vials. If the membrane can be removed from the microplate, the complete membrane (96 samples) can be

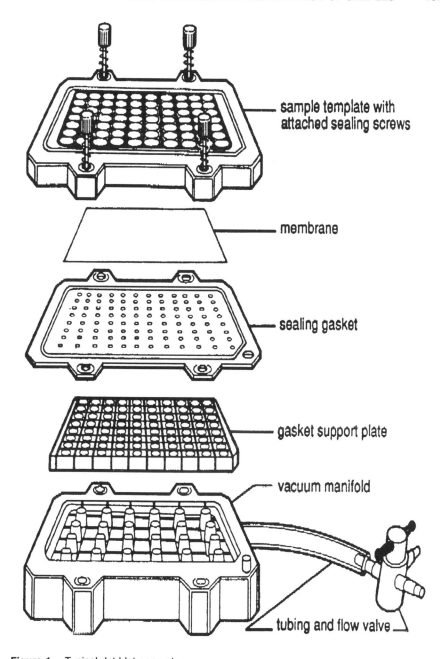

Figure 1. Typical dot blot apparatus.

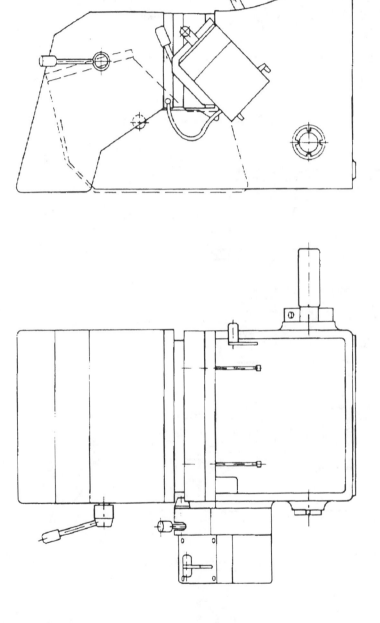

Figure 2.   Packard Micromate™ 196, 96 sample microplate harvester.

analyzed on the new Packard Matrix 96 radioactivity reader. This stripping of the membrane from Microplate is illustrated in Figure 3. At present these strippable membrane bottom plates are manufactured by Pall Biosupports.

In summary, three separate methods exist for preparing samples on a solid support media in the microplate format. The first is the direct method of spotting as exemplified by the dot blot application. The second is the harvester applications which aspirate the cell/particulate from the microplate onto a filter media and remove the unincorporated/unbound radioactivity by extensive washing steps. The third is the use of membrane bottom microplates which allow the incubation, harvesting and washing of the assay components in a single microplate. All three methods can be used to prepare samples for quantitation using the Matrix 96 radioactivity reader. (Figure 4).

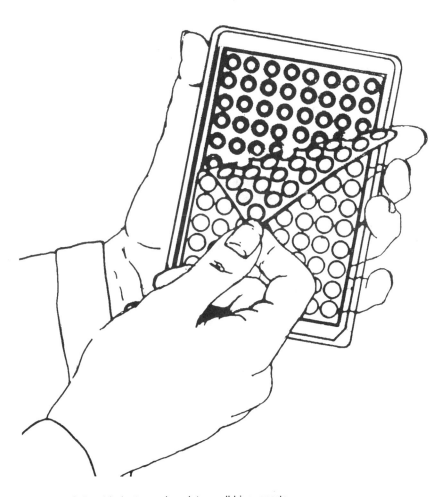

Figure 3. Strippable bottom microplates pall biosupports.

Figure 4. Packard Matrix™ 96, 96 sample direct beta counter.

QUANTITATIVE TECHNIQUES

Four separate techniques for quantitating samples on a solid support exist in the microplate format. These include standard liquid scintillation counting, multi detector liquid scintillation counting, position sensitive proportional counter scanning, and multi-detector (96) avalanche gas ionization detector quantitation. Each technique will be evaluated in detail with the number of steps, time, and cost of each presented. The first technique is liquid scintillation counting. This technique is the gold standard by which all others are measured. This technique involves over 500 steps from the sample harvesting to the actual quantitation. The first step is harvesting the samples from the microplate 6 samples at a time. This procedure is repeated 16 times in order to harvest all of the samples from the 8×12 microplate. Each of the individual samples, which are harvested and washed onto a filter, is punched out into one of the 96 individual scintillation vials which were previously labeled to prevent a sample mix up. Scintillation fluid (4–7 mL) is added to each of the 96 scintillation vials one sample at a time. Each of the 96 samples is manually capped, transferred to the samples cassettes for the counter, placed in the liquid scintillation counter, and analyzed for 1 min each. The data is removed from the counter and processed using an external computer system. The over 500 sample handling steps from harvesting to sample analysis, takes over 3 hr to complete.

The second technique is harvesting the sample on a specially prepared filter mat using a multidetector (6) liquid scintillation counter. This technique harvests 12 samples at a time using a manual harvester and a special filter mat in the 6×16 format. Thus the harvesting time is similar to that of the liquid scintillation counter technique. Once the samples have been harvested, the entire filter mat is dried and placed into a special chemically resistant plastic bag. To this bag is added 10 to 15 mL of a hydrophobic scintillation cocktail. The bag is sealed to prevent the cocktail from escaping and rolled to insure the cocktail completely saturates the filter mat. This sealed bag is placed into a special holder, and 6 samples are analyzed at a time on the multidetector (6) liquid scintillation counter. This reduces the counting time by one sixth that of conventional liquid scintillation counting. Once the entire plate has been quantitated the data must be converted from the 6×16 format to the 8×12. This technique reduces the total time of harvesting and analysis to one third that of liquid scintillation counting. It takes over 60 minutes for a microplate of 96 samples and involves 18–20 steps.

The third technique is position sensitive proportional counter scanning. The scanners were originally designed to detect radioactivity on a flat surface for such thing as TLC or paper chromatography. The technique uses a position sensitive wire detector to count 12 samples (in a single row) simultaneously and scans lane by lane over the 8×12 matrix (8 separate scans). This method has three major disadvantages.

First, the detectors are not highly sensitive to low energy beta emitter (i.e., ^3H). Second, the detectors are subject to high amounts of cross talk when high energy isotopes (i.e., ^{32}P) are analyzed. Third, the detectors are not uniform across the entire length of the wire. This is critical because the wire detectors must be able to locate and quantitate the radioactivity on the solid filter matrix. These factors make the technique unsuitable for quantitative applications.

The fourth technique is quantitating with a specially designed 96 detector quantitation system. It uses avalanche gas amplification detectors with collector/cathode voltage bias operating in the Geiger-Muller voltage region. This system is capable of analyzing 96 samples simultaneously with 96 individual detectors in the 8×12 format. The technique uses an open ended avalanche gas ionization detector and is capable of quantitating ^3H, ^{32}P, ^{35}S, ^{14}C, ^{125}I and many other isotopes which produce ionizing radiation. The only steps involved in using this technique is harvesting 96 samples simultaneously with the Micromate 196, drying the sample filter, and quantitating all 96 samples simultaneously in the Matrix 96. This entire process from harvesting to quantitation, requires less than 12 min/microplate of 96 samples, and there is no liquid radioactive waste to dispose of at the end of the experiment, only a filter membrane. This technique does not destroy the sample either, so the sample can be removed from the filter mat and analyzed further (NMR, Mass spectrometry, DNA sequencing, etc.). The filter mat can then be analyzed by the Matrix 96 and placed in a plastic bag for storage or liquid scintillation counting for quantitation at a later date.

Table 1. Cost of Analysis of 500 Microplate/Year for Mixed Lymphocyte Cultures Assays by Various Methods

		LSC[a]	MD-LSC[b]	MD-AGD[c]
1.	Scintillation Vials	$2500	0	0
2.	Scintillation Cocktail	$2500	$250	0
3.	Glass Fiber Filters	$350	$1000	$350
4.	Special Sample Filter Bags	$0	$350	$0
5.	Cost/Technician Time	$25,000	$8500	$1700
	(sample preparation, harvesting, counting)			
6.	Waste Disposal Costs	$500	$100	$10
	TOTAL	$30,850	$10,250	$2060

[a]LSC = liquid scintillation counting.
[b]MD-LSC = multi-detector LSC.
[c]MD-AGD = multi-detector avalanche gas detector.

Now that the steps involved in each technique have been shown, what about the cost involved in analyzing a series of microplates? If 10 microplates/week were analyzed over the period of a year then approximately 500 plates would be quantitated in one year. The cost for each of these three methods is based on the time required to prepare and analyze the samples and the cost of chemicals and supplies required. The first technique of liquid scintillation counting requires scintillation vials, scintillation cocktail, filter mats, and time to prepare and dispose of the 96 samples obtained from each plate. If 500 plates were analyzed by this method it would require over $30,000/year as shown in Table 1. For the second technique of multidetector liquid scintillation counting, the scintillation vials have been eliminated, but more expensive special filter mats are required. The cost of the filter bags, cocktail, special filter mats, and labor to prepare and dispose of these samples is over $10,000 for the same 500 plates (Table 1). For the third technique which uses the Micromate 196 cell harvester, to harvest all 96 samples simultaneously, and the Matrix 96 reader, to quantitate all 96 samples simultaneously, the only cost is the filter mat (the same type used on the manual harvester) and the labor costs. The cost of labor and materials is approximately $2000 for the 500 microplates or 15 times less than the standard liquid scintillation counting technique.

APPLICATIONS

Several applications exist involving radioactivity quantitation on a solid support in a microplate format. These include dot blots, ^{3}H-thymidine cell proliferation assays, receptor binding assays, radioimmunoassays, DNA polymerase spot assays, broken cell assays (fungi), and many others. Three specific assay types will be evaluated in detail with reference to the three quantitation techniques described earlier.

The first application is dot blots. This technique specifically identifies a sequence of DNA or RNA of interest in a specific disease or DNA/RNA fragment. The technique involves the following basic steps. The DNA or RNA of interest is bound to a special membrane in a dot blot device (Figure 1). A special radiolabeled DNA/RNA probe, complementary to the region of DNA/RNA of interest is prepared. The radiolabeled probe is added to various DNAs to locate the sequence of interest. The noncomplementary DNA is removed from the membrane by washing, and the radioactivity on the membrane is quantitated. This quantitation is normally accomplished by using X-ray film exposure, because the radionuclide is high energy ^{35}S or ^{32}P. The radioactivity on the film is determined by densitometry. The alternative method is the use of the Matrix 96 direct beta counter for these samples. A comparison of the quantitation of the densitometry and the Matrix 96 is shown in Figure 5. As can be clearly seen, the densitometry has a small dynamic range because the X-ray film becomes saturated (50 to 100 fold range). On the other hand the Matrix 96 has a dynamic range of over 10^5 for the ^{32}P dot blots. In addition to

MATRIX 96 VS DENSITOMETRY
32P DOT BLOT

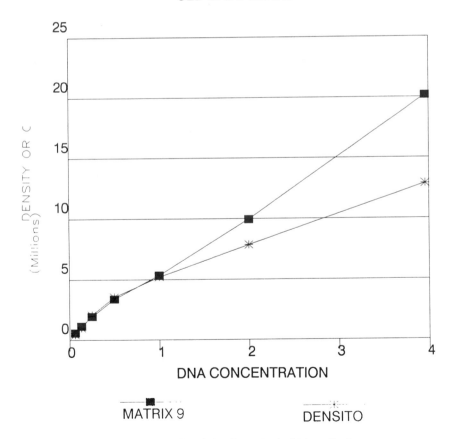

Figure 5. Comparison of Matrix 96 and densitrometry for blot applications.

the larger dynamic range the Matrix 96 is able to analyze the dot blots in 1/50 to 1/100th the time of the X-ray film method. A correlation between the Matrix 96 and liquid scintillation counting on the same dot blot samples was determined to be an R^2 of 0.949 (Matrix Application Note).

The second application is ^{3}H-Thymidine cell proliferation assays. These assays use the tritiated thymidine which becomes specifically incorporated into DNA as a measure of cell growth or proliferation in culture. This proliferation assay is used to test toxic substances, potential cancer drugs, AIDS drugs, and other important naturally occurring and synthetic substances. The conventional method of analysis is cell harvesting with a manual harvester and analysis by liquid scintillation counting. A series of samples with various radioactivity incorporated into the cellular DNA are analyzed using both the Matrix 96 reader and the conventional liquid scintillation counting techniques. The data from this experiment is shown in Figure 6. The correlation of the data was performed and an R^2 of 0.999 was calculated (Matrix Application Note). This clearly demonstrates that this technique provides results which are as accurate as the liquid scintillation counting technique in one tenth the time and at one tenth the cost.

The third application is the radiolabeled receptor binding assays. This technique involves using either a specific type of cell culture or a tissue homogenate preparation of a specific animal tissue which contains the receptors of interest. The substrate for the receptor is radiolabeled and a competitive binding assay is performed with unlabeled substrate. The number of receptors and the binding constant can be determined using this technique. The application can be analyzed by one of the three quantitative methods described earlier. If the multidetector liquid scintillation counter is compared to the Matrix 96 radioactivity reader, the results of the two techniques can be correlated. The R^2 for these two techniques was calculated to be 0.993 (Figure 7). A similar correlation for liquid scintillation counting and the Matrix 96 was also obtained. In addition to performing receptor binding assays with ^{3}H, the radionuclide ^{125}I can be performed and quantitated on the Matrix 96 (Matrix Application Note).

Several other applications (Matrix Application Notes) can be performed which use quantitation on a solid support in the microplate format. These include radioimmunoassays with either ^{3}H or ^{125}I. Special DNA polymerase reactions which using spotting in a microplate format can be used. Initial experiments indicate that chromium release studies can be performed with the Matrix 96 detectors.

SUMMARY

This article presents three different methods for preparing samples on a solid support in the microplate format. These include dot blotting or direct spotting, cell harvesting in a manual or Micromate 196 96 sample microplate

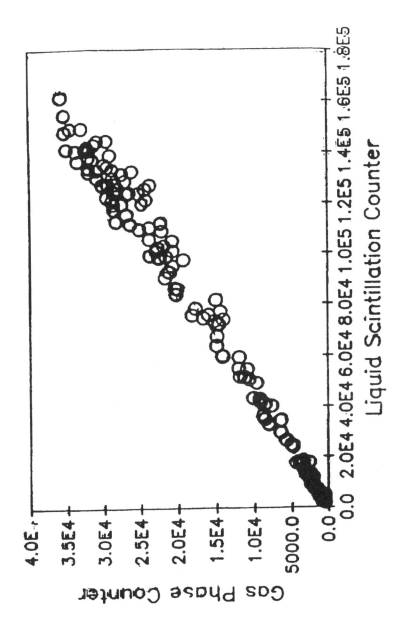

Figure 6. Comparison of MLC samples analyzed on Matrix 96 and liquid scintillation counter.

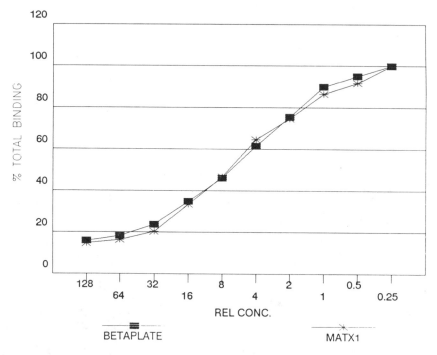

Figure 7. Receptor binding application comparison of Matrix 96 and betaplate.

harvester, and the use of a membrane bottom strippable microplate which allows incubation, washing, and harvesting in a single microplate. The strippable bottom allows the membrane to be analyzed directly on the Matrix 96 radioactivity reader.

Four separate methods of quantitating these harvested samples are presented and compared. The liquid scintillation counting technique requires about 3 hours of sample preparation/microplate and cost about $60/plate. The second technique is multidetector liquid scintillation counting which requires about 60 min/microplate and cost about $20/plate. The third technique is scanning with position sensitive proportional counter. This method improves the throughput but suffers from the disadvantages of cross talk, low ^{3}H-efficiency, and wire detector nonuniformity. The fourth technique, and the most efficient, is the use of the Micromate 196 96 sample cell harvester and the Matrix 96 96 detector direct beta counter. This technique requires about 12 min/microplate and cost $4/microplate. In summary, this new technique provides a method of rapidly analyzing 96 samples simultaneously from microplates into a solid support, and quantitating the radioactivity directly, without any liquid scintillation counting waste. The Matrix technique uses *No Vials, No Cocktail, No Bag, No Sample Destruction, No Special Filters*, and *No Waste* (liquid) to analyze the 96 samples on the 8×12 Microplate.

CHAPTER 15

Dynodic Efficiency of Beta Emitters

F. Ortiz, A. Grau, and J.M. Los Arcos

INTRODUCTION

In liquid scintillation counting measurements[1-3] of beta emitters, contrasted models give us efficient detection through a study of internal processes in the vial and the response of the photocathode. Many of these models use a figure of merit of the system as a parameter.[4-8]

Other models explain the dynodes response in a photomultiplier as a set of electrons reaching the first dynode. In this way we can get the electron distribution at the output of any dynodic stage in the photomultiplier.[9-12]

For each electron set that enters the first dynode, however, there is always a nonzero probability of nondetection in the dynodes. This leads to a lower efficiency at the dynode output than at the photocathode.

The present paper has studied the efficiency loss through the dynodic stages for ^{3}H and ^{14}C and applied it to the standardization of ^{14}C via the efficiency tracing method.[7-8] First, a model for the electron multiplication loss is described, then the efficiency loss through dynodes is computed for ^{3}H and ^{14}C, for a single tube and for two tubes in coincidence.

Experimental quench curves for ^{14}C have been compared to the theoretical predictions.

NON DETECTION PROBABILITY MODEL

It is assummed that the electron production process, at the output of the photocathode, is governed by the Poisson statistics. If N electrons are expected in the average, the probability of obtaining n electrons is:

$$P(n) = \frac{\overline{N}^n \, \overline{e}^N}{n!} \quad (1)$$

The number **n** of electrons that leaves the photocathode is multiplied through sucessive dynodes, the amplification being typified by means of a free parameter, **G**, the dynodic gain. This parameter is defined as the number of electrons that come out from a dynode when a single electron comes into the dynode.

Statistically, there is a nonzero probability of having no electrons at the output of the Kth dynodic stage when **n** electrons arrive at the first dynode. This nonzero probability follows this iterative expression:[12]

$$P_K(n,0) = \exp(-G \cdot n) \exp[G \cdot n \cdot P_{K-1}(1,0)]; \quad K > 1 \tag{2}$$

$$P_K(n,0) = \exp(-G \cdot n) ; \qquad\qquad K = 1 \tag{3}$$

where   n = the number of electrons that leaves the photocathode
K = the order of dynodic stages
G = the dynodic gain

The nondetection probability leads to a dynodic detection efficiency lower than the photocathode efficiency, because when **n** electrons leave the photocathode with probability 1, the total sum of the emission probabilities at the Kth dynodic output is $1 - P_K(n,0)$.

DETECTION EFFICIENCY

Single photomultiplier tube

Beta emitters have an emission spectra given by the Fermi distribution:

$$N_i(E) = C \cdot F(Z,E) \cdot (E_i - E)^2 (E + 1)(E^2 + 2E)^{1/2} \tag{4}$$

where $N_i(E)$
= the number of particle beta with energies between E, E + dE
E_i = the maximum kinetic energy of beta particles in the ith energy band
C = the shape factor[13]
F(Z,E) = the relativistic expression of the Coulomb factor[14] (all in m_0c^2 units)

When a radionuclide is dissolved in a scintillation liquid, its beta particles transfer almost all their energy to the scintillator, that is, convert to light. This light emission is partially detected by the photocathode of the photomultiplier tube. The number of electrons that leave the photocathode is a function of the beta particle energy.

The average number of electrons that can be expected when a beta particle is emitted in the energy band (E, E + dE), is:

$$N(E) = \frac{E \cdot Q(E)}{M} \tag{5}$$

where E = the beta particle energy
   $Q(E)$ = the ionization quench factor
   M = the figure of merit, or energy necessary to produce an electron at the photocathode

A good aproximation for $Q(E)$ is Equation 7:

$$Q(E) = 1 - 0.9624 \; e^{-0.5457}; \; E > 10 \text{ keV} \tag{6}$$

$$Q(E) = 0.1253 \; Ln(E) + 0.4339; \; 0.1 \text{ keV} < E < 10 \text{ keV} \tag{7}$$

According to that, mean value r electrons could be obtained with probability:

$$P(r,E) = \frac{N(E)^r \; e^{N(E)}}{r!} \tag{8}$$

The detection efficiency at the output of the photocathode for a single tube and for a unique beta particle in the ith band is:

$$E_{f1} = \sum_r P(r,E) \tag{9}$$

For all the particles in the ith energy band, the probability for having electrons is:

$$P_i(r) = P(r,E) \; N_i(E) \tag{10}$$

and the detection efficiency for the ith energy band at the output of the photocathode is:

$$E_{fi} = \sum_r P_i(r) \tag{11}$$

Therefore the efficiency for the whole spectrum at the output of the photocathode for a single photomultiplier tube is:

$$E_f = \sum_i E_{fi} \tag{12}$$

Now the detection efficiency for a beta emitter of the Kth dynodic stage output of a single photomultiplier tube can be analyzed. When a beta particle in the ith energy band is emitted, the r electrons that leave the photocathode with probability $P(r,E)$ will have a nonemission probability at the Kth dynode given by:

$$P_{K1}(r,0) = P_K(r,0) \, P(r,E) \qquad (13)$$

and summing up for all the electrons that leave the photocathode when a beta particle in the ith energy band is emitted, we arrive at:

$$P_{K1}(0) = \sum_r P_K(r,0) \, P(r,E) \qquad (14)$$

Therefore, the efficiency at the output of the Kth dynode for this single beta particle in the ith band is:

$$E_{K1} = \left(\sum_r P(r,E) \right) - P_{K1}(0) \qquad (15)$$

For the $N_i(E)$ particles in the ith band, the nondetection probability at the output of the Kth dynode is:

$$P_{Ki}(0) = P_{K1}(0) \, N_i(E) \qquad (16)$$

and the efficiency of detection for this band will be:

$$E_{Ki} = E_{fi} - P_{Ki}(0) \qquad (17)$$

Finally the detection efficiency of the Kth dynodic stage output of a single photomultiplier tube for the whole beta emitter spectrum is:

$$E_K = \sum_i E_{Ki} \qquad (18)$$

Two Photomultiplier Tubes in Coincidence

The previous procedure can be easily extended for a system composed of two photomultiplier tubes in coincidence. Both tubes are assummed to be identical so the detection efficiency at the output of the photocathodes for a beta particle emitted in the ith-energy band for that coincidence system is:

$$E'_{f1} = \left(\sum_r P(r,E) \right)^2 \qquad (19)$$

and for $N_i(E)$ particles in the energy band we will have:

$$E'_{fi} = E'_{f1} \cdot N_i(E) \qquad (20)$$

Therefore the photocathode efficiency for the whole electron spectrum at the output of the system of two tubes in coincidence is:

$$E'_f = \sum_i E'_{fi} \tag{21}$$

Now the detection efficiency at the output of the Kth dynode for two tubes in coincidence can be analyzed. For a single beta particle in the ith energy band the detection efficiency is:

$$E'_{K1} = (E_{K1})^2 \tag{22}$$

and for all electrons in the ith energy band we will have:

$$E'_{Ki} = E'_{K1} \cdot N_i(E) \tag{23}$$

Finally, the detection efficiency for a beta spectrum at the output of the Kth dynodic stages for two tubes in coincidence will be:

$$E'_K = \sum_i E'_{Ki} \tag{24}$$

RESULTS

Non detection probability

The nondetection probability has been studied for the following parameters:

The dynodic gain varying, from 1 to 10 with step 1, the number of dynodic stages 12, and the number of electrons injected into the first dynode, from 1 to 19 with step increment 2.

Table 1 shows the nondetection probability for the 12 dynodic stages and for the 1, 3, 5, 7, 11, and 19 electrons injected into the first dynode. The dynodic gain is 1 or 2.

In Figure 1a, the nondetection probability is plotted vs the dynodic stage. For 1, 5, 11, or 19 electrons injected and a dynodic gain 1, in Figure 1b, the nondetection probability is plotted vs dynodics gain.

Detection Efficiency for Beta Emitters

Two different photomultiplier tubes with 12 dynodes have been simulated:

1. low gain, G has a value of 7 in the first dynode and a value of 2 in the others.
2. high gain, G has a value of 30 in the first dynode and a value of 3 in the others.

Table 2 gives the efficiency at the output of the 12th dynode for ^{3}H, for high gain and two photomultiplier tubes in coincidence.

Table 1. Nondetection Probability in Successive Dynodic Stages and Different Number of Electrons Injected into the First Dynode

STAGE	1 ELECT.	3 ELECT.	5 ELECT.	7 ELECT.	11 ELECT.	19 ELECT.
01	0.37	0.05	0.01	0.00	0.00	0.00
02	0.53	0.15	0.04	0.01	0.00	0.00
03	0.63	0.24	0.10	0.04	0.00	0.00
04	0.69	0.32	0.15	0.07	0.01	0.00
05	0.73	0.39	0.21	0.11	0.02	0.00
06	0.76	0.45	0.26	0.15	0.03	0.01
07	0.79	0.49	0.31	0.19	0.05	0.01
08	0.81	0.53	0.35	0.23	0.07	0.02
09	0.83	0.57	0.39	0.26	0.10	0.03
10	0.84	0.60	0.42	0.30	0.13	0.04
11	0.85	0.62	0.45	0.33	0.15	0.05
12	0.86	0.64	0.48	0.36	0.17	0.06

DYNODIC GAIN: 1

STAGE	1 ELECT.	3 ELECT.	5 ELECT.	
01	0.13	0.00	0.00	
02	0.17	0.01	0.00	Nondetection probability < 1.E-4
03	0.19	0.01	0.00	
04	0.20	0.01	0.00	
05	0.20	0.01	0.00	
06	0.20	0.01	0.00	
07	0.20	0.01	0.00	
08	0.20	0.01	0.00	
09	0.20	0.01	0.00	
10	0.20	0.01	0.00	
11	0.20	0.01	0.00	
12	0.20	0.01	0.00	

Figures 2a, 2b, 2c, and 2d show the plot of the efficiency vs figure of merit at the photocathode, at the 12th dynode with high gain and with low gain. Figures 2a and 2b are for ^{3}H and 2c and 2d are for ^{14}C.

Figures 3a, 3b, 3c, and 3d plot the difference between the efficiency at the output of the photocathode and the efficiency at the output of the 12th dynode as a function of the figure of merit for high and low gain.

Table 3 shows the efficiency variation from the photocathode to the 12th dynode for different values of the figure of merit at low and high gain, for ^{3}H, for a single photomultiplier tube, and for two photomultiplier tubes working in coincidence.

Figure 4 plots the composed and the experimental efficiency vs quenching for ^{3}H and ^{14}C.

Table 4 contains the numerical values shown in the Figure 4.

Figure 5 shows the efficiency for low quantum yield at the photocathode for ^{14}C . The points are computed and the line corresponds to the experimental data.

The experimental values data have been obtained with a set of quenched standards for ^{3}H and ^{14}C, in a L.K.B. 12 19 Rackbeta spectral liquid scintillation spectrometer.

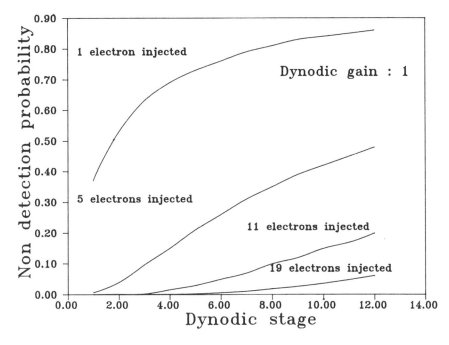

Figure 1a. Nondetection probability vs dynodic gain.

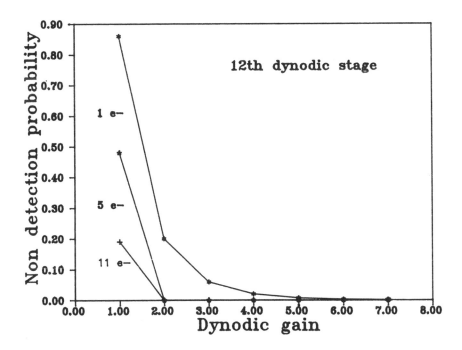

Figure 1b. Nondetection probability vs dynodic stage.

Table 2. Total LSC Efficiency Calculation for H-3 at the 12th Dynode, High Gain Two Photomultiplier Tubes in Coincidence

Fig. Merit	Efficiency (%)		Fig. Merit	Efficiency (%)		Fig. Merit	Efficiency (%)		Fig. Merit	Efficiency (%)
0.2500	84.093	*	12.7500	2.5803	*	25.2500	0.75114	*	37.7500	0.35151
0.5000	73.011	*	13.0000	2.4946	*	25.5000	0.73746	*	38.0000	0.34712
0.7500	63.529	*	13.2500	2.4131	*	25.7500	0.72417	*	38.2500	0.34278
1.0000	55.439	*	13.5000	2.3355	*	26.0000	0.71121	*	38.5000	0.33855
1.2500	48.576	*	13.7500	2.2616	*	26.2500	0.69862	*	38.7500	0.33436
1.5000	42.766	*	14.0000	2.1911	*	26.5000	0.68637	*	39.0000	0.33028
1.7500	37.845	*	14.2500	2.1238	*	26.7500	0.67441	*	39.2500	0.32628
2.0000	33.665	*	14.5000	2.0596	*	27.0000	0.66278	*	39.5000	0.32232
2.2500	30.100	*	14.7500	1.9983	*	27.2500	0.65142	*	39.7500	0.31846
2.5000	27.044	*	15.0000	1.9397	*	27.5000	0.64038	*	40.0000	0.31464
2.7500	24.411	*	15.2500	1.8836	*	27.7500	0.62958	*	40.2500	0.31092
3.0000	22.131	*	15.5000	1.8299	*	28.0000	0.61909	*	40.5000	0.30726
3.2500	20.146	*	15.7500	1.7784	*	28.2500	0.60886	*	40.7500	0.30363
3.5000	18.410	*	16.0000	1.7291	*	28.5000	0.59885	*	41.0000	0.30010
3.7500	16.883	*	16.2500	1.6818	*	28.7500	0.58911	*	41.2500	0.29660
4.0000	15.535	*	16.5000	1.6364	*	29.0000	0.57958	*	41.5000	0.29320
4.2500	14.339	*	16.7500	1.5928	*	29.2500	0.57031	*	41.7500	0.28985
4.5000	13.273	*	17.0000	1.5510	*	29.5000	0.56125	*	42.0000	0.28652
4.7500	12.320	*	17.2500	1.5107	*	29.7500	0.55238	*	42.2500	0.28329
5.0000	11.465	*	17.5000	1.4720	*	30.0000	0.54375	*	42.5000	0.28007
5.2500	10.695	*	17.7500	1.4348	*	30.2500	0.53529	*	42.7500	0.27694
5.5000	9.9984	*	18.0000	1.3990	*	30.5000	0.52706	*	43.0000	0.27387
5.7500	9.3673	*	18.2500	1.3645	*	30.7500	0.51898	*	43.2500	0.27081
6.0000	8.7936	*	18.5000	1.3312	*	31.0000	0.51111	*	43.5000	0.26784
6.2500	8.2704	*	18.7500	1.2992	*	31.2500	0.50343	*	43.7500	0.26488

6.5000	7.7923	*	19.0000	1.2683	*	31.5000	0.49588	*	44.0000	0.26200
6.7500	7.3541	*	19.2500	1.2384	*	31.7500	0.48853	*	44.2500	0.25917
7.0000	6.9515	*	19.5000	1.2097	*	32.0000	0.48132	*	44.5000	0.25636
7.2500	6.5809	*	19.7500	1.1818	*	32.2500	0.47429	*	44.7500	0.25362
7.5000	6.2390	*	20.0000	1.1550	*	32.5000	0.46742	*	45.0000	0.25089
7.7500	5.9228	*	20.2500	1.1291	*	32.7500	0.46066	*	45.2500	0.24823
8.0000	5.6300	*	20.5000	1.1040	*	33.0000	0.45408	*	45.5000	0.24562
8.2500	5.3582	*	20.7500	1.0798	*	33.2500	0.44760	*	45.7500	0.24302
8.5000	5.1056	*	21.0000	1.0563	*	33.5000	0.44130	*	46.0000	0.24049
8.7500	4.8703	*	21.2500	1.0336	*	33.7500	0.43510	*	46.2500	0.23797
9.0000	4.6508	*	21.5000	1.0116	*	34.0000	0.42906	*	46.5000	0.23552
9.2500	4.4458	*	21.7500	0.99030	*	34.2500	0.42314	*	46.7500	0.23310
9.5000	4.2540	*	22.0000	0.96964	*	34.5000	0.41731	*	47.0000	0.23070
9.7500	4.0743	*	22.2500	0.94966	*	34.7500	0.41163	*	47.2500	0.22836
10.0000	3.9057	*	22.5000	0.93026	*	35.0000	0.40603	*	47.5000	0.22602
10.2500	3.7473	*	22.7500	0.91148	*	35.2500	0.40059	*	47.7500	0.22375
10.5000	3.5983	*	23.0000	0.89326	*	35.5000	0.39525	*	48.0000	0.22152
10.7500	3.4580	*	23.2500	0.87555	*	35.7500	0.38998	*	48.2500	0.21928
11.0000	3.3257	*	23.5000	0.85839	*	36.0000	0.38485	*	48.5000	0.21711
11.2500	3.2009	*	23.7500	0.84169	*	36.2500	0.37978	*	48.7500	0.21495
11.5000	3.0829	*	24.0000	0.82551	*	36.5000	0.37485	*	49.0000	0.21284
11.7500	2.9713	*	24.2500	0.80976	*	36.7500	0.37002	*	49.2500	0.21077
12.0000	2.8657	*	24.5000	0.79449	*	37.0000	0.36524	*	49.5000	0.20869
12.2500	2.7655	*	24.7500	0.77965	*	37.2500	0.36059	*	49.7500	0.20668
12.5000	2.6705	*	25.0000	0.76518	*	37.5000	0.35599	*	50.0000	0.20466

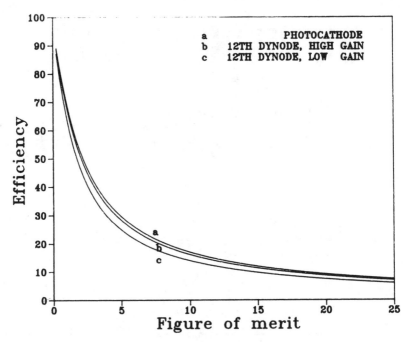

Figure 2a. ^{3}H: Efficiency vs Figure of merit, a single photomultiplier tube.

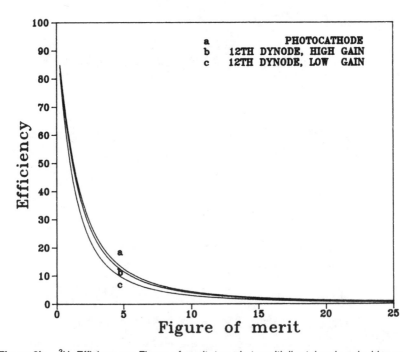

Figure 2b. ^{3}H: Efficiency vs Figure of merit, two photomultiplier tubes in coincidence.

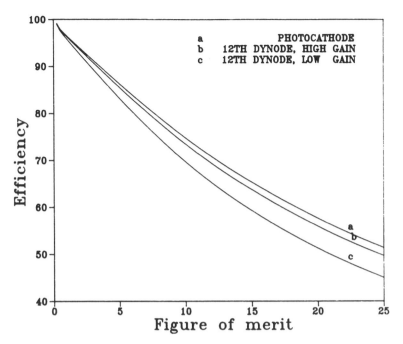

Figure 2c. ^{14}C: Efficiency vs Figure of merit, a single photomultiplier tube.

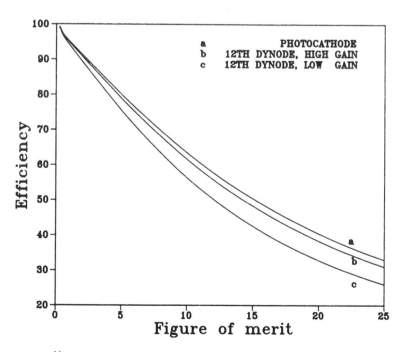

Figure 2d. ^{14}C: Efficiency vs Figure of merit, two photomultiplier tubes in coincidence.

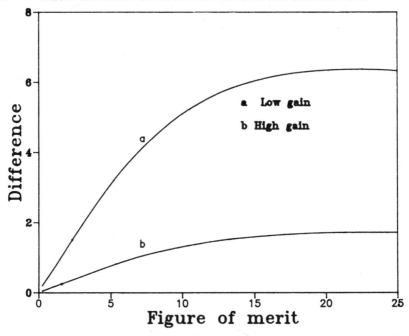

Figure 3a. Difference of efficiencies at the photocathode and at the 12th dynode for ³H, a single photomultiplier tube.

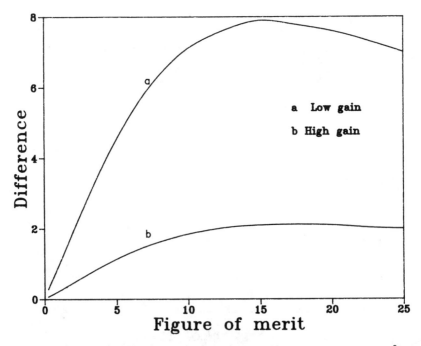

Figure 3b. Difference of efficiencies at the photocathode and at the 12th dynode for ³H, two photomultiplier tube.

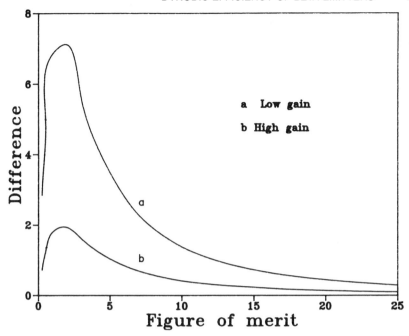

Figure 3c. Difference of efficiencies at the photocathode and at the 12th dynode for ^{14}C, a single photomultiplier tube.

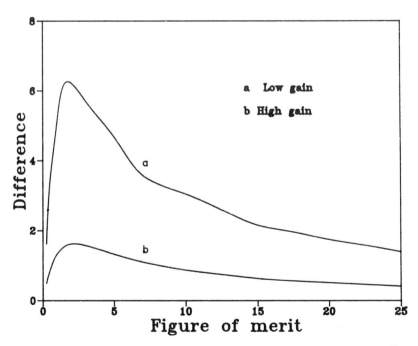

Figure 3d. Difference of efficiencies at the photocathode and at the 12th dynode for ^{14}C, two photomultiplier tube.

Table 3. Efficiency for ^{3}H in Photocathode and Successfive Dynodes

STAGE	FOT.	2TH DY.	4TH DY.	6TH CY.	12TH DY.	
Two Photomultiplier Tubes in Coincidence						
F.M.						
0.25	84.81	84.21	84.10	84.09	84.09*	
1.00	57.25	55.74	55.45	55.44	55.44*	
5.00	12.49	11.63	11.47	11.46	11.46*	
10.0	4.32	3.97	3.91	3.91	3.91*	HIGH GAIN
15.0	2.16	1.97	1.16	1.15	1.15*	
20.0	1.29	1.18	1.16	1.15	1.15*	
25.0	0.86	0.65	0.58	0.56	0.56*	
0.25	84.81	83.04	82.13	81.99	81.96*	
1.00	57.25	52.89	50.77	50.44	50.38*	
5.00	12.49	10.16	9.19	9.04	9.01*	
10.0	4.32	3.39	3.01	2.96	2.95*	LOW GAIN
15.0	2.16	1.66	1.47	1.45	1.44*	
20.0	1.29	0.99	0.87	0.86	0.85*	
25.0	0.86	0.65	0.58	0.56	0.56*	

STAGE	FOT.	2TH DY.	4TH DY.	6TH CY.	12TH DY.	
A Single Photomultiplier Tube						
F.M.						
0.25	89.03	88.61	88.53	88.53	88.53*	
1.00	69.44	68.33	68.11	68.11	68.10*	
5.00	29.54	28.44	28.23	28.22	28.22*	
10.0	16.93	16.21	16.07	16.06	16.06*	HIGH GAIN
15.0	11.85	11.31	11.21	11.21	11.21*	
20.0	9.11	8.69	8.61	8.60	8.60*	
25.0	7.39	7.05	6.98	6.98	6.98*	
0.25	89.03	87.80	87.16	87.06	87.04*	
1.00	69.44	66.20	64.60	64.35	64.30*	
5.00	29.54	26.47	25.09	24.89	24.85*	
10.0	16.93	14.23	14.05	13.92	13.89*	LOW GAIN
15.0	11.85	10.38	9.75	9.65	9.63*	
20.0	9.11	7.95	7.46	7.38	7.37*	
25.0	7.39	6.44	6.04	5.98	5.97*	

Table 4. Quenching Curve for ^{14}C Theoretical Values

Q(E)	EFFICIENCY 3H	FIGURE OF MERIT			EFFICIENCY FOR ^{14}C		
		FOT.	H.G.	L.G.	FOT.	H.G.	L.W.
451.61	58.33	0.97	0.91	0.77	95.23	95.24	95.25
434.61	54.18	1.11	1.04	0.89	94.67	94.69	94.66
394.12	47.15	1.40	1.31	1.11	93.54	93.57	93.57
351.88	38.05	1.85	1.74	1.47	91.79	91.79	91.82
309.94	27.41	2.61	2.47	2.10	88.88	88.82	88.79
250.72	15.86	4.19	3.94	3.33	82.95	82.95	82.99
206.16	09.03	6.27	5.90	4.99	75.47	75.45	75.49

PHO: Photocathode; H.G.: 12th Dynode high gain; L.W.: 12th Dynode low gain

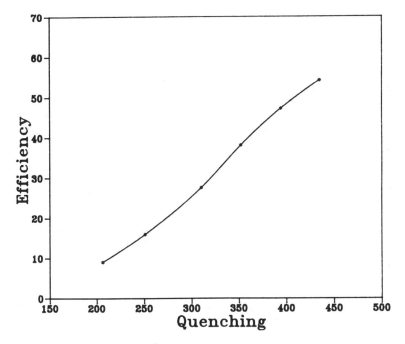

Figure 4a. Quenching curve for [3]H.

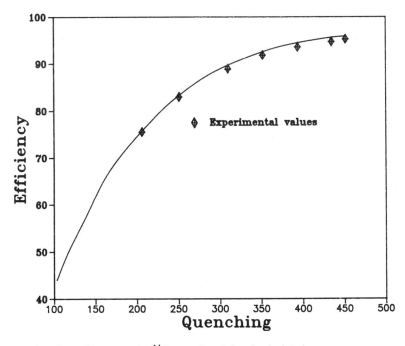

Figure 4b. Quenching curve for ^{14}C, experimental and calculated.

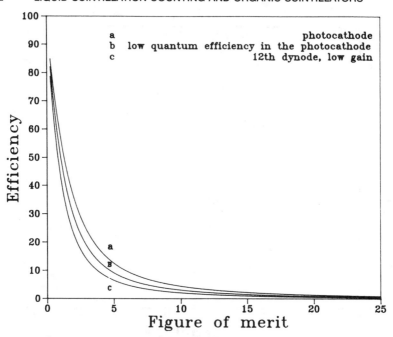

Figure 5a.   ³H efficiency vs Figure of merit, two photomultiplier tubes in coincidence, low quantum efficiency at the photocathode.

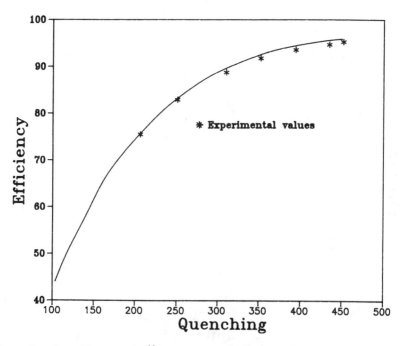

Figure 5b.   Quenching curve for ¹⁴C, experimental and calculated, low quantum efficiency at the photocathode.

DISCUSSION

The results of the previous section show that the nondetection probability falls down very quickly when the number of electrons reaching the first dynode grows. It also falls with the increase of the dynodic gain. However, this nondetection probability is greater for the latest dynodic stages.

The model developed allows efficiency computation in any dynodic stage that is lower than the efficiency at the photocathode, as a function of the figure of merit for a single photomultiplier or for a two photomultiplier system working in coincidence. The computed efficiencies do not vary significantly after the 4th or 5th dynodic stage.

The model permits obtaining the quenching curve of a nuclide from the experimental quenching curve of another nuclide. It does this by using the figure of merit efficiency function for both nuclides, at the photocathode or at the output of any dynodic stage. The computed values are the same regardless of the case considered.

REFERENCES

1. Horrocks, D.L. *Applications of Liquid Scintillation Counting*, (New York: Academic Press, 1965).
2. Gibson, J.A.B. "Modern Techniques for Measuring the Quenching Correction in a Liquid Scintillation Counter", *The Int. Conf. on Liquid Scintillation Counting* (San Francisco: Aug. 1979) p. 21.
3. Horrocks, D.L. and M.H. Studier. "Determination of the Absolute Disintigration Rates of Low Energy Beta Emitters in a Liquid Scintillation Counting," *Analyt. Chem.* 33: 615 (1961).
4. Gibson, J.A. and H.J. Gale. "Absolute Standardization with Liquid Scintillation Counters," *J. Scint. Instrum.* ser. 21 1 199(1968).
5. Grau, A. and E. Garca-Toraño. "Evaluation of Counting Efficiency in Liquid Scintillation Counting," *Int. J. Appl. Radiat. Isot.* 33:249 (1981).
6. Garca-Toraño, E. and A. Grau. "EFFY, a New Program to Compute the Counting Efficiency of Beta Particles in Liquid Scintillators," *A Comp. Phys Commun* 36:307 (1985).
7. Houtermans H. "Probability of Non Detection in Liquid Scintillation Counting," *Nucl. Instr. and Methods* 112:121 (1973).
8. Jordan, P. "On Statistics of Coincidence Detection Efficiency in Liquid Scintillation Spectrometry," *Nucl. Instr. and Methods* 97 1 07 (1971).
9. Lombard, F.J. and F. Martin. "Statistics of Electron Multiplication," *Rev. Sci. Instr.* 32:200 (1971).
10. Gale, H.J. and J.A. Gibson. "Methods for Calculating the Pulse Height Distribution at the Output of a Scintillation Counter," *J. Sci. Instr.* 43:225 (1965).
11. Ortiz, J.F. and A. Grau. "Estadística de la Multiplicación de Electrones en un Fotomultiplicador: Métodos Iterativos," JEN report 574 (Madrid, Spain: Junta de Energía Nuclear).
12. Prescott, J.R. "A Statistical Model for Photomultiplier Single-Electron Statistics," *Nucl. Instr. and Methods* 39:173 (1966).

13. Konopinski, E.J. *The Theory of Beta Radioactivity*, (Oxford U.K.: Oxford University Press 1966), p. 280.
14. Rose, M.E. *Beta and Gamma-Ray Spectrometry* (Amsterdam, North-Holland: Ed K. Siegbahm, 1055), p. 280.

CHAPTER 16

Solid Scintillation Counting: A New Technique—Theory and Applications

Stephen W. Wunderly and Graham J. Threadgill

ABSTRACT

Solid scintillation counting is a new technique for measuring radioactive samples that previously had to be counted in liquid scintillators. The solid scintillator, XtalScint™, produces a light output brighter than standard organic scintillators. It is nonhazardous, nonodorous, and insusceptible to impurity (chemical) quench. It can be attached to either porous or nonporous surfaces for solution and filtration applications. Under optimum conditions, tritium efficiency approaches 60%. This report will discuss the theory and applications of XtalScint to solid scintillation counting techniques.

INTRODUCTION

Recent developments in liquid scintillation counting have focused on improving liquid scintillation solution safety by decreasing the volatility of the solvent (Kellogg,[1] Kalbhen and Tarkkanen,[2] Reed,[3] and Lin and Mei[4]). A solid, fine, scintillator powder that replaces the solvent would be the ultimate in low volatility. The powder, supported on either a porous or nonporous carrier, could be placed in a vial and counted in a liquid scintillation counter. The scintillation performance of the solid scintillation system would be equivalent to or superior to counting with liquid scintillation solutions. Also, in contrast to liquid scintillation solutions, the counting vials could be reused; the labeled samples could be recovered from the solid scintillators. Solids also have health and disposal advantages that make this new technique potentially revolutionary to the industry.

This report will discuss, in depth, the theory and applications of such a solid scintillation counting system using Beckman's Ready Cap and Ready Filter.

THEORY

There are five important properties of a scintillator. First, it must efficiently convert the energy of the radiation decay into measurable light. Second, it must be chemically inert to the conditions of measurement. Third, it must have low noise, or background. Fourth, it must be able to interact with low energy betas and augers. Finally, it should present minimal hazards to the user. XtalScint by Beckman meets all these criteria.

Conversion Efficiency

Figure 1 demonstrates that the light output for XtalScint is much greater than even an unquenched liquid scintillation calibration standard. Since these emission spectra were generated by the same isotope, this implies that XtalScint has a much greater light output per KeV of excitation energy than the best liquid scintillators. The wavelength of the emission maxima for XtalScint is 395nm, ideal for maximum sensitivity of current photomultiplier technology. The decay time of emission is 80 to 120 nsec. While this is slower than most traditional liquid scintillators, it is well below the minimum microsecond resolving power of current liquid scintillation counters and therefore very acceptable.

Chemical Properties

XtalScint is a solid with a melting point above 1000°C. It is chemically inert to aqueous buffers, aqueous bases, and organic solvents. It partially dissolves in concentrated acid, however, it can withstand moderate exposure to 1 M hydrochloric acid and 0.1 M sulfuric acid. Since XtalScint is completely impervious to organic reagents it is immune to impurity (chemical) quenching interferences. See Table 1 for data. It is also immune to chemiluminescence caused by common chemical reagents, such as base and peroxide.

Low Noise (Background)

XtalScint as used in Ready Cap and Ready Filter applications has essentially the background of an empty vial. Background generated from cosmic particle interaction with 10 mL of liquid scintillator is greatly reduced with Ready Cap or Ready Filter (see Table 2) because of the small scintillator target. Gillespie[5] has reported an 8- to 10-fold improvement in the signal to noise ratio for measurements of ^{32}P and ^{125}I with Ready Cap compared to liquid scintillator cocktail.

Detection Capability

While plastic scintillators and crystal scintillators are not new solid scintillators, they have found very little acceptance for measuring biological samples, most of which are labeled with tritium and ^{14}C. Although effective as scintilla-

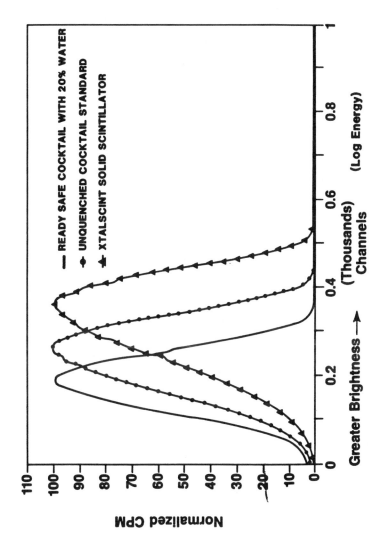

Figure 1. Light output (Tritium Spectra) with various scintillator forms.

Table 1. Effect of Chemical Quenching on Tritium Efficiency, at 20 nsec, of inulin in Liquid and Solid Scintillators

		Scintillator	
Quench Agent	Volume	Liquid	Xtalscint
None	—	45.5%[a]	39% (DRY)[b]
Isopropanol[c]	100µL	45.0%	28.5% (WET)
Nitromethane[d]	100µL	2.9%	28.5% (WET)

[a]Tritiated inulin in water added to 5 mL of ready protein+.
[b]Tritiated inulin in water added to ready cap, dried and counted.
[c]Isopropanol, 100µL, added to both cocktail and ready cap preparations. Samples were counted without removing isopropanol.
[d]Nitromethane, 100µL, added to both cocktail and ready cap preparations. Samples were counted without removing nitromethane.

tors, their poor contact with the labeled substrate has meant poor detection capability.

Two important developments make the XtalScint scintillator system effective. First, XtalScint is a fine (3 to 8 micron) powder providing enormous surface area for the intimate interaction between the labeled substrate and the scintillator. Second, Seltzer and Berger[6,7] and Unak[8] have calculated that by replacing the liquid media between scintillator particles with air, one increases the distance an electron can travel by a factor of 1000. Thus, removing water from an aqueous solution of [3]H 5-fluorouracil on ready cap resulted increased the tritium count rate from 13,700 to 75,900 cpm. This is a fivefold increase.

The performance of XtalScint may also be enhanced by altering the coincidence gate on the LS instrument. By increasing the coincidence gate time one can improve the tritium count rate by 25 to 40%. Enhancement of more energetic isotopes is not as significant (see Table 3).

Table 2. Background for XTALSCINT and Ready Safe

Sample	[3]H 0–400	Window ^{14}C/[3]H 401–670	>[14]C 671–1000
Empty Plastic Vial	14.43	14.17	0.33
10 mL Ready Safe in Plastic Vial	26.10	13.90	17.77
XTALSCINT, 35 mg, in Plastic Vial	14.55	14.06	0.53

Table 3. Radioisotope Efficiencies of XTALSCINT vs Ready Protein+

Radioisotope	Substrate	Solvent	XtalScint 20 nsec GATE	XtalScint XTAL GATE	Ready PROTEIN
[3]H	Palmitic Acid	2-Propanol	41.5%	57%	44%
[14]C	Glutamic Acid	Water	93.4%	97.7%	94%
[125]I	Triodothyroxane	2-Propanol	75%	78%	78%
[32]P	Phosphoric Acid	Water	94%	98.5%	100%

Health and Safety Aspects

XtalScint has certain inherent laboratory safety advantages over liquid scintillators. XtalScint has no vapor, protecting the user from exposure to bad odors or toxic vapor inhalation. XtalScint is not absorbed through the skin, saving the user from possible internal organ damage. XtalScint does not need to be used in a hood, providing additional hood space previously occupied by liquid scintillators. Spilled XtalScint, should it come free from its carrier, can be cleaned up with a broom and dust pan rather than spill pillows and absorbent. XtalScint will not burn, in contrast to most liquid scintillators which are classified as flammable or combustible. XtalScint with carrier is smaller than liquid scintillators, making storage more efficient. XtalScint with carrier may be stored in any quantity; storage of hazardous liquid scintillators is proscribed by regulations.

Disposal Aspects

Solid scintillators such as XtalScint are much easier to dispose of than are liquid scintillators. Liquid scintillator waste must be packaged with an absorbent adequate to hold the spilled liquid in the event the scintillation vials or other waste containers should leak or break. As a result, a drum of liquid scintillator waste contains about 2/3 absorbent and only 1/3 actual vials. XtalScint, in contrast, is classified as dry waste and requires no supplemental absorbent. Because of its compactness (small size) and dry form, one drum of XtalScint samples on its carrier is equal to 30 to 300 drums of the same number of liquid scintillator samples in vials. This represents tremendous cost savings as well as stress relief for disposal sites and the environment in general.

SAMPLE PREPARATION AND COUNTING PROCEDURE

Ready Cap

Ready Cap is the cap of a small plastic vial coated with a layer of XtalScint[9] (see Figure 2). It is designed for counting nonvolatile, radiolabeled substrate in a volatile solvent. The liquid sample, less than 200 μL, is pipetted onto the scintillator surface in the Ready Cap. The volatile liquid is evaporated by numerous methods (evaporation in hood, heat lamp, hot air blow dryer, microwave oven, or vacuum centrifuge), leaving the labeled, nonvolatile substrate deposited on the surface of the scintillator. The Ready Cap is placed in an LS vial and counted on an LS counter. The Ready Cap can subsequently be removed from the vial and the vial reused. In many cases the radiolabeled substrate may be extracted and recovered for further analysis; this is not possible with liquid scintillator solutions.

The position of the cap in the vial makes very little difference to the measured count rate (see Table 4).

The labeled substrate may be any nonvolatile material, such as proteins,

Figure 2. READY CAP: A small plastic container coated on the bottom with XtalScint.

peptides, amino acids, oligionucletides, steroids, etc., dissolved in a volatile liquid.

There are three limitations on the use of Ready Cap. First, the volume of the sample is limited to 200 μL. Second, because of the small target area of scintillator, it is difficult to generate a spectrum from an external gamma source for quench determination. Finally, noncorrectable quench is introduced by codeposition of nonradioactive, nonvolatile substances such as salts

Table 4. Ready Cap Position in Vial: Effect on Efficiency[a]

Ready Cap Position		Count Rate (CPM)	Relative to Bottom Down Configuration
Bottom Down	⌐	239,000	100%
Bottom Up	⌐	239,000	100%
Side Up Bottom Parallel to PMT Face	○	245,500	102.7%
Side Up Bottom Perpendicular to PMT Face	]	236,300	98.9%

[a]Single ready cap labeled with ^{3}H-5-Fluorouracil counted on four possible positions resting on bottom of plastic 20 mL Poly Q Vial.

Table 5A. Effect of Buffers on Ready Cap ^{3}H Counting Efficiency; Values Reported are Counting Efficiencies (%) at 20 nsec

Liquid Carrier for ^{3}H-Palmitic Acid	Volume of ^{3}H-Palmitic Acid Added			
	25 μL	50 μL	100 μL	200 μL
D.I. Water	41.8	39.6	35.8	33.0
0.1 M NaCl	42.7	40.3	37.1	33.2
1.0 M NaCl	41.7	38.0	32.8	23.3

(greater than 1 M) or high boiling solvents like glycerol, which absorb the beta decay before it reaches the scintillator (see Table 5a and 5b).

Ready Filter

Ready Filters are glass fiber filters coated on one side with XtalScint and designed to be used in a manner similar to conventional glass fiber filters. Ready Filter comes in two formats: filter mats for automated cell harvesters, and 25 mm filter circles for manual filtration experiments. Particulate suspensions are filtered onto the XtalScint side of the filter. To inactivate nonspecific binding, it may be occasionally necessary to presoak the filter in unlabeled substrate.

The flow rate of water through the Ready Filter is about 70% the flow rate through S&S #32 glass fiber filter. The retention of BSA precipitate for the two filters is greater than 99%.

Counting efficiency comparisons were made between Ready Filter and S & S #32 glass fiber filter counted in Ready Organic. The samples used contained 250 μg of BSA or DNA, labeled with different isotopes. The results of this comparison, reported in Table 6, show that the efficiencies of the two filters are similar to each other for ^{3}H, ^{14}C, ^{125}I, and ^{32}P labeled precipitates when counted on conventional liquid scintillation counters. Using a counter with an optimized coincidence gate yields superior tritium efficiency for the Ready Filter. Other isotopes are not as dramatically affected. Both filter types show similar patterns of efficiency dependence on substrate identity (Table 7) and similar weight of precipitate deposited (Table 8). Efficiency for Ready Filter was not affected by the vacuum caused increased flow rate through the filter (Table 9).

After drying Ready Filter, it may be counted in any vial, but the orientation of the filter in the vial makes a difference in counting efficiency (Table 10).

Table 5B. Effect of Buffers on Ready Cap Counting Efficiency; Values Reported AE Counting Efficiencies (%) at 20 nsec

Liquid Carrier for Amino Acid Mixture	Volume of ^{14}C-L-Amino Acid Mixture Added			
	50 μL	100 μL	10 μL	200 μL
D.I. Water	83.0	78.2	78.1	77.4
0.1 M NaCl	83.2	78.8	78.2	77.4
1.0 M	80.5	76.2	72.4	67.0

Table 6. Counting Efficiency of Common Isotopes: Ready Filter vs Standard Filter in Cocktail

Isotope	Substrate	Efficiency		
		Glass Filter[a]	Ready Filter[b]	Ready Filter[c]
^{3}H	BSA	32.6	30.1	36.6
^{14}C	BSA	96.9	87.3	92.0
^{125}I	BSA	57.8	62.2	70.8
^{32}P	LYSATE	101.6	87.35	95.5

[a]S & S #32 glass fiber filter counted in Ready Organic.
[b]Ready filter counted with 20 nsec gate.
[c]Ready filter counted with Xtal gate.

Table 7. Counting Efficiency as a Function of Substrate: Ready Filter vs Standard Filter in Cocktail

Substrate	Efficiency		
	Glass Filter[a]	Ready Filter[b]	Ready Filter[c]
^{3}H BSA	32.2	30.1	36.6
^{3}H Lysozyme	27.4	24.9	30.7
^{3}H DNA	29.2	30.2	36.3

[a]S & S #32 glass fiber filter counted in Ready Organic.
[b]Ready filter counted with 20 nsec gate.
[c]Ready filter counted with Xtal gate.

Table 8. Efficiency vs Weight of Precipitate: Ready Filter vs Standard Filter in Cocktail

Weight of ^{3}H BSA (μg)	Efficiency	
	Glass Filter[a]	Ready Filter[b]
5	35.1	36.5
50	43.1	41.0
100	37.4	37.2
250	35.8	35.5
500	32.0	32.0
750	31.5	29.6
1000	31.0	30.5

[a]S & S #32 glass fiber filter counted in Ready Organic.
[b]Ready filter counted with 20 nsec gate.

Table 9. Efficiency vs Flow Rate: Ready Filter vs. Standard Filter in Cocktail

Vacuum Measured in Inches of Hg Drop	Efficiency of ^{3}H BSA	
	Glass Filter[a]	Ready Filter[b]
6.8	30.6	33.1
10.0	29.3	30.0
15.0	30.3	30.4
20.0	32.9	32.9
25.0	32.2	32.2

[a]S & S #32 glass fiber filter counted in Ready Organic.
[b]Ready filter counted with 20 nsec gate.

Table 10. Ready Filter Position in Vial Effect on Efficiency[a]

Position of XTAL in Vial	Vial	Count Rate (CPM)	Relative to Maxi Up Configuration
Up	Maxi	52,500	100%
Down	Maxi	21,000	40%
In	Maxi	41,500	79%
Out	Maxi	39,000	75%
In	Mini	44,500	84%
Out	Mini	41,500	79%
In	Bio	38,000	72%
Out	Bio	39,500	75%

[a]Single ready filter with ^{3}H BSA precipitate counted in different orientations in three different vial types.

The best position for counting Ready Filter is flat on the bottom of the vial with the XtalScint side up. As with Ready Cap, the analyte can be recovered by extraction from the XtalScint and used for further analysis.

Ready Filter should not be used for counting nonvolatile substrate in true solution. The glass fiber support for Ready Filter is capable of strong, nonspecific binding. Upon drying, the nonvolatile substrate can be bound to the nonscintillating support and give reduced scintillations. An example of this is reported in Table 11. Precipitates on the other hand, are captured in the solid scintillator layer and give the high counting efficiencies reported. External quench monitors are not recommended with Ready Filter due to the time required to obtain the Compton spectra from such a small target.

CONCLUSIONS

XtalScint is an effective solid scintillator ideally suited for measuring ionizing radiation from radioactive decay. It may be measured on conventional liquid scintillation instrumentation, performing similar to conventional liquid

Table 11. Risks of Counting Labeled Substrate in Solution on Porous XtalScint Base (Ready Filter)

Solution	Efficiency[a]	
	Ready Filter	Ready Cap
^{3}H Uracil	38.8%	39.1%
^{3}H Histamine	6.1%	39.5%

Scrap XtalScint from Ready Filter surface
Extract Filter Base and XtalScint with Ready Protein+

	% Total DPM Extracted	
	Uracil	Histamine
XtalScint from Ready Filter	52.4	7.4
Filter Base from Ready Filter	47.6	92.6

[a]Efficiency measured at 20 nsec gate.

scintillators, or measured on instruments with optimized coincidence circuitry (such as the Beckman LS 6000 series), performing better than conventional liquid scintillators. Its low background and high signal make it especially attractive for those experiments requiring high sensitivity.

REFERENCES

1. Kellogg, T.F. "Progress in the Development of Water-Miscible Non-Hazardous Liquid Scintillation Solvents," in *Advances in Liquid Scintillation Counting*, S.A. McQuarrie, C. Ediss, and L.I. Wiebe, Eds. (Edmonton, Alberta: University of Alberta, 1983) pp. 387–393.
2. Kalbhen, D.A. and V.J. Tarkanen. "Review on the Evolution of Safety, Ecology and Economical Aspects in Liquid Scintillation Counting Materials and Techniques," in *Advances in Liquid Scintillation Counting*: S.A. McQuarrie, C. Ediss, and L.I. Wiebe, Eds. (Edmonton, Alberta: University of Alberta, 1983) pp. 66–70.
3. Reed, D.W. "Triton x-100 as a Complete Liquid Scintillation Cocktail for Counting Aqueous Solutions and Ionic Nutrient Salts," *Int. J. Appl. Radiat. Isot.* 35(5): 367–370 (1984).
4. Lin, C.Y. and T.Y.C. Mei, "The BAM Scintillators for the Measurement of Radionuclides," *Int. J. Appl. Radiat. Isot.* 35(1): 25–38 (1984).
5. Solomon, R., J. Thompson, and D. Gillespie, "Low Background Counting with Ready Cap," Technical Information Bulletin, T-1688-NUC-89-26 (Beckman Instruments: Fullerton, CA, 1989).
6. Seltzer, S.M. and M.J. Berger. "Evaluation of the Collision Stopping Power of Elements and Compounds for Electrons and Positrons," *Int. J. Appl. Radiat. Isot.* 33: 1189–1218 (1982).
7. Seltzer, S.M. and M.J. Berger, "Improved Procedure for Calculating the Collision Stopping Power of Elements and Compounds for Electrons and Positrons," *Int. J. Appl. Radiat. Isot.* 35(7): 665–676 (1984).
8. Unak, T. "A Practical Method for the Calculation of the Linear Energy Transfers and Ranges of Low Energy Electrons in Different Chemical Systems," *Nucl. Instrum. Methods Phys. Res.* A255, 274–280 (1987).
9. Wunderly, S.W. "Solid Scintillation Counting: A New Technique for Measuring Radiolabeled Compounds," *Int. J. Appl. Radiat. Isot.* 40(7): 569–573 (1989).

CHAPTER 17

Photon Scattering Effects in Heterogeneous Scintillator Systems

Harley H. Ross

SUMMARY

This paper describes a new experimental approach that reveals the individual contributions of sample geometry and scattering phenomena in heterogeneous flow-cell detectors. The experimental detector responses obtained using scintillating polystyrene beads with optically smooth surfaces are compared with those obtained using similar beads with highly diffuse surfaces. These comparisons are carried out for both alpha- and beta-emitting nuclides. The experimental detection efficiencies are compared to Monte Carlo simulations of the detection process. Also, a new technique will be described for the fabrication of scintillating beads.

INTRODUCTION

Flow cell scintillation detectors are used extensively for monitoring alpha- and beta-emitting nuclides in flowing aqueous or organic streams. Major applications of such cells include liquid chromatography detectors, in-line process monitors, and a variety of environmental measurement and control devices.

For low-energy emitters, two quite different approaches are used: the heterogeneous and the homogeneous flow cell. Both, however, use a scintillation process. Homogeneous flow cell systems operate by mixing a suitable liquid scintillator with all or a portion of the flowing stream. Several mixing and flow control units are used to establish stable, fixed measurement conditions. The

*Research sponsored by Office of Energy Research, U.S. Department of Energy under contract DE-AC05-84OR21400 with Martin Marietta Energy Systems. "The submitted manuscript has been authored by a contractor of the U.S. Government under contract No. DE-AC05-840R21400. Accordingly, the U.S. Government retains a nonexclusive, royalty-free license to publish or reproduce the published form of this contribution, or allow others to do so, for U.S. Government purposes."

195

cell is optically coupled to one or two photomultiplier tubes which are used to detect the scintillation output. Although this type of device can exhibit high sensitivity, it suffers from a number of problems: it uses large amounts of expensive liquid scintillator, it must provide safe disposal of its scintillator effluent, its quenching effects are unstable and unpredictable, it is difficult to maintain (mechanical and reagents), and its resolution is often poor. For these reasons the heterogeneous detector cell is usually preferred for most on-line measurement tasks.

Heterogeneous detector cells[1-4] are characterized by contacting the flowing liquid (to be measured) with a solid phase scintillator, most often in powder form. Such a flow cell must fulfill certain physical requirements for efficient operation. Obviously, detection geometry must be high in relation to the energy of the nuclide(s) to be measured. Cell volume must be as large as possible (for maximum sensitivity), consistent with the required temporal resolution of the application. The cell must be designed for minimal mixing and virtually no dead space to prevent any hold up. There may also be requirements for the shape of the cell, the surface to volume ratio, and the optical demands of coupling photodetectors.

The most efficient heterogeneous units have a tube or column that contains a solid scintillator (organic, plastic, or glass); the sample flows within the interstices of the scintillator particles or fibers. While the geometry of this device is not as good as that of the homogeneous detector, commercially available cells that use finely divided organic or glass scintillators show that it can count ^{14}C at moderate efficiency. The tritium efficiency is, however, quite low. Other powdered materials used in cells of this type include europium-activated calcium fluoride and scintillating plastics.

Powdered scintillator detector cells also exhibit deficiencies characteristic of their design. For example, radiation generated by weak emitters degrades significantly in its passage from the sample liquid phase to the solid fluor. If the sample absorption path length is sufficiently large, and the radiation energy sufficiently low, total absorption takes place, and the event is lost to the system. Attempts to use even finer particles to improve the geometry have not improved detection efficiency as expected and routinely have made other cell characteristics worse. For example, back pressure of fine powder cells can exceed the working limits of conventional flow systems. Also, such cells with their high surface to volume ratio exhibit strong and sometimes irreversible memory effects.

Rucker et al.[5,6] have attempted to overcome these difficulties by designing a cell that uses aligned scintillator fibers rather than the usual powder. Back pressure and memory effects improved significantly, but detection efficiency increased only slightly over previous designs. However, the results of this work prompted the idea that photon scattering within the detector, in addition to the geometry parameter, was crucial in establishing the ultimate efficiency of these devices. While smaller fluor particles lead to improved geometry, they also generate increased light scatter. The reason for this is that most of the fine-particle scintillators have very poor surface characteristics and simultaneously exhibit extensive scattering within the particles themselves. This scatter leads to elevated absorption of the generated photons with subsequent loss of pulse height, detection efficiency, and spectral resolution. Although it appears clear that both geometry and photon scatter can diminish detector performance, it is not obvious which parameter is most important in a specific application, nor has it been demonstrated how the effects of these parameters change with different cell designs. What is clear, however, is that attempts to optimize one of the parameters often result in a degradation of the other; small particle size scintillators usually exhibit maximum scattering. Practical flow-cell implementations are often a compromise between these two, conflicting optimization constraints.

On the basis of the above cited work and other work, it is clear that scintillation efficiency, geometry, and light collection are the major parameters that determine flow cell radiation response to low-energy emitters. It also appears that using conventional powdered materials for the solid scintillator cell phase cannot improve detector geometry and reduce photon scatter simultaneously. This apparent stalemate is unfortunate since improved flow cell detectors would have a broad range of important applications. Thus, while numerous different flow radiation detectors are used daily, throughout the world, for monitoring energetic nuclides, virtually none exist for low-energy emitters. This is particularly disturbing when one notes that many of these weak emitters have significant biological application.

In order to design better flow cell radiation detectors, it became clear that more fundamental information was needed about the effects of geometry and light scatter, and the relationships between the two. This investigation was initiated to hopefully develop such information.

EXPERIMENTAL

Equipment and Reagents

All scintillation measurements were carried out using a Packard Tri-Carb liquid scintillation counter, model 4530. This unit contains a low-resolution multichannel capability used solely for a visual display of the pulse-height spectrum. Two of the spectra shown in this paper are simply photos of this display. When detailed spectral information was needed, appropriate linear signals were taken from the Packard counter and fed to a Nuclear Data multichannel analyzer system. That system included all of the standard MCA input/output and spectrum storage features. As high resolution was not needed for these studies, data were collected in a minimal 128 channels.

The clear, polystyrene spheres (three size distributions) were synthesized in the Department of Chemistry, University of Tennessee, Knoxville. These spheres were processed with conventional scintillation fluors and organic solvents to produce the scintillating beads used here. The ^{14}C test solution (as aqueous carbonate) was standardized by liquid scintillation counting using the NBS hexadecane as an internal standard. ^{241}Am (as nitrate, in dilute nitric acid, pH 2 to 2.5) was assayed via 2π gas counting and confirmed in a liquid scintillation counter. All other reagents and fluors used in this study were reagent or scintillation grade.

Monte Carlo calculations were carried out on three different personal computers that all used the conventional MS-DOS operating system. Each computer was fitted with a numeric coprocessor chip (8087 or 80287) to speed the large number of floating point calculations simulation requires. Source code was written in Turbo Pascal (Borland International) that was compiled to executable files using the Turbo compiler, versions 3.01 and 4.0.

Preparation of Scintillating Beads

The untreated polystyrene beads had certain physical and optical properties that were crucial to the success of this project. First, they exhibited an extremely clear internal structure that was virtually fracture free. Also, the surface of each bead looked mirror smooth. Finally, the majority of beads assumed an almost perfect spherical shape. The visual effect through an optical microscope was similar to looking at drops of water. Although there were some spheres that were undeniably poor in optical quality, their number was small. The removal of such beads did not seem to be a viable task.

The beads were received already divided into three size distributions. There were many more beads of the larger size; these were selected for the preliminary tests directed toward endowing the beads with efficient scintillating properties. It was known that the polystyrene beads, when placed in toluene, would swell to several times their dry size. An obvious first approach was to dissolve scintillation fluors in toluene, place the beads in this solution, let them swell, filter the expanded beads, rinse them with ethanol, and air dry them so they shrink to their original size. The idea was that the solid fluors trapped within the beads would serve as efficient scintillating centers when excited by ionizing radiation. After some experimentation with different fluors and different fluor concentrations, the soak/dry technique did yield efficient scintillating beads.

A 1 g portion of dimethyl-POPOP was added to 100 mL of toluene and was allowed to mix for 24 hr at room temperature. About 3 g PPO were added to the saturated solution and the mixing was continued for 12 hr. The doubly saturated solution was separated from the excess fluors by filtration. The polystyrene beads were allowed to soak and swell in this solution for about 12 hr. The swelled beads were separated by filtration, washed with several portions of absolute ethanol, and air dried. This was the procedure used to treat all of the beads used in this study. (Some tests carried out after the start of this work have indicated that scintillation efficiency of the beads could be improved slightly by incorporating naphthalene in the soak solution. However,

Table 1. Average Bead Sizes and Size Distributions

Size	Ave. Radius (μm)	Std. Dev. (μm)
Small	28.6	2.8
Medium	68	7
Large[a]	295	36

[a]This bead size was not used for any of the nuclide measurements. It was used during the bead processing phase and is reported here for completeness.

to keep results consistent, this was not done for any of the beads studied here.)

Samples of each size distribution were examined by scanning electron microscopy (SEM) to evaluate the range of sizes in each distribution and to reveal clearly the surface characteristics and bulk structure of the beads after treatment. The size data are shown in Table 1. Figure 1 is a 300 × photo of the 28.6 micron cut; the highly damaged bead is obvious.

Some of the measurements carried out in this study required surface alteration of the scintillating beads from very smooth to thoroughly diffuse. A piece of fine sandpaper (8 × 10 in) was glued to a piece of plate glass, rough side exposed. The sample of beads to be etched was placed on the sandpaper surface along with 1 to 2 mL of distilled water. A second piece of plate glass was placed on top of the beads and gently rotated by hand to work the beads

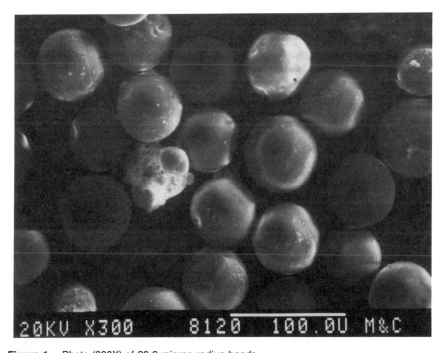

Figure 1.   Photo (300X) of 28.6 micron radius beads.

against the abrasive surface. As the beads are relatively soft, only two to three minutes of grinding were needed. The beads were washed into a small gas bubbler and separated from the sandpaper debris by flotation. The beads were inspected with an optical microscope; the surfaces were sufficiently etched such that it was not possible to see inside the beads. Also, no changes in the bead size could be discerned, although this was not rechecked with SEM photo.

Radionuclide Measurements

All activity measurements using the scintillating beads were carried out using an assembly similar to that shown in Figure 2. A glass tube of approximately 5 mm. diameter was cut to an appropriate length and flame sealed at one end. A small amount of white, room-temperature curing silicone rubber was placed on the inside bottom of a standard LSC glass vial. The sealed end of the tube was placed in the center of the vial and pressed into the rubber. The silicone was allowed to cure for 24 hr. For use, a selected sample of scintillating beads was transferred into the central glass rod and vibrated to aid packing. The activity to be measured was pipetted onto the beads. Water was added to the space between the vial and rod (to act as a light coupler), the vial was capped, and the assembly was gently centrifuged to draw the sample into the beads and eliminate air voids. This device was transferred to the Packard instrument for counting.

The Monte Carlo Simulations

Spheres of uniform size can be loaded into a container in two different close packed arrangements of identical efficiency: hexagonal and face centered cubic. In both cases, the spheres fill about 74% of the container, which leaves 26% void volume. It is interesting to note that 74% is not the maximum volume of space which can be occupied in the packing of spheres, although it is the maximum for symmetrical periodic arrangements. For irregular packing it can be shown that the maximum must be less than 78%.

Although crystallographic studies tend to focus on the arrangements of packed spheres, this investigation gives major importance to void geometry. Spheres that are hexagonally close packed create a "unit" void that is illustrated in Figure 3. The shaded portion is the void created by four spheres in two layers having mutual contact. When another layer of spheres is added (three total layers), the total void created is made up of two unit voids as seen through a single plane. All of the total voids in hexagonal close packing are of this type. Cubic close packing exhibits a symmetry that requires four layers rather than three. The effect on the void structure is that two different arrangements of the unit void are created. These are usually referred to in the literature as tetrahedral and octahedral interstitial holes. Both hole structures must be considered in the cubic pack simulation.

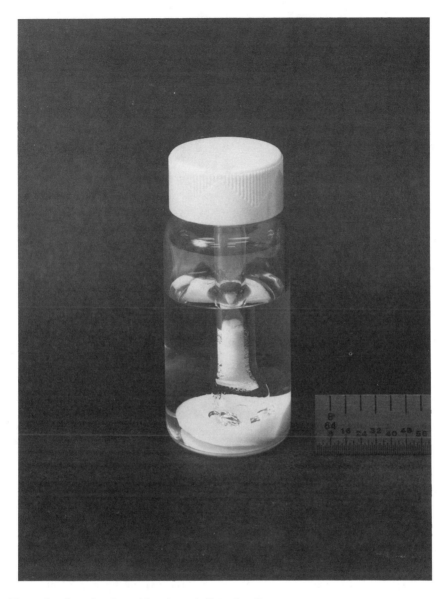

Figure 2.   Counting assembly using scintillator beads.

The Monte Carlo simulation is designed to answer the following question: if scintillating spheres are close packed and if a liquid containing a radioactive emitter fills the void volume, what fraction of the decay particles are expected to reach the solid spheres? This is, of course, the geometry. Several factors must be considered in designing the simulation. These include:

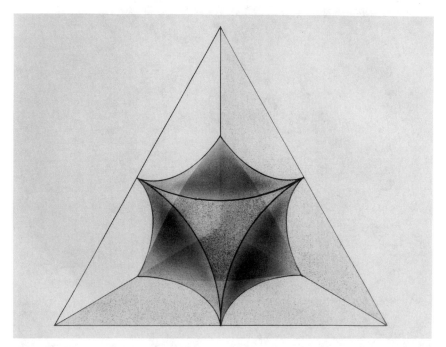

Figure 3.   Diagram of void volume in hexagonal close packed spheres.

1. the emitted particle (alpha or beta)
2. the decay energy
3. range-energy relationships and absorption characteristics
4. the absorber composition
5. the sphere size
6. the detailed geometry of the voids

Four major simulations (two different but related programs) were used to develop geometry information as a function of sphere size. These were for ^{241}Am alphas and ^{14}C betas for both the hexagonal and cubic close packing arrangements. After the appropriate program is loaded, the user is asked to supply: (1) the decay energy (an average of 5.46 MeV was used for ^{241}Am, (2) the absorber density and effective atomic mass, (3) the hexagonal or cubic close packing, and (4) the number of decay events to evaluate (2000 to 5000). At this point the simulation starts and continues until the final result is obtained.

The geometrical aspects for both alpha and beta simulations are very similar. Using a fixed coordinate system, a random point is selected and tested to verify its position within the liquid phase (the void volume). This point is used as the origin of the decay event. Next, a random direction is selected to create a linear displacement from the origin point. Although these two operations appear to be relatively straightforward, much of the program run time is used up here. The complex geometry and extensive spacial testing cause this. Also, for cubic packing, even more verification must be carried out. Processing from this point is quite different for alphas and betas.

Simulations of alphas are simplified by two important factors. Alpha radiation is monoenergetic and the emitted alpha particles travel in virtually straight lines. After the emission origin and random direction are determined, an alpha range corresponding to decay energy helps to create a vector from the event origin. The alpha range is calculated from experimental air measured alpha ranges combined with the Bragg-Kleeman[7] rule (to correct for absorbers other than air). Finally, the vector is tested for intersection with any of the void-surrounding spheres. If intersection is not observed, and the alpha range could go to a second layer of surrounding spheres, than another algorithm is used to evaluate intersection. All events that intersect any sphere are totaled; the result is divided by the total of events tested. The outcome is the predicted geometry.

A convenient simplification can also be employed for evaluations of beta simulation response. Although betas are not monoenergetic and can travel in very convoluted paths, an empirical relationship that exits between E_{max} and linear distance can be used to evaluate the response of an entire beta spectrum. This requires that the maximum beta range of the simulated nuclide be sufficiently long to allow at least some intersection of the beta energy distribution (from the origin of the decay event) with the solid scintillator. The technique used is described in detail in Rucker *et al.*;[6] the geometry studied, however, is quite different from that examined here. After the degree of intersection of each beta distribution is calculated, the individual percent values are averaged to predict the geometry.

RESULTS AND DISCUSSION

The first series of tests were designed to evaluate the basic quality of the scintillating beads. The two major factors involved are photon yield and photon transmission. The baseline values of these parameters are important. If the bead emission and transmission characteristics proved substantially inferior to a quality liquid scintillator, it would have been necessary to devise an alternate procedure for bead preparation.

Aqueous samples of the ^{241}Am tracer were mixed with different 15 mL. aliquots of Insta-Gel and Opti-Fluor liquid scintillation cocktails contained in standard size vials. The tracer was also added to bead scintillator samples contained in the assembly described above. A pulse-height spectrum was obtained for each sample. Figure 4 shows such a spectrum obtained in Insta-Gel and Figure 5 a spectrum using 68 micron beads. The marker is at the same position in both spectra.

Although the bead spectrum shows considerable energy degradation due to absorption in the liquid phase, it is quite clear that the full energy peak is both sharper and at a higher pulse height then that seen in the liquid scintillator

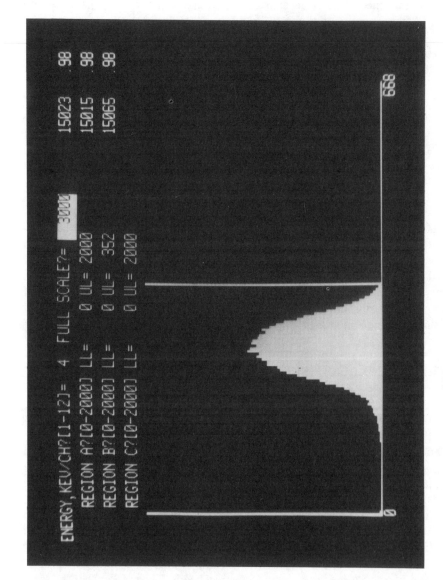

Figure 4. Pulse-height spectrum of ^{241}Am in Insta-Gel.

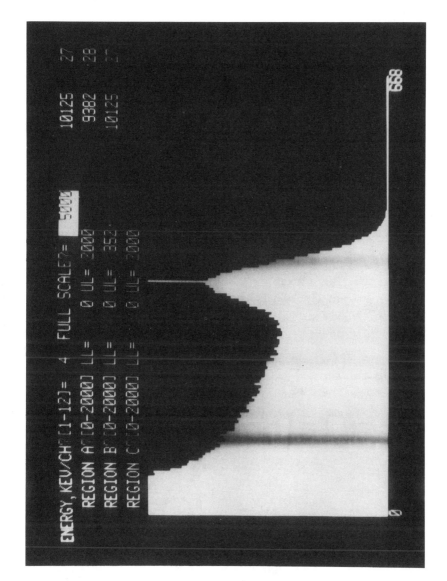

Figure 5. Pulse-height spectrum of ^{241}Am using 68 micron beads.

spectrum. This was a most surprising result. It is not really possible to determine the individual quantitative effects of photon yield and transmission from these data, but I believe it is fair to conclude that the bead scintillators do extremely well in both of these parameters. One might argue that the bead results could arise from an outstanding photon emission combined with a somewhat attenuated photon transmission characteristic. While such a conclusion might be suggested, the observed optical properties of the beads clearly argue against it.

Similar tests were made with ^{14}C tracer. Again bead results were very good, although they were more difficult to observe because of the continuous nature of the beta spectrum.

Figures 6 and 7 illustrate the results obtained with the four Monte Carlo simulations. The second order regression lines in both figures are only plotted to show point continuity. It is not suggested that they represent an analytical solution to the simulation. As might be expected, all four of the simulations tend to a geometric efficiency of one as the sphere radius goes to zero. The responses of the alpha regressions show less sensitivity to sphere radius when the radius is small; the opposite is true for the ^{14}C beta simulations. The values obtained for cubic packing are less than those obtained for hexagonal packing for both nuclides.

The Monte Carlo simulations only give an expected geometric efficiency for a given sphere size. This efficiency can be thought of as the maximum detection efficiency that could possibly be observed in the system. Of course, in real counters, results are always less than the geometry because of factors such as the minimum energy required for photon generation, presence of nonradiative

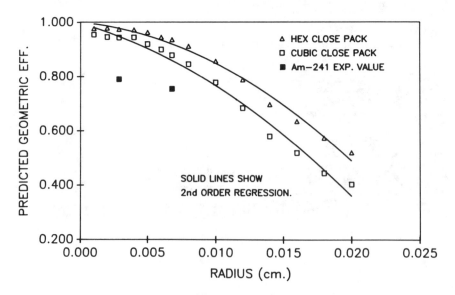

Figure 6. Monte Carlo simulation of [241]Am and experimental data.

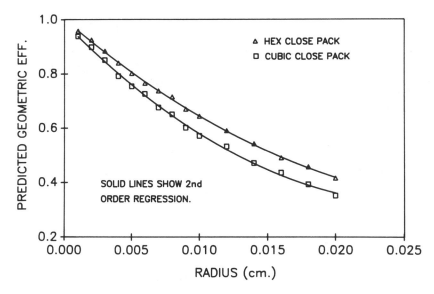

Figure 7. Monte Carlo simulation (linear plot) of ^{14}C.

absorption processes, photon collection efficiency, and electronic thresholds in the counter. Also, it would be unreasonable to assume that close packing would be of one type only or that close packing could even be achieved throughout the majority of detector volume. However, if all of these factors remain essentially constant within a given system, the validity of the simulation can be verified by comparing an experimental curve with the appropriate simulation. A shift to lower efficiency would be expected but the shape of the curves should match.

Figure 6 shows two experimental points for ^{241}Am obtained with two different size smooth-bead samples. While it is not possible to verify the curve with only two points, the experimental values do exhibit the trend of the simulation. Figure 8 shows the ^{14}C simulations replotted in a semilog format along with four experimental points. The average experimental values for the smooth beads lend additional support for the simulation results. Figure 6 also includes data obtained with etched beads that exhibit a high degree of light scattering. Along with the drastic drop in counting efficiency, one sees a surprising reversal in the response as a function of bead size. Table 2 summarizes the counting results for data shown in Figures 7 and 8.

CONCLUSIONS

Monte Carlo simulations for heterogeneous flow cell radiation detectors (fabricated with spherical scintillator beads) show somewhat different geometric response functions for alpha and beta emissions. For alpha radiation, the

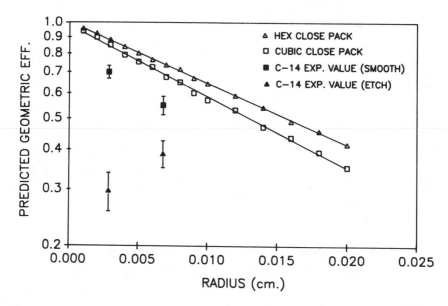

Figure 8. Monte Carlo simulation (semi-log plot) of ^{14}C and experimental data.

response is relatively flat up to about a 40μ bead radius. This implies that particles smaller than this value probably will not materially improve alpha detection efficiency in practical implementations. The experimental data observed with ^{241}Am support the shape of the observed simulation function which further strengthens the idea that ultrasmall particles are not needed. The overall maximum pulse height and spectral resolution (not counting efficiency) exceed that obtained with a typical homogeneous cocktail for aqueous samples.

Conversely, the ^{14}C response functions exhibit an increasing slope as bead radius drops. Obviously, much more attention must be directed toward bead size for low-energy beta emitters. The experimental data with smooth and rough bead surfaces demonstrates, however, that significant photon scattering is an extremely important limitation in heterogeneous cells. The minimal conditions investigated in this study show that these cells may be 3 to 4 times more sensitive to photon scatter and absorption than to geometry. The conclusion is

Table 2. Counting Efficiency of ^{241}Am and ^{14}C With Scintillating Beads

Nuclide	Bead Radius (μm)	Bead Surface	Counting Eff. (ave.)
Am-241	28.6	Smooth	0.79
Am-241	68	Smooth	0.75
C-14	28.6	Smooth	0.70
C-14	68	Smooth	0.55
C-14	28.6	Etched	0.30
C-14	68	Etched	0.39

clear that the development of high-sensitivity heterogeneous flow cell detectors may ultimately be determined more by the optical rather than physical properties of the scintillator cell.

ACKNOWLEDGEMENTS

The author would like to thank Professor Spiro Alexandratos, Department of Chemistry, University of Tennessee, for providing the polystyrene beads used in this study; Ms. Lisa Rozevink, Science Alliance Summer Participant, University of Tennessee, for writing major pieces of the Pascal source code used in the simulations, executing some of the alpha simulations, and making several of the alpha response measurements with the beads; and Dr. R. G. Haire, Transuranium Research Laboratory, Oak Ridge National Laboratory, for supplying the ^{241}Am tracer.

REFERENCES

1. Mackey, L.N., P.A. Rodriguez, and F.B. Schroeder. *J. Chromatogr.*, 208, 1 (1982).
2. Everett, L.J. *Chromatographia*, 15, 445 (1982).
3. Harding, N.G.L., Y. Farid, M.J. Stewart, J. Sheperd, and D. Nicoll. *Chromatographia*, 15, 468 (1982).
4. Frey, B.M. and F.J. Frey. *Clin. Chem.*, 28, 689 (1982).
5. Rucker, T.L., H.H. Ross, and G.K. Schweitzer. *Chromatographia*, 25, 31 (1988).
6. Rucker, T.L., H.H. Ross, and G.K. Schweitzer. *Nucl. Instr. Methods in Physics Research*, A267, 511 (1988).
7. Evans, R.D. "The Atomic Nucleus," McGraw-Hill, 1955, p.652-3.

CHAPTER 18

Some Factors Affecting Alpha Particle Detection In Liquid Scintillation Spectrometry

Lauri Kaihola and Timo Oikari

ABSTRACT

Pulse shape analyzers have been introduced in commercial liquid scintillation counters, such as the Wallac 1220 QuantulusTM and 1219 SM, enabling separation of alpha particle spectra from other types of radiation in a single measurement and making ultra sensitive alpha counting possible. Background count rates as low as one count per day for [214]Po in a teflon vial are achievable while still retaining nearly 100% counting efficiency.

The number of photons per decay event however, is relatively low in liquid scintillation counting, leading to energy resolution of the order of 10%. Energy resolution has been found to improve with low water to cocktail ratios and opaque vial types.

Easy and fast sample preparation together with the ultra low background make liquid scintillation spectrometry a competitive method in environmental counting of alpha particles.

INTRODUCTION

Liquid scintillation counting (LSC) is a well established method for beta particle counting. Its use in alpha particle counting has suffered from large background count rates in energy range due to the cosmic ray and environmental radiation caused events in the cocktail. Background reduction for alpha counting requires identification of the origin of the background. It has long been known that the relative amounts of the prompt and delayed components of fluorescent decay depend on the specific ionization, hence the type of particle causing it. Typically, the heavily ionizing particles, such as neutrons and alpha particles, produce more delayed component than beta particles, thus longer pulses. This phenomenon is the basis for the pulse shape analysis (PSA) as particle identifier, and it makes possible rejection of the beta-like background component in alpha counting.

The method has been applied in nuclear physics for neutron counting in the presence of gamma radiation since the '50s. In alpha counting it has also been applied with good success and is well reviewed.[1,2]

Experimental liquid scintillation counters, developed in the '70s, used zero crossing techniques for pulse shape discrimination.[3,4] These devices contained silicon oil as a light guide, improving the coupling from the sample to a single photomultiplier tube, thus giving an excellent energy resolution.

Wallac 1220 Quantulus, an ultra low background liquid scintillation spectrometer, has contained a pulse shape analyzer since 1987. Being a general purpose alpha-beta counter with an automatic sample changing mechanism, the optical coupling described above was not applied, but a standard detector configuration with two PMTs was maintained.

The Wallac pulse shape analyzer uses a concept quite similar to the one adopted by, e.g., Brooks.[5] It initiates the integration of the pulse tail after 50 nsec from the leading edge and compares it with the integral of the total pulse. Surface mounted electronic components manufacture the high speed analog circuitry that produces the pulse shape related signal without amplitude dependence. By setting the PSA level through software commands it is possible to route alpha-like events into one half of the MCA and beta-like events (due to beta particles, Compton recoil electrons, Cerenkov radiation, and X-rays) into the other half in a single measurement. Maximum resolution of 1024 channels is available with a logarithmic energy scale.

The Quantulus is equipped with a guard counter, set in anticoincidence with the sample detector PMTs, to minimize the background component caused by environmental radiation, cosmic flux, and gamma flux from the bedrock and building materials. Further, a user programmable pulse amplitude comparator is available for reduction of cross talk events.[6]

Pulse shape separation characteristics depend on cocktail. Adding naphthalene to conventional toluene or xylene based cocktails enhances their pulse shape separation. Modern Hisafe cocktails by Wallac contain naphthalene derivatives as solvent and offer excellent pulse shape characteristics without any further naphthalene addition.

PULSE SHAPE ANALYSIS

Since the pulse shape characteristics vary between different cocktails, the pulse shape analyzer has to be set individually for each one to achieve the optimum separation of alpha and beta particle spectra. This can be done by preparing a pure alpha and beta sample whose amplitude spectra overlap, e.g., ^{36}Cl and ^{241}Am. The PSA level is then scanned for acceptable rejection of the radiation not of interest. With conventional cocktails one may allow say 5% loss of alpha particle counting efficiency to remove beta-like background, if alpha particle spectrum is only needed. If both particle types are to be measured, new cocktails containing naphthalene derivatives will give good separation (Table 1).

Table 1. Beta Residual, the Percentage of Betas Remaining Among Alpha-like Events. Normalized to 100% when PSA is off. The Residual is Given for ^{36}Cl Beta Particles in ^{241}Am Alpha Window for Corresponding Alpha Counting Efficiencies. The Sample is Contained in 1 mL Water with 10 mL Optiphase Hisafe 3 in Equilibrium with Air.

PSA	Alpha Eff %	Beta Residual %
0 = off	100	100
25	100	100
50	100	84.7
75	100	36.1
100	99.3	2.75
120	98.9	0.098
130	95.9	0.025
140	88.7	0.02
150	73.4	<0.02
160	54.4	

ENERGY RESOLUTION IN LIQUID SCINTILLATION SPECTROMETRY

Energy resolution is proportional to the number of photons created in the decay event.[7] With LSC one may not expect to achieve the energy resolution of solid state detectors, because number of liquid scintillator charge carriers is considerably greater than that of photons detected in a typical liquid scintillation event. Therefore, considerable improvement in energy resolution will only be achieved with a profound change in the light production properties of the cocktail.

Current cocktails have about 10% resolution in opaque vials at 5 MeV alpha energy (Figure 1). Corresponding resolution in standard glass vials is poorer, 20% (see item 2).

There are ways to improve the energy resolution:

1. High light output cocktails give better results.
2. Opaque vials transmit light with fewer losses than standard glass vials where photons can get trapped through total reflections in the vial wall. Teflon and etched glass have the best resolutions so narrower windows can be used to obtain smaller backgrounds with the same counting efficiency. Alignment of the liquid meniscus with the center axis of the PMT gives the best sensitivity and optimal resolution.
3. The lower the quench level, the better the photon emission; this leads to improved energy resolution. In practice this means having the smallest amount of water, contradicting the need to have as good a lower limit of detection as possible. In extractive methods, using two phase samples, radioactivity from water is transferred into the cocktail phase thereby eliminating the aqueous quench.
4. Nitrogen bubbling improves the performance further, due to the removal of quenching by dissolved oxygen.

[A]   4.888 CPM/ch   59.47 min   A:\DEMODATA\RA226\9815781M.888   SP#11
[B]   4.888 CPM/ch   59.47 min   A:\DEMODATA\RA226\9815781M.888   SP#12

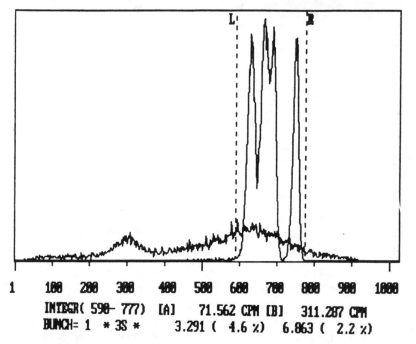

INTEGR(598- 777)   [A]   71.562 CPM [B]   311.287 CPM
BUNCH= 1 * 3S *   3.291 (4.6 %)   6.863 (2.2 %)

Figure 1. ^{226}Ra spectrum from 2 mL water sample with 18 mL Optiphase Hisafe 3 in 20 mL teflon vial. Alpha emission peaks from left to right are ^{226}Ra (4.78 MeV), ^{222}Rn (5.49 MeV), ^{218}Po (6.00 MeV), and ^{214}Po (7.69 MeV).

In the logarithmic amplitude scale, the alpha peak shape is quite close to being symmetric (Figure 2). At very high alpha particle energies, an asymmetric shape may result due to increased losses close to the vial walls, c.f. ^{214}Po, 7.7 MeV.

PERFORMANCE IN ALPHA COUNTING

Glass vials contain U and Th isotopes, their contribution results in a wide background spectrum under the alpha peaks (Figures 3 and 4). These alpha particles are emitted in a 2π geometry and are subject to losses due to their need to penetrate through a layer of glass in order to reach the cocktail. Background also contains slow fluorescence from the glass excited by the environmental radiation.

Attenuation of the beta-like decay events among alpha-like ones by pulse shape analyzer is given in Table 1. Typical background count rates and lower

[A] 1500.000 CPM/ch   89.17 min   A:\NOGUARD\9811501N.001   SP#12
[B] 2300.000 CPM/ch   2.84 min   A:\CLAM\9010101N.001   SP#12
[C] 1500.000 CPM/ch   89.17 min   A:\NOGUARD\9821601N.001   SP#12

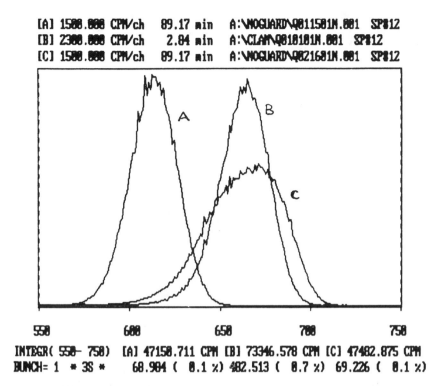

550   600   650   700   750
INTEGR(550- 750) [A] 47150.711 CPM [B] 73346.578 CPM [C] 47482.875 CPM
BUNCH= 1 * 3S *   68.904 (0.1 %) 482.513 (0.7 %) 69.226 (0.1 %)

Figure 2. ^{241}Am alpha emission spectrum (5.5 MeV) in teflon vial (A), etched glass vial (B), and in standard glass vial (C).

limits of detection are given for ^{241}Am (5.5 MeV) and ^{214}Po (7.7 MeV) in two instruments.

CONCLUSIONS

There are several advantages to alpha particle counting with the LS method compared to solid state detection, gridded ion chambers, or the nuclear track method, especially when high energy resolution is not required. The LSC performance does not involve a variable geometrical factor, variable counting efficiency, self absorption, and detector contamination. A counting efficiency of at least 98% with a background of less than 0.1 cpm can easily be achieved. Sample preparation is simple; saving time in low level counting. Sample sizes considerably larger than in solid state alpha spectrometry can be analyzed, and enrichment is a simple procedure. Spectrum analysis allows early recognition of alpha-emitting nuclei in the samples and thus time for decision making if any further analyses are required.

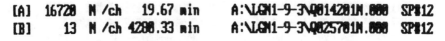

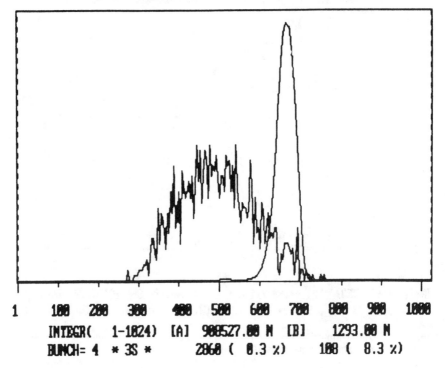

Figure 3. [241]Am alpha emission and background spectra in a standard glass vial (1 mL water + 9 mL Lumagel + 20% w/v naphthalene). Background count rate in full window is 0.38 cpm and 0.002 cpm in [214]Po window.

Table 2. Alpha Peak Background (cpm) and Lower Limit for Detection (LLD, Bq/L) for 100 min Counting Time in the Wallac 1220 Quantulus and 1219 SM (which contains no guard counter). Sample is 2 mL water + 18 mL Lumagel + 20% w/v Naphthalene in Equilibrium with Air

Instrument	Vial	bkg	LLD	Comment
1219 SM	glass	0.23	1.2	Am-241
1219 SM	glass, op	0.11	0.8	Am-241
1219 SM	glass, op	0.03	0.08	Po-214
1219 SM	teflon	0.02	0.05	Am-241
1220	glass	0.03	0.08	Am-241
1220	teflon	0.005	0.01	Am-241
1220	teflon	0.001	0.002	Po-214

[A] 19918 N /ch   14.75 min   A:\LGN1-9-3\Q834381N.888 SP#12
[B]   4 N /ch 4288.35 min   A:\LGN1-9-3\Q845881N.888 SP#12

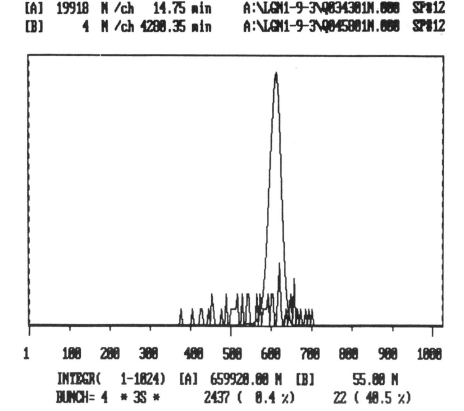

1   188   288   388   488   588   688   788   888   988   1888

INTEGR(1-1824) [A] 659928.88 N [B]   55.88 N
BUNCH= 4 * 3S *   2437 (8.4 %)   22 (48.5 %)

Figure 4. ^{241}Am alpha emission and background spectra in a teflon vial (1 mL water + 9 mL Lumagel + 20% w/v naphthalene). Background count rate in full window is 0.013 cpm and 0.001 cpm in ^{214}Po window.

REFERENCES

1. Horrocks, D.L. "Pulse Shape Discrimination with Organic Liquid Scintillator Solutions," *Applied Spectroscopy*, 24:397 (1970).
2. Brooks, F.D. "Developments of Organic Scintillators," *Nucl. Instr. Methods*, 162:477 (1979).
3. McDowell, W.J. "Alpha Counting and Spectrometry Using Liquid Scintillatio Methods," Monograph, NAS-NS-3116 (DE86 007601) (publ. by Technical Information Center, Office of Scientific and Technical Information, US DOE), 1 January 1986.
4. McKlveen, J.W. and W.J. McDowell, "Liquid Scintillation Alpha Spectrometry Techniques," *Nucl. Instr. Methods Phys. Res.* 223:372 (1984).
5. Brooks, F.D. "A Scintillation Counter with Neutron and Gamma-Ray Discriminators," *Nucl. Instr. Methods* 4:151 (1959).
6. Kaihola, L. "Liquid Scintillation Counting Performance Using Glass Vials with the Wallac 1220 Quantulus^{TM}," paper presented at tis conference.

7. Oikari, T., H. Kojola, J. Nurmi, and L. Kaihola, "Simultaneous Counting of Low Alpha-and Beta-Particle Activities with Liquid-Scintillation Spectrometry and Pulse-Shape Analysis," *Appl. Radiat. Isot.* 38A(10):875–878 (1987).

8. Salonen, L. "Simultaneous Determination of Gross Alpha and Beta in Water by Low Level Liquid Scintillation Counting," paper presented at the 2nd Karlsruhe International Conference on Analytical Chemistry in Nuclear Technology, Karlsruhe, FRG, June 5–9, 1989.

CHAPTER 19

Modern Techniques for Quench Correction and dpm Determination in Windowless Liquid Scintillation Counting: A Critical Review

Staf van Cauter and Norbert Roessler, Ph.D.

ABSTRACT

With the advent of readily available computing power, it seems reasonable to describe the probability of the liquid scintillation process replacing quench curves. The pulse height spectrum and the counting efficiency can be expressed as functions of the scintillator photon yield as measured by the external standard Compton spectrum. The energy dependence of the photon yield can be accounted for by combining information from the sample spectrum and the Compton spectrum. A variety of numerical techniques can then be applied to describe the spectral response and to calculate dpm values of an unknown from appropriate reference spectra.

These methods for determining dpm values have recently been applied to liquid scintillation counters equipped with multichannel analyzers. Although these "windowless" dpm determinations are very convenient to the user, many inherent characteristics of the scintillation process may render results invalid under certain experimental conditions.

Indeed, the photon yield at a particular energy is significantly affected by the microscopic environment of the nuclide and the composition of the sample, expressed by the concept of ionization quenching. As a further complication, the prompt and delayed components of the fluorescence depend on these parameters as well. As a consequence, theoretical expressions proposed to account for ionization quenching must remain approximations; their validity needs to be verified for each cocktail-sample combination. Due to the unstructured nature of LSC pulse height spectra, it is difficult to accomplish this verification using spectral overlay and comparison techniques. Pulse height discrimination is often required after the fact to make results less dependent on varying experimental conditions.

Numerical methods applied to windowless dpm counting will be reviewed, and the results will be compared to methods based on quench compensated discrete window settings.

INTRODUCTION

In the field of optical spectroscopy, energy dispersion is obtained using filters or monochromators that can be calibrated independently of the sample using absolute physical methods. In liquid scintillation counting, this is not the case. Although the response of the PMT and its related pulse shaping and aplification circuits can be determined independently of the sample, quench compensation is required to correlate the pulse height measured in the multichannel analyzer with the electron energy of the nuclear event. Most commercial systems monitor the quench level using the mean sample pulse height (SIS) and the Compton edge from an external source of gamma radiation (tSIE or H#). Horrocks has presented the results of extensive studies determining the energy needed to produce a photoelectron in a liquid scintillation counter.[1] This energy depends both on electron energy and on quench level. These functions are not linear. Rundt et al., investigated theoretical expressions for the effect of ionization quenching and found them all dependent on adjustable parameters which vary with the solvent.[2] The theoretical functions are valid for chemically quenched homogeneous samples only.

The reference method of compensating for quench effects, the method of standard addition, is too labor intensive for routine use. Usually, "quench curves" are constructed using a series of standards at various quench levels. Ideally, the standards are prepared using the same cocktail and quench material present in the sample. A variety of mathematical methods is then used to interpolate counting efficiency values for the measured value of the quench indicating parameter.

In practice, commercially prepared sets of quench standards in a toluene based cocktail, yield satisfactory results for ^3H and ^{14}C. In no small part, this is due to the large body of experimental information available on the use of these two nuclides. Indeed, hardware, software, and cocktail formulations have been optimized to ensure reliable results for the broadest possible range of applications. Especially important are experiments using emulsifier cocktails to count aqueous samples.

The use of liquid scintillation counters equipped with multichannel analyzers and the computing power to automatically analyze the results offers considerable convenience for the user. Channel selection in the MCA has replaced adjustment of analog threshold values, and spectral storage features even allow for post run analysis. This has allowed the development of windowless protocols for dpm determinations that are very general and almost as easy to use as "absolute" instruments such as monochromator based spectrophotometers. Nevertheless, the physics of scintillation counting has not changed.

This presentation will focus on quench correction and dpm determination of dual labeled samples containing ^3H and ^{14}C because these are the two nuclides most commonly counted. The discussion can easily be generalized to other nuclides or to the triple label case.

Figure 1 is a schematic drawing of the beta spectra of ^3H and ^{14}C, as well as a composite of the two. The implicit assumption made in the drawing is that the spectra are from a sample at the same quench level. As is well known, the shape of a spectrum is distorted as quench changes. Consequently, accurate quench correction is a prerequisite for accurate dpm calculation, whether of single or multilabel samples.

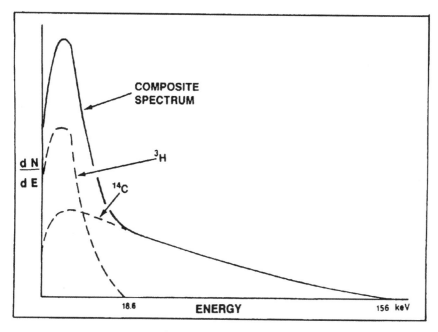

Figure 1. Schematic spectra for ^{3}H, ^{14}C, and dual label sample.

FIXED WINDOW SETTINGS

When using fixed window settings (see Figure 2), the multichannel analyzer is used in a quasianalog fashion. The energy scale is divided up into two regions, A and B, and the counts from these channels are summed up as cpm_A and cpm_B. According to the principle of radionuclide exclusion, the lower limit of region B is chosen above the endpoint energy of the low energy nuclide. This allows direct calculation of the high energy nuclide dpm_H without interference from the low energy nuclide. The efficiency for the high energy nuclide in channel B, E_{HB}, is interpolated from a quench curve.

$$dpm_H = \frac{cpm_B}{E_{HB}} \qquad (1)$$

When calculating dpm_L of the low energy nuclide, the counts need a spill down correction due to the high energy nuclide in the low energy channel, E_{HA}. The efficiency for the low energy nuclide in the A channel, E_{LA}, is also needed.

$$dpm_L = \frac{cpm_A - dpm_H \, E_{HA}}{E_{LA}} \qquad (2)$$

The relative precision for the low energy nuclide in the dpm value can be expressed as proportional to factor analogous to the figure of merit (E2/B) where B is the spill down (= $dpm_H \, E_{HA}$).

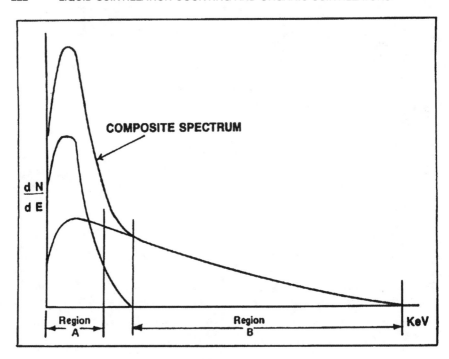

Figure 2. Fixed window region settings.

$$\% \text{ Precision} \approx \frac{E^2_{LA}}{dpm_H \, E_{HA}} \left(\approx \frac{E^2}{B} \right) \tag{3}$$

The relative precision is highly quench dependent because as quench increases, the ^3H efficiency decreases and the spill down from the ^{14}C increases. The dotted lines in Figure 3 represent quench curves obtained using the fixed window method, while the solid lines represent quench curves obtained by automatically adjusting the windows as a function of quench. In our case the quench level is determined by the quench indicating parameter of the external standard tSIE. This method is also called automatic efficiency control (AEC) because the efficiency of the high energy nuclide in the low energy channel is kept constant. As a result, the relative precision is independent of quench level.

The absolute precision of dpm_L, however, still depends on the relative activity of the high energy nuclide, $R = dpm_H/dpm_L$.

$$\text{Precision} \approx \frac{E^2_{LA}}{\dfrac{dpm_H}{dpm_L} \, E_{HA}} = \frac{E^2_{LA}}{R \, E_{HA}} \tag{4}$$

As R increases the precision decreases due to increasing spill down, which leads to an activity ratio dependency.

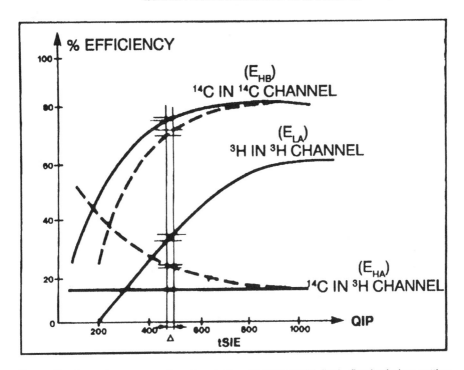

Figure 3. Quench curves: automatic window tracking (solid line), fixed window setting (dashed line).

FULL SPECTRUM DPM

One method of calculating dpm without windows and using the information in the full spectrum is to separate out the counts due the two nuclides with the spectral index of the sample (SIS) (see Figure 4). Since the spectral index of the composite, SIS_T, is a cpm weighted average of the spectral indices of the low energy nuclide, SIS_L, and the high energy nuclide, SIS_H, it is possible to assign the correct proportion of the total count rate, cpm_T, to the low and high energy nuclide, cpm_L and cpm_H.

$$cpm_L = \frac{SIS_H - SIS_T}{SIS_H - SIS_L} \quad cpm_T \qquad (5a)$$

$$cpm_H = \frac{SIS_T - SIS_L}{SIS_H - SIS_L} \quad cpm_T \qquad (5b)$$

While SIS_T is measured, the SIS_H and SIS_L values are interpolated from a quench curve (see Figure 5), as are the full spectrum efficiencies used to calculate dpm values.

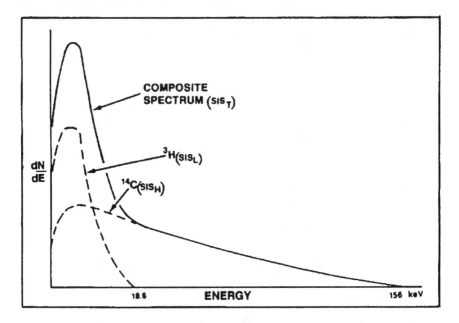

Figure 4. Full spectrum dpm.

$$dpm_L = \frac{SIS_H - SIS_T}{SIS_H - SIS_L} \frac{cpm_T}{E_L} \qquad (6a)$$

$$dpm_H = \frac{SIS_T - SIS_L}{SIS_H - SIS_L} \frac{cpm_T}{E_H} \qquad (6b)$$

We see from the curve that both the efficiency and the SIS interpolations depend on the accuracy of the external standard quench indicating parameter since the curves have a slope. Furthermore, the leveraging effect in Equation 6 on the difference between SIS_T and SIS_H (and to a lesser extent SIS_L) magnifies the effect of any inaccuracies on calculating dpm_L. The dpm calculation is now quite sensitive to experimental conditions affecting the QIP_{ES} differently from SIS.

Figure 6 contains the results of SIS measurements of tritium standards made up in a variety of cocktails. We see that the relationship between the quench indicating parameters of the external standard and the internal standard is similar for all cocktails, but not identical. Using the tSIE (see Figure 7) brings some improvement, but a discrepancy remains.

Since the early work on homogeneous solutions as presented in Birks's[3] monograph reviewed at the International Symposium on Organic Scintillators at Argonne,[4] the photochemical literature has expanded into the area of colloid solutions such as those used in emulsifier based cocktails. Interest in laser dyes, which include some common scintillators, has led to a better understanding of the prompt and delayed components of scintillator emissions. It turns

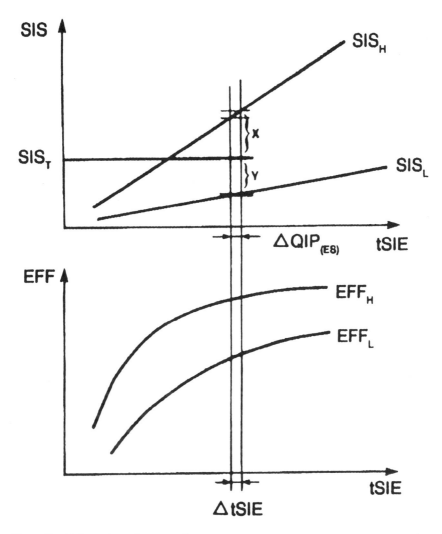

Figure 5. Full spectrum dpm quench curves.

out that the microscopic environment of the nuclide in the sample varies with sample load and sample composition. Furthermore, this effect is different for the prompt and delayed component of the scintillation pulse. Tracks from high energy electrons pass through the different phase regions of emulsifier cocktails so that their effect tends to be averaged out. Low energy electrons, such as those emitted by tritium, do not. This makes it difficult to correlate scintillation efficiencies of high energy electrons with the scintillation efficiency of low energy electrons. Unfortunately, it is precisely this correlation that is at the heart of all spectral interpolation schemes, whether they use the sample spectrum alone or a combination of sample and external standard spectra. This

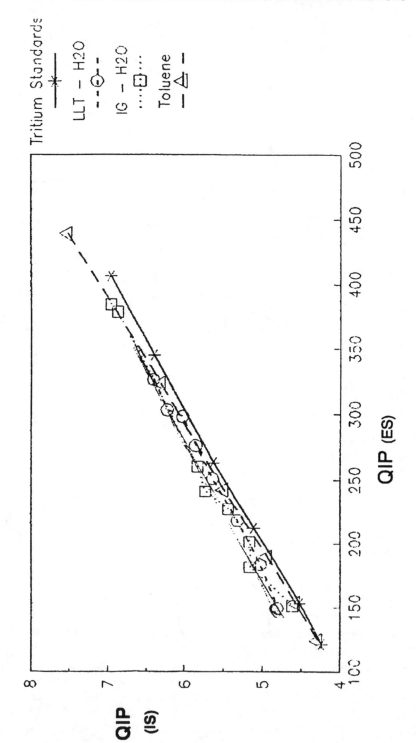

Figure 6. Effect of various cocktails on the relationship QIP$_{(IS)}$ = f(QIP$_{ES}$).

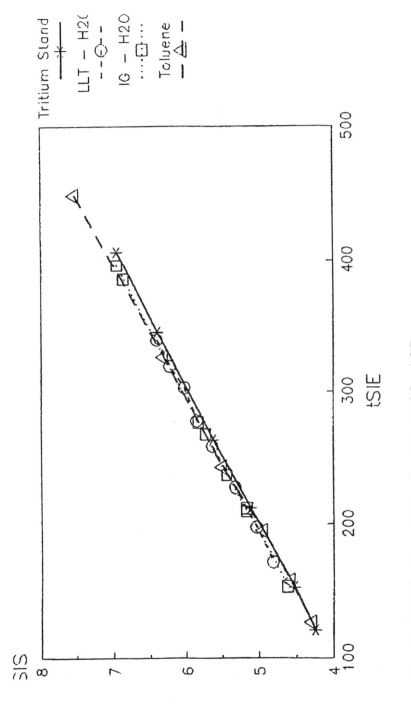

Figure 7.   Effect of various cocktails on the relationship SIS = f(tSIE).

problem becomes even more critical when quench standard and sample are counted in different cocktails. As a consequence, theoretical expressions accounting for ionization quenching break down in the critical low energy region. They are approximations whose validity needs to be verified for each cocktail/sample combination.

SPECTRUM OVERLAY TECHNIQUES

An alternative method of windowless counting that does not reduce the spectrum to a single parameter is the spectral overlay technique (see Figure 8). In this technique, the QIP of the external standard is measured and used to interpolate the spectra of quench standards channel by channel to the quench level of the sample. The spectra are normalized to separate the shape function from the counting efficiency.

The cpm values of the nuclides are adjusted by a method of fitting the least square of the standard spectra to the composite spectrum of the sample. When the best fit is obtained, the cpm values for the nuclides are known, and the dpm values can be calculated directly using the efficiencies.

$$dpm_L = \frac{cpm_L}{E_L} \tag{7a}$$

$$dpm_H = \frac{cpm_H}{E_H} \tag{7b}$$

This numerical method is derived from fields such as optical spectroscopy, which allow for absolute energy calibration. The error in the energy value is at least constant if not negligible. This allows for curve fitting routines to evaluate the goodness of fit based only on the variability of the photon count rate. The energy is assumed to be known exactly. In LSC count spectra, this assumption is not valid. As a result, these algorithms tend to give spurious results when challenged by errors in the energy value.

One way to overcome this problem is to allow the fitting procedure to "adjust" the quench level at which the standard spectra are calculated. Now the quench level is no longer a measured value; it becomes the third parameter in a least square fit of the interpolated spectra to the measured composite spectrum. The added parameter improves the fit. However, the algorithm assumes that the physics of the sample is identical to that of the standards. That is, the effect of sample preparation is not accounted for.

Figure 9 illustrates the consequences of adjusting QIP_{ES}. Two single label tritium standards are shown in toluene and a gel phase cocktail (Insta-Gel). The Insta-Gel spectrum is slightly deformed due to beta self-absorption in the gel phase. In order to improve the least square fit, an overlay technique will raise the external standard QIP to broaden, and thus deform, the whole spectrum.

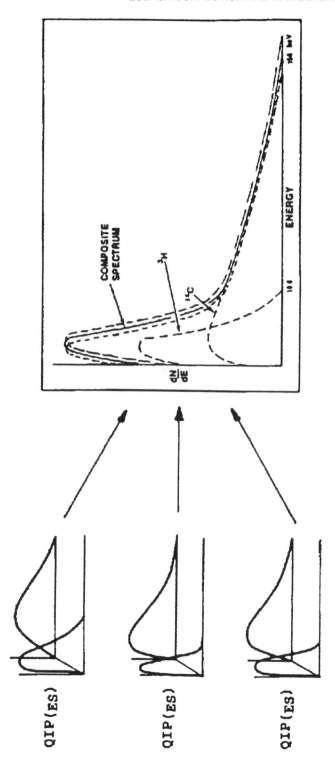

Figure 8. Spectrum overlay technique.

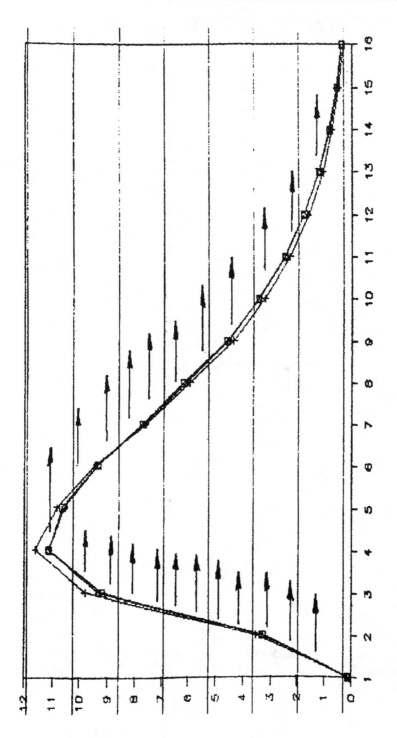

Figure 9.   Consequences of adjusting QIP_ES on spectrum overlay technique.

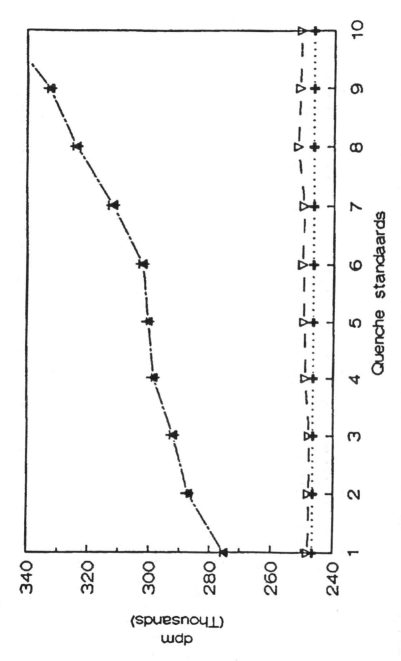

Figure 10. dpm Recovery of toluene samples using OptiFluor standards. + added dpm, ▲ DOT-dpm with OptiFluor library, ▽ AEC.

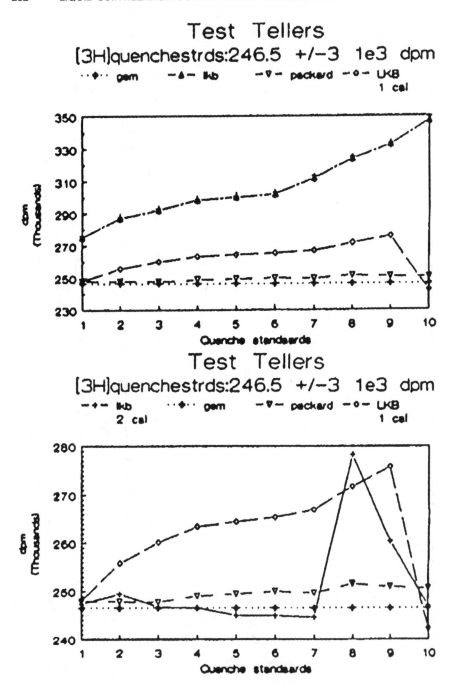

Figure 11.   Effect of "fine tuning" the spectrum library on dpm recovery using the spectrum overlay technique: with one quench standard (top), with two quench standards.

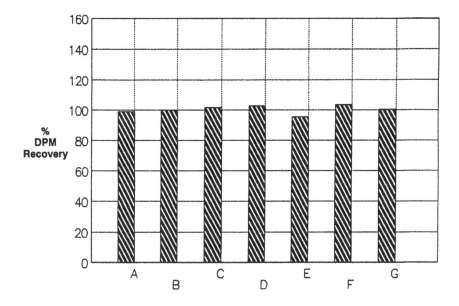

Figure 12.   Effect of sample preparation on % dpm recovery of single label ^{3}H samples using AEC.

Figure 10 contains the test results of toluene single label samples using an Optifluor standard. While the automatic window racking method (AEC) results in values close to the dpm added, the spectral overlay technique leads to spuriously high dpm values. Clearly this technique is not reliable because of its sensitive to sample composition.

One way to improve the performance of the technique is to "fine tune" it by adding a standard prepared in the same cocktail as the sample. This ameliorates the problem (see Figure 11a), but even using a second quench standard (see Figure 11b) does not eliminate it. These results show that most reliable results are still obtained using several standards on a quench curve.

Figure 12 contains the data from a test of tritium samples prepared using different vial types and cocktail compositions. We see that the dpm recovery of Insta-Gel is under 100% due to self-absorption. The values from all other samples are quite close to 100%. In Figure 13, we see the performance of the

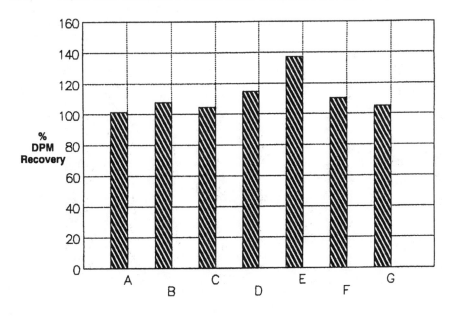

A= 20mL glass Ultima Gold
B= 20mL plastic Ultima Gold
C= 20mL glass Opti-Fluor
D= 20mL glass Opti-Fluor O
E= 20mL glass Insta-Gel
F= 20mL glass Toluene
G= 6mL glass Ultima Gold

Figure 13.   Effect of sample preparation on % dpm recovery of single label ^{14}C samples using spectrum overlay technique.

spectrum overlay technique. The error in the Insta-Gel sample is particularly large due to the effect of self-absorption, but by almost any standard of comparison, the overlay technique does not perform as well as automatic window tracking.

POST COUNT dpm WITH AUTOMATIC WINDOW TRACKING

In order to obtain the most reliable performance and maximum precision it is possible to combine automatic window tracking with automatic region adjustment. In this method, the full spectra of all quench standards are stored in the computer. This allows the software to create quench curves with any region settings, saving the user from having to recount the samples. When the sample is counted, the energy limits of region A are adjusted to obtain the best precision for the activity ratio of that particular sample (see Figure 14a). Once

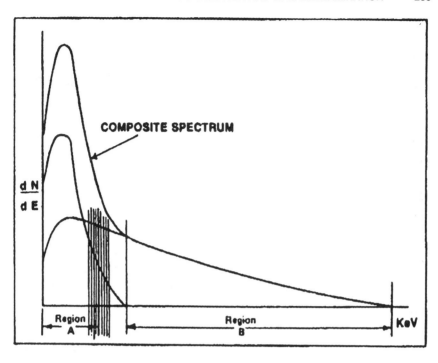

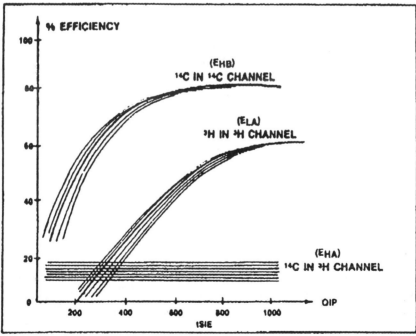

Figure 14. Automatic optimization of region settings for quench and activity ratio using Uniquench-dpm.

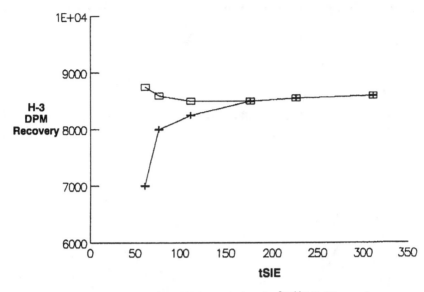

Figure 15.   dpm recovery of AEC and Uniquench-dpm for ³H:¹⁴C (1:15) sample.

the region is determined, the stored quench spectra are used to create a full quench curve under the optimum conditions for that sample (see Figure 14b).

The improved performance obtained by adjusting the region settings is shown in Figure 15. We see that the regions set for the unquenched standards no longer give good results at higher quench levels (lower tSIE). Adjustment of the regions extends the range of good performance considerably.

CONCLUSION

In this review of quench correction methods, we see that addition of computing power to a state of the art liquid scintillation counter will result in improved performance and more convenience for the user. However, the physical problems connected with the relationship between external standard and sample spectra still remain. Software techniques based on a probabilistic analysis of instrument performance and sample composition now allow us to extract the maximum possible amount of information from the sample spectrum. However, for general purpose instruments, they are no substitute for an appropriately chosen set of quench standards and algorithms that respect the limitations on precision set by counting statistics and our limited knowledge of nuclear track photochemistry.

REFERENCES

1. Horrocks, D.L. "Energy per Photoelectron in a Liquid Scintillation Counter as a function of Electron Energy," in *Advances in Liquid Scintillation Counting*, S.A. McQuarrie, C. Ediss, and L.I. Wiebe, Eds., (Edmonton, Alberta, Canada: University of Alberta, 1983), p. 16.
2. Rundt, K., H. Kouru, and T. Oikari. "Theoretical Expressions for the Counting Efficiency and Pulse Height Distribution Obtained for a Beta-Emitting Isotope by a Liquid Scintillation Counter," in *Advances in Liquid Scintillation Counting*, S.A. McQuarrie, C. Ediss, and L.I. Wiebe, Eds., (Edmonton, Alberta, Canada: University of Alberta, 1983), p. 30.
3. Birks, J.B. *The Theory and Practice of Liquid Scintillation Counting* (New York: Macmillan, 1964).
4. Horrocks, D.L., Ed. *Organic Scintillators* (New York: Gordon and Breach, 1966).

CHAPTER 20

Multilabel Counting Using Digital Overlay Technique

Heikki Kouru and Kenneth Rundt

ABSTRACT

The development of electronics has brought multichannel analyzers (MCA's) into liquid scintillation counting. However, the data analysis methods are not on the same level. The currently used window-based calculation methods cannot use all the information measured from the sample. In this presentation a new method called digital overlay technique (DOT) is introduced. This method efficiently uses the information available from the MCA.

In DOT, the shape of the sample spectrum is used for resolving multicomponent samples, by fitting the spectrum of each component to the measured composite spectrum. Furthermore, the fitting of a reference spectrum to the measured spectrum enables the monitoring of sample quality.

In this report the basic principles of DOT are illustrated. Measurements made by a liquid scintillation counter, the Wallac 1410, which uses this principle, are presented. Comparisons are made with a conventional window-based method. The results show that DOT gives the best precision irrespective of the activity ratio of the isotopes.

INTRODUCTION

Dual label liquid scintillation counting is currently a relatively common discipline for which almost all commercial instruments provide some kind of solution. Traditionally, liquid scintillation counting of dual-label samples is performed in two counting windows. The windows may either be fixed, moving,[1] or automatic. Fixed window counting is a fairly uncomplicated procedure accurate enough for samples of moderate quench and activity ratios. If the quench level of the samples within one batch varies immensely, fixed windows have to be exchanged for moving or automatic windows. In the method with moving windows the windows are first set by the user for an unquenched sample. The counter thereafter automatically moves the window limits for each sample prior to counting, according to the quench level of the sample. In the method with automatic windows, there are no counting windows during counting, but only a multichannel analyzer (MCA). The user need not specify any window limits, only isotopes. After counting, the windows are selected from a table of window limits as a function of an external standard quench index for different isotope combinations. Moving windows and automatic windows have the advantage of providing quite robust methods for a wide quench range. However, window methods are not particularly suitable for counting multilabel samples containing more than two isotopes, as the windows have to be defined either by the user (moving windows) or by the manufacturer (automatic windows) through a tedious trial-and-error procedure. Furthermore, separate quench curves for different labeling situations are needed as the windows may be set differently. This means that ordinary single label quench curves just cannot be combined for multilabel counting when using moving or automatic windows.

During the last few years, three new methods for dual label counting have been introduced in commercial counters: Full Spectrum DPM by Packard,[2] Three-Over-Two introduced by Wallac Oy in the RackBeta LS counters in 1987, and Digital Overlay Technique introduced in 1988 by Wallac Oy in the new LS counter 1410. Three-Over-Two counts dual labeled samples with three fixed windows. The count rate of each isotope is determined by solving a set of three equations in two unknowns by using, e.g., the method of least squares. For dual label counting the number of windows with this method is three, or one more than the number of isotopes. This method can, in analogy, be extended to multilabel counting of n isotopes as well by always having the number of windows greater than the number of isotopes n. The number of windows may naturally be much higher than $n + 1$. The ultimate solution is when the number of windows approaches the resolution of the MCA; this is then equal to Digital Overlay Technique (DOT).

DOT is based on a methodology similar to spectral analysis in other branches of spectroscopy and analytical sciences. Oller and Plato[3] were the first to mention spectrum analysis in connection with liquid scintillation counting. In their original paper they did not take quench into consideration though, nor did they describe how the composite spectrum was resolved mathematically. DOT makes use of a multichannel analyzer having 1024 channels.

Mathematically, DOT can be described as follows. Let S_1 and S_2 denote the normalized spectra of isotopes 1 and 2 respectively, and $s_{i,j}$ the normalized count rate of isotope i (i = 1 or 2) in channel j. A normalized spectrum is a spectrum for which the total count rate is equal to 1 cpm. For a dual-label test sample, let U denote the predicted spectrum and u_j the predicted count rate in channel j. Finally, let c_1 and c_2 denote the unknown, *true* count rates of the two isotopes in the sample. Then the following linear equation is valid for each channel:

$$u_j = c_1 \, s_{1,j} + c_2 \cdot s_{2,j} \qquad j = 1 \ . \ . \ m \qquad (1)$$

or, for n isotopes

$$u_j = \sum_i^n c_i \cdot S_{i,j} \qquad\qquad j = 1 \mathinner{\ldotp\ldotp} m \qquad\qquad (2)$$

The number of equations that describe the whole spectrum of m channels is equal to m. For the general case of n isotopes it is most convenient to introduce matrix notation. Let S denote an n by m matrix comprising the n spectra S_1 to S_n and let C denote a vector comprising the n count rates c_1 to c_n. For the spectrum U the equation

$$U = S \cdot C \qquad\qquad (3)$$

is then valid. Let Y denote the measured composite spectrum and, y_j the measured count rate in channel j. In the least squares solution the count rates c_i^* are found by minimizing the squared difference between the measured and the predicted spectrum:

$$\chi^2 = \sum_{j=1}^m (y_j - \sum_{i=1}^n c_i \cdot s_{i,j})^2 \qquad\qquad (4)$$

which is the least squares solution without weights. In matrix notation, the solution is

$$C = (S^T \cdot S)^{-1} \cdot S^T \cdot Y \qquad\qquad (5)$$

The count rates c_i can thereafter be converted to activities a_i (or dpm_i) by using the familiar relation

$$a_i = c_i/e_i \qquad\qquad (6)$$

where e_i is equal to the counting efficiency of isotope i, which must separately be determined from a quench curve. Although Equations 4 and 5 would be accurate enough in favorable situations, they should not be used when the count rates are low and the spectra overlap considerably. The next step is to introduce weighted least squares fit. Let W denote a weight vector comprising the, so far unknown, weight values. Equation 4 then becomes

$$\chi^2 = \sum_{j=1}^m w_j \cdot (y_j - \sum_{i=1}^n c_i \cdot s_{i,j})^2 \qquad\qquad (7)$$

and the matrix solution becomes

$$C = (S^T \cdot W^{-1} \cdot S)^{-1} \cdot S^T \cdot W^{-1} \cdot Y \qquad\qquad (8)$$

W now denotes a weight matrix in which all elements are equal to zero except the diagonal ones, which are equal to the weight values in vector W. The weight values can be chosen in many ways, but the optimal values can only be found through an iterative procedure. A better procedure in this case is to make full use of the maximum likelihood technique,[4,5,6] of which least squares fit is a subclass. The maximum likelihood technique gives the minimum variance solution. The principle of MLT applied to LS counting is shortly as follows: radioactive decay is governed by Poisson statistics and the counts y_j in channel j are Poisson distributed (y_j is the mean value). The probability p_j for detecting k_j counts in channel j is:

$$p_j = (y_j)^{k_j} \cdot e^{-y_j} / k_j! \tag{9}$$

For a spectrum S the joint probability l is the product of p_j for all m channels:

$$l = \prod_{j=1}^{m} p_j$$

Combining Equations 2 and 10 leads to the conclusion that the probability l is a function of the component spectra S_i and the count rates c_i, or $l = l(S_i, c_i)$. In DOT, the shape of the spectra S_i are fixed, but the count rates are iteratively determined so that l achieves a maximum value. The spectra S_i are retrieved from the spectrum library as soon as the quench level of the sample has been determined with the external standard. In the LS counter 1410 there is a spectrum library comprising spectra that cover a large quench region of both chemical and color quench for six common isotopes. This library can be used as such or extended with the users own calibrations or "fine-tunings."[7] The influence of color on counting efficiencies and spectrum shape, and the Wallac color quench monitor, has been dealt with elsewhere by the present authors.[8]

EXPERIMENTAL

This report is an account of a comparison between the traditional two fixed window method and DOT, under similar conditions and in the same instrument. Organic tritium and ^{14}C were used in this study. Two calibration standards with the pure isotopes were prepared for fine tuning the Wallac library. Unknown samples were prepared with the activities shown in Table 1. The scintillation liquid was OptiScint HiSafe, a high flashpoint cocktail. The samples were quenched with carbon tetrachloride so that the counting efficiency of tritium was around 23% and the counting efficiency of ^{14}C around 86%.

The samples were measured repeatedly 120 times and the counting time was one minute. All measurements were made on a typical 1410 LS counter. As this instrument is not equipped with any traditional window methods, the cpm

Table 1. Activity of the Two Isotopes of the Samples used in this Work. "dpm$_{exc}$" is the Activity of the Isotope in Excess and "dpm$_{less}$" is the Activity of the Other Isotope. "Ratio" is the Ratio Between These Two

dpm$_{less}$	dpm$_{exc}$	Ratio
95000	1500	64/1
95000	3000	32/1
95000	6000	16/1
95000	12000	8/1
95000	23800	4/1
95000	47500	2/1
95000	95000	1/1

values in five different windows for each isotope were printed out and converted to dpm off-line.

RESULTS AND DISCUSSION

The aim of this test was to find experimental evidence for our theoretical assumption that DOT is at least as good as or even better than any window method. For this reason, the standard deviation of the 120 dpm values was computed. The results are presented in Figures 1 to 4 as relative standard deviation (rDev) as a function of the activity ratio. rDev is defined according to the equation:

$$rDev = Dev_W / Dev_{DOT}$$

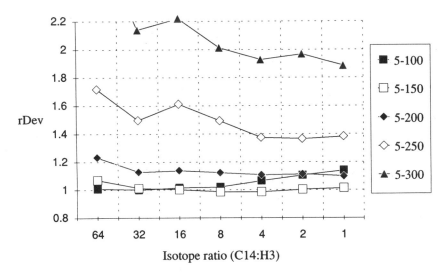

Figure 1.   The relative standard deviation of dpm for ^{14}C as a function of the activity ratio for a number of low window limits. rDev is defined in the text.

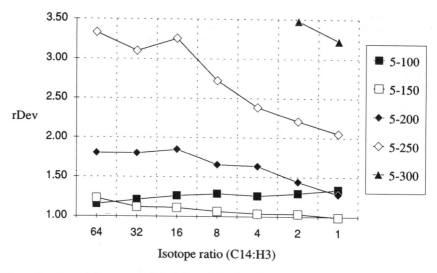

Figure 2. The relative standard deviation of dpm for tritium as a function of the activity ratio for a number of low window limits. rDev is defined in the text.

wherein Dev_W and Dev_{DOT} are equal to the standard deviations of the 120 dpm values for the two isotopes computed by using a windows method and DOT, respectively. In Figure 1, ^{14}C is in excess, and rDev is computed for ^{14}C while in Figure 2, ^{14}C is also in excess, but rDev is now computed for tritium. Accordingly, in Figures 3 and 4, tritium is in excess and rDev is computed for ^{14}C and tritium, respectively. These figures also mentioned the window limits of the

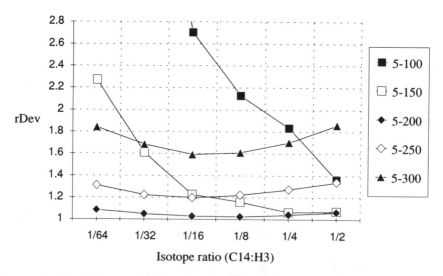

Figure 3. The relative standard deviation of dpm for ^{14}C as a function of the activity ratio for a number of low window limits. rDev is defined in the text.

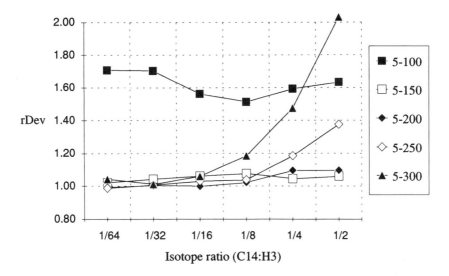

Figure 4. The relative standard deviation of dpm for tritium as a function of the activity ratio for a number of low window limits. rDev is defined in the text.

lower (A) of the two windows A and B. Window B extends from the upper limit of A up to channel 650.

The data in Figures 1, 2, and 4 states that, at the quench level assessed here, the narrow A window 5-150 gives a standard deviation almost as low standard deviation as DOT. However, Figure 3 shows that the window 5-150 is not the best when tritium is in excess. The data in these figures disclose one of the limitations of the traditional two-window method: there is no single window setting that gives as low standard deviations ad DOT for a wide range of activity ratios.

One important conclusion that can be drawn from these figures is that selecting the correct limits for window A is very important when using two windows. If using an automatic method like moving windows, the unquenched windows must be set correctly and the user can do no more than hope that the program always chooses the best windows settings at different quench levels. With DOT, the user has no such worries, as the method always gives lower or equally low deviations as any window method.

SUMMARY

The advantages of DOT are the following:

1. Statistically, DOT is always better than or equally good as any combination of moving or automatic windows.
2. The calibrations are of general characters and may be used in any connection, single label or multilabel.

3. Counting is optimized for all conditions, such as varying quench levels or activity ratios.
4. The method is similar in the general multilabel case as in the special case of dual label.
5. The reference spectra are also used in a quality control manner to ensure that the unknown sample does indeed contain the assumed radioisotopes; they are also used to determine, in a qualitative manner, what radioisotopes are present in completely unknown samples.

REFERENCES

1. Nather, R.E., U.S. Pat. No. 4,029,401 (1977).
2. van Cauter, G.C., L.J. Everett, and S.J. DeFilippis, European Patent Appl. No. 0 202 185 (1986).
3. Oller, W.L. and P. Plato, "Beta Spectrum Analysis: A new Method to Analyze Mixtures of Beta-Emitting Radionuclides by Liquid Scintillation Techniques," *Int. J. Appl. Rad. Isot.*, 23:481-485 (1972).
4. Orth, P.H.R., W.R. Falk, and G. Jones. "Use of the Maximum Likelihood Technique, for Fitting Counting Distributions," *Nucl. Inst. and Meth.*, 65:301-306 (1968).
5. Ciampi, M., L. Daddi, and V. D'Angelo. "Fitting of Gaussians to Peaks by a Maximum Probability Method," *Nucl. Inst. and Meth.*, 66:102-104 (1968).
6. Awaya, T. "A New Method for Curve Fitting to the Data with Low statistics Not Using the χ^2-Method," *Nucl. Inst. and Meth.*, 165:317-323 (1979).
7. Kouru, H. "A New Quench Curve Fitting Procedure: Fine-Tuning of Spectrum Library," in *New Trends in Liquid Scintillation Counting and Organic Scintillators*, (1989).
8. Rundt, K., T. Oikari, and H. Kouru. "Quench Correction of Colored Samples in LSC," in *New Trends in Liquid Scintillation Counting and Organic Scintillators*, (1989).

CHAPTER 21

A New Quench Curve Fitting Procedure: Fine Tuning of a Spectrum Library

Heikki Kouru

ABSTRACT

In liquid scintillation counting the counting efficiency and the spectrum of the sample depend on the quench level; quench calibration is needed to establish the true activity of the sample. The quench curve relates the counting efficiency to an appropriate quench indicating parameter. The quench curve is formed by measuring a number of reference samples with known activity and varying quench levels. Typically six to ten reference samples are required in quench calibration.

This presentation shows that the number of reference samples can be reduced by using model based quench curve fitting. The model uses a large set of reference samples, which form the spectrum library of the isotope. The fine tuning of the spectrum library is the model adjustment procedure in which the spectrum library data is modified to fit to the data of the new calibration.

The principles of the fine tuning are illustrated and a theoretical example is calculated. Measurements made by a liquid scintillation counter that uses the described principle, the Wallac 1410, are presented and a comparison with a conventional counter is made.

INTRODUCTION

The quench level of liquid scintillation counting samples is often variable. Quenching decreases the number of photons detected after one radioactive decay. In instruments the variation of quench level is seen as variations of the counting efficiency and shifts of the pulse height energy distribution from sample to sample of the same isotope. The reduction in the amount of photons shifts the pulse height energy distribution or the spectrum to the smaller pulses and decreases the counting efficiency. Figure 1 shows three ^{14}C spectra of different quench levels.

To relate the disintegration rate of a sample to the count rate it is necessary to determine the quench level of the sample and correlate it to the counting efficiency. There are different methods to estimate the quench level of a sample, but most of them rely on the effect of quenching on the position of the

247

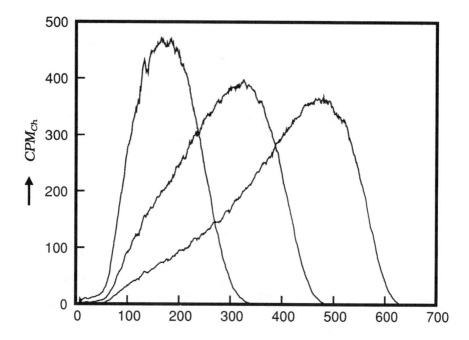

Figure 1. Three ^{14}C spectra of different quench levels.

spectrum on the amplitude scale. A common method uses an external gamma ray source or an external standard and measures the shift in pulse amplitudes of the spectrum resulting from Comptom electrons scattered by this external standard. The end point of an external standard spectrum ($Sqp(E)$) can be used as a quench indicating parameter.

The relationship between a quench indicating parameter and the counting efficiency of a radionuclide is established by measuring reference or standard samples of known activity and expressing the counting efficiency of these samples as a function of the quench indicating parameter. The whole procedure is called quench calibration and the resulting function, which correlates counting efficiency and the quench indicating parameter, is called a quench curve.

Figure 2 shows tritium quench curves for different scintillation solutions. Because typical quench curves are not linear but curved, normally six to ten reference samples have to be prepared and measured to make a quench calibration that covers the desired quench level range. Figure 2 shows that even if the quench curves differ from each other, the basic shape of the curves is the same. This is especially true for one radionuclide, but the same shape can be used as a good approximation for isotopes with energy of the same order of magnitude.

Traditionally quench curves have been formed by fitting a polynomial or a spline function to the quench calibration data or by making a linear interpola-

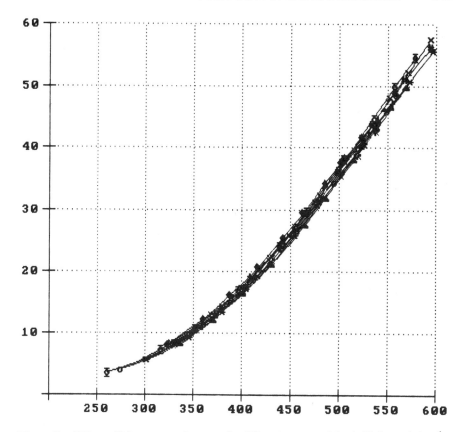

Figure 2. Tritium efficiency quench curves for different commercial scintillation solutions[1].

tion between the data points. In this presentation a quench curve producing method, which uses the known form of the quench curve is introduced; the method is called model based quench curve fitting.

This introduction is concerned with the efficiency quench curve, but the spectrum can be treated in an analogous way. The spectrum can be expressed as parametric form, and the parameter values can be expressed as a function of the quench indicating parameter. Another possible way to treat spectrum data is to normalize the spectrum and follow the variation of the spectrum proportion in each channel.

The preservation of the quench curve form can be realized in many ways. It is possible to develop a theoretical quench curve model with a few parameters and find the solution of these parameters for the particular quench data. This kind of procedure can preserve the right form of the curve if the number of free parameters is low enough. Another more straightforward possibility is to use an experimental model. The model adjustment should fulfill two demands: flexibility and stiffness. The procedure should be flexible enough to follow the natural variation and stiff enough to preserve the right shape of the curve. One possible solution is to make a transformation of the variables, make the fit on them, and then return to the real values by inverse transformation.

THEORETICAL EXAMPLE

As a theoretical example, we may consider a situation where the true efficiency quench curve is assumed to be a second degree polynomial. The quench curve model is then also a second degree polynomial as shown in Figure 3. Figure 3 also shows the measured counting efficiency and sample quench index value pairs for two reference samples, which are assumed to have 4 and 5% higher counting efficiency values than predicted from the model. Figure 4 shows the same model curve and reference data points after a transformation on the efficiency scale. Here the curve fitting is not made on efficiency values but on the relative deviations of efficiency from the model. A linear interpolation is used as a fitting method. Figure 5 shows the corresponding fit in untransformed coordinates; it also shows a normal linear interpolation without transformation. The comparison between model based fitting and normal fitting is shown in Figure 6, where the relative dpm errors of both systems are given as a function of counting efficiency. The error in normal fitting is three times greater than the one using model based estimation.

EXPERIMENTS

The Wallac 1410 counter uses a model based fitting principle for quench calibration. As a model, the 1410 has built in spectrum libraries for the most common isotopes. The spectrum library includes counting efficiencies, whole sample spectra and external standard spectra, external standard based total quench indication parameters, and external standard based color quench indication parameters of reference samples. The data is retrieved from the spectrum library using both quench parameter as keys. The quench calibration in the 1410 effects both counting efficiency and spectra, and it is called fine tuning of the spectrum library.

A tritium quench series of nine samples was made using Hisafe liquid scintillation cocktail and nitro methane as quenching agent. Using two samples of the series as references, a quench calibration has been made in the 1410 and a RackBeta 1219. In all measurements 0.5% one sigma counting statistics were used. RackBeta 1219 uses linear interpolation, as only two reference samples are used. Figure 7 shows the efficiency quench curve made in the 1410. Figure 8 shows the relative error of activities in the quench series samples, which have been measured as unknown samples. The systematic errors are one order of magnitude lower in System 1410 when using the fine tuning principle.

To illustrate the robustness of the 1410 fine tuning principle an experiment was conducted, where the spectrum library of another isotope is used as a

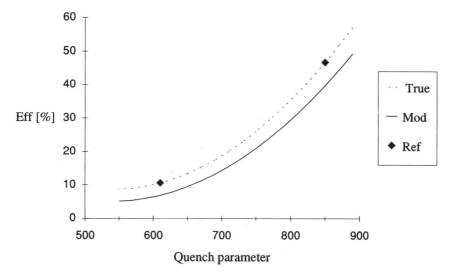

Figure 3. Mod is a tritium efficiency quench curve model, Ref is used for measured quench parameter—counting efficiency values of reference samples—and True is the real efficiency quench curve corresponding to the reference samples.

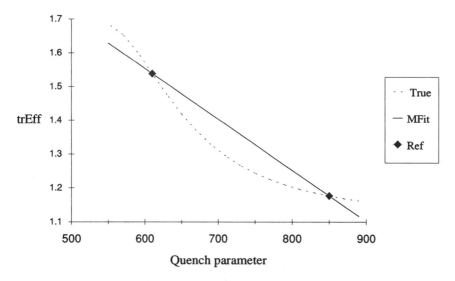

Figure 4. Transformed efficiency quench curves. The values of trEff-axes are calculated by dividing the efficiency with the efficiency of the model curve of the corresponding quench parameter value. MFit is the model based quench curve in this coordinate system. This quench curve was made using a linear interpolation of the efficiency quench curve in this coordinate system. Symbols True and Ref are defined in Figure 3.

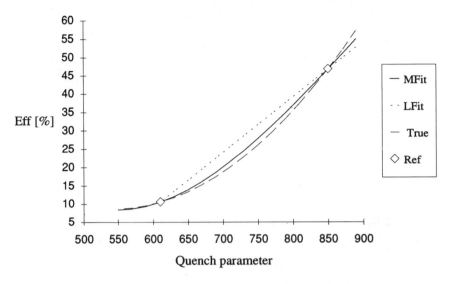

Figure 5. LFit is a quench curve made using a linear interpolation without the model. Symbols MFit, True and Ref are defined in Figures 3 and 4.

model in quench calibration. A quench series of nine ^{14}C samples was made using Hisafe cocktail and nitro methane as a quenching agent. Quench was calibrated in the 1410 using three ^{14}C samples, having the ^{45}Ca spectrum library as a model. Figure 9 shows the efficiency quench curve. All nine samples were measured as unknowns, and the dpm errors are shown as a function of sample

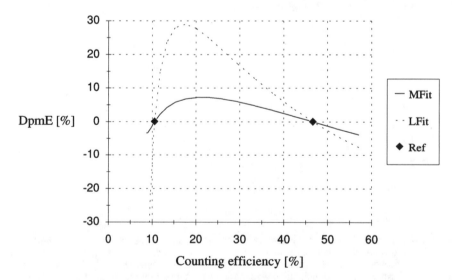

Figure 6. Relative DPM errors, DpmE as a function of counting efficiency for the model based quench curve and for the quench curve not based on a model.

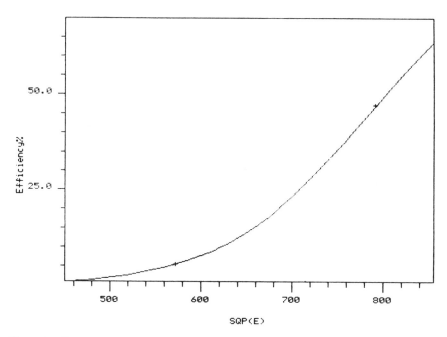

Figure 7.   Tritium efficiency quench curve of the 1410 using two reference samples.

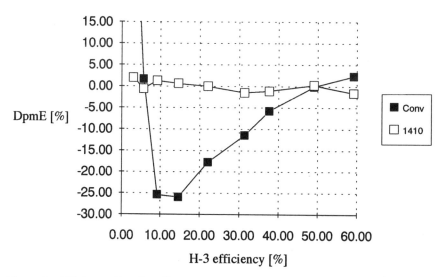

Figure 8.   DPM error, DpmE for ^3H samples as a function of the counting efficiency of the samples when two reference samples were used in quench calibration. The counting efficiencies of the reference samples were 6.7% and 47.3%. *1410* is the 1410 and Conv is the 1219.

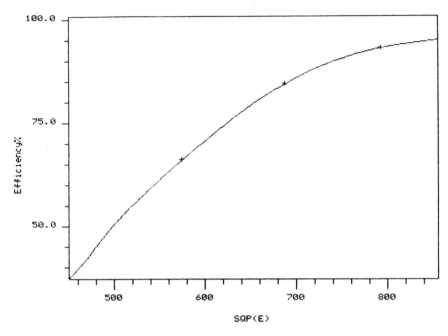

Figure 9.   Carbon efficiency quench curve of the 1410 using three reference samples and ^{45}Ca spectrum library.

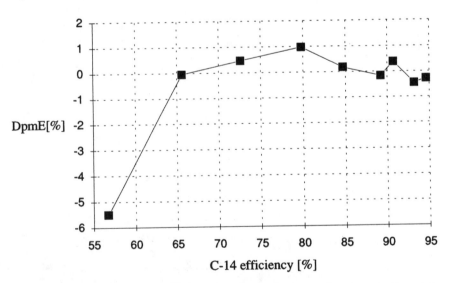

Figure 10.   DPM error, DpmE for ^{14}C samples as a function of the counting efficiency of the samples when three reference samples and ^{45}Ca library were used in quench calibration. The counting efficiencies of the reference samples were 93%, 84%, and 65%.

quench level in Figure 10. In every case the interpolation error is under 2%, hence a model could also be selected from a spectrum library of some other isotope. Of course, the library of a radionuclide with a similar energy would be preferred.

CONCLUSIONS

Using the model based quench calibration or spectrum library fine tuning considerably increases the accuracy of the results when only a few reference samples are used.

REFERENCES

1. Rundt, K. "On the Determination and Compensation of Quench in Liquid Scintillation Counting," PhD Thesis, Åbo Akademi, Turku, Finland (1989).

CHAPTER 22

The Effect on Quench Curve Shape
of the Solvent and Quencher
in a Liquid Scintillation Counter

Kenneth Rundt

ABSTRACT

When preparing quench calibration curves, commercial sealed standards are often used, although it is advisable to use the same composition, scintillation liquid, and sample as the unknowns. Commercial sealed standards usually contain a toluene or xylene based cocktail and carbon tetrachloride or nitromethane as quencher. However, the unknown samples may contain another solvent and other chemical quenchers. The solvent has become more significant with the introduction of new "safe" solvents that behave differently from the traditional toluene solvent.

As reported in earlier literature, different solvents and chemical quenchers lead to different quench curves, but no physical explanation has been given. In this work, quench curves for several commercial cocktails with chemical quenchers and one color quencher have been measured. The counting efficiency and the mean pulse height of the isotope (^{3}H) were recorded for different amounts of quencher and with varying coincidence resolving time of the counter. The differences in quenching characteristics can be explained as differences in the scintillation pulse shapes. The shape of the scintillation pulse is primarily dependent on the way the excitation energy is deposited in the solvent and on how the chemical quenchers act on the excited solvent molecules. The data indicate that there are mainly two different chemical quenching mechanisms: quenching of the higher excited states, which are precursors of the triplet states, and quenching of the lower singlet states.

INTRODUCTION

As liquid scintillation counting is frequently used for quantitative analytical measurements of low energy isotopes, such as tritium and ^{14}C, it is extremely important to know the counting efficiency of the samples. For that purpose, most counters today have automatic methods for determining the quench level and the counting efficiency. Generally, an empirical relationship between the counting efficiency and the quench index must be established at first, e.g., by measuring a set of quenched standards. A question that often arises is how to prepare this set of quenched standards; which scintillation cocktail and which

257

quencher substance should be used in the standards? The usual advise to users who want to suppress the risk of systematic errors is to use the same composition in the standards as in the unknowns. This means using the same vials, same scintillation cocktail (same volume), and same quencher. Using the same vials is mostly not a problem, but the two other factors are worth some consideration.

During the last few years the number of new solvents used in liquid scintillation cocktails has increased remarkably through the introduction of high flashpoint solvents. Typically these solvents have a flashpoint above 60°C. The purpose of these solvents is to make scintillation counting less hazardous. The most common scintillation liquids today are still based on toluene or xylene, both of which are rendered as toxic and hazardous chemicals. The new solvents not only have a low vapor pressure, they are generally considered less toxic than toluene. In a laboratory, cocktails based on both traditional solvents and on the new safe solvents may be in use at the same time. Commercial sealed calibration standards usually contain a traditional solvent. Because of this, a natural question that arises is whether or not quench curves based on traditional cocktails may be used together with these new safe cocktails. In current literature, there is very little or no information available on the quenching behavior of these new solvents as compared to the traditional ones. There are also remarkably few reports in current literature comparing quench curves from different quenchers. Wunderly[1] has investigated a number of substances in relation to the sample channels ratio (SCR) and the H-number used in Beckman Instruments counters as quench level index. The H-number is equal to the shift of the ^{137}Cs compton edge as compared to the unquenched position. Wunderly found differences between the assessed quenchers in terms of counting efficiency vs. SCR or H-number. He could not explain this nor could he propose any solution on how to avoid it. It is also well known at Wallac Oy that users of the Wallac Rackbeta LS counters have encountered this problem without being able to explain it or reduce the errors involved. Furthermore, the new safe cocktails introduce one more source of uncertainty. The present work has been undertaken in order to cast some light on these problems and show how the errors can be minimized by improving the LS counter. The aim of this work has also been to increase the general knowledge of the physics behind liquid scintillation and quenching.

EXPERIMENTAL

Two different measurements were performed. At first, quench curves for five commercial cocktails were recorded. The cocktails were: xylene based Lipoluma from Lumac AG, Netherlands, toluene based OptiScint 'T', pseudo cumene based OptiScint Safe, di-isopropylnaphthalene based OptiScint HiSafe, and OptiPhase HiSafe II (containing emulsifiers, but no added water).The last four are trademarks of Wallac Oy. The isotope spectra were recorded at three coincidence resolving times (15, 72, and 800 nsec). The isotope mean pulse height of the isotope spectrum, SQP(I), was used as the quench index. The quenching agent was carbon tetrachloride.

In the second set of measurements, the quench behavior of eight quenchers was recorded using the toluene based cocktail (OptiScint 'T', Wallac Oy). The uncolored substances were: nitromethane, carbon tetrachloride, methanol, acetone, acetophenone, dibutylamine, and methyl benzoate. One colored substance, the commercial yellow dye Sudan 1, was also used. All quenchers were of *pro analysi* quality. A dilution series was also made by diluting the OptiScint 'T' with pure toluene, resulting in a decreased concentration of the fluors.

Normal 20 mL glass vials where used, with 10 mL of scintillation liquid. All samples were in equilibrium with air; thus containing the same amount of oxygen-quench. Wallac Internal Standard Capsules 1210–120, containing tritium labeled cholesterol, were used for dispensing the activity. All measurements where done on a prototype of the new 1410 LS counter from Wallac Oy, which has an adjustable coincidence resolving time.

RESULTS

The results of the solvent measurement are shown in Figures $1-3$. Figure 1a shows the quench curves for all the cocktails at normal coincidence resolving time (15 nsec), and Figure 1b shows the relative error in DPM when the toluene based calibration curve was used as a reference for all cocktails. The DPM error is defined according to the equation

$$DPM_{Err} = 100 \% \cdot (DPM_0 - DPM_1) / DPM_1$$

wherein DPM_1 is the activity predicted by the toluene calibration qurve and DPM_0 is the true activity. Figures 2 and 3 show the DPM errors at longer coincidence resolving times.

A quencher other than CCl_4 may lead to quench curve behavior different than that described above. The next figures show the influence of quencher on the quench curve shape in a toluene based cocktail. Figure 4a shows a plot of all nine quench curves for tritium in the case of normal coincidence resolving time, while Figure 4b shows the same data plotted as relative error in DPM when a CCl_4-based calibration curve is used as a reference. The curves in Figures 4a and 4b show that at normal coincidence resolving time, and in the case of tritium, the nine quenchers can be divided into two groups. One group comprises methanol, carbon tetrachloride, nitromethane, and dibutylamine situated above, and another group, comprises the toluene dilution, acetone, methyl benzoate, acetophenone, and Sudan 1 situated below. The toluene dilution shows the greatest deviation. From these curves one can see that nitromethane or carbon tetrachloride, which are often used for preparing quenched standards, may not be the correct choice when the unknown samples contain ketones like acetone or acetophenone.

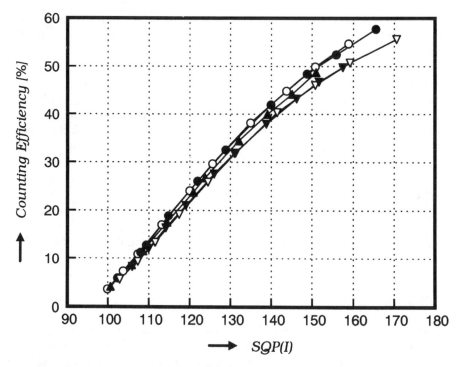

Figure 1a.   Tritium quench curves (counting efficiency as a function of quench index) for the 5 different cocktails and a coincidence resolving time of 15 nsec.

The relative error for some of the quenchers depends very much on the coincidence resolving time. Figures 5 and 6 show the relative DPM error as a function of SQP(I) for two more resolving times, 40 and 72 nsec. At 72 nsec all of the curves join except the color curve and the toluene dilution. The color curve does not in fact depend very much on the coincidence resolving time.

In normal commercial instruments the coincidence resolving time has to be quite short (≈ 15 nsec) in order to decrease the number of random coincidences from, e.g., chemiluminescence. As can be seen from Figure 1b, in single label counting, the relative error in the DPM will be a constant value independent of quench level when measuring unknown samples containing other solvents than toluene against a toluene based calibration curve in a normal instrument using SQP(I) as the quench index. This can be satisfactory if the *absolute* activity is of minor interest and the results are used for comparison only. With different quenchers this is not so as the error varies with the quench level from 0 to –40% for a sample quenched with a ketone measured against a CCl_4 calibration curve. When measuring dual labeled samples, the errors in both cases will be even bigger, as the shape of the isotopic spectra differ and the spill over curves look completely different. Figures 7a and 7b

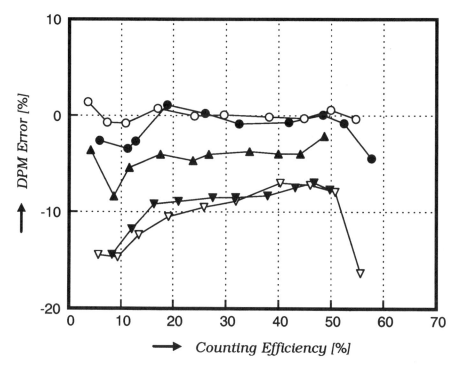

Figure 1b. The *DPM* error as a function of the tritium counting efficiency when using the toluene based quench curve (OptiScint 'T') for computing DPM values. The symbols are: ○ = OptiScint 'T', ● = Lipoluma, ▲ = OptiScint Safe, ▽ = OptiScint HiSafe, ▼ = OptiPhase HiSafe II.

shows the influence of coincidence resolving time on spectral shape for 1. an OptiScint 'T' sample and 2. an OptiScint HiSafe sample.

DISCUSSION

From the curves in Figures 1–6 one general conclusion can be drawn: the shorter the coincidence resolving time, the bigger the difference between the various solvents and quenchers. Why then is the coincidence resolving time critical to the quench curve behavior? The lowest excited singlet state of the aromatic solvent and most fluors have quite short lifetimes and will result in prompt light emission, a few nanoseconds after the disintegration. Triplet states, superexcited states, and ionized states have, in comparison to the singlet states, long lifetimes. These states may either relaxate, through internal conversion, to the ground singlet state and liberate energy in the form of heat, or they may end up in an excited singlet state which can lead to light emission several hundreds of nanoseconds after the disintegration. The first light burst is usually called the prompt pulse, and the second emission is called the

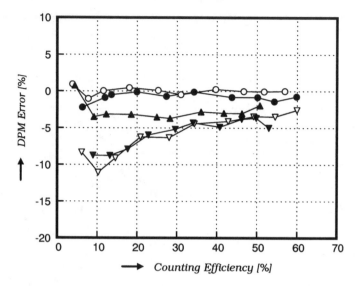

Figure 2. The DPM error as a function of the tritium counting efficiency when using the toluene based quench curve and a coincidence resolving time of 72 nsec. The symbols are the same as in Figure 1.

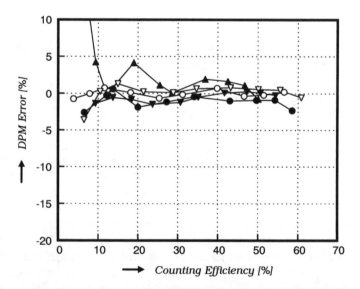

Figure 3. The DPM error as a function of the tritium counting efficiency when using the toluene based quench curve and a coincidence resolving time of 800 nsec. The symbols are the same as in Figure 1.

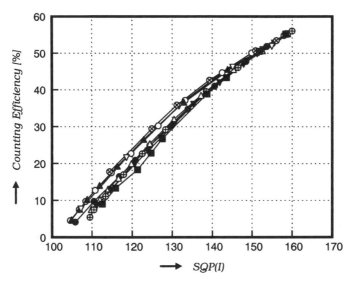

Figure 4a.   Tritium quench curves for OptiScint 'T' and 9 different quenchers and a coincidence resolving time of 15 nsec.

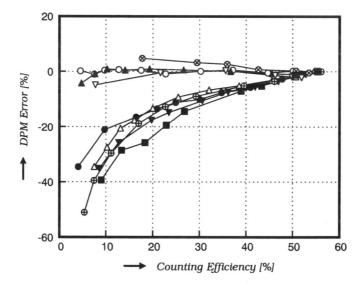

Figure 4b.   The DPM error as a function of the tritium counting efficiency when using the carbon tetrachloride quench curve for computing DPM values. The symbols are: ○ = methanol, ○ = carbon tetrachloride, ▲ = nitromethane, ▽ = dibutylamine, ■ = toluene dilution, ● = acetone, ▼ = methyl benzoate, △ = acetophenone and ○ = Sudan 1 (a yellow dye).

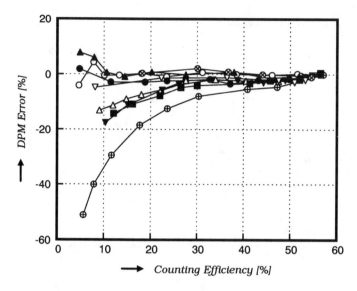

Figure 5. The DPM error as a function of the tritium counting efficiency when using the carbon tetrachloride quench curve and a coincidence resolving time of 40 nsec. Symbols as in Figure 4.

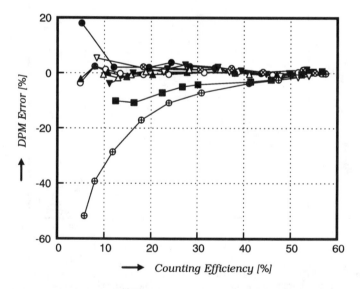

Figure 6. The DPM error as a function of the tritium counting efficiency when using the carbon tetrachloride quench curve and a coincidence resolving time of 72 nsec. Symbols as in Figure 4.

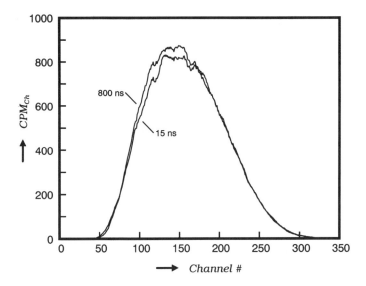

Figure 7a. The tritium spectrum of an air-quenched OptiScint 'T' sample at two different coincidence resolving times.

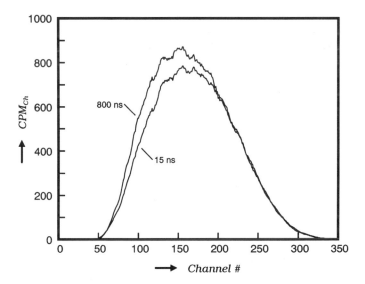

Figure 7b. The tritium spectrum of an air quenched OptiScint HiSafe sample at two different coincidence resolving times.

delayed pulse (see, e.g., Fuchs, et al.[2,3]). When the coincidence resolving time is so short that a noticeable part of the delayed component is situated after the coincidence analyzer closes, one would expect to find mainly a reduced number of small pulses comprising only a few photoelectrons. The reason for this is as follows: when the number of photoelectrons produced decreases, the spacing in time between the photoelectrons increases if the overall pulse length remains constant (compare randomly dividing a distance into ten parts or two parts). Thus the chance increases that the time delay between the first electron, starting the coincidence analyzer, and the second electron, causing a valid coincidence pulse, will be longer than the coincidence resolving time. Hence, for a sample at fixed quench level, as the number of small pulses is decreased, the mean pulse height is shifted upwards and the counting efficiency downwards. This behavior can be observed in all the data presented here and especially in Figure 7.

Solvents

As all investigated cocktails may be expected to contain principally the same fluors, like PPO and bis-MSB, the reason for the difference in quench curve shape lies in the behavior of the solvents. It is known that naphthalene crystals have very long scintillation pulses[4] and that solvents based on naphthalene derivatives have longer pulses than mono-aromatics.[5] There may be three reasons for the longer pulse:

1. The electron may cause more extensive ionization in the naphthalene based solvent than in xylene. Ionization leads to formation of triplets through ion and electron recombination. The triplet states may eventually form singlet states through the triplet-triplet annihilation process.
2. Longer lifetimes of the excited singlet states may cause longer pulse decay times.
3. The bulkier molecule may slow the diffusion rates of the excited molecules.

Even at the longest coincidence resolving time (800 nsec), there is still some difference between the curves. But this difference may be due to other factors than the pulse shape. In general, with "infinitely" long coincidence resolving time, both the counting efficiency (E) and the mean pulse height (F) may be considered as functions of three parameters:[6]

$$E = E(a,b,X) \text{ and } F = F(a,b,X)$$

The parameter a is equal to the scintillation efficiency of the liquid, and thus depends on both the solvent and the concentration of quenching agent, while the parameters b and X are dependent on solvent only. The parameter b reflects the degree excited states are quenched by ionized states in the vicinity, while X, the mean excitation potential, has a direct effect on the stopping power of the solvent, and thus also on the density of ionized and excited states.

Variations in the mean excitation potential X and the ionization quenching parameter b result in variations of the quench curve shape.

Quenchers

What then is the explanation for the behaviors depicted in Figures 4–6? The answer might lie in the relation between the energy levels of a certain quencher and the solvent (toluene). This phenomenon has been extensively studied by, e.g., Fuchs, et. al.,[2,3] who have proposed the theory that certain quenchers, which can be classified as "electron scavengers" (like oxygen and carbon tetrachloride) and have their excited levels above the energy levels of the solvent, may interact with the higher energy states of the solvent, and thus cause a reduced relative contribution of the delayed component. Compounds having their singlet energy levels just below the corresponding energy levels of the solvent may interact by accepting the excitation energy from the singlet states of the solvent and decrease the intensity of both the prompt component and the delayed component. The different forms of "impurity" or chemical quenching has also been discussed extensively by Birks.[7]

The fact that the color quench curve deviates from all the others, at all coincidence resolving times, is related to the special characteristics of color quench.[8] Apart from the other quench modes, color quench depends on the spatial coordinates of the disintegrations. Sudan 1 also has a small amount of chemical quench, otherwise it would behave exactly like the toluene dilution, but the main amount of quenching comes from the color. The results show that color quench can not be handled the same as other forms, just by prolonging the coincidence resolving time.

CONCLUSION

Different quench curves are produced by different solvents and quenchers, when using a quench index reflecting the position of the pulse height spectrum of the dissolved isotope (as the mean pulse height or samples channels ratio), this can be explained on the basis of the shape of the scintillation pulse. The pulse shape is mainly dependent on two factors: 1. the distribution of different excited states of the solvent molecules when a β-electron traverses the solution and 2. the probabilty of quenching certain excited states in the solution. If the probabilty of forming higher excited states (e.g., Rydberg states) or ionized states is large, then the scintillation pulse will be streched out in time. Quenchers capable of quenching the triplet and higher excited states, combining with ionized states, or acting as electron scavengers very effectively quench the delayed component of the pulse. Substances capable of quenching only the lowest singlet levels however, quench both the prompt and the delayed component to an equal degree. The differences in the quench curves caused by this effect can be reduced by increasing the coincidence resolving time of the counter to above 40 nsec.

REFERENCES

1. Wunderly, S. W. "Evaluation of Liquid Scintillation Counting Accuracy: Effect of Various Chemical Quenching Agents and Effect of Milky Sample Preparations," in *International Conference on Advances in Scintillation Counting*, S. A. McQuarrie, C. Ediss, and L. I. Wiebe, Eds. (Edmonton, Canada: University of Alberta, 1983), pp. 376–386.
2. Fuchs, C., F. Heisel, R. Voltz, and A. Coche. "Scintillation Decay and Absolute Efficiencies in Organic Liquid Scintillators," in *Organic Scintillators and Liquid Scintillation Counting*, D. L. Horrocks and C.-T. Peng, Eds. (New York: Academic Press, 1971), pp. 171–186.
3. Fuchs, C., F. Heisel, and R. Voltz. "Formation of Excited Singlet States in Irradiated Aromatic Liquids," *J. Phys. Chem.*, 76(25):3867–3875 (1972).
4. Birks, J. B. and R. W. Pringle. "Organic Scintillators with Improved Timing Characteristics," *Proc. Roy. Soc. Edinburg (A)*, 70:233–244 (1971/72).
5. Koike, Y. "Measurement of Decay Times of Loaded Liquid Scintillators," *Nucl. Instr. Meth.*, 97:443–447 (1971).
6. Rundt, K., H. Kouru, and T. Oikari. "Theoretical Expressions for the Counting Efficiency and Pulse Height Distribution Obtained for a Beta-Emitting Isotope by a Liquid Scintillation Counter," in *International Conference on Advances in Scintillation Counting*, S. A. McQuarrie, C. Ediss, and L. I. Wiebe, Eds. (Edmonton, Canada: University of Alberta, 1983), pp. 30–42.
7. Birks, J. "Impurity Quenching of Organic Liquid Scintillators," in *Liquid Scintillation Counting, Vol 4*, M. A. Crook and P. Johnson, Eds. (London: Heyden & Son Ltd, 1977), pp 3–14.
8. Rundt, K., T. Oikari, and H. Kouru. "Quench Correction of Colored Samples in LSC," in *New Trends in Liquid Scintillation Counting and Organic Scintillators*, 1989.

CHAPTER 23

^{67}Ga Double Spectral Index Plots and Their Applications to Quench Correction of Mixed Quench Samples in Liquid Scintillation Counting

Shou-li Yang, Itsuo Yamamoto, Satoko Yamamura, Junji Konishi, Ming Hu, and Kanji Torizuka

INTRODUCTION

To prepare samples suitable for liquid scintillation counting various materials are added to scintillator solutions. These materials and the sample itself will alter the counting efficiency to some degree; this effect is known as quenching. Chemical quenching arises in an energy transfer process prior to creating luminescence, whereas color quenching is a phenomenon occurring after the luminescence production. Color quenching effect diminishes the mean free path of the fluorescence photons. It is emphasized that chemical quenching is really a physical effect rather than a chemical effect.[1] The pulse height distribution of a color quenched sample differs from that of a chemically quenched sample, even if both samples have identical radioactivity and counting efficiencies. Therefore, chemical and color quench correction curves are also different. For strongly quenched samples there is a large difference between the two kinds of quench correction curves.[2] Significant errors can be encountered if a quench correction curve obtained from chemically quenched standards is applied to color quenched samples. Thus, it is advantageous to determine the type of quench in a sample before quench correction is applied.

Let us discuss some mixed quench samples where quenching is produced by the mixed solutions of a given chemical quencher (CCl_4) and a color quenching solution at different ratios. Since the counting efficiency of the sample with known radioactivity (dpm) can be calculated, its quenching type is easily determined according to the location on the chemical and color quench correction curves. This method cannot be used, however, if sample activity is unknown. Therefore, a new parameter, quench type contribution factor (QTCF), was established to denote the degree of contribution from chemical or color quenching in a mixed quenched sample.[3,4] The mixed quenched sample could be corrected automatically by means of the chemical and color quenched standard samples, and QTCF value, and an on-line computer. Some authors have used the double ratio technique to determine the quench type.[5] This method was first used by Bush in 1968 to detect adsorption or precipitation of radioactive material from liquid scintillation solutions.[6] Earlier, an isolated internal standard method[7] and four curve method[8] were also tried for this purpose.

This paper describes a method called *67Ga Double Spectral Index Plots*, which can determine quench type of unknown samples and correct the quenching of mixed quench samples. A novel report about using ^{67}Ga as an internal standard in liquid scintillation counting was presented by McQuarrie and Noujaim in 1983.[9] Gallium is an element in Group IIIa of the periodic table. The physical half life of the radionuclide ^{67}Ga is 78 hours with major gamma emissions at 93 (40%), 184 (24%), and 296 (22%) keV. A variety of radiopharmaceuticals have been developed in the past two decades in an attempt to obtain tracers for tumor imaging, among these ^{67}Ga-citrate has found widespread use[10] thus, ^{67}Ga can be obtained easily from clinical departments for use in liquid scintillation counting.

MATERIALS AND COUNTING CONDITIONS

- scintillator solution: aquasol-2, a commercially available xylene based cocktail (New England Nuclear)
- chemical and color quenchers: carbon tetrachloride (CCl_4), saturated solution of methyl red (reagent grade) in acetone (reagent grade), a specially prepared reagent. All quenchers and their solvents, products of Nakarai Chemicals Ltd. (Japan)
- labeled compounds: ^{67}Ga-citrate injection (Medi-physics Co. Japan) as ^{67}Ga source, Thymidine (6-^{3}H), >15 Ci/mmol, aqueous solution (New England Nuclear), n-(1,2(n)-^{3}H)hexadecane (TRR-6) and n-(1-^{14}C)hexadecane (CFR-6), manufactured by Amersham International, used as standards in preparations of ^{3}H and ^{14}C quenched series
- instrument and counting conditions: Packard Tri-Carb (Model No.460CD) liquid scintillation spectrometer; spectral index (SIS and SIE), sample channels ratio (SCR), and spectra, as counting conditions used as listed in Table 1

All counting was done at 16°C and all samples were contained in 20 mL standard glass vials in which 10 mL cocktail was added.

EXPERIMENTS AND RESULTS

Quenched ^{67}Ga Spectra

A series of quenched samples containing the same amount of ^{67}Ga-citrate (5 μL aqueous solution) and varying amounts of carbon tetrachloride (chemical

Table 1. The Selection of Counting Conditions

	Energy Range (keV)	SCR	Spectra (keV) Window Width	Spectra (keV) Range
^{3}H	0–19	counts (4–19) keV / counts (0–19 keV)	1	0–20
^{14}C	0–156	counts (10–156) keV / counts (0–156 keV)	1 / 5	0–20 / 21–160
^{67}Ga	0–3,5,7,10,19,156 0-variable upper limit 10–100,15–100	counts (0–19 keV) / counts (0–156 keV)	1 / 5	0–20 / 21–156

quencher) were prepared and tested. Increasing the amounts of the quencher caused a shift of the observed ^{67}Ga pulse height spectrum in the direction of lower energy (Figure 1). These spectra of ^{67}Ga are characterized by two peaks, i.e., lower and higher energy peaks corresponding to Auger electrons at 8 keV and conversion electrons from the 90 keV level, respectively. Though the spectral shift of both lower and higher energy peaks is concentration dependent, the counts at the lower energy peak were sequentially reduced by increasing the quencher. In contrast, the counts at the higher peak were sequentially increased. This response is similar to the spectral shift of ^{3}H and ^{14}C quenched series (Figure 2). The energies of two peaks of ^{67}Ga are closely related to the average beta decay energies of ^{3}H and ^{14}C, thus, a single or dual labeled sample containing ^{3}H and/or ^{14}C would behave similarly to the corresponding peaks of ^{67}Ga at the same level of sample quench. McQuarrie and Noujaim[9] developed a technique using parameters related to the pulse height and number of events in each peak of ^{67}Ga. With it they may obtain a mathematical relationship that monitors the degree of sample quench.

Spectral Index of ^{67}Ga

Spectral index of samples (SIS) and spectral index of external standard (SIE) are two of the current quench monitoring techniques. The spectral index is a unique number for the given total spectrum and is related to the radionuclide and its level of quenching. If only a part of the spectrum is analyzed, however, this number will not be unique and will depend on the counting conditions. This last point has not attracted sufficient attention until now. Variations of SIS in an unquenched ^{67}Ga sample for various pulse height ranges (i.e., energy range) are shown in Figure 3. With a region of interest set between a lower limit of 0 keV and an upper limit of 10 to 30 keV, a plateau is seen on the curve; if the upper range is greater than 60 keV, a second plateau is seen. Both lower and higher energy plateaus are related to two peaks of ^{67}Ga pulse height spectrum. The ^{67}Ga SIS values determined in the same region of interest will vary as the quenching increases (see Figure 4).

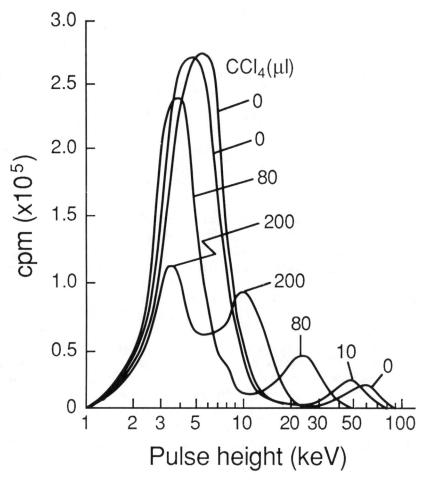

Figure 1. The unquenched and quenched (by carbon tetrachloride) spectra of ^{67}Ga.

Double Spectral Index Plots for ^{67}Ga

The plot of SIS (0 to 100 keV) vs SIE for chemically quenched ^{67}Ga is shown in Figure 5. The differences between the shape of chemical vs color quenched spectra of ^{67}Ga were also obtained (Figure 6); thus, an isolative curve from a chemical quenching curve was obtained for color quenched sample series that contained the same amounts of ^{67}Ga-citrate (5 μL aqueous solution) and varying amounts of color quenching solution (from 0 to 20 μL).

A plot of SIS in the lower energy region vs that of SIS in the higher energy region can be obtained when two regions of interest are used to determine ^{67}Ga SIS values for both chemical and color quenched series, the series must include the lower and higher energy peaks of ^{67}Ga (Figure 7a). Because the peaks of the color quenched sample were lower and broader in the higher energy side than

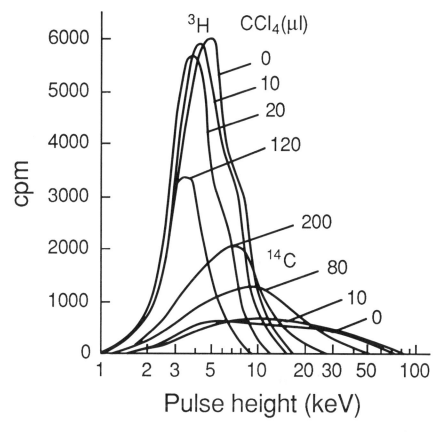

Figure 2.   The unquenched and quenched (by carbon tetrachloride) beta spectra of ^{3}H and ^{14}C.

the peaks of a chemically quenched sample for ^{67}Ga (Figure 4), in the same way as in Figure 5, two separate curves were given for chemical and color quenched series (Figure 7a). There is an inflection point on each curve and a crossover point at moderate quenching. When the regions of interest were changed to 7 to 100 keV and 0 to 19 keV, new curves were obtained (Figure 7b) in which there are two inflection points on each curve and a crossover point at a higher level of quench.

Mixed Quench Zone

Let us analyze the region between the chemical and color quenched curves in Figures 5, 7a, and 7b. Obviously, these are transitional regions in which samples will simultaneously contain both chemical and color quench; they are referred to as a mixed quench zone. In fact, we prepared samples containing

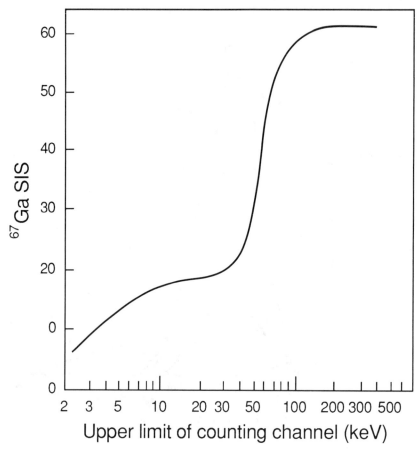

Figure 3. The variation of SIS of unquenched ^{67}Ga sample in various pulse height ranges.

varying amounts of chemical quenching agent (carbon tetrachloride) and color quenching agent (saturated solution of methyl red in acetone) (Table 2) and measured them at the counting conditions of Figures 5, 7a, and 7b.

These points of mixed quench samples fall in a zone; the positions in the zone are generally near chemical or color quenching curves, according to their ratio of chemical to color quenching. Similar phenomena have been observed and analyzed in the quench correction curves of sample channels ratio (SCR) and SIS, i.e., the curves of counting efficiency vs SCR or SIS.[3,4] Thus, we can identify the quench type of an unknown sample and calculate the degree of contribution of chemical or color quenching according to its location on the double spectral index plot (Figures 5, 7a, and 7b.) For example, if the quenching agent of a sample is not known, ^{67}Ga is added to the sample and measured at the above counting condition. A point in double spectral index plots can be obtained from the observed SIS and/or SIE values of the unknown sample.

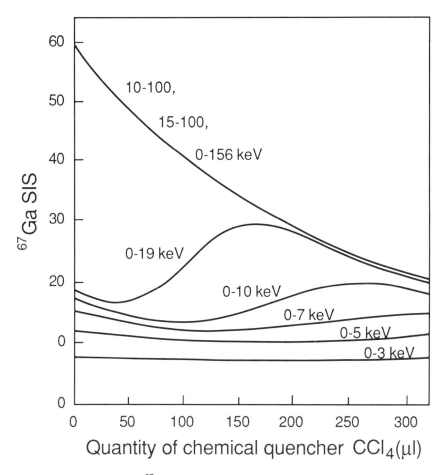

Figure 4. The variations of ^{67}Ga SIS with increasing quench.

The sample was quenched by chemical quenching when the point is on the chemical quench curve and, in contrast, by color quenching when the point was on the color quench curve. Those samples corresponding to points lying in a mixed quench zone can be considered to contain both chemical and color quenchers simultaneously.

The mixed quench zone was a function of counting conditions in the same way as double spectral index plots. Attention should be paid to the points located near the crossover. Two points [(40,2) and (50,2)] fall outside of the mixed quench zone in Figure 7a, and one point (30,6) is on the color quenched curve in Figure 7b. When counting conditions were changed the mixed quench zone also changed, and these points fell in the new mixed quench zone. Generally speaking, these double spectral index plots are not sensitive to the degree of quench near their crossover point.

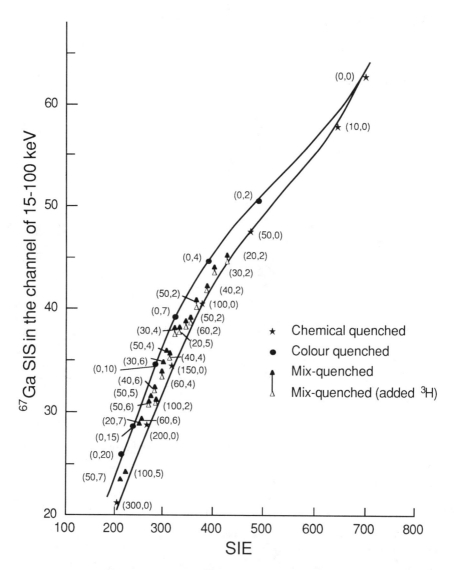

Figure 5. Double spectral index plots of ^{67}Ga SIS in the channel of 15 to 100 keV vs SIE. The figures of the first and second groups in the parentheses represent the amounts (μL) of chemical quencher (CCl4) and color quencher (saturated solution of methyl red in acetone).

Reproducibility of the Point Positions for Mixed Quench Samples

We have given two kinds of double spectral index plots for ^{67}Ga, i.e., SIS vs SIE, and SIS in a lower energy region of interest vs SIS in a higher energy region of interest. Generally, these two kinds of curves also can be used as a tool to identify the quenching type of an unknown sample. A more important

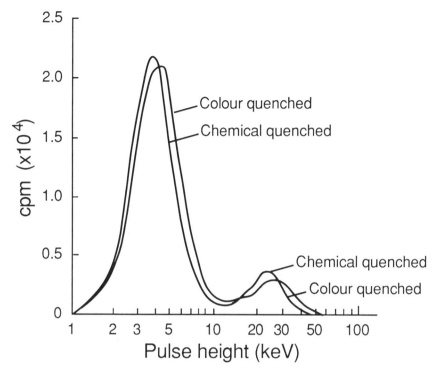

Figure 6.   The differences of the shapes of ^{67}Ga pulse height spectra for chemical and color quenching samples containing the same degree of quenching (relative counting efficiency was 68%).

problem for mixed quench samples is the reproducibility of the position in the mixed quench zone, because the zone is smaller than the total region of quench variation.

The main factors affecting the reproducibility of this point are: (1) variation of ^{67}Ga activity, (2) interference from ^{3}H and/or ^{14}C, and (3) selection and reproducibility of counting conditions. The selection of counting conditions has been discussed under "Spectral Index of [67]Ga." The reproducibility of counting conditions in this instrument are satisfactory as they are set automatically. To observe the effect of ^{67}Ga radioactivity on the reproducibility of SIS, a series of quenched samples with three different quantities of [67]Ga (5,10, and 20 μL) were prepared and measured repeatedly. The standard deviations of measurements caused by the variation of ^{67}Ga activity were $(0.60 \pm 0.20)\%$ (mean $\pm$ S.D.) for the region of interest of 0 to 7 keV and $(1.14 \pm 0.66)\%$ for the region of interest 15 to 100 keV which were similar to the standard deviation for 10 measurements of a single sample were similar $(0.80 \pm 0.26)\%$. There was no significant difference between either. Changing the quantity of ^{67}Ga from 5 to 20 μL affected the reproducibility of SIE $2.1 \pm 0.8\%$.

If unknown samples containing ^{3}H and/or ^{14}C only make these points shift a

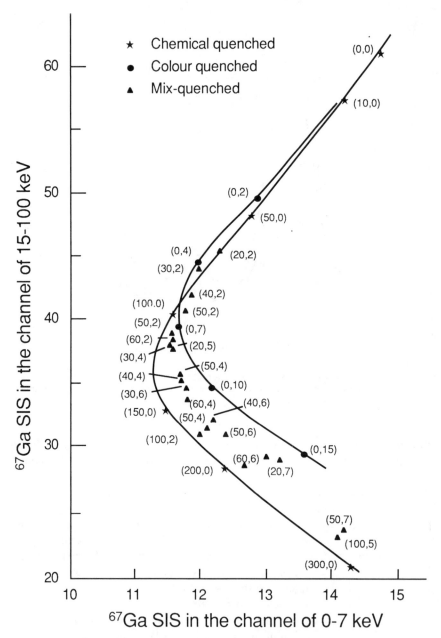

Figure 7a. ^{67}Ga double spectral index plots of SIS in the channels of 15–100 keV to 0–7 keV. The illustration of the figures in the parentheses is shown in Figure 5.

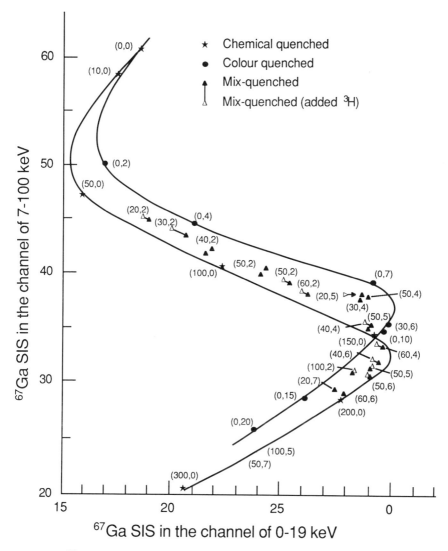

Figure 7b. ^{67}Ga double spectral index plots of SIS in the channels of 7–100 keV to 0–19 keV. The illustration of the figures in the parentheses is shown in Figure 5.

little (Figures 5, 7a, and 7b), their effect on the identification of quenching type and on quench correction can be ignored, because ^{67}Ga has a sufficiently high counting rate. If ^{3}H and/or ^{14}C in the samples have higher counting rates, usual double ratio techniques[5] and SCR quench correction methods[3,4] can also be used to identify and correct the quench.

Table 2. The Components and Parameters of Mixed Quench Samples

Batch	CCl_4 (μL)	Color Solution[a] (μL)	Equivalent Counting Efficiency (%)		QTCF[b]	F[c]
			[3]H	[14]C		
	20	2	63.5	95.0	0.615	0.407
		5	34.2	85.6	0.381	0.219
		7	22.7	77.7	0.313	0.185
1	50	2	47.9	92.6	0.706	0.607
		5	25.8	83.5	0.353	0.385
		7	17.1	75.8	0.278	0.337
	100	2	28.1	85.9	0.892	0.724
		5	15.2	77.4	0.695	0.516
		7	10.1	70.3	0.497	0.464
	30	2	57.8	94.6	0.308	0.500
		4	38.4	88.7	0.273	0.327
		6	25.8	80.8	0.188	0.267
	40	2	52.8	93.6	0.759	0.564
		4	34.8	87.8	0.523	0.360
		6	23.5	80.0	0.417	0.320
2	50	2	47.9	92.6	0.531	0.607
		4	31.8	86.9	0.445	0.428
		6	21.4	79.1	0.333	0.359
	60	2	43.7	91.2	0.625	0.639
		4	29.0	86.0	0.488	0.462
		6	19.6	78.3	0.387	0.392

F = 0.1296 ± 0.6280 QTCF   r = 0.7967

[a]Saturated solution of methyl red in acetone.
[b]Experimental QTCF of chemical quench from Figure 3.
[c]Calculated contribution factor (F) of chemical quench from [3]H.

QUANTITATIVE QUENCHING CORRECTION OF MIXED QUENCHED SAMPLES

The color quencher used in this study was a solution of methyl red in acetone (Table 2), but the acetone itself also is a chemical quenching agent. In order to decrease the effect of chemical quenching of acetone, we prepared a saturated solution of methyl red in acetone. The maximum quantity of the color solution used in the work was 20 μL, which decreased the relative counting efficiency of ^{3}H to 3.5%, and that of ^{14}C to 32% (Figure 8), relative to the unquenched sample. Under this condition the relative quenching contribution factor (F_c) of chemical quenching of acetone was only about 0.04 for [3]H. The formula for calculating F_c factor is as follows:

$$F_c = \frac{1 - E_{rc}}{(1 - E_{rc}) + (1 - E_{rcl})} \tag{1}$$

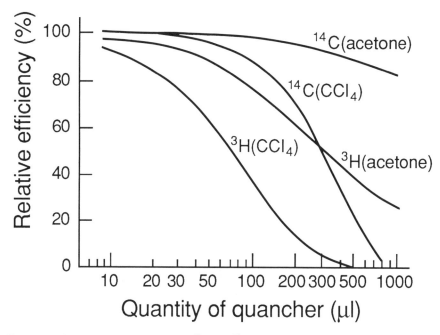

Figure 8. Chemical quench curves for ^3H and ^{14}C (quenching agents, acetone and carbon tetrachloride).

where E_{rc} and E_{rcl} are relative counting efficiencies caused by chemical quencher and color quencher in the mixed quench sample. The chemical quenching effect of methyl red was not considered.

The decrease in relative counting efficiencies of ^3H and ^{14}C with increasing amounts of carbon tetrachloride and acetone are shown in Figure 8. Likewise, changes in relative counting efficiencies due to the addition of increased quantities of saturated methyl red solution in acetone are shown in Figure 9. The equivalent relative counting efficiency of mixed quench samples can be represented by the product of the relative counting efficiencies of chemical and colour quenching in these samples (Table 2). In addition, F_c of chemical quenching can be calculated according to Equation 1 (Table 2). The quenching contribution factor for color quenching (F_{cl}) equals $1 - F_c$.

Obviously, Equation 1 is usable only for the samples with known quenching components. When the quenching components are unknown, a new parameter, quenching type contribution factor (QTCF), has been defined by Yang[4] to represent the relative position of mixed quench samples in a mixed quench zone:

$$QTCF_{cl} = \frac{E_c - E_s}{E_c - E_{cl}} \qquad QTCF_c = \frac{E_s - E_{cl}}{E_c - E_{cl}} \qquad (2)$$

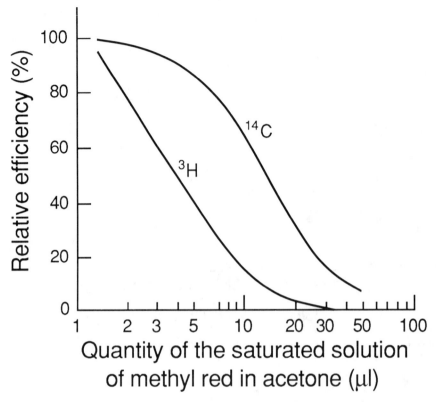

Figure 9. Color quench curves for ^3H and ^{14}C (quenching agent saturated solution of methyl red in acetone).

where E_c and E_{cl} are the counting efficiencies on the chemical and color quenching correction curve corresponding to the QIP (quenching indicating parameter) of the mixed quench sample. E_s is the counting efficiency of a mixed quench sample. In contrast, when QTCF is known we can calculate the counting efficiency of unknown mixed quench samples according to its QIP and QTCF:

$$E_s = E_c - QTCF_{cl} (E_c - E_{cl})$$
$$E_s = E_{cl} + QTCF_c (E_c - E_{cl}) \tag{3}$$

The QTCF value of the sample with unknown quenching components cannot be determined by Equation 2 as the counting efficiency E_s is unknown. The double vial method[4] has been used to solve this difficulty: $1/m$ of total volume is taken from an unknown sample after the first counting and is diluted to original volume with the same cocktail, then it is recounted with the same conditions. Suppose QIP_1 and QIP_2 were obtained from the first and the second measurements, respectively; thus,

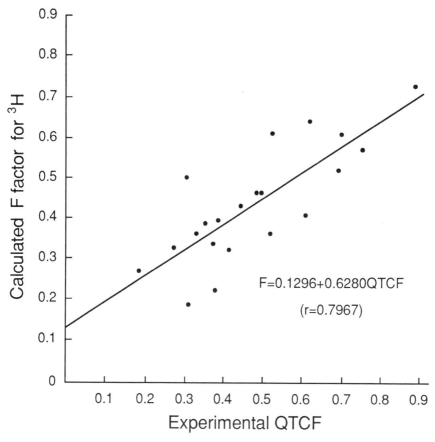

Figure 10. The relation of calculated F factor to experimental QTCF. These parameters were obtained from mixed quench samples.

$$QTCF = \frac{mrE_c'' - E_c'}{mr\ E_{cc}'' - E_{cc}'} \qquad (4)$$

where E_c' and E_c'' are the counting efficiencies from the chemical quenching correction curve corresponding to QIP_1 and QIP_2, respectively. Therefore, $E_{cc}' = E_c' - E_{c11}'$ and $E_{cc}'' = E_c'' - E_{c1}''$; and E_{c1}' and E_{c1}'' are the counting efficiencies on the color quenching correction curve corresponding to QIP_1 and QIP_2, respectively; where $r = n_2/n_1$, and where n_1 and n_2 are the counting rates of the first and the second measurements, repectively. If the amount of the second sample was m times that of the first sample, then

$$QTCF = \frac{mE_c'' - rE_c'}{m\ E_{cc}'' - r\ E_{cc}'} \qquad (5)$$

The reason for using the double vial method to determine QTCF is that the position of an unknown sample in a mixed quench zone cannot be determined by a single count, as in the usual quench correction method. Now, in this experiment the difficulty was resolved by the addition of ^{67}Ga to the unknown sample. The quench correction curves mentioned above were replaced by double spectral index plots. The point distributive character of mixed quench samples in the mixed quench zone of double spectral index plots are the same as in quench correction curves. Therefore, QTCF can also be used, for example, when an unknown sample was measured and its SIE was determined. The chemical and color quenched SIS_s values in the energy range of 15 to 100 keV, corresponding to SIE in Figure 5, were SIS_c and SIS_{c1} respectively, while the SIS value of the sample itself was SIS_s. We can then obtain experimental QTCF values from Formulas 6 and 7 listed below:

$$QTCF_{c1} = \frac{SIS_c - SIS_s}{SIS_c - SIS_{c1}} \qquad (6)$$

$$QTCF_c = \frac{SIS_s - SIS_{c1}}{SIS_c - SIS_{c1}} \qquad (7)$$

where $SIS_c - SIS_{c1}$, $SIS_s - SIS_{c1}$, and $SIS_c - SIS_s$ are negative values because the color quenched curve was above the chemical quenched curve. However, the value of QTCF is always positive. $QTCF_{c1}$ and $QTCF_c$ characterize the degree of the contribution of color and chemical quench to mixed quench samples, respectively. The values of QTCF taken from Figure 7 are always the same as those taken from Figure 5, as only ^{67}Ga SIS_c and SIS_{c1} in the region of interest of 15 to 100 keV (corresponding to the SIE of a mixed quench sample) will change correspondingly to ^{67}Ga SIS_s in the region of interest of 0 to 7 keV. Thus, when we substitute QTCF into Equation 3, the counting efficiency of an unknown sample is resolved. The ^{3}H corrected results, which are obtained by using QTCF (from Figure 5) and SIS quench correction curves, are given in Table 3. Except for one sample, the QTCF method gave more accurate values than those obtained by using either the chemical or color quenching correction curve alone. The mixed quench zone is wider in the range of strong quenching; therefore, the employment of QTCF for quench correction in that range has particularly significance.

DISCUSSION

The experimental QTCF values of 21 mixed quench samples were obtained from Figure 3 and Equation 7 (Table 2). The correlation of the calculated F factor with experimental QTCF values for mixed quench samples was given in Fig. 10, in which $F = 0.1296 + 0.6280$ QTCF ($r = 0.7967$). It is illustrated that F factor is proportionally related to QTCF; in fact, this is the basis of the

Table 3. The Comparison Between Activities Obtained Using QTCF and Activities Obtained Using the Usual Quench Correcting Curves for Strongly Mixed Quench Samples[a]

Sample Number	SIS of 0–19 keV	Sample Counts (cpm)	Counting Efficiency			Activity (dpm)					
			Chem. (%)[c]	Color (%)[c]	QTCE (%)[d]	Chem. Curve	Error (%)[b]	Color Curve	Error (%)[b]	Using QTCF	Error (%)[b]
1	12.8	6290	24.5	20.4	22.9	25673	−14.5	30833	2.7	27467	−8.5
2	12.4	5610	21.9	17.4	18.8	25616	−14.6	32241	7.4	29840	−0.6
3	11.2	3204	13.2	9.1	10.2	24273	−19.1	35209	17.3	31412	4.7
4	12.1	5328	20.0	15.3	18.9	26640	−11.2	34824	16.0	28190	−6.1
5	11.1	2917	12.6	8.5	10.6	23151	−22.9	34318	14.3	27519	−8.3
6	11.8	4444	17.8	13.2	15.6	24966	−16.8	33667	12.2	28487	−5.1
7	10.8	2445	10.7	6.3	8.3	22850	−23.9	38810	29.3	29458	−1.9
							−17.6 ± 4.3		14.2 ± 7.8		−3.7 ± 4.4

[a]Each sample contained 20 μL aqueous solution of ^{3}H-thymidine (30014 dpm).

[b]Error (%) = $\dfrac{\text{actual activity} - \text{corrected activity}}{\text{actual activity}} \times 100$

[c]Taken from curves in Figure 7 (left)

[d]Counting efficiencies were obtained by QTCF from Figure 3.

quench correction method for mixed quench samples in this chapter. If we call the usual quench correction "first order quench correction," then QTFC quench correction in mixed quench zone can be called "second order quench correction." Obviously, the magnitude of the second order quench correction is less than the first order quench correction; therefore, it must have greater relative variability. Based on this point, in principle we can divide the mixed quench zone into subzones by the measurement of two parameters of the curve. All points falling in the same subzone were considered to have the same QTCF value which is more practical for application.

If a liquid scintillation counting system can automatically give two spectral indexes for two different parts of a sample (or external standard) spectrum at the same time, it is not only convenient for measurement of double spectral index plots, but also more usable to various new and existing applications. The authors had hoped to prepare double spectral index plots of external standards to correct mixed quench samples, but the commercial instrument used was unable to measure the SIE of different energy ranges.

The main advantages of using ^{67}Ga or ^{113m}In as an internal standard include, (1) the elimination of accurate pipetting of the internal standard, (2) the ability to recover the sample after the spiked activity has decayed, and (3) the high count rate.[11] Among these advantages the one with the greatest importance is probably the fact that the content of "standard" was changed from quantity of activity to energy or its spectrum shape. The mixed quench samples in this paper consisted of simple chemical and color quenchers. Practical samples are very complex, therefore, it is necessary to do further work. In addition, further work with the double spectral index plots or ratio of spectral index may yield some new applications.

ACKNOWLEDGMENTS

The authors wish to express their gratitude to Mr. Toru Fujita and Mr. Hideo Miyatade for kindly supplying ^{67}Ga-citrate and labeled hexadecane respectively.
Note: This manuscript has undergone extensive revision and condensation in order to meet the style requirements and space limitations of the proceedings. We sincerely hope that we have not changed the substance or connotation of any information intended to be conveyed by the authors. Eds.

REFERENCES

1. Dyer, A. *An Introduction to Liquid Scintillation Counting* (London: Heyden & Son, 1974), p. 57.
2. Parmentier, J.H. and F.E.L. Ten Haaf. "Developments in Liquid Scintillation Counting Since 1963," *Int. J. Appl. Radiat. Isot.* 20:305 (1969).
3. Yang, S.L. "Effect of Chemical and Color Quenching Ratios on Calibration Curve

in Liquid Scintillation Counting," *Chinese J. of Nuc. Med.* 1:92 (1981); C.A. 96, 170718 (1982).

4. Yang, S.L. "Automatic Quenching Correction Method for Samples with Both Chemical and Color Quenching in Liquid Scintillation Counting," *Acta Acad. Med. Sinicae* 3:114 (1981); C.A. 96, 42804 (1982).

5. Takiue, M., T. Natake, and M. Hayashi. "Double Ratio Technique for Determining the Type of Quenching in Liquid Scintillation Measurement," *Int. J. Appl. Radiat. Isot.*, 34:1483 (1983).

6. Bush, E.T. "A Double Ratio Technique as an Aid to Selection of Sample Preparation Procedure in Liquid Scintillation Counting," *Int. J. Appl. Radiat. Isot.*, 19:447 (1968).

7. Ross, H. H. "Color Quench Correction in Liquid Scintillator Systems Using an Isolated Internal Standard," *Anal. Chem.*, 37:621–623 (1965).

8. Lang, J.F. "Chemical vs. Color Quenching in Automatic External Standard Calibration. Application of Empirical Observation in a Computer Program," in *Organic Scintillators and Liquid Scintillation Counting*, D.L. Horrocks and C.T. Peng, Eds. (New York: Academic Press, 1971), p. 823.

9. McQuarrie, S.A. and A.A. Noujaim. "^{67}Ga: A Novel Internal Standard for LSC," in *Advances in Scintillation Counting*, McQuarrie, S. A., C. Ediss, and L.I. Wiebe, Eds. (Edmonton: University of Alberta Press, 1983), pp. 57–65.

10. Ell, J. and E.S. William. *Nuclear Medicine: An Introductory Text*. (Oxford, London: Blackwell Scientific Publications, 1981), p. 159.

11. Oldendorf, W.H. "Liquid Scintillation Quench Correction Using 113m-Indium Conversion Electrons as an Internal Standard," *Anal. Biochem.*, 44:154 (1971).

CHAPTER 24

Modem Applications in Liquid Scintillation Counting

Yutaka Kobayashi

New counters today are all controlled by sophisticated software which means that new liquid scintillation counters are now married to computers. Therefore, it seemed appropriate about two years ago to explore the applications of a modem to liquid scintillation counting technology. The person who aroused my interest in this study was Stan De Filippis, senior marketing applications specialist, who was then at the Packard Instrument Company. He envisioned that the modem could provide both the instrument manufacturer and the user with a very powerful tool. The major concern with this idea was reliability because of the well known problems associated with early attempts at data transmission with modems and personal computers.

To explore this concept, a modem was placed into my counter, a Packard Model 2200CA, located at the Worcester Foundation for Experimental Biology in Shrewsbury, MA. The Packard 2200CA is controlled by an IBM model PS/2–30 computer equiped with a 20 MB hard disk. The modem was a Leading Edge Model "L" set to run at 2400 baud/sec. An identical modem was used in a computer at the Packard Instrument Company located in Downers Grove, IL. Later in this study a Hayes Smartmodem 2400B modem was installed in my home computer in Wellesley, MA, about 25 miles from Shrewsbury.

Once the modems were in place, we needed to choose a communications package to make the system work. After trying a few commercial software packages, we settled on one called CLOSE-UP from Norton-Lambert Corporation. The CLOSE-UP system consists of two separate software packages: one called Support, which was used at the remote receiving stations located in Downers Grove and Wellesley and another called Customer, which was installed in the counter. At the Worcester Foundation all telephone calls are routed through a switchboard. Although the communication software is designed to operate under these conditions, we decided to install an outside line directly to the modem in the counter to avoid this high potential source of problems.

Table 1. Modem Applications in Liquid Scintillation Counting from a Remote Site Investigated in this Study

1. Monitor counter
2. Control/Operate counter
3. Trouble-shooting aid
4. Update software
5. Transfer data
6. Automatic data transfer

When the counter is called from Downers Grove, IL and contact is established, the screen of the counter appears on the computer screen in Illinois. From that moment, the counter can be controlled from the computer keyboard in Illinois just as if the operator were standing at the counter in Massachusetts. We have found the system to be very reliable and useful. In Table 1, all the applications successfully executed during this study are listed.

This system allows you to monitor the counter from a remote site. For example, if you wanted to check on the state of your counter from home, you can call your counter and examine the Status Page. The Status Page in a Packard 2200 counter tells you what program is being used and which sample is currently being counted. On the other hand, if you were scanning a group of samples to see the distribution of counts among them and wanted to change the counting time from, say, one minute to ten minutes, you can also do that from home without going to the laboratory. This has proven to be a convenient option.

Perhaps the most intriguing potential application of the modem is its use as a trouble shooting aid. This aspect of the modem, however, was not really tested because the counter did not have any serious problems during the study. In the case of the Packard 2000 series counters, the potential usefulness of the modem is there because of the diagnostic software contained within the counting system software. An engineer in Illinois can actually look at all the operating parameters of the counter in Massachusetts, such as the history of the high voltage applied to the photomultiplier tubes for the last 10 instrument normalizations. He can also check the performance history in terms of the check source counting efficiency for the last 90 normalization cycles; all of this is stored in memory. In terms of diagnosing problems, the only occasion in which the modem was used for this purpose was to send the puzzling data for a particular spectrum to the engineers in Illinois for analysis.

During the course of this evaluation, a software update for the Packard 2200 counter was released. The installation of the new software was done using the modem. This meant erasing the old software and installing the revised version. The entire process went without any problems. This demonstrated one operation which could save both the instrument manufacturer and the user a lot of time. In the case of the manufacturer, the need to schedule a service call for the installation of the new software is eliminated, and, in the case of the user, the entire software update can be scheduled after working hours when the instrument is not in use.

Table 2. Example of a Task File used in CLOSE-UP for Automatic Data Acquisition using a Modem

File Name: DATAOUT.TSK
1. CONFIG DIAL REDIAL = 3   WAIT = 3
2. WAIT UNTIL 23:45
3. BAUD 2400
4. PRINT LOG ON
5. DIAL 15558425555[a]
6. FETCH C:SDATA10.DAT TO A:
7. HANGUP
8. PRINT LOG OFF

[a]Fictitious telephone number.

The most useful feature of a modem-equipped counter is its ability to transfer data to a remote site. In the Packard 2200CA, it is possible to store any counting data onto the hard disk. This data can then be transferred to any remote site via modem. During this study, this was achieved as a routine option. The convenience of this option is that the researcher can now transfer his counting data to his home computer and analyze the data without going to the laboratory.

Perhaps the most unique feature of the communication software package, CLOSE-UP, is its ability to transfer counting data automatically. For example, the computer at home can be programmed to call the counter, say at midnight when the phone rates are low, and transfer counting data while I sleep. This is accomplished by using a simple routine called a task file. An example of a task file is shown in Table 2.

This simple, eight line test program was written using EDLIN which is a resident DOS text editor. This task file was called DATAOUT and all task files are designated by the extension, TSK. DATAOUT is stored in the subdirectory which contains CLOSE-UP. The first line instructs CLOSE-UP to make a total of three attempts to connect with the counter. If the first try is unsuccessful, it will wait three minutes before trying again. The next line instructs the program to make the first call at 11:45 PM; military time is used here. The next line sets the baud rate. The next line turns on the printer to type a log of all activities. The next line gives the phone number (the phone number listed is a fictitious number). After contact is made with the counter, the next line instructs the computer operating the counter to fetch a file from the C drive called sdata10.dat and copy it to the A drive of the home computer. When the task is completed, the program hangs up the phone and turns off the printer.

Table 3 shows the log generated by the task file. The program was initiated on September 14th and the name of the task file in control was DATAOUT-.TSK. The program was activated at 11:45:07 PM; connection was made on the first try at 11:46:32. The data file, sdata10.dat, was located on the C drive and copied to the A drive of the home computer. The counting data for 28 samples were counted under low-level conditions. The transferred data received and stored in the home computer is shown in Table 4. The log shows

Table 3. Example of a Log for Task File, DATAOUT.TSK

	—Sept. 14 1989 DATAOUT.TSK
23:45:07	5—dial 15558425555[a]
	Connect: 1 attempt
23:46:32	6—fetch c:sdata10.dat to a:
	Complete – 00:00:06
23:46:47	7—hangup
23:46:51	8—print log off

[a]Fictitious telephone number.

that the call was terminated at 11:46:47 for a total connect time of only 15 seconds and 4 seconds later, the printer was turned off. All of this was done without anyone being present. This sample test file was run to check out the system. Also, in actual practice, the file transfer system can be protected by requiring a password to allow only authorized access to the counter computer system.

In setting up this automated data collection system, it was reasoned that the counting data should be collected in the A drive of the computer controlling the counter rather than in the C drive. The removable diskette would be a backup system for the automated data processing system; should any problem occur with data transfer, the original data will always be saved on a removable

Table 4. Data File, SDATA10.DAT, Transferred by Modem from a Packard 2200CA Counter. Data Transferred, in Sequence, are: Sample Number, Time in Min.: Region A cpm, Region B cpm, Region C cpm, SIS and tSIE.

```
1,10.00,10.9000,4.2000,9.8000,47.29,0.00
2,10.00,8.8000,2.1000,7.7000,54.20,0.00
3,10.00,8.4000,3.9000,7.7000,60.33,0.00
4,10.00,10.4000,4.5000,8.8000,62.50,0.00
5,10.00,10.2000,4.3000,9.2000,66.46,0.00
6,10.00,11.1000,3.5000,9.7000,58.33,0.00
7,10.00,10.8000,3.8000,9.3000,55.33,0.00
8,10.00,14.1000,3.9000,12.2000,49.20,0.00
9,10.00,10.4000,4.7000,9.0000,69.42,0.00
10,10.00,9.3000,3.7000,7.6000,56.67,0.00
13,10.00,12.5000,3.8000,11.7000,54.37,0.00
14,10.00,12.2000,3.6000,9.4000,51.17,0.00
15,10.00,9.6000,3.5000,7.9000,57.22,0.00
16,10.00,9.9000,3.2000,8.9000,53.73,0.00
17,10.00,11.7000,3.7000,9.9000,51.33,0.00
18,10.00,8.8000,3.3000,7.8000,59.44,0.00
19,10.00,9.7000,4.4000,8.2000,61.98,0.00
20,10.00,10.9000,3.8000,9.9000,60.83,0.00
21,10.00,9.3000,3.0000,8.4000,60.77,0.00
22,10.00,12.5000,3.2000,10.0000,44.14,0.00
25,10.00,8.0000,0.7000,6.2000,21.22,0.00
26,10.00,31.7000,12.300,30.2000,46.69,0.00
27,10.00,30.9000,14.1000,29.1000,49.59,0.00
28,10.00,30.5000,15.9000,29.1000,52.94,0.00
30,10.00,76.0000,45.0000,74.2000,56.67,0.00
31,10.00,168.600,89.0000,164.9000,53.49,0.00
32,10.00,77.4000,46.4000,75.6000,57.51,0.00
33,10.00,15.200,0.4000,13.4000,15.63,0.00
```

diskette. This means that the Packard 2200 counter will always have a diskette in the A drive. This can be a problem in the event of a power failure, because the counting system is programmed to automatically reboot from the C drive after power is restored. The problem is that, in the IBM PS/2–30 computer used in the counter, the system always first looks at the A drive, for rebooting. This situation can be accommodated by putting all the counting system software on the A drive but this would reduce the data storage capacity of the data diskette. This problem was avoided simply.

First, the A diskette is formatted using the FORMAT A:/S command. The slash S calls for two hidden DOS system files to be copied onto the formatted diskette. Without these two hidden files, the system will not boot up properly from the A drive. Next, using EDLIN, a DOS text editor, write a one-line autoexec.bat file as follows:

C:\KOBY.BAT

All data diskettes used in the A drive must be formatted this way and contain the one line autoexec.bat file. Then, on the C drive, that is, on the hard disk in the DOS directory, write another batch file called KOBY.BAT using EDLIN. This file contains only two instructions:

1. C:
2. autoexec.bat

This little routine ensures that the counter will reboot normally after power failure when there is a data diskette prepared as described in the A drive.

In considering the practical aspects of modems in counting technology, I believe they have limited, but very useful attributes. From an instrument manufacturer's point of view, the possibility of updating software via modem directly from the plant should be an attractive prospect. This system would assure timely distribution of software to all customers and save a service call as well. Also, the ability to inspect a customer's counter performance remotely can be advantageous, especially after a service call was made or a new procedure is being evaluated. I can envision the time when a manufacturer can offer a preventive maintenance program which involves scheduled, invisible inspections of a working counter through a modem. The fact that a log of these invisible inspections can be generated would assure the customer that the service was indeed provided.

For the user, the modem does have limited applications at this time. However, in certain situations I believe the modem can be very useful. For example, any diagnostic laboratory running assays can have the completed results transferred to its various customers via modem automatically every night by using a task file. The customers will then have the results available before the start of the working day. In an industrial plant where the counting is done in a central analytical laboratory, the completed counting data can be sent directly to any laboratory on site provided the remote site, has a modem-equipped computer.

In my limited experience, I have found the modem to be a very convenient and useful option. Even though it has limited applications, the data transfer capability alone justifies its presence in my counter . What is attractive about this technique is that it is a relatively inexpensive option to own. Modems can be purchased for under $100 and good communications software is available in the public domain. For example, a highly regarded public domain software is PROCOMM. This software has been commercialized and is sold as PRO-COMM PLUS at nominal cost, especially at software discount houses. However, the original PROCOMM is available and should be more than adequate for this application. The advantage of using a commercially produced software is that these packages are easy to use and well documented.

CHAPTER 25

A New Procedure for Multiple Isotope Analysis in Liquid Scintillation Counting

A. Grau Carles and A. Grau Malonda

INTRODUCTION

Double and triple labeled radiative samples are interesting in biological tracing work. The procedures to obtain the activity of each one of the components in a multilabeled sample are: double or triple window measurementes,[1] spectral shift by quench variation,[2] rate counting at different times,[3] different counting methods,[4] different decay modes.[5]

All these procedures are based on the counting rate of the whole or part of the pulse height spectra. It seems interesting to look for new counting procedures in order to overcome some of the very known restrictions. This procedure could be based on using the differential shape of the pulse height spectra.

The aim of this chapter is to develop a procedure, based on spectra fitting, that obtains the activity of each one of the radionuclides in the sample.

The procedure will be applied to mixtures of pure beta-ray emitters, ^{3}H + ^{14}C, and ^{35}S + ^{14}C. The application to other kinds of decay schemes is possible, but they will not be studied here.

A code named DILATA has been developed in order to compute the spectral components and activities of radionuclides mixtures. In this chapter, only the mathematical procedure will be described.

FITTING PROCEDURE

The spectral issue of a liquid scintillation counter for beta-ray emission depends on the quenching degree of the sample. Two of the eight ^{14}C spectra obtained with a logarithmic amplifier have been plotted in Figure 1. The shape and pulse height of the spectra depend very strongly on the quenching of the samples.

The first problem to be solved is how to obtain the spectrum which corresponds to a given quench level. A direct interpolation can not be applied

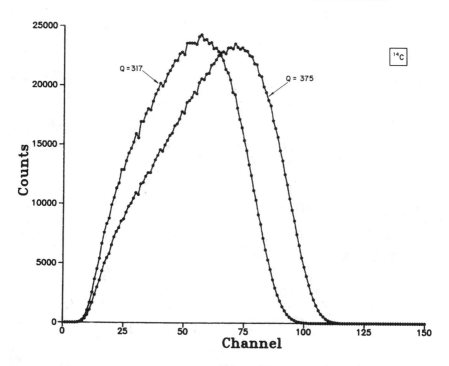

Figure 1.   Experimental beta-ray spectra of ^{14}C for two different quenching values.

because of the dispersion of the spectra. One possible interpolation method could be the interpolation by spectral dilatation.

The general procedure has the following steps:

1. A set of spectra with different quench is obtained for each nuclide.
2. The interpolation procedure is applied to each nuclide, to obtain the spectra corresponding to the quench level of the experimental double labeled sample.
3. A least square fitting is done with the interpolated spectra and the double labeled spectrum in order to compute the activities of each one of the nuclides.

The procedure can be applied to multilabeled samples, but for simplicity and clarity we only discuss double labeled samples.

Special Functions

The experimental spectra, see Figure 1, are histograms and have the inconvenience of being defined for discrete values of the channel spectra. It is interesting, from a mathematical point of view, to have a continuous function defining the spectra.

The spectral function has been obtained by fitting Fourier series to the experimental spectra. The spectral function f(w) is obtained from:

$$f(w) = \begin{cases} a + bw + \sum_{k=1}^{N} C_k \sin \dfrac{k\pi w}{M-1} & 0 \le w < w^* \\ 0 & w > w^* \end{cases} \quad (1)$$

where N is the number of Fourier harmonies and M is the number of experimental points of the spectrum. The first and last spectral points are $w = 0$ and $w = M - 1$. w^* is the end of the spectrum.

The constants a, b and C_k can be obtained from the following equations:

$$a = y_0 \quad (2)$$

$$b = \frac{y_{M-1} - y_0}{M - 1} \quad (3)$$

$$C_k = \frac{2}{M - 1} \sum_{j=0}^{M-1} \bar{y}_j \sin \frac{\pi k w_j}{M - 1} \quad (4)$$

where y_j the number of counts in channel j, and

$$\bar{y}_j = y_j - (a + bw_j) \quad (5)$$

$$w_j = j = 0, 1, 2, \ldots M - 1. \quad (6)$$

The calculation of the spectral function f(w) is interesting for two reasons: it can eliminate very properly the statiscal fluctuations of the spectra and permits a continuous function that allows application of the spectral dilatation interpolation method.

The first problem is defining the number of harmonics to be taken in the Fourier series. It is known that if the number of harmonics is equal or close to the number of experimental points of the spectrum, the function f(w) will follow all the experimental points. In this case the statistical fluctuations will be included. We are interested in having a certain smoothing of the spectrum in order to attenuate the effects of statistical fluctuations. However, if we take a very low value for N, the function f(w) can loose interesting or essential structures of data. That will produce a modification of the real shape of the spectrum.

Different values of N have been tried. $N = M/2$ seems to be a good compromise to obtain a moderate smoothing without loosing essential structures.

Spectral Dilatation-interpolation Method

The spectral dilatation-interpolation method is interesting in the case of logarithmic spectra, nevertheless it is also applicable to linear spectra.

The spectral function f(w) can be separated into two regions. Region 1 ranges from zero to the value that corresponds to the maximum of the spectrum. Region 2 is the remainder of the function.

The spectral dilatation method consists of applying a shift to region 1 of all the spectra that makes all the maxima coincide in the same point. The same procedure can be applied to region 2, however, the common point is w* in this case.

The transformation applied is a linear dilatation defined by:

$$w' = \frac{w_1}{w_k}w \tag{7}$$

where w_1 is the value of the common point after doing the dilatation and w_k is the position of this common point for each spectral function before carrying out the dilatation (k = 2,3 . . . v), where v is the number of spectra for each nuclide.

We see in Figure 2 that the common point for region 1 is the maximun position of the less quenched spectrum. The w value of the less quenched spectrum is the common point for region 2.

It is proved that the choose of the position of the inflexion in the spectra or the point y/r as the common point for region 2 also gives good results.

Once the dilatation is computed it is not difficult to obtain the interpolated spectrum which corresponds to a given value of the quench parameter. First of all, we carry out a channel by channel interpolation between the dilated regions of the spectral functions. Then we calculate a new spectral function for

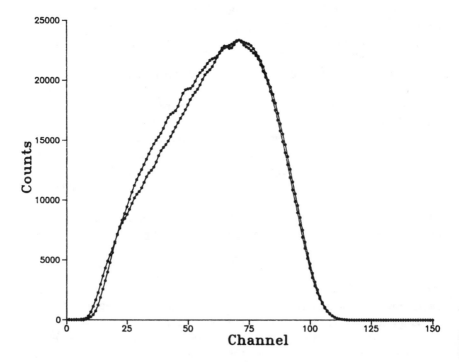

Figure 2.   Fourier spectra after shifting.

these points as we have done before for the experimental spectra. Finally, we apply the contraction:

$$w' = \frac{\bar{w}_2}{w_1} w \qquad (8)$$

where w_1 and $\bar{w}_2$ are the positions of the common point before and after the contraction is applied respectively. The next section shows how to calculate w_2.

Spectral Parameters Depending on the Quench

In the previous section we associated the maxima and the ends of the spectral functions with the common points that correspond to the maximum and the end of the least quenched spectrum.

It is interesting to see the variation of these quantities as a function of the quench parameter. A straight line can be obtained by least square fitting of the data. The maximum and the end of the spectral function for a given value of the quench is calculated from this fitting.

Special Fitting

Once the radionuclide spectra, corresponding to the quench parameter of the experimental mixture have been interpolated, obtaining the activity of each component of a radionuclide mixture is a problem of least square fitting.

Let y_i be the number of counts in the channel i of the mixture spectrum and y_{ij} be the number of counts in the channel i for the spectral component j. The intensities p_j for each nuclide are related to the spectral components by the equation:

$$y_i \simeq \bar{y}_i = \sum_{j=1}^{j=n} p_j y_{ij} \qquad (9)$$

where $\bar{y}_i$ is the number of counts of channel i in the computed spectrum and n is the number of spectral components. The values for p_j are obtained from the condition of minimum for the quantity:

$$G(p_1, \ldots, p_N) = \sum_{i=1}^{i=M} w_i [y_i - \bar{y}_i(p_j)]^2 \qquad (10)$$

where w_i is the weight for channel i. These weights have been taken inversely porportional to the statistical variance of the count number in each channel:

$$w_1 = \frac{1}{y_i} \qquad (11)$$

As the values of y_i depend linearly on the p_j, these can be computed by solving a system of n simultaneous linear equations.

The necessary and sufficient condition for the minimum of G is that the following equations hold,

$$\frac{\delta G}{\delta p_1} = \frac{\delta G}{\delta p} = \cdots \frac{\delta G}{\delta p_n} = 0 \tag{12}$$

these are equivalent to the system of equations:

$$p_1 \sum w_i y^2_{i1} + \cdots + p_n \sum w_i y_{in} y_{i1} = \sum w_i y_i y_{i1}$$
$$\cdots\cdots\cdots\cdots\cdots\cdots\cdots\cdots\cdots\cdots\cdots\cdots$$
$$p_1 \sum w_i y_{i1} y_{in} + \cdots + p_n \sum w_i y^2_{in} = \sum w_i y_i y_{in} \tag{13}$$

or in matricial form:

$$PQ = Y \tag{14}$$

where

$$P = (p_1 \cdots p_n) \tag{14}$$

$$Q = \begin{pmatrix} \sum wy^2_1 & \cdots & \sum wy_n y_1 \\ \cdots\cdots & \cdots\cdots & \cdots\cdots \\ \sum wy_1 y_n & \cdots & \sum wy^2_n \end{pmatrix} \tag{16}$$

$$Y = (\sum wyy_1 \cdots \sum wyy_n) \tag{17}$$

The index i has been omitted in order to lighten the expressions.
The p_i values are obtained from the equation:

$$P = YQ^{-1} \tag{18}$$

The activity A_i of the i-radionuclide is given by the formula:

$$A_i = \frac{p_i S_i}{\epsilon_i} \tag{19}$$

where S_i is the counting rate and ϵ_i the counting efficiency obtained from a calibration curve functioning as a quench parameter.

EXPERIMENTAL

The procedure described above can be applied to any mixture of radionuclides independently of the decay scheme. The limitations of the procedure will be given by the measurement data.

In principle the procedure can be applied to beta- and alpha-ray emitters and to electron capture nuclides. The multiple mixture of these nuclides and the

Table 1. Activity Discrepancies for ^{14}C + ^{3}H mixtures

Q	$^{14}C/^{3}H$	$\Delta^{14}C$ (%)	$\Delta^{3}H$ (%)
326.6	6.3	0.16	0.57
216.2	14.2	0.55	7.1
326.0	0.6	7.0	2.2

different quench and proportions give a great number of possible combinations to be checked. In this chapter the procedure will be applied only to some particular mixture in order to show its possibilities.

Three radionuclides have been used in our experiment: ^{3}H and ^{14}C (both as n-hexadecane), and ^{35}S (dioctyl sulphide).

All the solutions were standardized by LMRI (France) and dispersed homogeneously into 10 mL of toluene or dioxane based scintillator. The standard solutions were added gravimetrically to each vial using a Sartorious microbalance. Carbon tetrachloride was used as a quencher.

A LKB 1219 Rackbeta Spectral liquid scintillation counter has been used to obtain the spectra.

RESULTS AND DISCUSSION

Four different situations have been considered:

- Mixture of ^{14}C and ^{3}H with middle quench and 6.3 times more counting rate of ^{14}C than ^{3}H
- Mixture of ^{14}C and ^{3}H with high quench and 14.2 times more counting rate of ^{14}C than ^{3}H
- Mixture of ^{14}C and ^{3}H with moderate level of quench and very poor counting statistics.
- Mixture of ^{14}C and ^{35}S with moderate level of quench and sim ilar counting rates for both radionuclides.

Figure 3a shows the experimental spectrum of the ^{14}C and ^{3}H mixture. The quench paramenter for this sample is Q = 326.6, this corresponds to a counting efficiency for ^{3}H of about 0.29%. The counting ratio relationship between ^{14}C and ^{3}H was about 6. The spectrum of residuals Figure 3b shows a balanced dispersion. Table 1 shows that the discrepancies between experimental and computed activities are 0.16% for ^{14}C and 0.57% for ^{3}H.

Figure 4a shows the experimental spectrum of a ^{14}C and ^{3}H mixture with a quench parameter Q = 216.2, the corresponding counting efficiency is 0.11%. The counting ratio relationship between ^{14}C and ^{3}H is 14.2. The spectrum of residuals, Figure 4b, shows some predominant positive values due to the oscillations of the experimental spectrum close to the maximum. Table 1 shows that the discrepancies between computed and experimental activities are 0.55% for ^{14}C and 7.1% for ^{3}H. A better stability of the experimental spectrum would improve the discrepancies.

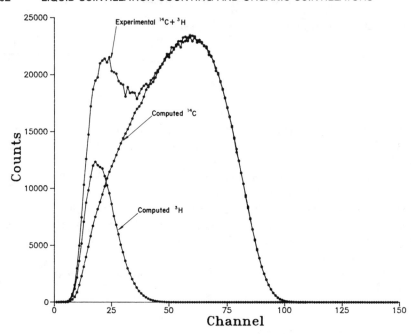

Figure 3a.   Experimental spectrum of a mixture of ^{14}C + ^3H and spectral decomposition for middle quenching.

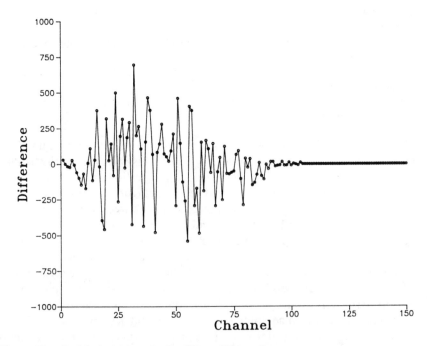

Figure 3b.   Residual spectrum for the fitting of Figure 3a.

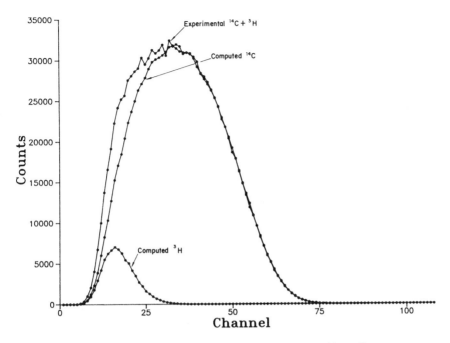

Figure 4a. Experimental spectrum for a high quenched mixture of ^{14}C + ^3H.

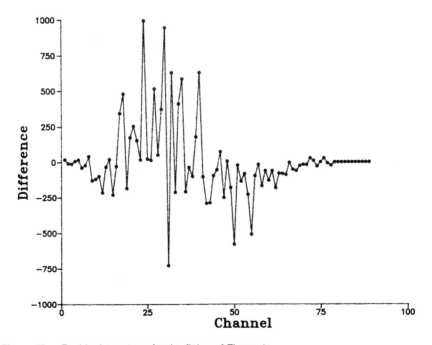

Figure 4b. Residual spectrum for the fitting of Figure 4a.

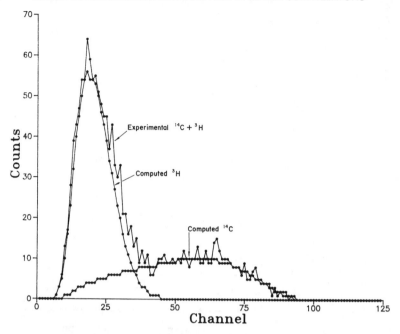

Figure 5a.   Experimental spectrum of a mixture of ^{14}C and ^{3}H with very poor statistics and spectral components.

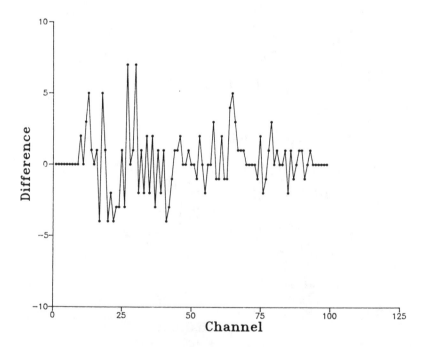

Figure 5b.   Residual spectrum for the fitting of Figure 5a.

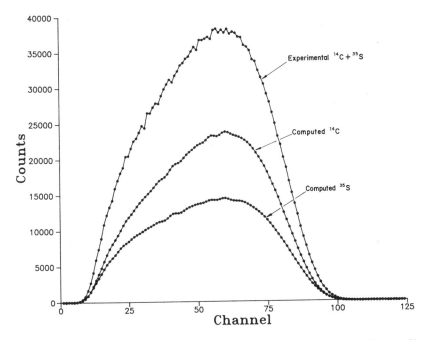

Figure 6a. Experimental spectrum and spectral components for a mixture of ^{14}C and ^{35}S.

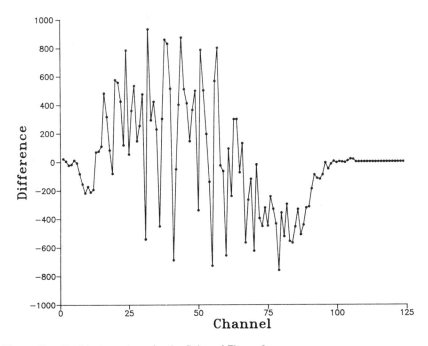

Figure 6b. Residual spectrum for the fitting of Figure 6a.

Figure 5a shows the experimental spectrum of ^{14}C and ^{3}H mixture with a quench parameter $Q = 326$. The measurement time was very short (1.5 seconds). That means that the counting rate is very low, like a low level activity sample, but without background. The counting rate between ^{14}C and ^{3}H is about 0.6. Figure 5b shows the residuals of the fitting. Systematic positive or negative values are clearly obtained, but the balance between both is quite acceptable. Table 1 shows that the discrepancies between computed and experimental activities are 7.0% for ^{14}C and 2.2% for ^{3}H.

Figure 6a shows the experimental spectrum of ^{14}C and ^{35}S. The quench is $Q = 330$, and the counting ratio relationship is $^{14}C/^{35}S = 1.5$. The discrepancies for ^{35}S and ^{14}C are 3.4% and 2.2%, respectively. Figura 6b shows the residuals of the fitting. It can be seen that for the low and high region of the spectrun there are several residuals unbalanced. The analysis of ^{35}S and ^{14}C is a very difficult problem because the maximum beta-ray energies for both radionuclides are very close, but the differences in the spectral shape permit the activities of each one of the radionuclides.

CONCLUSION

A computer code DILATA has been developed which is applicable to the analysis of radionuclide mixtures. The method is based on a new procedure to interpolate spectra and has been applied to obtain the activity of ^{14}C and ^{3}H mixtures in different conditions. The application to a mixture of ^{14}C and ^{35}S has shown the power of the method.

REFERENCES

1. Simonnet, G. and M. Oria. *Les Mesures de Radioactivité à l'Aide des Compteurs à Scintillateur Liquide*, (Paris: Eyrolles, 1977), pp. 89–95.
2. Martin-Casallo, M.T. and A. Grau Malonda. "Un Nuevo Procedimiento de Calibración de Muestras Doblemente Marcadas Basado en el Método del Trazador," CIEMAT Report (in press).
3. Horrocks, D.L. *Applications of Liquid Scintillation Counting*, (New York: Academic Press, 1974), p. 229,
4. Brown, L.C. "Determination of Phosphorus-32 and 33 in Aqueous Solution," *Anal. Chem.*, 43(10): 1326–1328 (1971).
5. Horrocks, D.L. and M.H. Studier. "Determination of Radioactive Noble Gases with a Liquid Scintillator," *Anal. Chem.*, 36(11): 2077–2079 (1964).

CHAPTER 26

On the Standardization of Beta-Gamma-Emitting Nuclides by Liquid Scintillation Counting

E. Garcia-Toraño, M.T. Martin Casallo, L. Rodriguez, A. Grau, and J.M. Los Arcos

INTRODUCTION

The $4\pi\beta$(LS) efficiency tracing method with ^{3}H has been successfully applied to the standardization of pure beta emitters.[1-5] In this method, a ^{3}H standard is used to calibrate the measuring system, and a combination of experimental measurements and theoretical computations standardizes the unknown nuclide. We present here the extension of this method to the case of any β-γ-emitting nuclide.

The basic diagram of the efficiency tracing method is shown in Figure 1 and a complete description can be found in previous papers.[1,2] A set of vials containing a ^{3}H standard is measured and a distribution of efficiency vs a quench parameter is obtained. On the other hand, the theoretical efficiency can be computed as a function of the figure of merit M, that we will define here as the energy in keV required to produce one photoelectron at the first dynode of the phototube.[4] The next step is obtaining the distribution of the figure of merit vs the quench parameter, which is independent of the radionuclide. The last distribution characterizes the measuring system. If we can compute the theoretical overall efficiency of the unknown nuclide vs the figure of merit, then we can calculate its activity from the precedent distributions.

COMPUTATION OF THE THEORETICAL COUNTING EFFICIENCY

In order to extend the method to any beta-gamma emitter, the counting efficiency must be computed. This implies the computation of all the possible ways the nuclide can decay to the ground state. There are three different processes that must be taken into account: beta emission, gamma emission, and electron conversion. First we will study the efficiency computation for these individual processes and then we will show how they can be combined to obtain the overall efficiency.

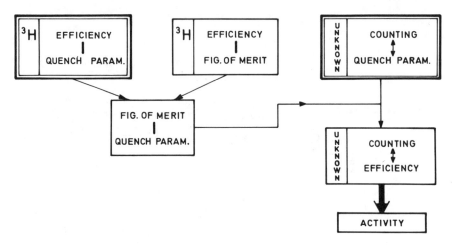

Figure 1.   Block diagram of the efficiency tracing method with ³H by LSC. Boxes drawn in thick lines correspond to experimental measurements.

Beta Efficiency

Taking into account shape factors, the beta spectra are calculated from the Fermi theory of beta decay. According to this theory, the number of particles of energy between E and E + dE is given by the equation:

$$\frac{dN(E)}{dE} = C(E)\ F_o(Z,E)\ L_o(Z,E)\ (E_m - E)^2\ (E + 1)\ p, \qquad (1)$$

where E_m is the maximum kinetic beta particle energy in m_oc^2 units, and F_o and L_o are given in reference.[4] C(E) is the shape factor, and its value[6] is given in the Table 1. In the case E = 0, the equation is:

$$\frac{dN(E)}{dE} = 8\pi\ C(o)\ (2R)^{2\gamma-1}\ (\alpha Z)^{2\gamma-1}\ \frac{E^2\ L_m(Z_\delta o)}{|\Gamma(2\gamma + 1|^2} \qquad (2)$$

where R is the nuclear radius of the residual nuclide and α is the fine structure constant. A more detailed description of the spectrum calculation can be found in References 1 and 4.

Table 1. Values of the Prohibition Parameter (q is the Neutrino Momentum in m_oc^2 Units)

Prohibition	Value
Allowed and first forbidden	1
Unique-first forbidden and second forbidden	$p^2 + q^2$
Unique-second forbidden and third forbidden	$p^4 + \dfrac{10}{3}p^2 \cdot q^2 + q^4$
Unique-third forbidden and fourth forbidden	$p^6 + 7p^2q^2\ (p^2 + q^2) + q^6$

If the beta spectrum is normalized to be unity, the counting efficiency is given by the expression:

$$\epsilon = \int_{0}^{E_m} N(E) \left(1 - \exp \left[- \frac{E.Q(E)}{2.M} \right] \right)^2 dE \tag{3}$$

Here, Q(E) is the ionization quenching correction factor given by:

$$Q(E) = \frac{1}{E} \int_{0}^{E_m} \frac{dE}{1 + k.B \left(\frac{dE}{dx} \right)} \tag{4}$$

In this work, we used the following approximate equation[7]:

$$Q(E) = \frac{.357478 + .459577*t + .159905*t^2}{1 + .0977557*t + .215882*t^2} \quad \text{with } t = \log_{10}E \tag{5}$$

The expression E.Q(E) in Equation 3 represents the amount of energy converted in photons and if the figure of merit (M) is considered, the probability of nondetection will be given by:

$$\exp \left(- \frac{E.Q(E)}{M} \right) \tag{6}$$

Hence, for a system with two tubes working in coincidence mode, the probability of detection will be given by:

$$\left\{ 1 - \exp \left(- \frac{E.Q(E)}{2.M} \right) \right\}^2 \tag{7}$$

Gamma Efficiency

To compute the γ counting efficiency, the interaction probability and the Compton spectrum distribution must be obtained. This is carried out by means of a Monte Carlo simulation. The emission point of the photon is drawn according to the vial dimensions, then three direction cosines and a pathlength are drawn to define the arrival point for this step. The type of interaction is decided by a random process, taking into account the cross sections for photoelectric and Compton processes; the trajectory of the photon is modified accordingly. The process finishes when the photon escapes out of the vial, its energy drops below 10 eV, or if the interaction is photoelectric. In any case, the amount of energy lost by the photon in the scintillator is considered to build up the electron distribution. The interaction probability is also obtained from the spectrum. A complete description of the process is given in García-Toraño and

Grau (1987), and a typical spectrum obtained in the simulation is shown in Figure 2.[8]

If N(E) is the distribution of energy of the electrons, the counting efficiency can be obtained from the expression:

$$\epsilon = \int_0^{E_\gamma} N(E) \left(1 - \exp\left[-\frac{E.Q(E)}{2.M}\right]\right)^2 dE \tag{8}$$

where E_γ is the gamma-ray energy.

Internal Conversion

Internal conversion transitions produce vacancies in the atomic shells. To compute the effective energy converted into light, one must first calculate the probabilities and effective energies for the different processes involved. Only K,L, and M internal transitions are considered. The complete expressions for all the processes have been detailed in Grau (1982).[9]

The internal conversion processes contributes to the efficiency as a function $F(\phi_1, n_1, E_{ij}, E_e)$, where ϕ_1 and n_i are the probability and number of emitted particles in the i-atomic rearrangement, E_{ij} is the energy corresponding to the j-th particle and E_e is the converted electron energy.

Overall Efficiency

Consider the case of a radionuclide which decays by beta emission and is followed by a cascade of n gamma rays to the ground state of the daughter nuclide. Although this is a very simplified model of β-γ emitter, the expression that gives the counting efficiency is as complex as:

$$\epsilon = \int_0^{E_\beta} \int_0^{E_{\gamma 1}} \int_0^{E_{\gamma 2}} \ldots \int_0^{E_{\gamma n}} N(E)S_1(E_1) \ldots S_n(E_n) \left(1 - e^{-E_\beta Q(E_\beta) - \sum E_{\gamma i} Q(E_{\gamma i})}\right)^2 dE_1 \ldots dE_n \tag{9}$$

where

$\qquad E_\beta$ = Maximum energy of the beta emitter
$\qquad E_{\gamma i}$ = Energy of the i-th gamma ray
$\qquad N(E)$ = Beta spectrum
$\qquad S_i(E_i)$ = Spectrum of Compton and photoelectric electrons

In a real case, some of the gamma transitions could be converted, and the appropriate conversion coefficients would affect Equation 9, which should be split in all the possible combinations of gamma and electron conversion processes. The argument of the exponential should also be modified in accordance

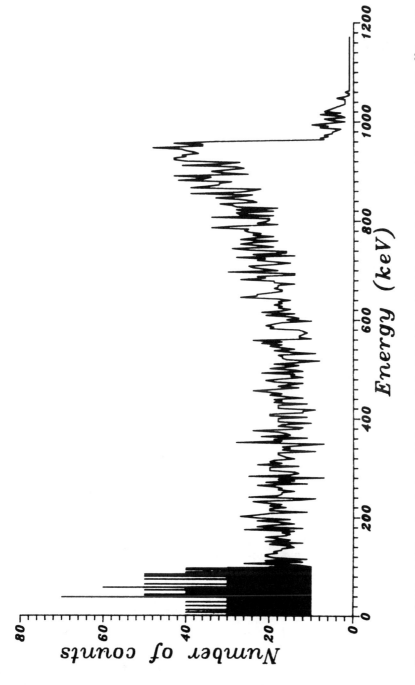

Figure 2. Spectrum of Compton electrons obtained in a typical simulation of the interaction between the 1.17 MeV γ ray of the ^{60}Co and the scintillator (dioxane-naphthalene). The volume of the vial is 10 ml and the radius is 1.25 cm.

with the internal conversion processes, and Equation 9 would become a sum of similar expressions, with the required normalizing factors.

If we adopt a general scheme, other β emissions can feed the intermediate levels, and some additional gamma transitions will appear. The general expression for the efficiency then becomes a combination of expressions like Equation 9; for instance, in the case of ^{60}Co, 15 expressions must be computed. Its numerical evaluation is not possible without some kind of factorization; a typical Compton spectrum has about 2000 points and the multiple integral could take excessive time.

It may be seen that the integral (Equation 9) can be factorized in terms of some factors which only depend on the individual processes. We have developed a FORTRAN program which builds up all the possible decay ways, factorizes the integrals, and takes into account the normalizing factors (branching ratios, conversion coefficients, interaction probabilities, etc.).[10] It has been used in the efficiency calculations needed in this work.

EXPERIMENTAL RESULTS

The method described in this paper has been applied to the standardization of ^{60}Co. This nuclide, which decays by β-γ emission to ^{60}Ni is usually standardized by the method of $4\pi\beta$-γ coincidence-anticoincidence. A simplified decay scheme is shown in Figure 3, and some characteristic nuclear data are presented in Table 2.[11] Although there are other transitions, only the most significant have been considered in this study.

Materials

A standard solution of n-hexadecane ^{3}H from Amersham, U.K., was used as the reference for the efficiency tracing method. The radioactive concentration was 50 kBq/mL and the uncertainty, taken as the addition of both systematic and random components, was certified to 3%. The ^{60}Co was also from Amersham, and its chemical form was Cl_2Co, 0.1 M. The certified uncertainty was 0.5%.

The scintillation solution was formed by naphthalene 60 gr, PPO 4 gr, Dimetil POPOP 0.1 gr, methanol 100 mL, etilenglicol 20 mL and Dioxane until 1 L.

Sample Preparation and Equipment

Two sets of seven identical vials were prepared, one for the ^{3}H, the other for the ^{60}Co. Each vial contained 10 mL of the cocktail solution. In each set Cl_4C was added in 5 μL increments to obtain different quench parameters. The radionuclide solutions were added gravimetrically to the vials. The stability was studied over a period of one week and proved to be good.

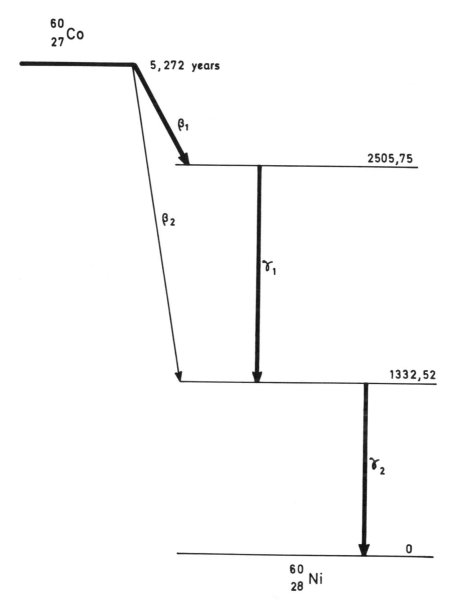

Figure 3. A simplified decay scheme of the ^{60}Co.[10]

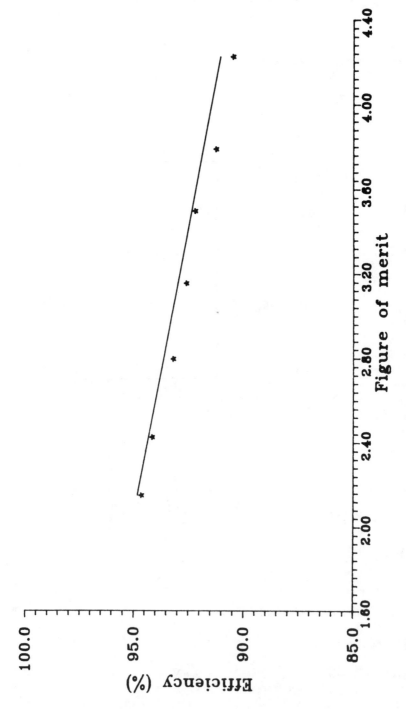

Figure 4. Measured (single points) and computed (full line) efficiencies in the standardization of ^{60}Co, as a function of the figure of merit.

Table 2. Selected Nuclear Data[10] from ^{60}Co

Transition	Energy (keV)	Intensity (%)
β^1	317.9 (0.3)	99.89 (0.06)
β^2	1491.1 (0.3)	0.10 (0.02)
γ^1	1173.22 (0.05)	99.89 (0.06)
γ^2	1332.51 (0.05)	99.993 (0.002)

Table 3. Measured and Computed Efficiencies for ^{60}Co

Sample	Quench Parameter	Figure of Merit	Measured Efficiency	Computed Efficiency	Diff (%)
1	344	2.1492	94.6194	94.8144	−0.20
2	327	2.4262	94.1109	94.2893	−0.19
3	306	2.7979	93.1427	93.5946	−0.48
4	289	3.1562	92.5480	92.9340	−0.41
5	275	3.4985	92.1405	92.3104	−0.18
6	262	3.7898	91.1859	91.7846	−0.66
7	249	4.2260	90.4283	91.0053	−0.63
					Average −0.4%

Measurements were carried out with a LKB RackBeta liquid scintillation counter, which was connected on-line to a personal computer.

Results

We present in Table 3 and Figure 4 the results for the measured and computed efficiencies. In the same table, values are also given for the quench parameter and the figure of merit. The differences between experimental and computed[11] efficiencies, also shown in the table, vary between 0.18 and 0.66%, with an average value of 0.4%.

The uncertainties estimated for the method are shown in Table 4. The most important are due to the counting statistics and the quench parameter determination; the contribution of the 3% uncertainty in the ^{3}H standard leads to only a 0.1% in the efficiency of ^{60}Co. We also considered the component due to the nuclear data, and finally, we numerically estimated the influence of the Monte

Table 4. Estimated Uncertainties in the Standardization of ^{60}Co by the LSC Efficiency Tracing Method with ^{3}H

Source of Uncertainty	Uncertainty (%)
Liquid scintillation counting of ^{3}H	0.2
Liquid scintillation counting of ^{60}Co	0.17
Quench parameter determination	0.1
Sample preparation	0.1
Nuclear data	0.09
Monte Carlo simulation	0.06
^{3}H standard	0.1
Combined uncertainty	0.33%
Overall uncertainty (three times the combined uncertainty)	0.99%

Carlo simulation as 0.06. The combined uncertainty resulted in 0.332% and for the overall uncertainty, taken as three times the combined, we found 0.99%. These values are in good agreement with the average differences found in Table 3, taking into account the uncertainties in the ^{60}Co standard.

In conclusion, a method based on the 4πLS efficiency tracing with ^{3}H has been developed which allows the standardization of β-γ emitters; the application to the case of ^{60}Co has given results which agree well with the values obtained for other methods.

REFERENCES

1. Grau, A. and E. García-Toraño. "Evaluation of Counting Efficiency in Liquid Scintillation Counting of Pure β-ray emitters," *Int. J. Appl. Radiat. Isot.*, 33:249–253 (1982).
2. Coursey, B.M., W.B. Mann, A. Grau, E. García-Toraño, J.M. Los Arcos, and D. Reher. "Standardization of Carbon-14 by $4\pi\beta$ Liquid Scintillation Efficiency Tracing with Hydrogen-3," *Appl. Radiat. Isot.*, 37(5):403–408 (1986).
3. Coursey, B.M., L.L. Lucas, A. Grau, and E. García-Toraño. "The Standardization of Plutonium-241 and Nickel-63," *Nucl. Instr. and Methods*, A279:603–610 (1989).
4. García-Toraño, E. and A. Grau. "EFFY, a New Program to Compute The Counting Efficiency of Beta Particles in Liquid Scintillators," *Comp. Phys. Comm.* 36:307–312 (1985).
5. Coursey, B.M., J.A. Gibson, M.W. Heitzman, and J.C. Leak. *Int. J. Appl. Radiat. Isot.*, 35:1103–1107 (1982).
6. Daniell, H. "Shapes of Beta-Ray Spectra," *Rev. of Modern Phys.* 40:659 (1968).
7. Los Arcos, J.M., A. Grau, and A. Fernández. "VIASKL: A Computer Program to Evaluate the LSC Efficiency and Its Associated Uncertainty for K-L-Atomic Shell Electron-Capture Nuclides," *Comp. Phys. Comm.*, 44:209 (1987).
8. García-Toraño, E. and A. Grau. "EFYGA, A Monte Carlo Program To Compute The Interaction Probability and The Counting Efficiency of Gamma Rays in Liquid Scintillators," *Comp. Phys. Comm.* 47:341–347 (1987).
9. Grau, A. "Counting Efficiency for Electron-Capturing Nuclides in Liquid Scintillator Solutions," *Int. J. Appl. Radiat. Isot.* 33:249–253 (1982).
10. García-Toraño, E., J.M. Los Arcos, and A. Grau. "MULBEGA: Counting Efficiency for β-γ Emitters in Liquid Scintillation Counting," to be published in *Comp. Phys. Comm.*
11. Lagoutine, F., N. Coursol, and J. Legrand. *Table de Radionucleides*, (Giff Sur Ivette, France: Laboratoire de Metrologie des Rayonnements Ionisants, 1983).

CHAPTER 27

The Standardization of [35]S Methionine by Liquid Scintillation Efficiency Tracing with [3]H

J.M. Calhoun, B.M. Coursey, D. Gray, and L. Karam

INTRODUCTION

[35]S ($T_{1/2}$ = 87.44 days) decays by beta-particle emission of maximum energy 166.74 keV and average energy 48.60 keV.[1] [35]S has been long used for labeling organic compounds for *in vitro* measurements.[2] It is now increasingly used for labeling sulfur-containing amino acids methionine (Met) and cystine (Cys) for protein sequencing studies, as well as in following the progress of amino acid incorporation during protein synthesis. Further potential applications of ^{35}S labeled compounds are in studies of the interactions of sulfur-containing carcinogens and mutagens (e.g., sulfur mustards, methyl methane sulfonate, ethicnine, etc.) with DNA, particularly with respect to the nature of the binding of such compounds to the macromolecule. Because of the low energy of the beta particles, ^{35}S is usually assayed by liquid scintillation counting.[3] Very few papers are available on the standardization of [35]S. Bryant et al., used the method of $4\pi\beta$-γ efficiency tracing with [60]Co.[4]

In previous work we have shown that efficiency tracing with ^{3}H can be used to standardize pure beta-particle emitters of low energy, e.g., nickel [63]Ni and [241]Pl,[5,6] and intermediate energy, e.g., [14]C.[5,6] These measurements were made, hwoever, under low quenching conditions. The present work was undertaken to see if the method could be used for ^{35}S labeled amino acids under high quench conditions such as are routinely encountered in assaying biological samples.

The basic principle of the efficiency tracing technique is to explicitly account for the beta-particle spectra of the radionuclide to be assayed (^{35}S) and the standard nuclide (^{3}H). The counting efficiency for a liquid scintillation system with two phototubes operating in coincidence, ϵ_c, is given by Coursey et al.[6]

$$\epsilon = \left\{ \int_0^{E_{\beta_{max}}} P(Z,E) \times [1 - \exp(-E\eta Q(E)W(E))]^2 \, dE \right\}$$

$$\times \left\{ \int_0^{E_{\beta_{max}}} P(Z,E) \, dE \right\} \tag{1}$$

where P(Z,E) dE is the Fermi distribution function,
η is the figure of merit, photoelectrons per keV,
Q(E) is the ionization quenching function to account for the differences in light yield for electrons as a function of energy,[7] and
W(E) is a wall-loss function, taken here as unity for these low-energy β particles.

The liquid scintillation counter is the first efficiency calibrated with ^{3}H standards and the optimum value for the $\eta Q(E)$ term in Equation 1 is obtained. The system may then be used to standardize for activity any other radionuclide for which the Fermi spectrum is known.

EXPERIMENTAL

Materials and Methods

^{35}S labeled Met (5 mCi/0.5 mL (185 MBq/0.5 mL)) was obtained from Nordion (Ottawa, Canada).* A carrier solution was prepared for all dilutions of the ^{35}S consisting of 10 mM beta-mercaptoethanol, 50-mM N-[tris (hydroxymethyl) methyl] glycine (tricine), and 0.1 mM stable Met. The ^{3}H water was a dilution of NIST Standard Reference Material (SRM) 4927C. Each scintillation vial contained 10 mL Beckman Readysolv HPb scintillator in a polyconeseal glass vial (Kimble). Two sets of identical vials were prepared; one for the ^{35}S and one for the ^{3}H standards. For each set, chloroform was added in 50 μL increments (0 to 200 μL) to simulate the quenching expected in biological samples.

The radionuclide samples (27 to 82 mg) were added gravimetrically to the vials containing scintillation cocktail.

Equipment

Measurements were made with a Beckman LS7800 liquid scintillation counter on line to a Charles River Data Systems supermicro data aquisition system.[8] Data were downloaded to an IBM XT PC for processing. Efficiencies

---

*Mention of commercial products does not imply recommendation or endorsement by the National Institute of Standards, nor does it imply that the products identified are necessarily the best available for the purpose.

and spectra were computed using the program EFFY2 implemented on the NIST PC by Eduardo García-Toraño in 1988.[9]

RESULTS AND DISCUSSION

The samples were measured over a period of 30 days. The ^{35}S-methionine proved to be very stable (less than 0.1% change), while for the ^{3}H samples, the count rate decreased by approximately 0.16% per day.[5] The tritiated water separates from the scintillator with time, while the ^{35}S labeled Met is apparently in the organic phase.

Figure 1 shows the Fermi spectra of ^{35}S (Figure 1a) and ^{3}H(Figure 1b). The measured LS spectra for the two nuclides for low quenching (Horrocks' H# around 70) and high quenching (H# around 150) are shown in Figures 1c and 1d for the ^{35}S and ^{3}H, respectively. The low quenching samples were used to standardize the ^{35}S by the efficiency tracing technique.

Figure 2 shows the relationship between the figure of merit, the quench parameter H#, and the counting efficiencies of ^{3}H and ^{35}S. Although the ^{3}H efficiency decreases with quenching from 49 to 24%, the ^{35}S efficiency only

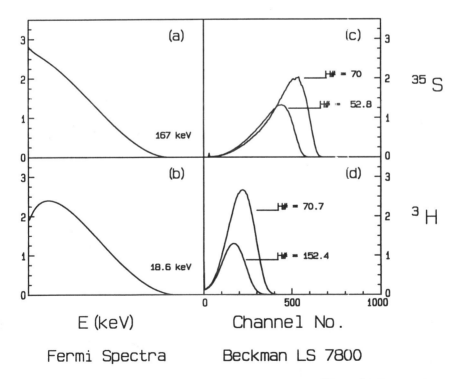

Figure 1.   Beta particle spectra and liquid scintillation spectra for ^{35}S and ^{3}H. The experimental spectra was obtained on a system with a logarithmic amplifier.

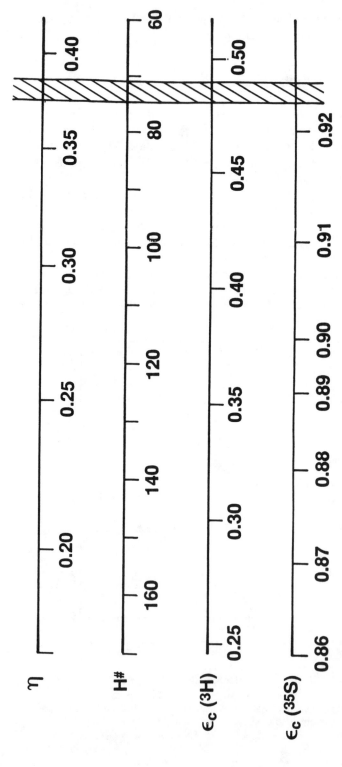

Figure 2. Relationships between the figure of merit (η), the quench paremeter (H#), and the two-phototube coincidence counting efficiencies for ^{3}H and ^{35}S. The shaded area is the unquenched region used for the most accurate standardization of the ^{35}S. This nomograph is strictly valid for only one scintillator and counting system.

Table 1. Results for Nordion ^{35}S Methionine as of 1200 EST February 24, 1989 (uncertainties are given as one standard deviation)

Vial No.	H#	ϵ_c	Predicted Activity (Bq/mg)
1	71.3	0.9256	69.43
2	70.5	0.9261	69.15
3	71.4	0.9255	69.14
4	72.8	0.9245	69.34
5	71.6	0.9253	69.23
			69.26 ± 0.12
6	104.7	0.9028	69.50
7	118.1	0.8940	69.81
8	131.4	0.8854	69.51
9	141.0	0.8793	69.45
10	152.8	0.8719	69.25
11	171.3	0.8604	68.75
			69.38 ± 0.36

drops from 92 to 87% over the same chloroform concentration range. The predicted values of the activity for all samples are shown in Table 1. The average for the quenched samples, 69.38 ± 0.36, compares favorably with the value obtained for the unquenched samples 69.26 ± 0.12 Bg/mg (uncertainties are one standard deviation). The estimated uncertainty in the ^{35}S activity concentration, which includes systematic as well as random components, is given in Table 2.

Since this standardization method can be extended to high quenching for ^{35}S, it should work as well for ^{14}C and ^{32}P. Figure 3 shows computed counting efficiency for all three radionuclides and ^{3}H as a function of quenching. Using the *"measured ^{3}H efficiency"* one can compute the efficiency of any of the other three to within 1 to 2%.

Table 2. Estimated Uncertainties in the Standardization of ^{35}S Methionine

	Percent (%)
a) Liquid scintillation measurements	0.18
b) ^{3}H reference beta-particle standard	0.21
c) Quenching in the liquid scintillation measurements	0.05
d) Source preparation	0.10
e) Uncertainty in efficiency curve fit	0.07
f) Scintillator stability	0.05
g) Uncertainty in numerical spectra integration	0.05
h) Dead time	0.03
Combined in quadrature	0.32
Overall uncertainty (× 3)	1.0

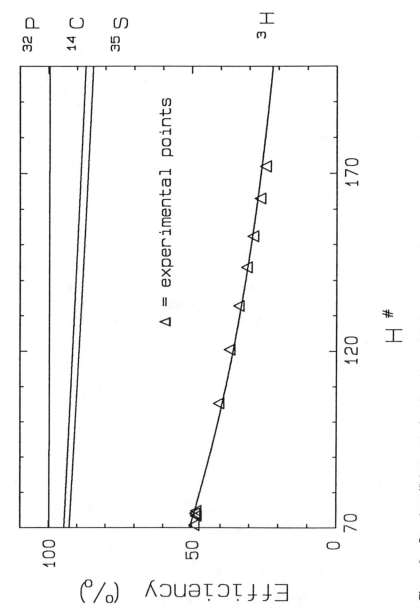

Figure 3. Counting efficiency as a function of quenching, with experimental points from this work. Curves computed using code EEFY2 (Grau Malonda and García-Toraño (1985)).

CONCLUSIONS

^{35}S labeled methionine has been standardized by the method of $4\pi\beta$ liquid scintillation efficiency tracing with ^{3}H. The method appears to work well for chemically quenched samples up to at least H# 171. Since this program is now implemented on a personal computer, it should be possible to adapt most commercial liquid scintillation counters to directly compute ^{35}S activities, providing a set of ^{3}H quenched standards is available to establish a quench curve vs a quenching parameter.

REFERENCES

1. *Handbook of Radioactivity Measurements Procedure 58*, Second ed. (Bethesda, MD: National Council on Radiation Protection and Measurements, 1985).
2. Gordon, B.E., H.R. Lukens, and W. ten Hove. *Int. J. Appl. Radiat. Isotp. 12*: 145 (1961).
3. Horrocks, D.L. *Applications of Liquid Scintillation Counting*, (New York: Academic Press, 1974).
4. Bryant, J., D.G. Jones, and A. McNair. in "Standardization of Radionuclides," proceedings of a symposium, International Atomic Energy Agency, Vienna, (1967).
5. Coursey, B.M., L.L. Lucas, A. Grau Malonda, and E. García-Toraño. *Nucl. Instrum. Meth. Part A.* in press, (1989).
6. Coursey, B.M., W.B. Mann, A. Grau Malonda, E. García-Toraño, J.M. Los Arcos, J.A.B. Gibson, and D. Reher. *Int. J. Appl. Radiat. Isopt. 37*, 403 (1986).
7. Grau Malonda, A. and E. García-Toraño. *Int. J. Appl. Radiat. Isopt. 33*: 249 (1982).
8. Coursey, B.M., M.P. Unterweger, L.L. Lucas, and E. García-Toraño. *Trans. Amer. Nucl. Soc. 55*: 54 (1982).
9. Grau Malonda, A. and E. García-Toraño. *Comp. Phys. Comm. 36*:307 (1985).

CHAPTER 28

A Review and Experimental Evaluation of Commercial Radiocarbon Dating Software

S. De Filippis and J. Noakes

ABSTRACT

New commercial radiocarbon dating computer software has been developed to provide on-line data analysis for Packard computer assisted counting systems and off-line data analysis for other instruments. Because different radiocarbon laboratories may have slightly different methods of calibration and data presentation, the software was designed based on the routine methods and calculations performed at the Center for Applied Isotope Studies (CAIS). Input from several other radiocarbon laboratories was also incorporated in the design of the computer software. This chapter presents a complete product critique, from a users' point of view, covering ease of use, calculations, and future applications of radiocarbon application software. Comparative data will be presented to show the differences between this radiocarbon software data analysis and the routine data analysis that is performed at the Center for Applied Isotope Studies.

The software is menu driven with on-screen programming and editing for a professional user interface environment. It also has the capability to save and analyze counting data directly from the Packard liquid scintillation system. The software can be programmed to accept oxalic calibration information and count data. The software data disk management develops count history files of a carbon reference standard and background over time. The user can archive data into a unique data subdirectory. Data can be selected to export collected and stored information for incorporation into other computer software such as Lotus^{R}1,2,3^R or database programs.

The software offers several correction features, including an account of benzene evaporation and Delta ^{13}C isotope value, and it can be programmed for the scintillation counter radionuclide efficiency.

INTRODUCTION

Laboratories that specialize in low level radiocarbon measurements as a part of their own research programs or as a commercial service to the scientific community have had to develop in-house computer software to automate these routine calculations. A question often raised by new laboratories that for the first time enter into low level radiocarbon measurements concerns the analysis and interpretation of the counting data. In many cases this question remains unanswered until a literature search through the annals of *RADIOCARBON* uncovers the commonly used calculations and methods of correction for radiocarbon age dating. It is not uncommon to discover a startup radiocarbon laboratory that has limited knowledge of these calculations and how to apply the correction schemes.

The application of laboratory computers used in routine data analysis for a specialized task like radiocarbon age dating has, in the past, required a programmer/scientist to develop and test the software used by the laboratory. In the absence of computer software, hand calculations would suffice.

When we learned that specialized application commercial software was being developed for radiocarbon age dating analysis, we became interested in evaluating and comparing the results of the new software package to our own radiocarbon analysis program.

BACKGROUND

The radiocarbon software was developed by Packard Instrument Company of Downers Grove, IL, for use with the computer assisted, as well as the noncomputer automated, line of liquid scintillation counters that the company manufactures. At the time of this independent product evaluation, the software has not been released for commercial distribution. The software provided for our studies was a prerelease version. The software as we know it was intended for on-line data analysis using the Packard Tri-Carb® series models 2260, 2250, 2200, 1050, 1000 liquid scintillation analyzers. Off-line data entry and subsequent analysis does not seem at this time to be an option of this software. The liquid scintillation analyzer used in our evaluation was a Packard Tri-Carb model 2050, predecessor of the model 2250, both are specially designed low background counting systems.

EVALUATION CRITERIA

This product evaluation was conducted solely on the basis of an individual end user. The software was evaluated in five areas of interest; (1) installation and documentation, (2) ease of use, (3) correctness of calculations, (4) user benefit features, and (5) future applications.

In order to compare results obtained from the new Packard software to the data analysis performed at CAIS, several wood samples from the radiocarbon laboratory at the University of Waikato in New Zealand were dated using both computer programs. The samples of wood ranged in age between 2,000 years before present (YBP) to approximately 50,000 YBP. All sample dates were corrected for carbon isotope fractionation. The opinions expressed here are those of the authors and do not represent those of Packard or the University of Georgia.

MATERIALS AND METHODS

Each wood sample was prepared for liquid scintillation counting by first cutting away the surface related wood material and then successively pretreating it with chemicals to remove possible contamination from modern carbon components. The treated wood sample was processed to pure benzene using a benzene synthesizer. A 3 cc sample of benzene was derived from wood combustion and mixed with scintillators; counting followed.

The samples were first counted using the CAIS scintillation counters, normally used to process routine samples. Each sample was counted for 2700 minutes in a Picker/Nuclear Liquimat 220 with a specially designed low background copper/Teflon®* shielded counting vial, and then the age date was calculated.

Originally, the counting experiment was designed for the CAIS system and the Packard software to be similar. It was later discovered that the Packard software was designed to count each sample once, unload, and cycle it around the sample changer before counting for a repeat time. To minimize statistical variations due to cycling, each sample was counted for a single 999 min interval, and then the date was computed using the Packard software. Normally at CAIS, each sample, background, and oxalic standard is counted in the same counting vial because each liquid scintillation counter is calibrated with its own vial to minimize experimental variations. Using the Packard system, the software is designed to count the sample in a multi-user environment. This means that the investigator should place the samples in a cassette in a desired sequence. This sequence can be programmed by the user into the software, so the software can recognize the difference between a standard, background, oxalic, and sample. However, this multi-user environment does not lend itself to counting each radiocarbon sample, oxalic, or background in the same counting vial. To accomplish this, the investigator must manually intervene to stop the counting system, prepare a new sample in the same vial, and start the counting again. This system works best when using a different vial for each sample. The counting vial used with the Packard counting system was a standard low ^{40}K borosilicate glass vial with an internal volume of 7 cc. Teflon cap liners were used to prevent solvent loss during the counting interval.

HARDWARE AND SOFTWARE REQUIREMENTS

The software was developed in the "C" programming environment and runs on an IBM Personal Computer (PC) or Professional Series (PS) microcomputer system and compatible. The computer in any case must be compatible with the interfacing to the Packard scintillation counter and the executing of the scintillation software. The radiocarbon application software is an add-on program that runs within the scintillation software operating environment

---

*Teflon is a Registered Trademark of E.I. DuPont de Nemours

on computer assisted counting systems. Using the Packard Datalink™ stand alone counting systems like the Tri-Carb model 1050 the investigator can compute age dates with the radiocarbon software. No additional computer equipment, other than what is typically provided with each Packard counting system, is required for operation. During the software installation, the color option can be activated for use with top of the line Packard couting systems, which are equipped with a color monitor.

Installation and Documentation

Since the software provided for evaluation was a prerelease version, the documentation supplied was not the information that would accompany the commercially released product. At this time, we cannot describe the quality or completeness of this material.

The program installation was complete. The original disk, provided in the 5.25 in. floppy format, contained an installation routine that prompted the user for the different Packard counting system models. Since the Packard system used was an earlier vintage, the use of subdirectories was not supported. The radiocarbon software was then installed to the root directory on the hard disk, which contained all other files used to operate and communicate with the scintillation counter. Newer models of Packard counters support the use of subdirectories, making the house keeping of different files of a given software package much easier.

Ease of Use

Once installed, the software was very easy to access through the Packard scintillation software user interface. The radiocarbon program was developed as a menu driven system with preprogrammed function keys for different operations and modules. The Packard operating environment allows application software, like the radiocarbon program to execute and run while the scintillation counter was performing the most recent instruction from the scintillation software. If the scintillation system was counting a sample when the user executed the radiocarbon software, the sample would continue to count until its time or statistical termination was satisfied. At that point, if the scintillation program was not returned to the main portion of memory, the counting system would wait until the user terminated conversation with the application program before the counter would go on to the next instructed task. A safety feature of the system is that it can sense keyboard activity. If the user executes the application program and is called away from the computer, the program will automatically terminate, saving the most recent information and returning to the operation of the scintillation counter within a preset period of time.

The use of eight preprogrammed function keys enables the user to navigate through the system as shown in the flow chart (Figure 1). There is a help line at

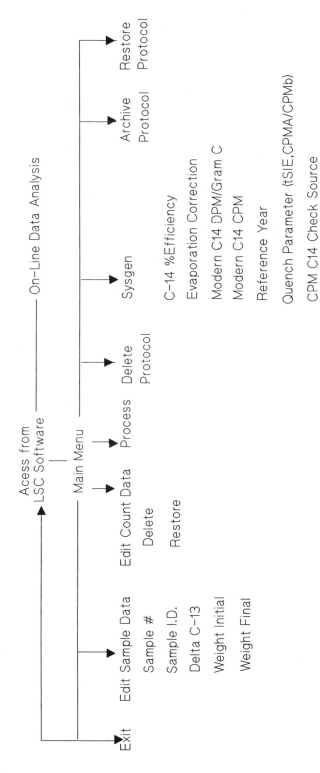

Figure 1. Radiocarbon software flow chart.

the bottom of each screen to provide expected responses for software prompts.

Once the counting protocol in the radiocarbon software has been predefined by the user, all that is required is counting the sample. The subsequent data analysis is automatic, a printed report follows. The protocol defined in the radiocarbon software must be identical to the counting protocol on the scintillation counter. The radiocarbon software can be user programmed through the scintillation software to process the sample data after each sample is counted (Sample Mode) or after a contiguous group of samples is counted (Batch Mode). However, in Batch Mode, the radiocarbon data analysis and report are not generated until the logical end of the group. So, if each sample is programmed to count for 1000 minutes, it could be some time before the radiocarbon age of the first sample is reported.

Overall, the software can automatically calibrate, based on the user providing the correct oxalic reference information, weight of the sample in terms of benzene, and Delta ^{13}C data. The user can edit previously counted data on-screen, should there be a need to delete or restore data. In addition, the user can achieve logical groups of raw sample count data and also export that information to a selected disk drive as an ASCII file for later import into other software packages.

SOFTWARE FEATURES

System Generation

One of the initialization functions the software provides the user is the ability to enter the externally computed ^{14}C counting efficiency, oxalic acid specific activity, and the count rate of a check source. Other information regarding the correction of benzene evaporation is activated within Sysgen. The Sysgen screen allows the user to select or enter their own specific activity for the oxalic acid standard. Presets were available for oxalic acid I (SRM 49) of 14.27 DPM/gr C and oxalic acid II (SRM 4990-C) of 18.46 DPM/gr C. The software was designed to automatically compute updated counting efficiency information, check source CPM, and oxalic acid CPM when these samples were identified by the user and counted.

This screen appears to be simple in design, but the prompts were somewhat vague and should be explained in the final documentation. The software does, however, allow the user to enter a nondefined specific activity for the oxalic acid in the event that the user is not calibrating on oxalic acid, but possibly some other reference material.

Edit Sample Data

The radiocarbon software provides lets user program each sample identification with a sample number corresponding to the position of the sample in the cassette, the Delta ^{13}C value, and the weight of sample benzene. When the evaporation question is activated through the Sysgen menu, another column of data appears on this screen, allowing the user to enter the final sample benzene weight. In order to correct the previous data for the loss of benzene during counting, the original count data must be manually reprocessed. The sample specific activity is corrected for the sample benzene loss by taking half of the difference between initial and final benzene weights. As a result of benzene loss, the percent difference between the initial and final weights are computed.

Once the sample data have been entered, the user can initiate counting and wait for the final results. This module of the software allows the user to program on what given date the calculation should begin.

Edit Count Data

Once the samples have been counted, a count history file is generated for each sample. When a particular sample is counted repeatedly, the total counts for each cycle are combined and a correlation value computed. If the sample has counted for more than 5 cycles, an additional correlation value over the most recent 5 cycles is computed. This provides the user with information regarding the reproducibility of the count results. If there is any bias in the data, the counts/day are computed and printed.

In this module the user can selectively delete one or more count repeats. The program flags the deleted point(s), but does not discard the information. If the user wishes to restore the deleted point(s), the Restore function can remove the delete flag so that the sample data can be incorporated into the final calculations.

COMPARATIVE RESULTS

In order to compare the radiocarbon age calculations between the CAIS routine method and the new Packard software, six different wood samples were processed to benzene and counted. These samples ranged in ages between approximately 2,000 YBP and 50,000 YBP. Also a modern oxalic standard was measured to see how the software would handle the 136% of modern sample. The benzene samples were measured in the standard CAIS counting vial when the sample was counted using the CAIS LSC spectrometers. The Packard software was designed to count each sample once and then cycle the sample, enabling the instrument to count another sample before repeating a given measurement. Because of this limitation, the same counting vial could not be used for the standard, sample, and background as was accomplished in the CAIS routine. Each benzene sample was transferred from the CAIS counting vial into a common 7 cc low ^{40}K glass vial.

The comparative results are presented in Table 1. It was apparent that on the

**Table 1. Radiocarbon Age Dates–Charcoal Samples
CAIS vs Packard Software**

Sample I.D. UGA	$\delta^{13}C$	CAIS YBP ± 1σ	PACKARD YBP ± 1σ	% DIFFERENCE
5894	−25.81	1,974 ± 50	2,019 ± 50	2.3
5897	−25.85	2,036 ± 52	1,894 ± 50	−7.0
5891	−24.96	3,232 ± 419	2,894 ± 50	−10.5
5903	−24.67	5,058 ± 65	4,695 ± 50	−7.2
5900	−24.83	49,033 ± 7569	38,635 ± 50	−21.2
5888	−25.56	>50,000	−6E38	−
Oxalic–4990C	−17.68	Modern	−2393 ± 50 (Modern)	−

basis of counting these radiocarbon samples there were differences between the CAIS method and the Packard software. Most of the age dates computed automatically with the Packard software demonstrated a negative bias ranging from approximately 2 to 21% from the CAIS corresponding values. The software was executed, with no correction schemes activated, to yield computed age dates that were not influenced by any possible errors in the code.

Upon discussing the results, Packard is still in a program developmental stage. CAIS endeavors to assist the program developers at Packard in identifying the program errors so this commercial software package will carry the full merit of radiocarbon age dating capability.

GENERAL COMMENTS

At present, the Packard software is in need of further debugging and testing before it is ready for commercial distribution. We believe that new startup radiocarbon laboratories will probably benefit the most from this on-line software analysis. Established laboratories such as CAIS and others around the world, most likely will contine to utilize their own developed radiocarbon software for computation of age dates. Since this software is closely related to the equipment of one particular manufacturer, it does provide the capability for new laboratories to obtain the counting equipment and analysis software from a single source and it is ready to use.

During our test and evaluation period, we had discovered several bugs in the program, some which prevented us from continuing with our evaluation. Packard responded quickly with programmer support to correct the apparent problems and provided us with another version. The program source code was not made availabe to us for modification, but we were pleased with the prompt attention from Packard.

FUTURE APPLICATIONS

During the recent year, other radiocarbon laboratories have developed radiocarbon age date and radiocarbon correction type application software.

The Centrum Voor Isotopen Onderzoek in Groningen, the Netherlands, under Professor Van de Plicht, has made available to the radiocarbon community computer software that will provide tree ring corrections to radiocarbon dates. This software was developed in the Pascal language. Recently, a computer program named "CALI" was published by the Laboratorio de Datacion por Carbono-14 at the University of Granada in Spain. This program was developed similarily, to provide a calibration of radiocarbon age dates.

Although at CAIS, we have not had an opportunity to fully evaluate and compare these newest radiocarbon software packages, there appears to be a growing interest among different laboratories in the development of computer software for this specific purpose.

We believe that the direction many radiocarbon laboratories are considering is one that will completely automate the radiocarbon sample calibration process. Ultimately, once the sample is loaded into the counter, the computer software that controls the scintillation counter or acquires the data from a stand alone counting system will capture the counting data from the equipment, compute the radiocarbon age of the sample, and provide the different types of correction schemes. At present, the Packard computer-assisted counting systems can execute user-programmed software to further analyze the counting data. The on-line radiocarbon software, once fully operational, will provide radiocarbon calibration of samples counted, but it does not completely include all of the different correction schemes. The next logical step in the sequence of software development will be the linking of on-line radiocarbon software to the other programs developed to correct the calibration dates for a complete data presentation.

CHAPTER 29

Efficiency Extrapolation Adapted
to Liquid Scintillation Counters

Charles Dodson

ABSTRACT

Efficiency extrapolation applied to β emitters has been adapted previously to conventional liquid scintillation counters. This concept has been extended to cover broad quench ranges, it has also been extended to cover tritium with a small decrease in accuracy compared with *a priori* approaches.

INTRODUCTION

The disintegration rate of an unknown β-emitting nuclide may be obtained by using a standard β-emitting nuclide to determine the overall system efficiency of a counter. Initial work by Gibson and Gale[1,2] was followed by Malonda and colleagues,[3-5] and used by Coursey et al.[6] who reviewed it, as did NCRP Report Number 58.[7]

This approach computes the detector efficiency for a two-phototube coincident counting system. An efficiency curve of a standard is measured experimentally in terms of a selected quench parameter (Q). That permits calculation of the corresponding figure of merit (M) of the system via the Fermi β-ray spectral distribution for the standard. The distribution is corrected for two energy exchange processes that do not contribute to photon production: ionization quenching and the wall effect. These corrections provide a figure of merit independent of the detectable energy of the counting system. Therefore the figure of merit can be expressed as a function of the selected quench parameter:

$$M = f(Q) \tag{1}$$

With that independence, the efficiency, $\epsilon(\text{unk})$, of an unknown beta emitter can be computed from the Fermi spectral distribution corrected for ionization quenching and the wall effect:

Table 1. Summary of Variables Studied

Isotopes: ^{3}H, ^{63}Ni, ^{14}C, ^{35}S, ^{22}Na, ^{36}Cl, ^{32}P, ^{241}Am
Cocktails: Toluene, Xylene and Pseudocumeme based gel, Non-emulsifier, Biodegradable
Volume: 0.4–16 ML.
Count rates: 1000–6 Million
Quench range: 0–475 H#
Reference standards: ^{3}H, ^{14}C
          ^{3}H, ^{63}Ni, ^{14}C, ^{36}Cl, and ^{241}Am   (NIST - SRMs)
          ^{35}S, ^{22}Na and ^{32}P   Not as well characterized
                                monitored by quench curves

$$\epsilon(\text{unk}) = f(M) \tag{2}$$

From Equations 1 and 2,

$$\epsilon(\text{unk}) = f(Q) \tag{3}$$

is available. The quench curve of the unknown has been calculated from the experimentally measured quench curve of the standard. Details of this approach are available in References 3 through 7 and their bibliographies.

An experimental approach initiated by Ishikawa and collegues[8-10] calibrates the counting system efficiency by determining the efficiencies of a standard for six defined pulse height regions. The count rates of an unknown are obtained for the same spectral regions as the standard. This provides integrated (or cumulatively summed) count rates of the unknown as a function of the system integrated efficiency determined by the standard. The resulting linear or quadratic functionality may be extrapolated to 100% efficiency to provide disintegrations of the unknown. Good results have been obtained for reasonable quench ranges.

EXTENDING THE QUENCH RANGE

First examine the use of the complete spectrum of the standard for system calibration and the complete spectra of unknown nuclides for count rate integration. Figure 1 illustrates one result based upon ^{14}C as the standard and quenched samples of ^{35}S as the unknowns. The counting efficiency range of the four ^{35}S samples taken from the same quench set is 60 to 95%. Figure 2 provides analogous data for ^{32}P and ^{36}Cl. Depending upon the quench level of the sample, the specific nuclide, and the resolution of the multichannel analyzer (MCA), tens to hundreds of channel data are available. A 4096 channel MCA provides about 300 channels for an unquenched ^{14}C spectrum.

Several hundred samples have been monitored. The variables examined (nuclides, cocktails, sample volume, count rates, quench levels, and quenching agents) are summarized in Table 1. Various curve fitting and extrapolation algorithms led to an iterative procedure which produces a least squares linear fit. It is subsequently extrapolated as illustrated in Figure 3. Criteria defining linearity are important but not hypercritical. Results reported here are based

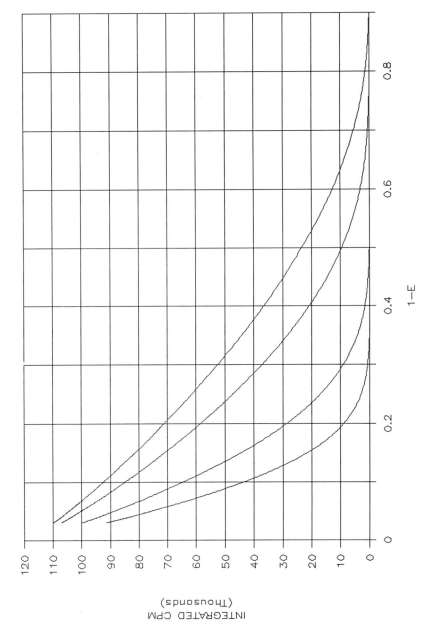

Figure 1. Integrated ^{35}S Spectra vs Integrated Standard ^{14}C Counting Efficiency.

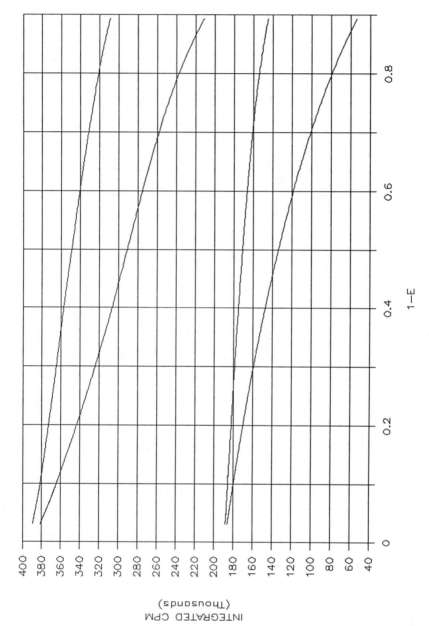

Figure 2. Integrated ^{32}P and ^{36}Cl Spectra vs Integrated Standard ^{14}C Counting Efficiency.

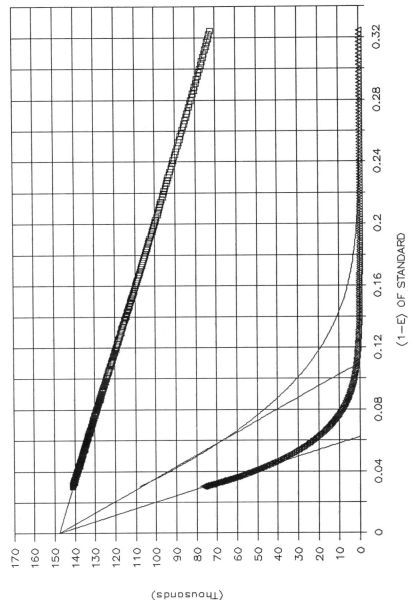

Figure 3. Integrated ^{14}C Spectra vs Integrated Standard ^{14}C Counting Efficiency.

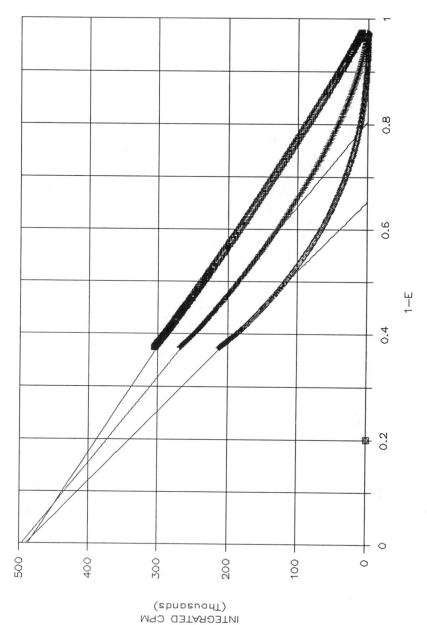

Figure 4. Integrated ³H Spectra vs Integrated Standard ³H Counting Efficiency.

Table 2. Results

For H# ranges of 0–350:	0.1 to 3.5% (except H3)
For H# range of 0–150 for H3 (65–30% counting efficiency)	0.1 to 8.1%

upon the square of the correlation coefficient (or coefficient of determination) exceeding 0.99. The three samples shown in Figure 3 were taken from a standard quench set containing 2470 Bq (148000 DPM). The raw data are represented by the thick line, the segment satisfying the linear constraint and subsequent extrapolation is the narrow line. These samples cover a counting efficiency range of 55 to 95% with an error in DPM recovery less than 2%.

APPLICATION TO TRITIUM

Direct application of the procedure to tritium provides useful results if the quench level remains small. For example, % error in DPM recovery for NIST SRMs with counting efficiency above 45% (or Horrocks' numbers less than 75) is less than 4%. However, if tritium unknowns are measured after calibration by tritium standards, recoveries better than 8% are obtained for counting efficiencies of 30%.

A summary of the results is provided by Table 2. The range of the % error in DPM recovery is 0.1 to 3.5% for all nuclides except ^{3}H for the volume, quench ranges, cocktails, and count rates presented in Table 1. For ^{3}H monitored for all the same variables but over a quench range of 65 to 30%, the % error range is 0.1 to 8.1%.

Lastly, ongoing work has established that recoveries better than 4% are possible for ^{14}C with efficiencies down to 20% (or Horrock's numbers of 450) if the system is calibrated by standard ^{3}H.

REFERENCES

1. Gale, H.J. and J.A.B. Gibson. *J. Sci. Instrument.*, 43:224 (1966).
2. Gibson, J.A.B. and H.J. Gale. *J. Phys.*, E1:99 (1968).
3. Malonda, A.G. and E. García-Toraño. *Int. J. Appl. Radiat. Isot.*, 33:249 (1982).
4. Malonda, A.G., E. García-Toraño, and J.M. Los Arcos. *Int. J. Appl. Radiat. Isot.*, 36:157 (1985).
5. Malonda, A.G. *Int. J. Appl. Radiat. Isot.* 34:763 (1983).
6. Coursey, B.M., W.B. Mann, A.G. Malonda, E. García-Toraño, J.M. Los Arcos, J.A.B. Gibson, and D. Reher. *Appl. Radiat. Isot.* 37:403 (1986).
7. Mann, W.B., Ed. *NCRP Report 58, A Handbook of Radioactivity Measurements Procedures*, 2nd edition, (Washington, DC: NCRP Publications, 1985) p. 199.
8. Ishikawa, H., M. Takiue, and T. Aburai. *Int. J. Appl. Radiat. Isot.* 35:463 (1984).
9. Takiue, M. and H. Ishikawa. *Nucl. Inst. Meth.* 148:157 (1978).
10. Fujii, H., M. Takiue, and H. Ishikawa. *Appl. Radiat. Isot.* 37:1147 (1986).

CHAPTER 30

Applications of Quench Monitoring Using Transformed External Standard Spectrum (tSIE)

Michael J. Kessler, Ph.D.

ABSTRACT

The transformed external standard spectrum is an effective quench monitor in quantitating samples in liquid scintillation analysis. It uses the Compton spectrum from the gamma source, ^{133}Ba, to monitor sample quench. This technique uses tSIE as an accurate method of quantitating the DPM in radiolabeled samples; it is shown to produce accurate DPM.

The use of tSIE is an accurate method of monitoring quench in a sample even under the following conditions:

1. a large dynamic range of tSIE (quench)
2. a large range of sample volumes
3. independence of wall effect, cocktail type and quenching agent
4. DPM determination without substantially increasing the count time of the sample

The use of tSIE parameter provides an accurate method for quench correction in the liquid scintillation analyzer. In addition, an automatic method (AEC) for adjusting the counting regions assists in DPM determination for both dual and triple labeled radioisotope samples.

INTRODUCTION

To understand the use of external standards for quantitating DPM in a radiolabeled sample, it is important that the basic principles of liquid scintillation process be understood. The entire process is to convert the energy from the beta particle into photons of light. This conversion must be efficient, and the intensity of the photons of light must be directly proportional to the energy of the beta particle. This process is illustrated in Figure 1.

The process of liquid scintillation involves the transfer of the energy of the beta particle to the solvent. Most all solvents used today for the liquid scintillation process contain an aromatic ring structure with π electrons. The solvent is chosen because it efficiently transfers the energy from the beta particles to the electrons of the solvent. The second step in the process is to transfer the energy from the activated solvent to a scintillator. This scintillator molecule becomes

343

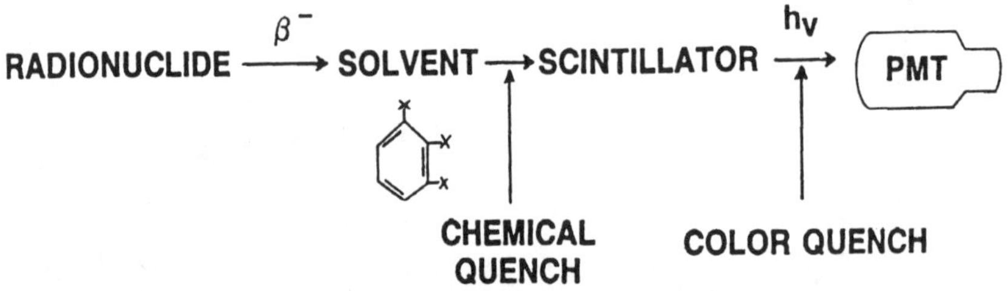

Figure 1. The basic liquid scintillation process.

excited, and when it returns to the ground state, it emits photons of light which are directly proportional to the energy of the beta particle. The photon (intensity of light) is then converted to a voltage pulse by the photomultiplier. The intensity or pulse height will be dependent on the number of photons entering the photo-multiplier. These voltage pulses (analog in nature) are then converted to a digital pulse for direct analysis by a multichannel analyzer (4096 channels) over the analysis time of the sample. The entire energy range which is analyzed, is 0–2000 keV. This is the energy range of all beta-emitting radionuclides.

The beta particle decay is a process in which a neutron from the nucleus is converted into a photon, an electron (beta particle), and a neutrino (mass-like particle). The energy loss in the decay is distributed between the beta particle and the neutrino. Thus, the beta particle can have any energy from zero to the maximum energy of the beta particle. For example, for ^{14}C, an energy from 0–156 keV can be seen in the spectrum. This is illustrated in Figure 2; it is a typical energy spectrum of an unquenched ^{14}C spectrum using the Spectralyzer of the Packard 2500TR liquid scintillation analyzer. The x-axis is a linear plot of the energy or light intensity of the beta particles, and the y-axis is the number of counts/unit time that a pulse of that intensity has been seen by the photomultiplier tube. If the area under the spectrum is integrated, then the total amount of radioactivity present in the sample, as measured by the counter (CPM) is obtained. The CPM value measures of how efficiently the energy transfers from the beta particle to the solvent, and how effeciently it transfers to the scintillator. It measures how efficiently the scintillator produces light, and how efficiently the instrument electronics convert the photons to voltage pulse. All of the steps together determine the sample CPM seen by the system. The CPM value is measured and compared to the actual DPM in the sample (obtained from the μCi in the sample). The efficiency of counting the sample is then calculated by the following equation:

$$\frac{cpm}{dpm} \times 100\% = \text{counting efficiency}$$

The difference between DPM and the CPM measured by the liquid scintillation analyzer, is a result of a process termed quenching. This quenching phenomenon is involved in the scintillation process at two points: energy transfer and light quantitation. The first process of quenching (the energy transfer) is called chem-

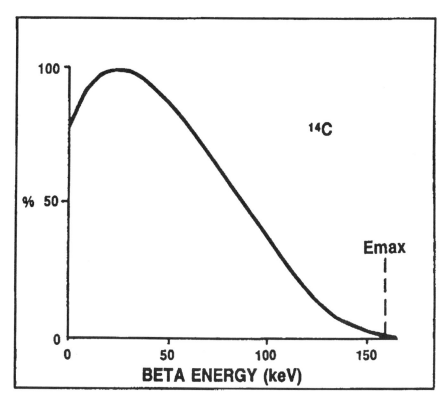

Figure 2. Energy spectrum of unquenched ^{14}C sample on Packard 2500TR liquid scintillation analyzer.

ical quenching. It involves the energy transfer of the beta particle to the solvent, or the transfer of energy from the solvent to the scintillator. Common chemical quenching agents (H_2O, nitromethane, $CHCl_3$, CCl_4, etc.) reduce the transfer of energy from the beta particle to the solvent. The quenching agent absorbs or reduces the energy of the beta particles transferred to the scintillator. The second process of quenching (photon reduction) is called color quenching. This phenomenon reduces the intensity of the scintillator produced photons seen by the photomultiplier. This phenomenon is similar to the use of a color filter on a photographic camera, which filters out certain colors of light on the photographs. The color in the scintillation process reduces the intensity of the photons seen by the photomultiplier. A typical spectrum for a ^{14}C sample, quenched by chemical and color quenching agents, is demonstrated in Figure 3.

The results of these quenching processes on the CPM can be extremely variable depending upon the quenching in the sample. For example, a sample prepared from a binding study may contain 2000 CPM, but a similar sample counted in solution might give 5000 CPM. In order to be able to compare these two samples, it is necessary to compare the DPM values (CPM compensates

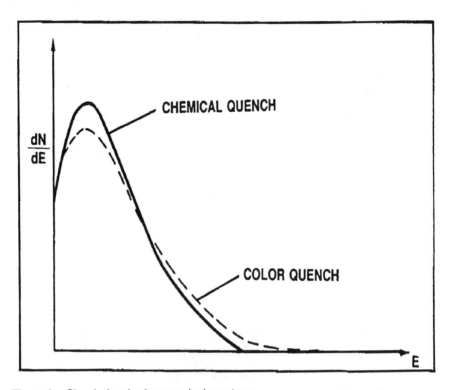

$\dfrac{dN}{dE}$

CHEMICAL QUENCH

COLOR QUENCH

E

Figure 3. Chemical and color quenched spectrum.

for quenching). Three methods of quench correction can be used to obtain accurate DPM values: internal standard, sample spectrum, and external standard spectrum analysis.

Transformed External Standard Spectrum—tSIE

Most of methods for DPM determination involve the preparation of a quench curve (% efficiency vs quenching level) for a set of standard quenched vials. The most accurate method of determining DPM presence in the sample is by using the external standard spectrum technique. This technique involves exposing each sample to an external gamma source (^{133}Ba). The gamma radionuclide creates a Compton spectrum by way of the Compton scattering phenomenon. The gamma ray interacts with an electron to create a new gamma ray with less energy and a Compton electron. This Compton electron is similar to a beta particle and creates a Compton energy spectrum, Figure 4, in the LSA.

The Compton spectral distribution of ^{133}Ba can be used to monitor quenching in the sample. The Compton spectrum is stored in the Spectralyzer so that various features of this spectrum can be monitored. After close evaluation of the Compton spectrum of various quenched samples, it was apparent that this spectrum could not be used directly as a measure of quench; volume, wall effect,

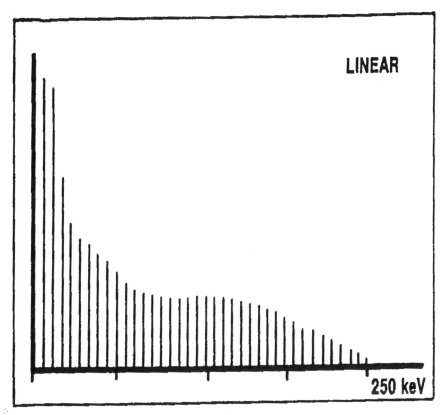

Figure 4. Compton spectrum of ^{133}Ba-linear.

and vial size changed the quenched indicating parameters in the Compton spectrum. After further investigation, it was found that if the integral spectrum of the ^{133}Ba external standard spectrum was computed, the end point of this transformed external standard spectrum (tSIE) could be used to measure sample quenching. The tSIE decreases as the quenching of the sample increases. This change of the tSIE with increasing quenching is illustrated clearly in Figure 5. This shows the Compton spectrum of two samples of different quench levels which have been transformed. The external standard end point of each sample is shown. This external standard end point is multiplied by a constant, such that the tSIE for an unquenched, argon purged ^{14}C sample is equal to 1000. The tSIE (quench indicating parameter−QIP) can be used as a method of relating the % efficiency of a standard quench set as a function of quenching the sample.

This plot of % efficiency can be used as an accurate method of obtaining DPM for samples with unknown DPM. This first requires the preparation of a standard quench curve (% efficiency vs tSIE) with a sealed set of standards containing the same radionuclide that is present in the unknown. Once each sample in the standard quench set has been counted, the % efficiency is deter-

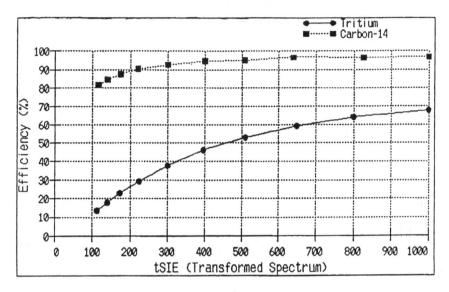

Figure 5. tSIE quench curves for tritium and ^{14}C.

mined. This is done by counting each sample and obtaining its CPM. Since the quenched set was made with a known amount (DPM) of radionuclides of interest, the % efficiency can be calculated by the following equation:

$$\frac{cpm}{dpm} \times 100\% = \% \text{ efficiency}$$

This is the y-axis number for each sample. The x-axis values are determined next. This is done by exposing each standard in the quenched set to the external standard (^{133}Ba). The Compton spectrum of each sample is then recorded in the Spectralyzer spectrum analyzer. This is converted to an integral spectrum. The tSIE value is determined from the extrapolated end point of this transformed spectrum. The tSIE value is then used as a measure of the quenching in the sample (QIP). Once the tSIE and % efficiency for each of the standard quench sets has been determined, it is plotted on a graph. The curve fit used to connect the points is a fixed point least square quadratic (FPLSQ). A typical quench curve for ^{14}C and ^{3}H is shown in Figure 5.

In summary, the following steps are used to prepare a quench curve for determining DPM:

1. prepare standards (two different methods can be used): (1) purchase sealed standards containing radioisotope of interest, (2) prepare and analyze set of standards of known DPM with different levels of quenching agent, and use a cocktail similar to that used in unknown sample
2. count standards in defined counting region determined by evaluation of sample spectrum of least quenched sample
3. count sample to determine CPM for each standard

4. compute % efficiency for each sample by using the following equation:

$$\% \text{ efficiency} = \frac{\text{cpm}}{\text{dpm}} \times 100$$

5. expose each sample to an external standard, ^{133}Ba, and obtain Compton spectrum. Transform spectrum to obtain the end point of this spectrum, and then determine tSIE value
6. plot % efficiency of each standard vs tSIE value
7. use fixed point least square quadratic curve fit (FPLSQ) to connect points and obtain quench curve for radionuclide of interest

Now that this quench curve has been obtained, how can it be used to obtain the DPM in an unknown sample? The first step in determining the DPM in the unknown sample, containing the same radionuclide as the nuclide used to prepare the standard curve, is to count the unknown sample under the same conditions used to create the standard curve. Then the CPM in the sample is determined. The sample is then exposed to the external standard, and the tSIE (quench indicating parameter) is determined. From the standard curve and the tSIE, the counting efficiency for the sample is extrapolated. The DPM for the unknown is then calculated from the equation:

$$\text{DPM} = \frac{\text{cpm}}{\% \text{ efficiency}}$$

In order to obtain accurate DPM values, it is necessary to have a quench indicating parameter (tSIE value) which has the following characteristics:

1. high dynamic range
2. reproducibly and accuracy at normal and high quench conditions
3. independence of sample volume
4. independence of "wall effect"
5. independence of vial type
6. independence of vial size
7. independence of cocktail density

The external standard used to determine the tSIE must have the following characteristics:

1. low radiation hazard not subject to stringent radiation safety regulation
2. the E_{max} low enough so as not to require excessive shielding which could increase background
3. maximum energy of Compton spectrum close to energy of ^{3}H and ^{14}C, most frequently used radionuclides in dual label counting
4. energy of external standard source sufficiently low to reduce or eliminate spectral distortions caused by gamma ray interaction with material in counting chamber environment
5. timely determination such that counting time of sample is not greatly increased in determining DPM

This chapter will evaluate each characterstic and present the data as to the accuracy of the DPM determination. This can be illustrated best by defining a term called % recovery. The % recovery of an experiment is the DPM determined by the liquid scintillation analyzer compared to the actual DPM in the sample times 100%.

$$\frac{\text{DPM counter}}{\text{DPM actual}} \times 100$$

For all characteristics of the tSIE, the % recovery is calculated and plotted.

The first characteristic of the tSIE is its volume independence. This can be illustrated by preparing a series of samples:

1. standard 20 mL vial — 19, 15, 12, 8 mL
2. miniature 7 mL vial — 6, 4, 3, 2, 1, 0.5 mL
3. microfuge tube — 0.4, 0.2, 0.1, 0.05, 0.025 mL

The samples were counted and a tSIE determined for each sample with a subsequent DPM printed out. The % recovery was then determined from the DPM of the counter, and the DPM actually present in the sample. A graph of the % recovery vs sample volume for ^{3}H is shown in Figure 6. For the 15 sample volumes, a straight line was obtained near 100% recovery. The exact statistical numbers of ^{3}H were a mean of 98.19%, a standard deviation of (SD) 1.58, and a % coefficient of variation of (%CV) 1.61. For ^{14}C, the numbers

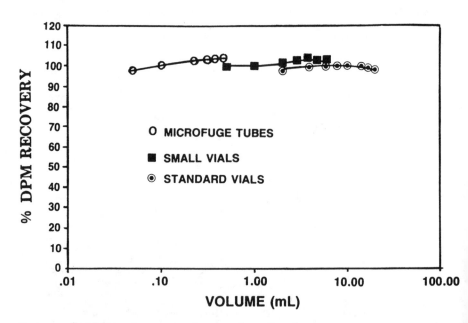

Figure 6. ^{3}H DPM % recovery as a function of sample volume.

Table 1. Quantitation of ^{3}H and ^{14}C and Dual Label Samples at Various Volumes Using tSIE as a Quench Monitor

Volume	Nuclide	% Recovery	tSIE
400 μL	^{3}H	98.9	663
200 μL	^{3}H	98.5	617
100 μL	^{3}H	97.6	555
50 μL	^{3}H	96.8	528
25 μL	^{3}H	96.4	453
400 μL	^{14}C	100.6	698
200 μL	^{14}C	101.5	630
100 μL	^{14}C	100.0	600
50 μL	^{14}C	98.3	538
25 μL	^{14}C	99.6	494
400 μL	^{3}H/^{14}C	93.3/99.6	677
200 μL	^{3}H/^{14}C	95.1/99.2	629
100 μL	^{3}H/^{14}C	95.3/104.5	610
50 μL	^{3}H/^{14}C	96.9/104.5	543
25 μL	^{3}H/^{14}C	99.2/98.5	475

were even closer to 100% with a mean of 99.51%, a SD of 0.917 and %CV of 0.922. This data clearly indicates that the DPM values obtained using the tSIE is volume independent. Because of the increasing use of microvolume counting, a special series of samples were prepared with volumes of 400, 200, 100, 50, and 25 μL containing ^{3}H, ^{14}C, or dual label ^{3}H/^{14}C. These volumes were chosen because they are commonly used in the microvolume counting procedure. The dual label samples were incorporated to determine if at low volumes, the DPM of a dual label sample could be determined accurately. Both single label ^{3}H and ^{14}C DPM % recoveries (Table 1) show excellent recoveries. The % recoveries are 97.6% (^{3}H) and 100% (^{14}C) with a small SD of 1.1 to 1.2% for these isotopes. For the ^{3}H/^{14}C dual label samples, the % recoveries are 96.56% for ^{3}H, and 101.2% for ^{14}C. The standard deviation for ^{3}H was 1.70, and for ^{14}C it was 2.94. The data in Table 1 clearly indicate that the tSIE is decreasing as the sample volume decreases. If the tSIE and % efficiency decreases, then the CPM/% efficiency increases, and the corrected DPM is determined accurately. As mentioned in the requirements for a DPM (quenching method), the counting time should not be substantially increased. Despite these low sample volumes, the time required to calculate the tSIE (external standard quench indicating parameter) is short (0.5 to 1.0 minutes).

The optimal geometry of the external source positioned directly below the sample and the ease of measurement of the Compton spectrum from the ^{133}Ba gamma source requires only a short time period for external source quench correction. The sample throughput, therefore, is not substantially affected by counting samples containing small volumes. These results are illustrated in Table 2.

As can be clearly seen, the tSIE is very reproducible over the large quench range, with the % efficiency for ^{3}H decreasing from 58.47 to 3.32%. The data shows that the % recovery is very close to 100%, with a maximum of %CV of 0.640 for the sample, with a tSIE = 83.59. The tSIE value is also very repro-

Table 2. Reproducibility of % Efficiency, % Recovery and tSIE for ^{3}H Quenched Samples, Counting Each Sample Ten Times

Sample	% EFF	% Rec.	% CV	tSIE	% CV/tSIE
1	58.47	99.94	0.362	815.2	0.412
2	55.78	99.77	0.200	725.9	0.334
3	44.49	100.23	0.224	472.2	0.240
4	24.77	99.82	0.304	231.1	0.246
5	18.07	99.42	0.303	177.1	0.179
6	10.81	99.6	0.318	119.8	0.352
7	6.15	99.29	0.640	83.59	0.480
8	3.32	99.00	0.381	58.75	0.230

ducible even for the most heavily quenched sample of the sample set tSIE = 58.75 (%CV = 0.230). The maximum %CV for any tSIE is 0.480.

Now that the tSIE has been shown to be reproducible at samples containing normal quenching values, what about samples that are severely quenched? Does the tSIE value become less reproducible as the quench becomes more severe? In order to assess this, a special set of ten samples was prepared. The tSIE values varied from 39 to 10, and the % efficiency varied from 1.388 to 0.011%. Each of these samples was counted 10 times with the statistical data presented in Table 3.

The samples (^{3}H-labeled) are severely quenched with efficiency decreasing to 0.011%. Even at this level, the % efficiency has a %CV of 3.31 which is extremely good for these heavily quenched samples. The tSIE values decrease to 9.956 with a %CV of 0.296. The table also shows the three tSIE values; the tSIE was identical each of the 10 times that the tSIE was determined. This data clearly indicates that the tSIE method of determining quench is very reproducible and has a large dynamic range. These experiments were also conducted with ^{14}C at similar quench level. Almost identical results for tSIE reproducibility were obtained. A graphic summary of the type of quench curve which could be obtained with these 18 standard quenched samples is shown in Figure 7. The major portion of the quench curve (% efficiency vs tSIE) is shown in the graph. The heavily quenched sample region of the curve has been exploded in order to assess the nature of these curves at the extremely heavily quenched levels.

Table 3. Reproducibility of % Efficiency, % Recovery and tSIE For Extended ^{3}H Quench Samples, Counting Each Sample Ten Times

Sample	% EFF	% CV/EFF	tSIE	% CV/tSIE
1	1.388	0.922	39.00	0.209
2	0.617	0.868	26.99	0.324
3	0.311	0.476	21.08	0.300
4	0.166	1.072	18.15	0.290
5	0.096	2.876	15.41	0.205
6	0.060	1.587	14.88	0.283
7	0.038	3.990	12.70	0.000
8	0.023	2.410	12.30	0.000
9	0.016	2.740	10.20	0.000
10	0.011	3.310	9.956	0.296

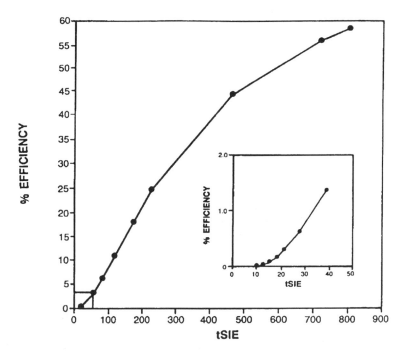

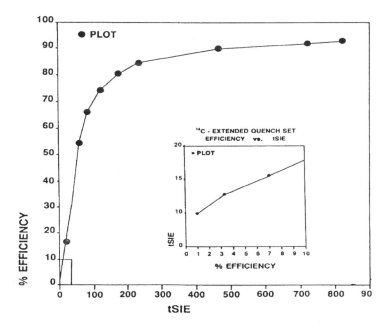

Figure 7. Quench curve for ^3H tSIE = 815-9.956 (top). Quench curve for ^{14}C for tSIE = 815-9.956.

The dynamic range for the tSIE method of determining efficiency can be defined by the following equation:

$$\text{Dynamic Range*} = \frac{\% \text{ eff. of unquenched sample}}{\% \text{ eff. of most highly quenched}}$$

Using this definition, the dynamic ranges are as follows:

$$^3H = \frac{63.57 \ (\text{tSIE} = 1000)}{0.011 \ (\text{tSIE} = 9.956)} = 5779$$

$$^{14}C = \frac{96.54 \ (\text{tSIE} = 1000)}{1.13 \ (\text{tSIE} = 9.956)} = 85.43$$

These results clearly indicate that the tSIE method of determining quenching in a sample have a very broad dynamic range with very reproducible values even at a heavily quenched sample.

The third characteristic of the tSIE is its ability to eliminate the "wall effect" problem, which may be present in samples prepared in liquid scintillation vials. The common organic solvent present in liquid scintillation solutions (toluene, xylene, pseudocumene) can readily penetrate the walls of plastic liquid scintillation vials. Two problems can be observed when using plastic liquid scintillation vials. First, the solvent can permeate the plastic wall of the scintillation vial. This gives in inside of the liquid scintillation counter a strong organic odor. The penetration of the solvent into and through the scintillation vials results in actual swelling of the vials; this makes it impossible to store and recount samples prepared in plastic vials. Second, not only the solvent but the scintillator can penetrate into the plastic wall. This can result in the plastic vial wall acting as a plastic scintillator. This "plastic scintillator wall" has a lower efficiency than the sample in solution. Thus, if the plastic vial with plastic scintillator wall is exposed to the external gamma source, extra low energy photons will be emitted which could affect the low energy external standard spectrum. This could thus affect the external standard quench indicating parameter and result in incorrect DPM values. The wall effect on the % recovery of a sample can easily be assessed by counting a sample initially at 12 hour periods. If the wall effect is affecting the DPM, then the % recovery would change as a function of time. This experiment was conducted for both the plastic standard (15 mL) and miniature vials (5 mL) over an 84 hour period.

The results in Table 4 clearly show that the % recovery of neither the standard vial nor the miniature vials change as a function of time. The %CV for this entire time period is 0.71 and 0.36 for the 5 and 15 mL samples respectively. This indicates that the wall effect commonly seen in some liquid scintillation counters has been completely eliminated by the Packard quench indicating parameter (tSIE).

The fourth characteristic of tSIE is its ability to give accurate DPM values

---

*% recovery of all samples must be 100 ± 5%.

Table 4. Effect of Possible "Wall Effect" on DPM Recovery in Plastic Vials Using ³H-Alanine in Instagel

Time (hr)	Volume (mL)	% DPM recovery
0	5	100.2
0	15	100.2
12	5	101.1
12	15	100.9
24	5	100.3
24	15	100.1
48	5	100.4
48	15	99.9
60	5	100.2
60	15	99.9
72	5	100.1
72	15	100.5
84	5	98.8
84	15	100.0

for different scintillation vial types (glass or plastic), different sample sizes (standard, miniature, and microfuge tubes), different scintillation solution types, and different quenching agents types (chemical and color). Let us divide these into two sections. The first will address the vial size and the vial type question. The second will cover the different types of scintillation solution, as well as various types of quenching agents. The results for different vial size and vial type is shown in Table 5.

The results in the table indicate the % recovery using glass or plastic, and any vial size is 100% for all of the samples assayed. This shows that the DPM using tSIE is independent of the type or size of vial used to hold the sample.

The second aspect to be investigated was the effect of various scintillation solutions and various quenching agents (color or chemical). The results are shown in Table 6.

The first four samples indicate the result of the various quenching agents (color and chemical) present in the sample. The results in Table 6 indicate that the three chemical quenching agents ($CHCl_3$ and CCl_3, and CH_3NO_2) do not affect the % recovery of the ³H-toluene in the sample. The addition of a red colored organic material, eosin, does not affect the % recovery with a 100.2% value obtained. This is further shown for various quench levels for ³H and ¹⁴C in Figure 8. The next three samples show that the sample density in g/cc does not

Table 5. Assessment of DPM/% Recovery–Various Vial Sizes and Types Using ³H-Toluene

Sample	Vial Type	% Recovery
1	20 mL glass	99.9
2	7 mL glass	99.6
3	6 × 50 mm Sealed	98.9
4	20 mL plastic	100.2
5	7 mL plastic	100.2
6	1.5 mL microfuge tube	99.8
7	0.5 mL microfuge tube	98.9

Table 6. Effect of Type of Quenching Agent on DPM Recovery for ³H

Treatment	Density (g/cc)	× 10²³/mL Elect Density	% Recovery
CHCl₃ (10 μL)	–	–	99.2
CCl₄ (10 μL)	–	–	98.5
CH₃NO₃ (10 μL)	–	–	99.6
Eosin Red (10 μL)	–	–	100.2
Toluene + PPO	0.863	2.82	99.9
Xylene + PPO	0.877	2.89	100.2
Pseudocume/PPO	0.860	2.84	100.8

affect the % recovery of the sample. In addition, the cocktail electron density of the samples was calculated. The cocktail electron density is the sum of the weight of each component, times the electrons per molecule, times 6.0238×10^{23} molecules/mol. The first units for the electron density is electrons/mL. The effect of this electron density on the tSIE/DPM and % recovery was evaluated and found not to affect the % recovery for the samples tested.

Now, the tSIE method of evaluating sample quenching has been shown to be independent of most of the factors which affect DPM determination in simple single radiolabeled analysis. What about its accuracy in counting the more complicated dual label and the most complicated triple label samples? First, let us evaluate the method used for quantitating dual labeled samples. The problem is how to obtain the true DPM for each of the two single labels in a dual label sample. This problem is shown in Figure 9 for dual labeled ³H/¹⁴C samples.

In order to obtain accurate DPM values for both radionuclides, the following steps should be taken:

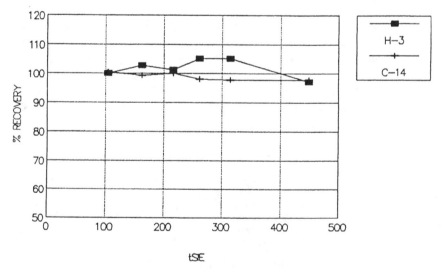

Figure 8. ³H and ¹⁴C recovery for colored samples at various tSIEs.

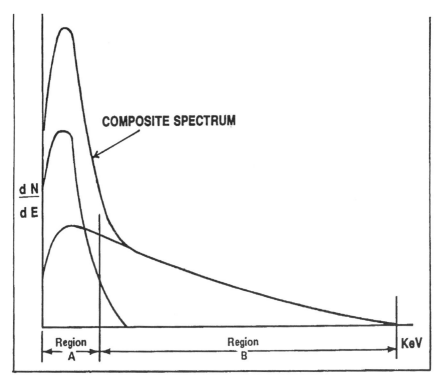

Figure 9. Dual label $^3H/^{14}C$ region settings for dual label counting region A = 0-12 keV, region B = 12-156 keV.

1. Prepare a set of ^3H samples, each containing a constant number of DPMs, with zero to a maximum amount of quenching agent. A commercial sealed set can also be obtained for each radionuclide.
2. Prepare a set of ^{14}C samples similar to the ^3H samples with a constant DPM in each and different quench levels.
3. Next, count each sample in the ^3H quench set to obtain DPM.
4. For each sample, calculate the % efficiency $= \dfrac{cpm}{cpm}$ equation is for both regions A and B.
5. Expose each sample to the built-in external gamma source, ^{133}Ba, with the tSIE being determined for each sample. This tSIE is independent of the radionuclide present or the region settings. The tSIE is an indicator of the quench level in the sample.
6. Obtain two plots of obtained—(Figure 10) of % efficiency, ^3H in region A and ^3H in region B, as a function of tSIE. From the plots, the % efficiency in both the A and B regions can be obtained.
7. Assay the ^{14}C standard in the same manner as the ^3H standards, with two additional plots being obtained (Figure 11). The plots are of % efficiency of ^{14}C in region A and ^{14}C in region B. From these plots, the % efficiency$_{CA}$ and % efficiency$_{CB}$ can be obtained using fixed point, at least square quadratic curve fit.

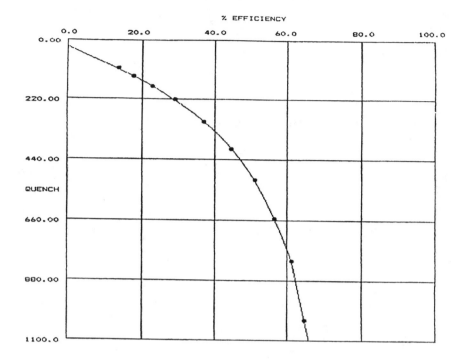

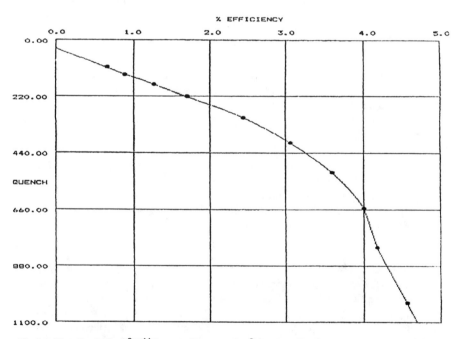

Figure 10.   Dual label ^{3}H/^{14}C quench curves for ^{3}H using tSIE for region A (top). Dual label quench curve for ^{3}H using tSIE for region B.

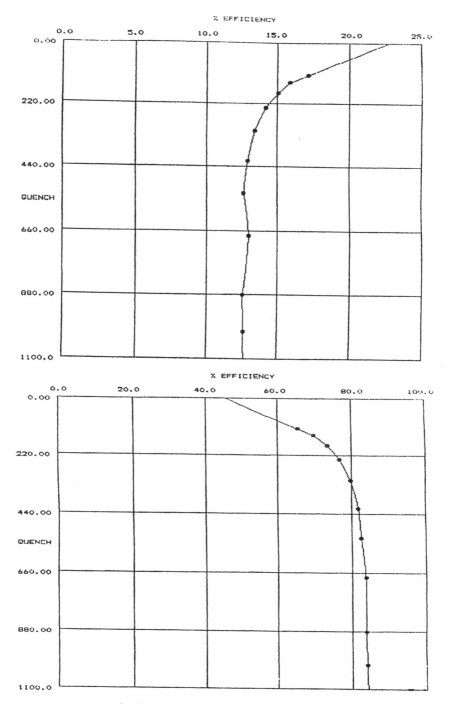

Figure 11. Dual label $^3H/^{14}C$ quench curves for ^{14}C using tSIE for region A (top). Dual label quench curve for ^{14}C using tSIE for region B.

8. Now the system is ready to determine DPM values for unknown samples. First the tSIE value is determined, and by using this value on the four plots above, four % efficiencies can be obtained (% efficiency$_{HA}$, % efficiency$_{HB}$, % efficiency$_{CA}$, and % efficiency$_{CB}$). The sample is then counted in each of the two regions with the CPM_A and CPM_B obtained. Then the following equation can be used:

$$CPM_A = (\% \text{ Eff.} \times DPM) + (\% \text{ Eff.}_{CA} \times DPM_C)$$

$$CPM_B = (\% \text{ Eff.}_{AB} \times DPM_H) + (\% \text{ Eff.}_{CB} \times DPM_C)$$

By rearranging and substituting into these equations, the DPM_H and DPM_C can be obtained.

$$DPM = \frac{(CPM_H \times \% \text{ Eff.}_{CB}) - (CPM_C \times \% \text{ Eff.}_{CA})}{(\% \text{ Eff.}_{HA} \times \% \text{ Eff.}_{CB}) - (CPM_C \times \% \text{ Eff.}_{CA})}$$

$$DPM = \frac{(CPM_C \times \% \text{ Eff.}_{HA}) - (CPM_H \times \% \text{ Eff.}_{HB})}{(\% \text{ Eff.}_{HA} \times \% \text{ Eff.}_{CB}) - (\% \text{ Eff}_{HB} \times \% \text{ Eff.}_{CA})}$$

As can be seen from the equations and plots, two of the most important factors are the % efficiency and carbon in region A. The % efficiency$_{CA}$ (crossover of ^{14}C into ^3H) increases drastically when the sample becomes more quenched. This could cause considerable problems when a large ratio of ^{14}C to ^3H exists in the sample. Therefore, a special feature termed AEC (Automatic Efficiency Control) was implemented in the system to help overcome this problem. This feature automatically tracks the theoretical end point of both the ^3H and ^{14}C spectrum in the sample. By doing this, the % efficiency of ^{14}C in the A region is kept relatively constant, as shown in Figure 12. This enables accurate DPM determinations for $^3H/^{14}C$ dual labeled samples at various quench levels and at various radionuclide ratios. This technique can be used for any dual labeled sample separated by at least 175 keV of energy.

In order to assess the performance of the tSIE as a quench indicating parameter, a DPM for dual labeled samples and a series of samples were prepared. These samples contained $^3H/^{14}C$ ratios of 1:1, 1:5, 1:10, 1:20, and 1:50. Each of these samples were prepared at various quenched levels. The DPM for each of the dual labeled samples was determined using the previously described procedure with AEC activated (Figure 13). The results, shown in Table 7, indicate that for the 14 samples evaluated, the ^3H recovery was 100.31 and the ^{14}C was 100.85%. These % recoveries are extremely stable, even for samples at high $^{14}C/^3H$ ratios of 20:1 and 50:1. It is also clear from the table that a quench range from very small to moderate to severe, that the % recovery of both radionuclides is close to 100%. The results are also very reproducible with a %CV of 0.887 for ^3H, and 0.886 for ^{14}C.

In order to evaluate the DPM determination for a dual labeled sample of $^3H/^{14}C$ samples of a 1:20 ratio were prepared and quantitated. A plot of %

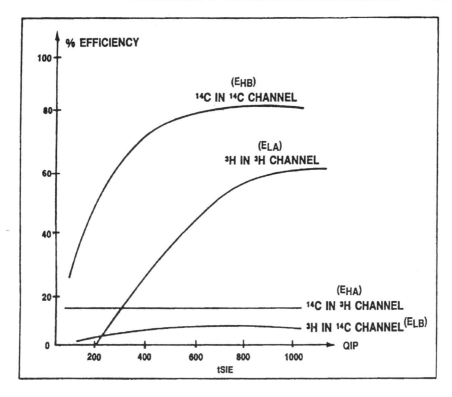

Figure 12. Dual label with AEC ^{3}H and ^{14}C in regions A and B.

recovery of both ^{3}H and ^{14}C is shown in Figure 10. The quenched levels range from a tSIE of 800 to 83.9% to 62.1% for ^{14}C. The results in the graph indicate a stable % recovery of ^{3}H and ^{14}C, even at a high ratio of 20:1.

Now that the dual label samples have been shown to use tSIE to determine accurate DPM values, what about the ultimate analysis of a triple label sample containing ^{3}H/^{14}C/^{32}P? This triple label sample analysis is very similar to the dual labeled analysis except that instead of % efficiency vs tSIE curves, 9 are required. Three for each radionuclide in each of the three regions of interest. From these 9 % efficiencies, the three CPM values for regions A, B, and C, and the DPM of each of the three radionuclides can be quantitated by an inverse matrix analysis procedure. In order to completely evaluate the triple label DPM procedure, four different ratios of ^{3}H/^{14}C/^{32}P were used at various quenched levels. The DPM was determined for each radionuclide. The statistics, mean, SD and %CV were also calculated.

The results clearly indicate that the % recoveries are very close to 100% (99.65, 101.46, and 100.15), and all samples have a %CV of less than 1.36 over the entire quenched range. This would be appropriate if all experiments were done at a ratio of 1:1:1, but most scientists use various ratios of the three radionuclides present. How do the tSIE perform under conditions when one of

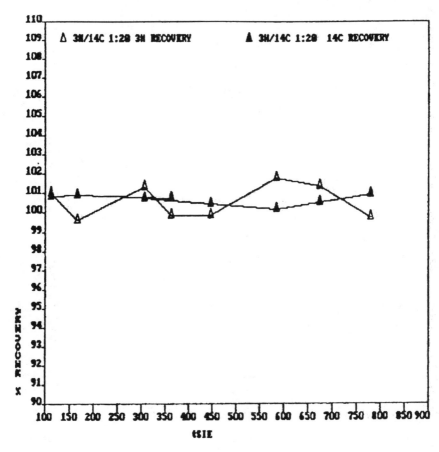

Figure 13.   % recovery of ^{3}H and ^{14}C for dual label with AEC ^{3}H/^{14}C ratio 1:20.

Table 7. % Recovery of Dual Label ^{3}H/^{14}C at Various Quench Levels and Isotope Ratios

Ratio ^{3}H/^{14}C	^{3}H % REC	^{14}C % REC	tSIE
1:1	100.6	101.0	808
1:1	100.6	101.3	473
1:1	100.2	98.3	161
1:5	99.9	101.8	828
1:5	102.3	100.5	463
1:5	100.5	102.0	174
1:10	100.1	101.3	786
1:10	99.7	101.1	464
1:10	100.8	101.5	173
1:20	99.7	100.9	780
1:20	99.8	100.5	448
1:20	99.5	100.9	166
1:50	100.9	100.6	786
1:50	101.5	100.2	681

Table 8. % Recovery of ^{3}H/^{14}C/^{32}P at Ratios of 1:1:1 at Various Quench Levels

	^{3}H	^{14}C	^{32}P	tSIE
	98.90	101.89	100.04	811
	98.12	100.78	99.16	724
	98.79	100.57	100.01	622
	98.98	101.78	100.51	499
	99.23	101.14	100.35	385
	99.80	101.04	101.13	329
	101.9	101.94	100.10	185
	101.50	101.78	99.87	117
Mean	99.65	101.46	101.15	
SD	1.35	0.56	0.62	
% CV	1.36	0.55	0.62	

Table 9. % Recovery of ^{3}H/^{14}C/^{32}P at Ratios of 10:1:1 at Various Quench Levels

	^{3}H	^{14}C	^{32}P	tSIE
	100.5	99.05	97.50	817
	100.1	94.71	98.90	753
	100.2	95.54	100.9	639
	100.1	98.23	98.96	572
	101.1	98.54	100.62	498
	100.5	97.80	99.2	420
	102.3	96.50	99.33	262
	102.5	97.40	102.50	178
Mean	100.91	97.24	99.68	
SD	0.97	1.52	1.41	
% CV	0.96	1.57	1.41	

the radioisotopes is in a tenfold excess over the other two radioisotopes present in the sample. This can be demonstrated by analyzing the data in Tables 9 to 11. Table 9 shows the results with an excess ($10\times$) of the lower energy ^{3}H in the sample.

The results show that the recoveries are very close to 100%. The mean % recoveries are 100.91 (^{3}H), 97.24 (^{14}C), and 99.68 (^{32}C). All show a standard deviation of less than 1.52. Table 10 displays the results for a sample contain-

Table 10. % Recovery of ^{3}H/^{14}C/^{32}P at Ratios of 1:10:1 at Various Quench Levels

	^{3}H	^{14}C	^{32}P	tSIE
	100.24	99.88	99.70	813
	103.70	99.54	100.02	738
	98.71	101.56	101.11	646
	105.17	100.14	100.18	572
	99.81	100.28	101.43	496
	103.40	100.48	100.73	435
	104.03	100.45	99.98	361
	103.21	100.78	104.67	265
	97.89	100.75	99.53	179
Mean	102.44	100.42	100.81	
SD	2.56	0.59	1.58	
% CV	2.51	0.59	1.57	

Table 11. % Recovery of $^3H/^{14}C/^{32}P$ at Ratios of 1:1:10 at Various Quench Levels

	^3H	^{14}C	^{32}P	tSIE
	96.72	99.89	100.49	817
	98.51	108.28	99.67	751
	102.19	95.53	101.38	655
	98.40	99.99	102.04	605
	95.92	106.95	99.97	504
	96.66	102.09	101.36	443
	103.73	91.66	101.24	269
	103.55	107.47	100.29	183
Mean	99.45	101.45	100.83	
SD	3.20	5.90	0.80	
% CV	3.22	5.82	0.80	

ing a $^3H/^{14}C/^{32}P$ ratio of 1:10:1. The results show a mean of 102.44 (^3H), 100.42 (^{14}C), and 100.81 (^{32}P). These results are over the entire quench range analyzed by the liquid scintillation analyzer. The %CV are all less than 2.51.

Finally, the most difficult triple label sample ratio to analyze is at a $^3H/^{14}C/$ ^{32}P ratio of 1:1:10. The reason that this is the most difficult is that both ^{14}C and large amounts of ^{32}P must be compensated for within the ^3H region. The results for this ratio are in Table 11. The statistical analysis of all of the quenched samples indicates extremely good % recoveries (99.45 [^3H], 101.45 [^{14}C], and 100.83 [^{32}P]. Because of the high ratio of ^{32}P in the sample, and consequently the high crossover of ^{32}P into both ^3H and ^{14}C, the %CV for these isotopes increased to 3.22 for ^3H, and 5.82 for ^{14}C. These numbers are extremely accurate considering that % efficiencies and three CPM values are used to calculate the DPM of each of the three radionuclides. Thus, the tSIE is an accurate method of determining the DPM in a triple label sample at various quench levels and with various isotope ratios.

SUMMARY

This paper definitely demonstrates that the use of the quenched indicating parameter, tSIE, is a rapid and accurate method of determining DPM of single, dual, and triple labeled samples. The DPM obtained with this technique is independent of sample volume, extent of quenching, wall effect, vial size, vial type, cocktail density, radionuclides number being analyzed (maximum = 3), and quenching agent type.

Accurate DPM can be obtained for dual label samples at 100:1 to 1:50 ratios and at various quenched levels. The DPM for triple label samples can be determined accurately, and with a small % coefficient of variation for various ratios of radionuclides. In conclusion, the use of tSIE as a quench indicating parameter results in accurate DPM values independent of most interferences found in liquid scintillation counting.

CHAPTER 31

Scintillation Proximity Assay: Instrumentation Requirements and Performance

Kenneth E. Neumann and Staf van Cauter

ABSTRACT

Recently, a new immunoassay and receptor binding analysis technology has been introduced. This methodology, scintillation proximity assay (SPA), belongs to a family of ligand binding techniques known as "sandwich assays." SPA technology employs a scintillation microsphere as the solid support, but requires no separation step. In addition, no conventional liquid cocktail is needed, minimizing safety hazards and disposal costs.

The unusual demands this technology places on currently available liquid scintillation instrumentation will be discussed. We will concentrate on the ability to accurately quantitate low levels of a low energy isotope, such as tritium, with adequate sensitivity. Since the recommended total sample volume is no more than 0.4 mL, acceptable sensitivity requires excellent light collection efficiency. Additionally, this technology is intended for high volume applications. Throughput, to a significant degree, is influenced by counting efficiency, and hence, instrument performance. Finally, the unique scintillation properties of the fluor material may affect instrument counting efficiencies, leading to poor quality data. Because assay sensitivity may be affected by the instrument employed, it is important that the technology is evaluated on several types of liquid scintillation counters. The results of such a study are presented. Alternative instrumental methods for counting SPA samples, providing increased sample throughput, are discussed.

INTRODUCTION

Classical immunodiagnostic and ligand-binding methodology using beta- and/or gamma-emitting radionuclides, is widely recognized in the industry for its sensitivity and specificity for the analyte of interest.[1,2] Several general techniques, such as direct and competitive assays, exist for a variety of applications. A third category is typically referred to as a sandwich assay.[3] Here, the analyte (usually an antigen) is bound to two different antibodies. One antibody is labeled with a radioactive tracer (typically ^3H or ^{125}I), while the other is permanently bound to a solid support structure such as a polymeric microsphere. When incubated together, these three elements form an Ab-Ag-Ab complex bound to the microsphere. The fraction of radiotracer bound to the ligand can be separated from the tracer in free solution via a number of techniques including centrifugation, filtration, charcoal adsorption, or magnetic separation. One significant disadvantage in these RIA methods is the time and effort required for this separation step.[4] A key challenge over the past 20 years has been to simplify the sample preparation steps, especially separation, required prior to assaying the tubes or vials.

A new technique, scintillation proximity assay (SPA), has recently been introduced for immunodiagnostic and ligand-binding assays. This method exploits the very low radioactive energy of the common beta emitting radionuclide tritium (^{3}H) or the Auger electrons emitted by ^{125}I. Because tritium decays energy on the average of 6 keV, beta electrons formed during the decay process travel only a short distance (4 μm in water). Therefore, in order to create a scintillation event, a tritium label must be in intimate contact with the scintillating medium.[5]

SPA technology employs a scintillating microsphere as the solid support structure. Formation of the Ab-Ag-Ab complex, which requires the presence of Ag in free solution, brings the radiolabeled antibody close to the scintillating particle, excites it, and causes photon emission. If there is little or no Ag in free solution, the complex will not be formed to any large degree. In this situation, radiolabeled Ab will not be bound close to the scintillating microsphere, and therefore, will not cause appreciable excitation of the fluor,[6,7] thus, the number of detected scintillation events is directly related to the amount of Ag in the sample. This effectively separates the bound fraction from the free fraction, without any need for manual separation steps.[8] Samples can be assayed directly in a conventional liquid scintillation counting system, following an appropriate incubation period. As a result, assay precision is ultimately improved due to fewer sample processing steps.

While SPA technology has obvious advantages in the areas of sample preparation, incubation, and separation, it also places unusual demands on commercially available liquid scintillation counting instrumentation. Most important is accurate quantitation of low levels of tritium with adequate sensitivity for routine immunoassay and receptor binding applications. Acceptable sensitivity requires both excellent tritium counting efficiency and low background count rate.

A primary factor influencing these requirements is the fact that typical SPA samples have a total volume of 0.4 mL and are prepared in 7 mL LS vials. A compounding factor is that the bound (scintillating) fraction consists of only a few milligrams of material which rapidly fall to the bottom of the vial. Sample geometry, as presented to the liquid scintillation counter, is therefore quite poor. This will tend to limit the photon collection efficiency, and thus the effective radionuclide counting efficiency. A low overall counting efficiency reduces the statistical accuracy with which one can measure a sample at a given count time. In order to compensate, the investigator or clinician must increase the sample count time or the sample volume to achieve better statistics. This either decreases sample throughput or increases assay cost. Furthermore, overall counting efficiency is critical in the determination of the signal to noise (S/N) ratio for the instrument employed in the assay. Poor efficiencies will limit the S/N ratio. As a result, assay sensitivity might be adversely affected.

A third factor influencing the net performance of this technology is the ability of a liquid scintillation counter to directly assay samples contained in 7 mL vials. Most currently available instrumentation is capable of loading and counting these vials directly. However, older counters may only be capable of counting samples in 20 mL vials. This necessitates placing the small SPA vial within an adaptor and assaying it as a large vial. This significantly reduces sample throughput.

Another factor influencing SPA performance is the fluor employed as the solid support medium. Experimental evidence indicates that the fluorescent emission of the scintillating microsphere occurs over a relatively long period of time compared to the emission from conventional LS samples. In addition, pulse height distributions obtained from typical SPA samples show that the fluorescent emissions average a slightly higher energy level than those usually encountered from conventional scintillation fluors. Currently available liquid scintillation systems incorporate various pulse discrimination schemes, based on pulse height and width, in order to optimize instrument performance. These systems have been optimized for use with existing scintillation chemicals.

The novel pulse decay characteristics of the scintillator used in SPA technology has broad implications for the signal processing techniques that discriminate true beta decay events from PMT thermal noise. Conventional liquid scintillation counters accomplish this via coincidence circuitry in conjunction with narrow pulse width scintillators. The design of conventional LSCs has been based on, and limited by, these requirements. By using a scintillator with a wider decay pulse, as shown below, patented time-resolved pulse discrimination techniques, using only a single PMT per detector, can discriminate against noise events. Radiochromatography counting systems based on this technology are well known.

The experiments detailed in the following section describe recent work done to better characterize SPA technology performance on existing liquid scintillation counting equipment. In addition, the evaluation has been carried out on experimental time-resolved counting equipment, using single PMT detectors. The results of these experiments are also presented.

EXPERIMENTAL

All experiments described herein were performed using either the Thromboxane B2 (code # TRK.951) or 6-Keto-prostaglandin F1a (Code # TRK.952) SPA kits available from Amersham Corporation. These kits contain all of the reagents necessary to prepare SPA samples. The components are listed in Table 1.

Table 1. Contents of SPA Reagent Kit

Item	Component	Volume (μL)
1	Assay buffer (PBS + gelatin + thimerosal)	100
2	Tracer (H-3 labeled antibody)	100
3	Antiserum	100
4	SPA protein a reagent (coupled to scintillating microspheres)	100
5	Standards (a through E, with varying amounts of antibody)	100 each

The instructions included with the reagents describe two sample preparation protocols — one day and one overnight. A preliminary study was done in which these two protocols were directly compared, to find the optimum preparation protocol. Triplicate sets of standard, NSB, and Bo tubes were prepared per the instructions provided for each protocol. All samples were prepared in 7 mL polyethylene LS vials (Packard, #6000192), and incubated with mixing for the appropriate period using a commercially available multi tube vortexer. All tubes were then assayed in a Packard Tri-Carb 2500TR liquid scintillation analyzer using a region setting of 0 to 200 keV. The results of this experiment, illustrated in Figure 1, indicated that few significant differences in sample count rate exist between the two protocols. For this reason, and for reasons of convenience, the overnight procedure was chosen as the preparation protocol to be used in all further experiments.

Because most current-generation liquid scintillation counting systems are

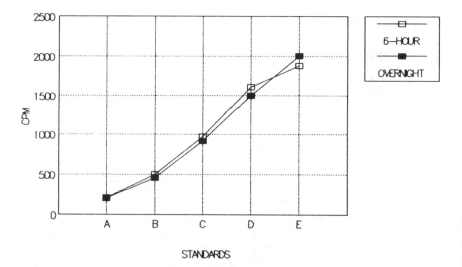

Figure 1. Comparison of 6 hr and overnight SPA incubation protocols.

equipped with MCA technology, counting regions can be optimized for the particular radionuclide/fluor combination being assayed. SPA technology employs a unique combination of radiolabeled tracer and solid scintillator. This can be illustrated by comparing an oscilloscope trace for a typical SPA sample to a conventional LS sample, as illustrated in Figure 2. Therefore, an initial experiment was performed to evaluate the spectral characteristics of a typical SPA sample. A Bo tube was prepared, incubated, and assayed in the Tri-Carb 2500TR LSA. A representative sample spectrum was collected and plotted (Figure 3). This plot indicates that preset ^{3}H counting regions will not capture the entire SPA sample spectrum. It is necessary to manually set the counting window to a region encompassing 0 to 30 keV.

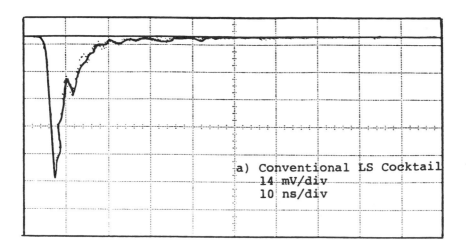

a) Conventional LS Cocktail
14 mV/div
10 ns/div

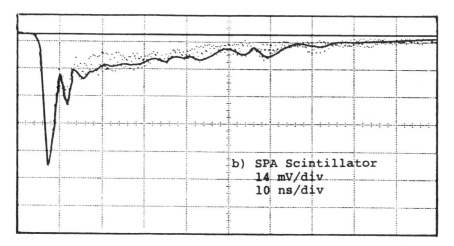

b) SPA Scintillator
14 mV/div
10 ns/div

Figure 2. Pulse shape characteristics: a) conventional LS cocktail, b) typical SPA sample.

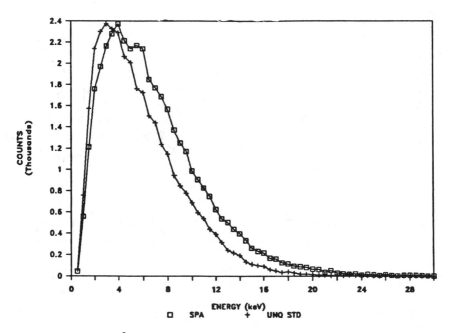

Figure 3. Normalized ³H spectra.

The first set of experiments was designed to determine the long-term stability of typical SPA samples after preparation. Triplicate sets of standard, NSB, and blank tubes were prepared per the previously qualified overnight protocol. Following incubation, all tubes were repeatedly assayed in the aforementioned liquid scintillation analyzer for a period of 60 hr. The data obtained from these assays indicated that the count rates of the samples, as a group, decreased by a significant amount over the 60 hr period. Figure 4 illustrates this for standards A-E. Furthermore, it was noted that the rate and function of this decrease varied with the initial count rate of the sample.

Because of the time dependency of the sample count rates, it becomes necessary to normalize the results of any one assay to an appropriate point in time relative to the incubation period. Therefore, a mathematical function was developed for each sample type (Stds A-O, and NSB) that relates the sample CPM to the time after incubation, when the sample was assayed. These functions were found to be fourth-order polynomials. These functions were used to calculate normalization factors based on sample type and time of assay.

A counting system using the aforementioned single-PMT discrimination scheme has been developed and constructed.[9] Experiments performed during development have demonstrated that efficiencies and backgrounds approaching those of conventional LSC can be achieved using a fluor with a relatively long decay constant. SPA technology employs such a scintillator. Therefore, a series of experiments was performed to evaluate SPA technology using this type of counter. Triplicate sets of Thromboxane B2 standards were prepared

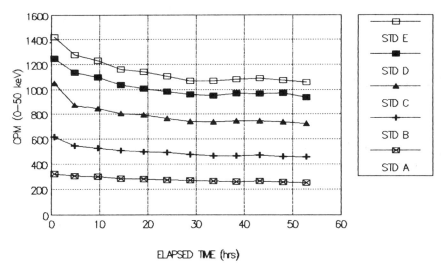

Figure 4. SPA sample stability after overnight incubation.

per the overnight protocol and assayed in the Tri-Carb 2500TR LSA. All samples were then assayed in the experimental counting system. Raw count results were normalized to time zero using the relationships described above. The normalized results for each test instrument were then compared to the reference instrument (the Tri-Carb 2500TR) by dividing each result by the count rate obtained in the reference system. The resulting CPM ratios are illustrated in Figure 5.

The data obtained from the above trials were also used to calculate figures

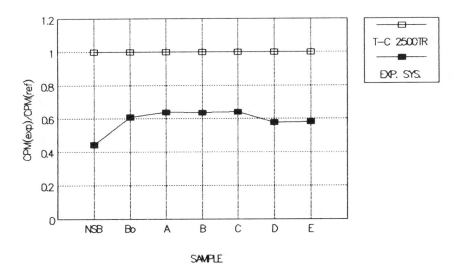

Figure 5. Normalized CPM ratios for reference and experiemental counters.

Table 2. Figures of Merit for SPA Counting Systems

Parameter	TRI-CARB 2500TR	Experimental System
CPM, Bo tube	1911.3	1160.0
CPM, blank tube	20.1	3.1
FOM $\left(\dfrac{\text{Bo-BLK}}{\text{BLK}}\right)$	94.1	373.2

of merit for each instrument. To evaluate the sensitivity of the counter, the following equation was employed:

$$\text{FOM} = \frac{\text{CPM(Bo)} - \text{CPM(blank)}}{\text{CPM(blank)}}$$

The value of this parameter is directly proportional to the dynamic range of the instrument for this application; that is, a maximum value is indicative of excellent performance. The values for each of the instruments employed in this study are displayed in Table 2.

Finally, standard curves were plotted for each of the trials, according to the format recommended in the SPA instruction booklet (Figure 6). Note that while count rates for the experimental system were generally about 40% lower, the resulting standard curves are almost identical.

The count rates obtained for each of the runs were then plotted against each other in order to directly compare instrument performance. Figure 7 illustrates these results. Here, we observe a high degree of correlation between the experimental system and the liquid scintillation counter. This suggests that a commercial system of the type described above would be ideally suited for SPA applications.

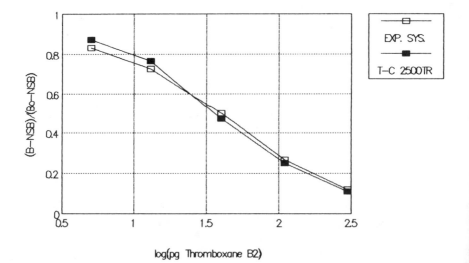

Figure 6. Thromboxane B2 standard curves.

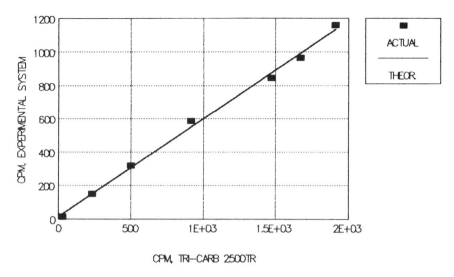

CPM, TRI-CARB 2500TR

Figure 7.   SPA count rate correlation.

RESULTS AND CONCLUSION

SPA technology represents a significant development in the field of radio-metric assay. Because of the lack of a separation step in the sample prepara-tion, labor costs for laboratories running routine ligand binding assays can be markedly reduced. Although most current assay kits are based on gamma-emitting radionuclides such as ^{125}I, the use of a beta-emitter such as ^3H is attractive due to its longer shelf life and lesser radiological hazard. Previous ^3H assays required separation steps and conventional liquid scintillation cocktails. These too represent both a cost and a hazard to many laboratories.

Scintillation proximity assay addresses these issues by eliminating both of them. In doing so, however, the technology forces a critical evaluation of the instrumentation required to perform these assays. The nature of the samples, and particularly the scintillator employed in them, is unique in comparison to classical LS samples and cocktails. While most commercially available liquid scintillation counting systems are capable of counting samples of this type, it is important to realize that SPA technology creates somewhat unusual counting requirements.

The data presented above clearly indicate that a typical current-generation liquid scintillation counter can effectively analyze SPA samples with adequate efficiency and sensitivity. Commercially available immunoassay data reduc-tion packages such as RiaSmart (Packard Instrument Company) can readily be joined to the LSC to provide an integrated immunoassay environment. One critical drawback, however, is the net sample throughput. Scintillation proxim-ity assay is intended for high volume applications such as receptor binding and drug screening. In addition, the homogeneous nature of SPA technology theoretically allows the user to perform kinetic measurements rather than endpoint determinations, effectively reducing incubation times. Conventional LSCs, possessing only a single detector, limit the number of samples that can be analyzed in a given period and make it impossible to perform kinetic measurements.

It has also been demonstrated that unique instrument technologies can count SPA samples. While classical coincidence counting, using two PMTs, is quite effective, new techniques, based on the unique scintillation properties of the fluor, are equally if not more effective for this type of sample. Single-PMT noise discrimination schemes can offer comparable counting efficiencies with reasonable background levels. One such scheme, used in the above experiments, resulted in an approximate quadrupling of the figure of merit. This ultimately results in greater assay sensitivity.

The effectiveness of time-resolved single PMT designs has major implications for the instrument technologies suited for scintillation proximity assay. Because the samples are of limited volume, large detectors are not required. In addition, the noise discrimination circuitry employed in the above experiments does not require massive lead shielding. These factors can lead to the development of extremely compact dedicated instrumentation. In addition, multiple detector instrumentation, which uses this technology, can address the problem of sample throughput and enhance assay performance through kinetic measurements.

REFERENCES

1. Yallow, R.S. and S.A. Berson. *J. Clin. Invest.* 39:1157 (1960).
2. Roitt, I.M., J. Brostoff, and D.K. Male. *Immunology*, (New York: Gower Medical Publishing Company, 1989), pp. 25.5–25.6.
3. Odell, W.D. and P. Franchimont. *Principles of Competitive Protein-Binding Assays*, (New York: John Wiley & Sons, 1983), pp. 243–254.
4. Chard, T. *An Introduction to Radioimmunoassay and Related Techniques* (1982).
5. Kirk, G. and S.M. Gruner. *IEEE Trans. Nucl. Sci.*, NS-29, p. 769 (1982).
6. Udenfriend, S., L. Gerber, and N. Nelson. "Scintillation Proximity Assay: A Sensitive and Continuous Isotopic Method for Monitoring Ligand/Receptor and Antigen/Antibody Interactions," *Anal. Biochem.*, 161:494–500 (1987).
7. Udenfriend, S., L.D. Gerber, L. Brink, and S. Spector. "Scintillation Proximity Radioimmunoassay Utilizing ^{125}I-Labeled Ligands," *Proc. Natl. Acad. Sci. USA*, 82:8672–8676 (1985).
8. U. S. Patent Number 4,568,649, Immediate Ligand Detection Assay.
9. U. S. Patent Number 4,528,450, Method and Apparatus for Measuring Radioactive Decay.

CHAPTER 32

The LSC Approach to Radon Counting in Air and Water

Charles J. Passo, Jr. and James M. Floeckher

INTRODUCTION

Various methods exist to monitor ^{222}Rn in air. There are seven commonly used types of detection: alpha track, activated charcoal adsorption, continuous radon monitoring, continuous working level monitoring, grab radon sampling, grab working level sampling, and radon progeny integrated sampling. Of these methods, the passive diffusion activated charcoal canister technique has been the method of choice when screening for radon. The liquid scintillation method is an improvement of the activated charcoal canister technique. Specially designed air detectors containing a small amount of activated charcoal and desiccant serve as the collection and counting vial. Because the liquid scintillation method offers the advantages of higher sensitivity, higher throughput, and shorter exposure times, interest in the method of measuring radon in air has increased.

Radon in water has been measured by both the Lucas cell[1,2] and liquid scintillation methods.[3] The liquid scintillation technique combines the advantages of minimal sample preparation time (1 min/sample), small sample sizes (10 mL), and high sensitivity (200 pCi/L).

This chapter will describe a liquid scintillation method for measuring radon in both air and water—the PICORAD™ system from Packard Instruments. Data comparing air measurements made with typical charcoal canisters and the liquid scintillation detectors will be presented.

RADON IN AIR USING LIQUID SCINTILLATION

The Detector

The PICORAD detectors are passive devices requiring no power. They are integrating detectors that determine the average radon concentration in the air they are placed.

The detector consists of a porous canister welded securely near the top of a plastic liquid scintillation vial. The porous canister contains a bed of a controlled weight of charcoal (1.3 g) and silica gel desiccant (0.9 g). The vial has a removable cap and an O-ring seal to prevent moisture or radon from entering the vial during storage and after exposure.

Exposure Procedure

To initiate the exposure, the vial is uncapped to allow radon laden air to diffuse into the charcoal. The nominal measurement time is 48 hr, at which point the radon accumulation has reached 95% saturation. A testing time of 24 hr is recommended in the summer when moisture problems can be severe. A 2 day exposure at 100% humidity will result in a 5% weight gain and a subsequent 10% loss of maximum counts. The radon accumulation reaches 80% of its saturation value in 24 hr. The rate of radon accumulation in the PICORAD detectors, empirically determined over a span of 9 hr to 5 days at room temperature, is shown in Figure 1. At the end of the exposure time, the detector is capped and sent to the laboratory for analysis.

Elution Procedure

In the laboratory, 10 cc of xylene based cocktail (InstaFlour™ Packard Instruments, Meriden, CT) is syringed or pipetted into the bottom of the detector, which is then recapped. With the cocktail below the charcoal canister, the desorption takes place through the vapor phase. Desorption is more than 80% complete after 3 hr (the time for full equilibrium of the decay product). The count rate reaches its maximum value in about 8 hr. The elution curve vs time, generated at room temperature, is shown in Figure 2. This curve is in good agreement with the elution curve determined by other investigators using a similar airborne radon detector of their own design.[4]

Analysis of Results

A calibration constant of 37 cpm $(pCiL^{-1})^{-1}$ has been empirically determined at room temperature. Counts per minute corrected for background, are multiplied by the calibration constant and by correction factors for the decay of radon, for adsorption time, and for elution time. The calculations are part of a computer program licensed through Niton Corporation in Bedford, Massa-

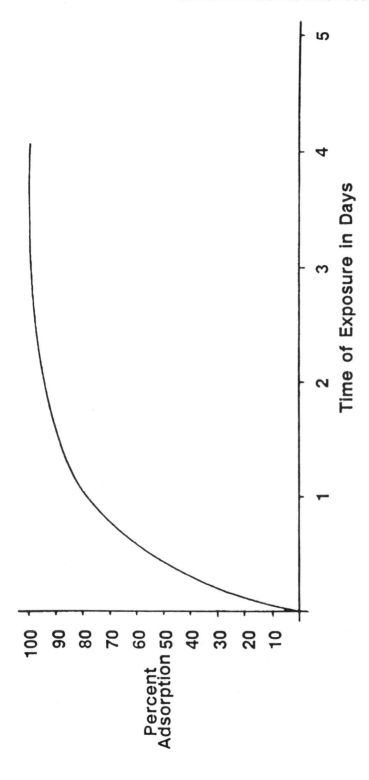

Figure 1. Adsorption of radon in air in PICO-RAD detectors.

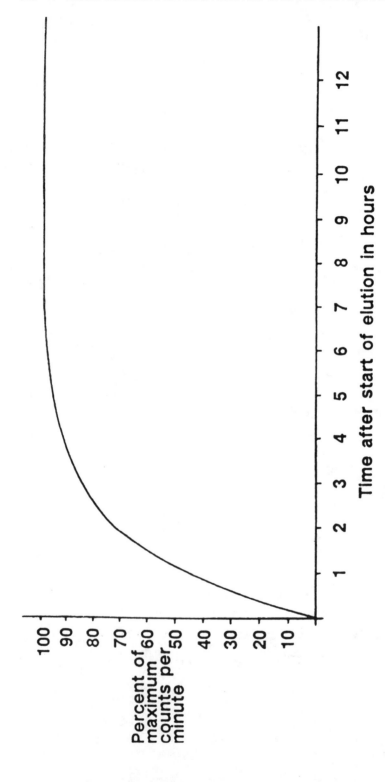

Figure 2. Desorption of radon from PICO-RAD detectors using a xylene based cocktail.

chusetts. The general form of the equation programmed into the computer is as follows:

$$I \text{ (pCi/l)} = K \; (N_1 - N_0 \; (e^{(T_4-T_2)/r_0)^*}$$
$$(1 - ep^{-(T_2 - T_1)/r_1)^{-1}} \; (1 - e^{-(T_4 - T_3/r_2)^{-1}}$$

where I = radon concentration in pCi/1

         K = normalization constant that converts net cpm to pCi/1

         N_1 = cpm obtained in the appropriate energy window of the LSC

         N_0 = mean value of the background counts obtained from many background samples

         r_0 = mean life of ^{222}Rn

         r_1 = mean time for the adsorption of radon into the activated charcoal

         r_2 = mean time for desorption of the radon out of the activated charcoal and into the liquid scintillation eluant

         T_1 = exposure start time

         T_2 = exposure end time

         T_3 = time the elutant cocktail was placed in the detector

         T_4 = analysis time

Comparisons with the Charcoal Canister Technique

A side by side study was performed to investigate the correlation of results obtained with the activated charcoal canister and the liquid scintillation method.

The charcoal canisters were obtained from Canberra Industries, Meriden, CT. The basic design and analysis of the canisters is the same as described by George.[5] An individual container is 10.2 cm by 3.3 cm in dimension and contains 150–200 g of activated charcoal. The container has no diffusion barrier. The canisters were weighed before and after exposure to determine the moisture content and correct for humidity.

The liquid scintillation detectors are the PICORAD detectors described earlier. The study consisted of exposing both types of detectors at the same location in the basement of eight private homes.

Results

Figure 3 is a bar graph illustrating the results of side by side exposures done in eight private homes. From the graph, it is apparent that the liquid scintillation results compare favorably to those obtained by the activated charcoal canister method over a range of radon levels. This relationship is shown in Figure 4. The data plotted are the mean values of 2 to 6 individual determinations for each test. The individual results for tests number 7 and 8 are pre-

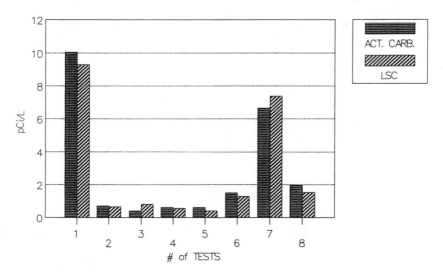

Figure 3. Method comparison of radon levels in eight private homes.

sented in Table 1. Six determinations with each method were made for these tests, because these were the homes found to have significant radon levels from an earlier screening. Student t tests were done to determine if the difference between the mean exposures for each method is significant. As reported in Table 1, the p values for test 7 and 8 indicate that in test 7, the mean values did not achieve a level of significance, while in test 8 they did. Even though the mean values for test 8 seem close, the standard errors around the means are

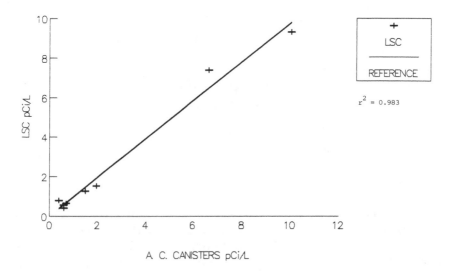

Figure 4. Relationship of radon measurements made with the activated charcoal and liquid scintillation methods.

Table 1. Individual Data and Statistics for Tests Numbered 7 and 8[a]

| | Test 7 | | Test 8 | |
Sample	A.C.	LSC	A.C.	LSC
1	7.0	8.9	1.9	1.8
2	6.8	6.4	2.3	1.6
3	6.3	6.8	1.7	1.2
4	7.1	8.1	2.0	1.5
5	6.5	7.3	2.0	1.5
6	6.2	6.8	1.9	1.6
Mean	6.65	7.38	1.97	1.53
S.D.	0.37	0.94	0.197	0.197
%CV	5.6	12.8	10.0	12.8
Student t	1.77		3.88	
	10 df		10 df	
	$p > 0.1$		$p < 0.01$	

[a]Radon Levels pCi/L.

small, accounting for the level of significance achieved. Ideally, a t test run on a larger sample size would be more appropriate.

RADON IN WATER

The Collection Technique

The method of collection is important to insure against loss of radon from the water sample. The water collection procedure used here is based on EPA documents and individual test results. The collection technique is as follows:

> If taking water from a faucet, remove the aerator and draw the water for five to ten minutes for the sample to be fresh. The water drawn should be a cold water sample since radon is less soluble in warm water. If taking water from a lake or stream, a direct sample can be taken. The recommended container is a standard EPA type water collection bottle with a volume of at least 20 mL. These vials have rubber-teflon plenums and prevent leakage of radon from the vial. Tests of radon leakage from water in such bottles indicate that the mean leakage time exceeds 3000 hr. The bottle should be filled such that a minimum amount of air is retained, preferably less than 1 cc, since radon prefers to be in air rather than water in the ratio of 4:1. As a result, a 1 cc bubble of air in a 20 mL bottle will yield an apparent radon concentration in the water which is 9% lower than the true concentration.

The Elution Procedure

The elution procedure for radon in water begins with a 20 mL syringe filled with 10 mL of Opti-Flour® O (Packard Instruments, Meriden, CT.). The cap of the water collection bottle is removed, and 10 mL of water is drawn into the syringe from the bottom of the bottle. The 20 mL mixture is then transferred to a liquid scintillation vial for counting. The syringe should be rinsed with "radon free" water between each test. The liquid scintillation vial is then shaken for approximately 5 sec (not critical) and set aside for equilibration of the radon into the scintillation cocktail. A typical count rate vs time curve such as the one reported by Prichard[3] is shown in Figure 5. As can be seen from the figure, a 3 hr equilibration time is appropriate. The elution will then be more than 95% complete.

Counting the Water Sample

Each decay of radon in the water will result in five detected counts: three alpha particles and two beta particles. The efficiency of counting the alphas and betas is close to 100%, with the lower level discriminator on the liquid scintillation counter set to 25 keV and upper value set to 900 keV. This value has been determined using a Bureau of Standards source of radium in water.

The number of pCi of radon in the water is calculated using the following formula:

$$pCi/l = (N - N_1) * 100 / (5 * 2.22 * 0.964) (D)$$

where N = cpm obtained from the liquid scintillation counter
              N_1 = background cpm
           100   = multiplication factor to give the concentration in pCi/l
             5   = number of counts per decay of radon
        2.22   = number of decays of radon per minute per pCi
     0.964   = correction factor applied for the radon in air vs water in the collection bottle
           D = radon decay correction factor

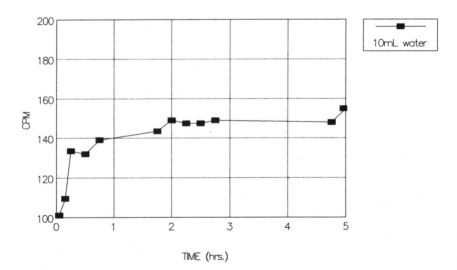

Figure 5. Count rate vs time for a 10 mL water sample in 10 mL of Opti-Fluor O.

A background sample can be prepared by from water drawn from a municipal source and either aerated for a number of hours, heated to 80°C for several hours, or allowed to stand idle for several months.

Comparisons with the Lucas Cell Method

A good correlation exists between the results obtained with the liquid scintillation and Lucas cell method. In early experiments, Prichard[3] reported a 9.34 counts/min/pCi calibration factor from a plot of the relationship of the liquid scintillation count rate vs the Lucas cell results.

The predicted value, based on the relative solubilities of radon and the volumes of water, scintillation fluid, and air in the collection bottle, was 9.75 counts/min/pCi. This relationship is shown in Figure 6.

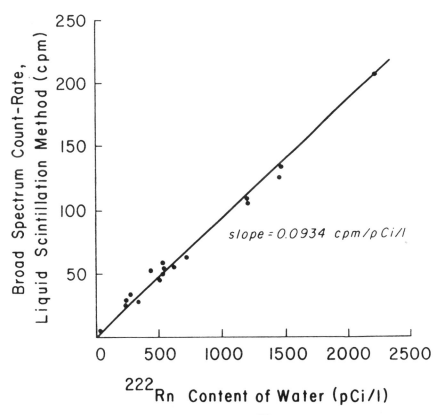

Figure 6. Count rate by liquid scintillation method vs ^{222}Rn content of water as determined by radon bubbler and Lucas cell. (This figure reproduced with the permission of Dr. Howard M. Prichard.).

Summary

The liquid scintillation method offers many desirable advantages for the measurement of radon: (1) Only one counting system is required to measure radon in both air and water. (2) The method is fast. A short exposure time of 24–48 hr for air measurements is ideal for short term screening. (3) The method is simple. Water samples can be counted directly while special air detectors serve both as a collection device and counting vial. (4) The method offers high throughput capabilities. Up to 200 samples can be counted and analyzed in a 24 hr period. (5) The method is accurate. It has passed EPA proficiency testing in rounds 4 and 5 and is currently a method listed by that agency for radon monitoring. (6) Measurements made with the liquid scintillation method show good correlation with established procedures for quantitating radon in air and water.

REFERENCES

1. Lucas, H.F. "Improved Low Level Alpha Scintillation Counter for Radon," *Rev. Sci. Instrum.*, 28:680 (1957).
2. Lucas, H.F. "A Fast and Accurate Survey Technique for Both Radon-222 and Radium-226," in *The Natural Radiation Environment* J.A.S. Adams and W.M. Lowder, Eds. (Chicago: University of Chicago Press, 1964), p. 315.
3. Prichard, H.M. and T.F. Gesell. "Rapid Measurements of ^{222}Rn Concentrations in Water with a Commercial Liquid Scintillation Counter," *Health Physics*, 33:577–581 (1977).
4. Schroeder, M.C., U. Vanags, and C.T. Hess. "An Activated Charcoal-Based, Liquid Scintillation Analyzed Airborne Rn Detector," *Health Physics*, 57(1):43–49 (1989).
5. George, A.C. "Passive, Integrated Measurement of Indoor Radon Using Activated Carbon," *Health Physics*, 46(4):867–872 (1984).

CHAPTER 33

Liquid Scintillation Screening Method For Isotopic Uranium in Drinking Water

Howard M. Prichard and Anamaria Cox

ABSTRACT

The radiation dose resulting from the ingestion of uranium is determined by activity rather than mass; isotopes such as [234]U can contribute significantly to the total uranium activity, without being detected by mass-based fluorometric techniques. As [234]U / [238]U ratios significantly in excess of 1 are frequently noted in groundwater, the determination of uranium on a mass basis and the assumption of isotopic equilibrium can lead to serious dose underestimation. Techniques for specific uranium isotope determination are well developed, but involve considerably more effort than fluorimetry. The relative ease with which uranium can be extracted from water into a liquid scintillation compatible solution provides an alternative approach for screening drinking water supplies. A technique originally developed for PERALS (Photon, Electron Rejecting Alpha Liquid Scintillation) systems has been modified for use in conventional LS systems as a screening test for isotopic uranium and other actinides in drinking water. While lacking the degree of alpha/beta discrimination and energy resolution available with PERALS systems, conventional units offer the advantages of large sample volume, high throughput, and widespread availability. This chapter demonstrates that with appropriate sample preparation, the resolution and background attainable with conventional systems is more than adequate to provide a simple, effective screening technique for uranium concentrations of significance to public health.

INTRODUCTION

Because of its long half-life and correspondingly low specific activity, [238]U is by far the most abundant uranium isotope in water on a mass basis, but the radiation dose resulting from the consumption of the water is determined on an activity basis. Other uranium isotopes, notably [234]U, can contribute significantly to the total uranium activity without being detected by existing mass-based fluorometric techniques.[1,2] The geochemical mobility of [234]U is often enhanced by the alpha recoil energy accompanying its formation, and [234]U/[238]U ratios significantly in excess of 1 are frequently noted in groundwater.[1,2] The determination of uranium on a mass basis and the assumption of isotopic equilibrium can therefore lead to a serious underestimation of the radiation dose from drinking water consumption. Techniques for specific uranium isotope determination are well-developed but involve considerably more effort than fluorimetry.

The high uranium extractibility from aqueous solutions into organic solvents is well known and has long been exploited in uranium recovery processes. Uranium in natural waters is nearly always in the form of the uranyl ion $(UO2)^{++}$, and is often complexed with carbonate or OH ions.[3,4] The complexes are destroyed below pH 4, leaving the simple uranyl ion in solution. At low pH, this ion is very readily extracted from water into toluene containing 60 g/L of HDEHP (bis[2-ethylhexyl] phosphoric acid), a scintillation compatible compound.[5] Liquid scintillation is therefore an attractive option which offers the possibility of a single step process: directly concentrating uranium into the counting medium.

The procedure described in this chapter is derived from a solvent extraction, alpha liquid scintillation technique developed at Oak Ridge National Laboratory and published in "Alpha Counting and Spectrometry Using Liquid Scintillation Methods."[5] The set of techniques outlined in that document rely on high resolution energy discrimination and pulse-shape discrimination available with the PERALS system, which uses small (ca. 1 mL) volumes of highly purified organic scintillators.

The high energy resolution and low backgrounds achievable with the PERALS system make this technology extremely attractive for many applications. However, the necessarily small liquid scintillator volumes and the amount of sample processing involved potentially limit the application of this approach to large scale, low specific activity drinking water screening programs. To meet this particular need, we have "scaled up" the extraction methods for use with conventional liquid scintillation systems, which have the advantages of large (20 mL) sample volume, automatic operation, and widespread distribution. The disadvantages of the commercial systems are low energy resolution, relatively high backgrounds, and less effective alpha/beta discrimination. However, in the restricted context of screening potable water for uranium, these disadvantages prove to be relatively unimportant. Furthermore, the energy resolution of commercial systems can be enhanced by a few simple steps to the extent that more definitive spectral analyses of positive samples can be performed.

SCREENING WATER SAMPLES FOR TOTAL URANIUM ACTIVITY

The main purpose of a screening program is to show that water is suitable for consumption, in other words, to demonstrate the absence of uranium above a particular concentration. This approach greatly simplifies analytical requirements in the (presumably) great majority of cases where gross extractable radioactivity is in fact below an established level.

Method

To screen for total uranium activity, a 1 L sample of water is treated by aeration or boiling to remove radon, which is highly soluble in toluene. The sample is then brought to approximately pH 1.5 with HNO_3 and agitated with 20 mL toluene containing 60 g/L of HDEHP and an appropriate fluor concentration. After a few minutes of contact, the phases are allowed to separate and the extractant scintillator drawn off for counting. If the initial solution had not been purged of radon three or more hours prior to extraction, the extracts should be held at least 3 hr before counting to allow for radon daughter decay. A screening count is then taken in a region of interest empirically set around the uranium alpha peaks (4.2 and 4.8 MeV).

Extraction and Counting Efficiency

Figure 1 shows the spectra of a gels incorporating 2 mL aliquots of a spiked sample containing 412 Bq/L of natural uranium alpha activity (A) and a second aliquot taken after liquid-liquid extraction (B). The net alpha count rate in the second gel indicated an activity of 21 Bq/L in the aqueous phase, or a deficit of 391 Bq/L. The net alpha activity of the extract was 376 Bq, with another 6 Bq recovered after a secondary organic stripping of the aqueous phase. The recovered activity is in good agreement with the deficit, indicating that the alpha counting is virtually quantitative in the extract, as might be expected from general principles. The ratio of activities in the aqueous phases after and before extraction is 21:412, or 5.1%, indicating a 95% recovery.

Background and Detection Limits

In a narrow region of interest around the uranium alpha peaks, the blank counting rate was found to be 3.2 cpm on our system. While this is still higher than that of the PERALS unit, the twentyfold increase in sample volume available with conventional LS systems produces adequate statistical reliability at concentrations of regulatory interest. (Some users may prefer a wider window with higher background but more tolerance for quench variation.) Under these conditions, the minimum true detectable activity (MDTA),[6] is given by:

$$MDTA = g(k_a + k_b)(CB)^{0.5}$$
$$g(K_a + K_b) C_B^{0.5}$$

where g = calibration factor (Bq/count),
  C_B = total measured background count,
  and k_a
  and k_b = number of normal standard deviations associated with the probabilities a and b of making Type I (false indication that activity is present) and Type II (false indication that activity is absent) errors.

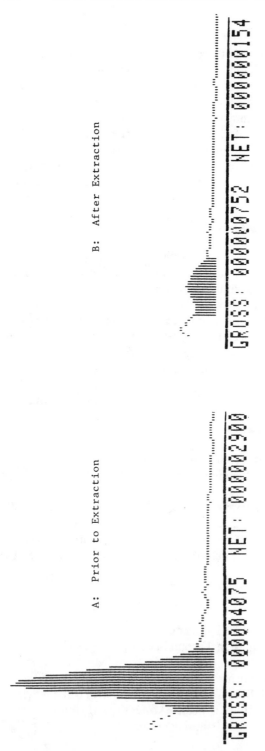

Figure 1. Aliquots of natural uranium spike solution in gel prior to liquid-liquid extraction (A) and after extraction (B). Counting times, scales, and regions of interest are identical.

The approximation holds if the background is stable and the square root of the total observed background counts is much greater than k_a. For screening measurements, it is more important to avoid a Type II error than it is to avoid a Type I error, so we might choose to set a = .05 and b = .025, with k_a = 1.645 and k_b = 1.960. If the sample volume is 1 L and combined extraction and counting efficiency is 90%, then a 10 min screening count would be expected to produce (10 × 60 × 0.9) or 540 counts per Bq/L, and g = 1/540 = 0.00185. The background count in 10 min, C_B, would on the average be (3.2 × 10) = 32 counts, and $k_a + k_b$ = 3.605. The minimum detectable true activity would then be:

$$MDTA = 0.00185 \times (3.605) \times (64)0.5 = 0.0377 \text{ Bq/L } (1.02 \text{ pCi/L})$$

SPECTROSCOPIC ANALYSIS OF POSITIVE SAMPLES

As noted above, the demonstration of the absence of activity poses less of an analytical challenge than the identification of the nuclide(s) producing an excess count rate. Several options are available for samples that exceed a predetermined screening level. One is to process the extract for high resolution alpha spectroscopy with either the PERALS system or solid state detectors. However, a relatively modest improvement in the energy resolution of conventional LS counting, combined with chemical and radiological considerations, is sufficient to identify and quantify the excess activity in many cases.

Resolution Enhancement

Because most liquid scintillation systems are multi-user devices, our efforts to further improve resolution have been directed to the vial and its contents, rather than on modifications to the instrument itself. As noted by McDowel,[15] a number of simple modifications can be made to improve the energy resolution potential of a liquid scintillator. Sparging the organic scintillator with inert gas reduces oxygen quenching, thus increasing both absolute light output and energy resolution. Total light yield and energy resolution are also increased by adding of 200 g/L of naphthalene to the primary solvent and by using PBBO (2-(4'-biphenylyl)-6-phenylbenzoxazole) instead of PPO as a fluor.

The substitution of a translucent for a transparent counting vial also has a marked effect on the alpha spectrum. (This was brought to my attention by Bernard Cohen[7] in the context of our long-standing interest in detecting radon in air by activated carbon adsorption and subsequent liquid scintillator desorption. A similar effect is noted in Donald Horrocks's 1972 text on liquid scintillation.) The diffuse transmission of light through the translucent material decreases spectral degradation due to photocathode nonuniformity, thus improving resolution. Figures 2 and 3 show the spectra of ^{238}U and ^{234}U in

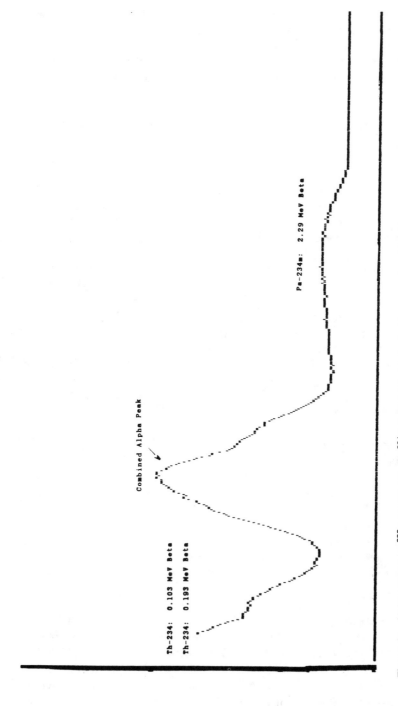

Figure 2. Alpha spectra of ²³⁸U (4.2 MeV) and ²³⁴U (4.7 MeV) obtained with conventional liquid scintillation (18 mL). The beta spectra of the short-lived intermediates are also shown.

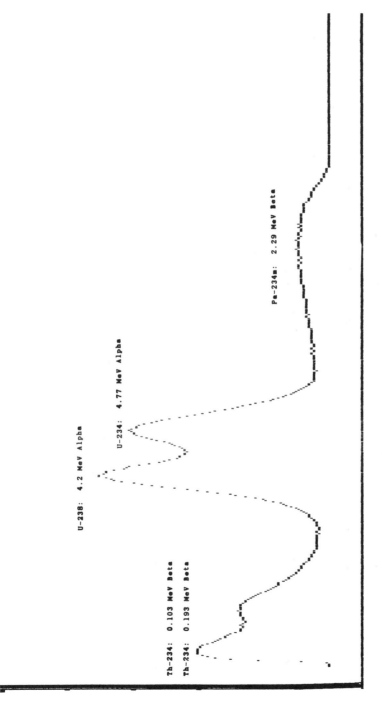

Figure 3.   Spectrum of the solution shown in Figure 2 after transfer to resolution enhancing vial.

NATURAL URANIUM, REAGENT GRADE

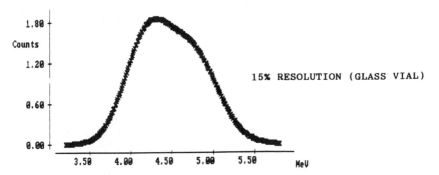

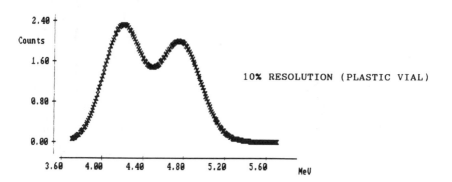

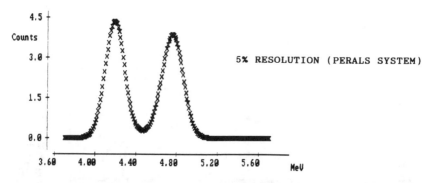

Figure 4.   Simulated alpha spectra at three levels of resolution.

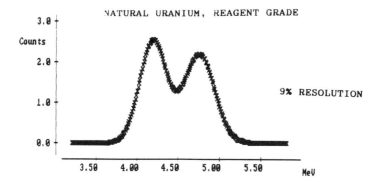

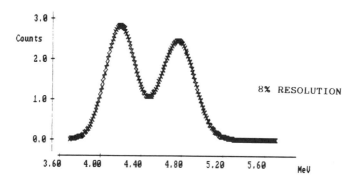

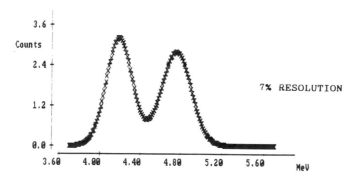

Figure 5. Simulated alpha spectra at three levels of resolution.

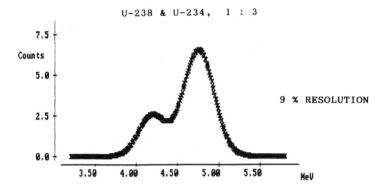

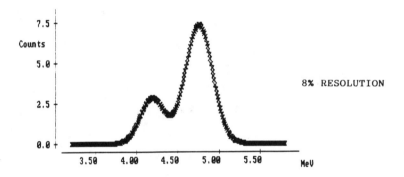

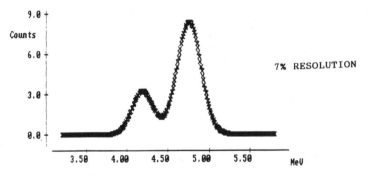

Figure 6. Simulated alpha spectra of ^{238}U and ^{234}U in disequilibrium at three levels of resolution.

argon-sparged extractant scintillator obtained in a glass and in a slightly modified polyethylene vial respectively. The beta spectra are largely unchanged, but the alpha resolution in the polyethylene vial is considerably improved. The alpha resolution is approximately 10% of the peak energy, just about midway between the resolution obtained in clear glass on a conventional system and the resolution available with a PERALS system, as shown in the simulated alpha spectra in Figure 4.

It is not yet known how much more resolution can be obtained on existing conventional LS systems by modifying the solvent and counting vial, but even if little additional resolution can be achieved, the level shown in Figure 3 is useful. While ^{238}U and ^{234}U are not completely resolved, the total alpha activity is obtained, and the degree of disequilibrium can be estimated from the relative peak areas. The importance of relatively small gains in energy resolution is illustrated in Figures 4 through 6. These figures depict synthetic alpha spectra generated by summing normal distributions with means corresponding to the ^{238}U, ^{235}U, and ^{234}U alpha peaks, full widths at half maximum at the stated percent of mean energy, and amplitude at the stated equilibrium ratios. ^{235}U is normalized to its natural abundance with respect to ^{238}U; while not resolved at any level shown, the presence of its alpha emission between that of the other two isotopes diminishes the resolution to some extent. Figure 4 shows the two main uranium alpha peaks unresolved at 15% percent resolution, which may be compared with Figure 2, the actual alpha plus beta spectrum of our uranium extract in a glass vial. Also shown is a simulation of the 5% resolution spectrum obtainable with the PERALS system and the 10% resolution obtained with a slightly modified high density plastic vial in our conventional LS system.

Figure 5 shows the projected effects of improving resolution to 9, 8, and 7%. Each incremental gain is seen to produce noticeable effects. The importance of gains of this magnitude is further illustrated in Figure 6, in which uranium in a state of disequilibrium is viewed with 9, 8, and 7% resolution.

Nuclide Identification

The 10% resolution currently available permits discrimination between uranium isotopes and many of the other alpha emitters of interest. Table 1 lists the nuclides of the ^{238}U and the ^{232}Th decay series that emit alpha particles with energies between 3.75 and 5.25 MeV. These nuclides could not be readily distinguished from $^{238}U - ^{234}U$ at the present level of resolution. However, the screening extractant HDEHP does not extract radium from the aqueous phase. Furthermore, except in one unusual well, ^{232}Th and ^{230}Th concentrations over 0.1 pCi/L have not been found in potable water.[2] In short, natural alpha activity from 3.75 to 5.25 MeV extracted from drinking water with HDEHP is quite likely to be due to uranium.

Table 2 lists common naturally occurring nuclides with alpha energies greater than 5.25 MeV. These nuclides are thus readily distinguishable from

Table 1. Natural Radionuclides with Alpha Emissions Between 3.75 and 5.25 MeV

Nuclide	Half-Life	Alpha Energy (Mev)	Abundance (%)
^{232}Th	1.4×10^{10} Y	3.95	24
		4.01	76
^{238}U[a]	4.5×10^{9} Y	4.15	25
		4.20	75
^{234}U[a]	2.5×10^{5} Y	4.72	28
		4.77	72
^{230}Th[a]	8.0×10^{4} Y	4.62	24
		4.68	76
^{226}Ra[a]	1.6×10^{3} Y	4.62	5
		4.78	95

[a] ^{23}U series nuclide.

uranium at the 10% energy resolution currently available. ^{222}Rn and its short-lived progeny can be easily excluded by appropriate sample handling. ^{224}Ra would not be extracted into HDEHP, but would build up with a 3.6 day half-life into any ^{228}Th extracted. A month after isolation, the ^{228}Th alpha emissions will be accompanied by several high energy alphas from its progeny, including one at 8.78 MeV. As noted above, however, it is quite unlikely that significant levels of ^{228}Th would be present in a drinking water sample, except through ingrowth (half-life = 1.9 years) into older samples containing ^{228}Ra. The presence of significant amounts of ^{228}Ra in even a fresh sample would cause a "false positive" in the initial screening test due to the rapid ingrowth of ^{228}Ac, a complex beta emitter with a major portion of its beta emission overlapping the alpha window. Subsequent medium resolution spectrometry would clearly dis-

Table 2. Natural Radionuclides with Alpha Emissions Above 5.25 MeV

Nuclide	Half-Life	Alpha Energy	Abundance (%)
^{210}Po[a]	138.4 D	5.30	100
^{228}Th	1.9 Y	5.34	28
		5.43	71
^{224}Ra[b]	3.6 D	5.45	6
		5.68	94
		6.05	25
		6.09	10
		6.29	100
		6.78	100
		8.78	36
^{222}Rn[a,b]	3.8 D	5.49	100
		6.00	100
		7.69	100

[a] ^{238}U series member.
[b] Includes radiations from short-lived daughters.

Table 3. Important Artificial or Enriched Alpha Emitters

Nuclide	Half-Life	Alpha Energy	Abundance (%)
^{235}U	7.1×10^8 Y	4.37	18
		4.40	57
		4.42	5
		4.56	5
^{237}Np	2.1×10^6 Y	4.8c[a]	86
^{242}Pu	3.8×10^5 Y	4.86	24
		4.90	76
^{239}Pu	2.4×10^4 Y	5.10	12
		5.14	15
		5.16	73
^{241}Am	458 Y	5.44	13
		5.49	86

[a]Complex peaks near 4.8 MeV.

tinguish between uranium isotopes and actinium. This, in effect, leaves ^{210}Po as the only common natural alpha emitter in this energy range likely to be present in a timely HDEHP extraction from potable water.

If artificial transuranic nuclides might be present in a sample, the situation becomes more complicated, as illustrated in Table 3. ^{241}Am alphas are energetic enough to be distinguished from ^{234}U at 10% resolution, but ^{239}Pu is only marginally distinguishable. ^{242}Pu and ^{237}Np would be interpreted as uranium unless higher resolution spectroscopic techniques or chemical separations are employed.

REFERENCES

1. Aieta, E.M., J.E. Singley, A.R. Trussel, K.W. Thorbjarnson, and M.J. McGuire. "Radionuclides in Drinking Water: An Overview," *J. Am. Water Works Assn.*, 79:144–152 (1987).
2. Hess, C.T., J. Michael, T.R. Horton, H.M. Prichard, and W.A. Coniglio. "The Occurrence of Radioactivity in Public Water Supplies in the United States," *Health Phys.* 48: 53–58 (1985).
3. Cordfunke, E.H.P. "The Chemistry of Uranium," (Amsterdam: Elsevier Publishing Co., 1969).
4. Sorg, T.J. "Methods for Removing Uranium from Drinking Water," *J. Am. Water Works Assn.*, 80L:105–111 (1988).
5. McDowell, W.J. "Alpha Counting and Spectrometry Using Liquid Scintillation Methods," *Nucl. Sci. Series*, NAS-NS-3116, (U.S. Dept. of Energy, January, 1986.)
6. Altshuler, B. and B. Pasternack. "Statistical Measures of the Lower Limit of Detection of a Radioactivity Counter," *Health Phys.* 9:293–298 (1963).
7. Cohen, Bernard, University of Pittsburg, Personal Communication, 1988.

CHAPTER 34

Assessment and Assurance of the Quality in the Determination of Low Contents of Tritium in Groundwater

C. Vestergaard and Chr. Ursin

INTRODUCTION

In the last few years it has become clear that groundwater resources are in danger of becoming unfit sources of drinking water as a result of human activities, e.g., waste dumps for hazardous materials and agricultural activities involving extensive use of fertilizers. In Denmark, the environmental protection authorities have launched a major program for monitoring and ensuring the quality of groundwater. The program involves periodic measurements of some 80 parameters of interest, e.g., nutrients, organic pollutants, trace and ultratrace metals, and tritium, in approximately 1000 positions covering the types of aquifers seen in the country.[1]

Tritium released into the atmosphere by the testing of thermonuclear weapons in the 1950s and early 1960s, and subsequently incorporated in atmospheric water and precipitation, has proven very useful in hydrological research over the last decades. In hydrological research the tritium content is often expressed in tritium units (TU), 1 TU being equivalent to 1 tritium nucleus per 10^{18} hydrogen nuclei and approx. 0.118 Bq/L. In many cases a profile of the tritium content in the groundwater, as shown in Figure 1, gives a quite precise dating of the groundwater and furthermore can give valuable hydrological information. Figure 1 shows tritium profiles from two geologically different locations common in Denmark. The vulnerability of a well in case of the threat of polution may be assessed by a single tritium measurement. A content of tritium approaching 1 TU, would indicate that the water originates from the time before the atmospheric bomb testing; consequently, that resource is relatively well protected. The interpretation of tritium data is beyound the scope of this chapter, but examples of such work may be found in the IAEA series of publications on isotope hydrology.[2]

Awareness of the need to ensure and document the quality of environmental

TRITIUM PROFILES

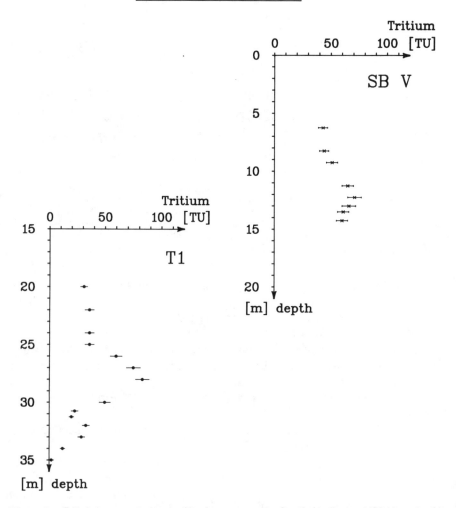

Figure 1. Tritium in groundwater profiles from an area by Syv Baek, Zealand (SB V), a glacial clay deposit (moraine), and an area by Rabis Baek, Jutland (T I) a sandy aquifer, 1988.

measurements is growing as the consequences of descisions based upon such measuremnets for human health, quality of life, and economy are becoming more widespread and complex. This chapter reports our attempt at assessing and assuring the quality of measurements we believe will be of growing importance in the work done to understand the mechanisms of groundwater pollution in the years to come.

Table 1. Summary of Method

Electrolytical Enrichment:
1. Distillation of water sample.
2. Addition of Na_2O_2.
3. Electrolysis in IAEA cell (Mild steel cathode, stainless steel anode), volume: 300 mL, 850 Ah, t = 0–5°C. Enrichment Factory (EF): approximately 14.
4. Distillation of enriched sample, volume: approximately 20 mL.

Liquid Scintillation Spectrometry:
– 12 mL sample in 10 mL Pico-Fluor LLT (Packard).
– Low potassium glass vials.
– Spectrometer: Packard 2000 CA/LL.
– Counting time: 300–500 min. t = 15°C.
– Efficiency (ϵ): approx. 0.18.
– Background (tritium free water): 2.4 CPM.

METHOD

The method for determining the content of tritium in groundwater is well established and commonly used. It is based on IAEA's method for the electrolytic enrichment of tritium in water,[3] followed by liquid scintillation counting. This method is regarded as applicable to the determination of tritium contents in water samples down to 1 TU.[4] Table 1 gives an outline of the method as it is routinely applied. It should be noted that multiple calibration standards are used and that the enrichment of samples with regard to tritium in the electrolysis is determined experimentally in each run by five identical samples of tritiated water. The liquid scintillation counting is done on uniformly quenched samples so that errors introduced by incorrect or poorly defined efficiency correction are avoided.

QUALITY ASSURANCE

Probably the most important aspect of any measurement quality is the reliability of its associated statement of uncertainty, which must be an integral part of reporting any result. This aspect has been discussed thoroughly in a book on neutron activation analysis in which it is shown that " . . . the precision of a single analytical result is determined by the method with which it is found. When all sources of variability are properly taken into account, the estimated precision will account for all the observed variability of analytical results."[5] This would seem to hold for most types of measurements and certainly for the measurements discussed here.

As can be seen in Table 1 several processes are involved in a measurement; that may contribute to the observed variability. The processes are the calibration of the spectrometer, the handling of samples, including the electrolytical enrichment, the determination of the enrichment factor, and finally the counting of the sample. This is also indicated by the expression used to convert counting results into tritium content (A), in the original sample:

$$A = (C-C_b) * \epsilon^{-1} * EF^{-1} * K$$

Where C is the observed count rate in the sample, C_b is the background count rate, ϵ is the the the efficiency of the spectrometer, EF is the degree of enrichment by the electrolysis, and K is a factor for unit conversion, decay time correction etc. In our method EF is determined by the actual enrichment of five identical samples of tritiated water in each run; thus:

$$EF = C_e * C_0^{-1}$$

Where C_0 is the mean count rate of the sample of tritiated water before electrolysis and C_e is the mean count rate of the same sample after electrolysis. An estimate of the standard deviation of an individual result, s_A may be calculated from this expression:

$$s_A = A*((s_c^2 + s_{Cb}^2)/(C - C_b)^2 + (s_\epsilon/\epsilon)^2 + (s_{c0}/c_0)^2 + (s_{Ce}/C_e)^2)^{1/2}$$

Contributions from the counting process, s_C and s_{Cb}, may be calculated from the counting statistics assuming identity between the total number of counts observed and variance. In estimating the contributions from the enrichment factor, s_{Ce} and s_{C0}, and calibration, s_ϵ, Shewart control charts of standard deviations are used.

Figure 2 shows such a control chart for the relative standard deviation of ϵ. It can be seen how a minor modification of the spectrometer, involving improved grounding, has improved the precision of the calibration; over the last 10 runs the precision has been homogeneous with a relative standard deviation of 0.4%.

In a similar way the contributions from the enrichment may be found. The result is summarized in Table 2. When the standard deviation of the single result has been estimated, the ability of this estimate to account for the observed variability may be tested by the analysis of precision, using the T statistic. In the case of independent duplicate measurements of M different samples, the statistic T is determined as:[5]

$$T = \sum_{i=1}^{M} ((A_{i1} - A_{i2})^2 / (s_{i1}^2 + s_{i2}^2))$$

Where A_{i1} and A_{i2} are the two independent results and s_{i1} and s_{i2} are the estimated standard deviations of the two results. T is approximated by an X^2 distribution with M degrees of freedom. Figure 3 shows T as a function of the number of duplicate determinations made over a 10 month period, in which approximately 10% of the determinations were done in duplicate. The individual results range from below 1 TU to 45 TU. As T is not significantly different from an X^2 distribution, it may be assumed that the estimated standard devia-

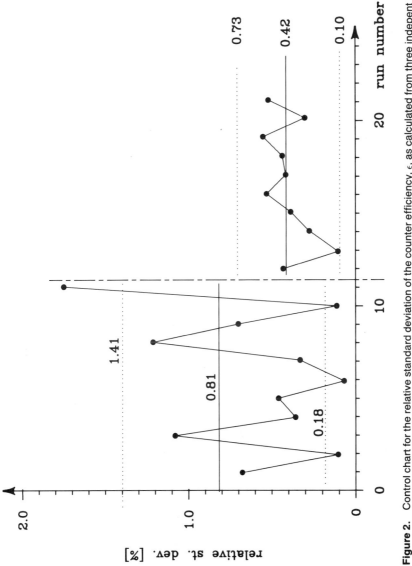

Figure 2.   Control chart for the relative standard deviation of the counter efficiency, ϵ, as calculated from three indepent calibration standards per run. The vertical dotted line coincides with the time of modification of the spectrom- eter. Control limits are given as 0.05 α.

Table 2. Individual Error Contributions

		Rel. Std. Dev. [%]
Efficiency		0.4
Tritiated water		1.2
-"-	after enrichment	4

tions of individual results do account for the observed variability in the duplicate results; consequently, no other significant source of variability is present.

In Figure 4 the calculated standard deviation for the samples is shown as a function of tritium content. s_A/A varies between approximately 0.5 for a content of 1 TU to approximately 0.05 at 45 TU. When the random error has been determined, what remains to be shown is that systematic errors or bias do not affect the results. An important source of bias would be the presence of a significant blank value in the determinations. This is checked by routinely measuring samples of supposedly tritium free groundwater. These measurements give results well below 1 TU and it is assumed that the blank can be neglected. The absence of systematic error can be demonstrated by measuring reference samples with certified tritium contents, similar to those usually measured in the laboratory, or by the participation in interlaboratory comparison exercises. The laboratory has participated in an intercomparision organized by the IAEA in which no systematic error was detected.[6]

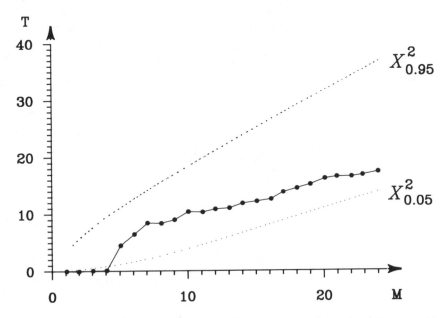

Figure 3. The statistic T, calculated from duplicate determinations plotted as a function of M, the number of duplicate determinations. Dotted lines represent the 5% and 95% fractiles in the X^2 distribution with M degrees of freedom.

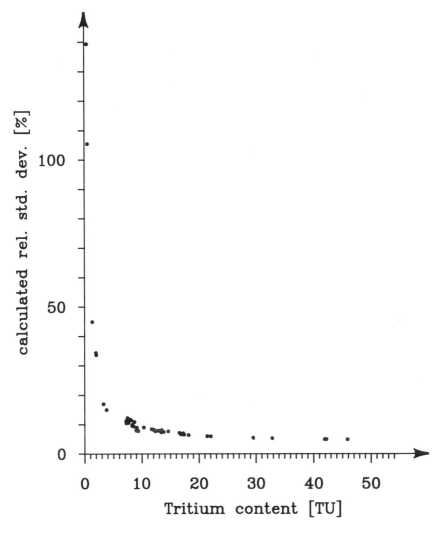

Figure 4. Calculated standard deviations as a function of tritium content.

CONCLUSION

The measurement quality of low tritium contents in groundwater can be controlled by comparing expected and observed variability in duplicate determinations of randomly selected samples. The analysis of precision shows that the identified sources of random error account for the observed variability, and the method may be said to be in statistical control.

ACKNOWLEDGMENTS

The authors wish to thank Ms. Lisbeth Just and Ms. Helle Schmidt Pedersen from the Danish Isotope Centre for performing the tritium determinations. The groundwater samples shown in Figure 1 were made available by the Geological Survey of Denmark, as part of a project sponsored by the National Agency of Environmental Protection, Denmark.

REFERENCES

1. "Monitoring Programme of the Aquatic Environment Plan," (in Danish) Environmental Project 115, National Agency of Environmental Protection, Denmark (1989).
2. Andersen, L.J. and T. Sevel. "Six Years Environmental Tritium Profiles in the Unsaturated and Saturated Zones, Gronhoj, Denmark," in *Isotope Techniques in Groundwater Hydrology 1974, Vol. I*, (Vienna: International Atomic Energy Agency, 1974).
3. "Enrichment of Tritium in Water by Electrolysis," IAEA Isotope Hydrology Laboratory Vienna (1972).
4. Florkowski, T. "Low-Level Assay in Water Samples by Electrolytical Enrichment and Liquid-Scintillation Counting in the IAEA Laboratory" in *Methods of Low-Level Counting and Spectrometry*, (Vienna: International Atomic Energy Agency, 1981).
5. Heydorn, K. *Neutron Activation Analysis for Clinical Trace Element Research, Volume I* (Boca Raton, FL: CRC Press, 1984), pp. 167–205.
6. Hut, G. "Intercomparison of Low-Level Tritium Measurements in Water," (Vienna: International Atomic Energy Agency, September 1986.

CHAPTER 35

Use of Liquid Scintillation in the Appraisal of Non-Radioactive Waste Shipments from Nuclear Facilities*

William L. McDowell

ABSTRACT

The generation of non-nuclear waste at the Savannah River Site (SRS) presents a special problem for analysis. As all waste must be suspected to contain radioactive materials prior to appraisal, large volumes of waste are retained on site prior to off-site shipment for incineration or burial. The amount of flammable waste solvent stored on site in late 1987 threatened to place the Savannah River Site in violation of EPA and OSHA regulations. One of the major contributors to this burden was some 2000 55-gallon drums of used paint solvent and waste paint. The Environmental Technology Section at the Savannah River Laboratory was charged with development of quick, reliable, and simple method for measurement of the maximum possible activity of paint waste samples. The <2 nCi/g guideline set by the Department of Transportation* is the upper limit of the radioisotope contamination for removal of the waste from the site.

Owing to the possible presence of tritium and the desire for a relatively quick analysis technique, liquid scintillation was used as the method of choice. Novel methods are presented for dealing with various phases (organic, aqueous, and solid) present in the samples and for determining the optimum dilution for severely quenched samples. Arguments are presented for the fragmentation of solid particles by ultrasonic vibration, and mathematical corrections are given for the escape fraction from solid particles.

INTRODUCTION

Nuclear facilities generate waste materials that must be removed from the site or disposed of on site. The characterization, isolation, and confinement of high level nuclear waste is currently the subject of much study. By contrast,

---

*This paper was prepared in connection with work done under Contract No. DEAC09-88SR18035 with the U.S. Department of Energy. By acceptance of this paper the publisher and/or recipient acknowledges the U.S. Government's right to retain a nonexclusive, royalty-free license in and to any copyright covering this paper along with the right to reproduce and to authorize others to reproduce all or part of the copyrighted paper.
*Code of Federal Regulations, 49CF173.403(g), June 1986.

relatively little attention has been given to non-nuclear waste from nuclear facilities.

On the Savannah River Site (SRS), nonradioactive or clean waste is normally assayed with a survey meter to appraise the extent of surface contamination. Because of the nature of the operations conducted at SRS, the majority of the waste generated is treated as contaminated waste until proven otherwise. Before such waste can be shipped off site, more rigorous assay must demonstrate compliance with government standards. In particular, it is necessary to appraise the samples with regard to the Department of Transportation (DOT) guideline for a total activity of 2 nCi/g of waste material.[1] Furthermore, the assay methods must be sufficient to comply with Environmental Protection Agency (EPA) regulations to prevent overcrowding of waste storage locations; thus, the present work developed efficient liquid scintillation methods for meeting these requirements.

BACKGROUND

The waste material at SRS is accumulated in 55-gallon drums and stored in interim storage facilities. These drums must be assayed in a relatively short period of time so that off-plant shipment can proceed in an orderly fashion, and the storage inventory complies with EPA regulations. All of this material was considered incinerateable nonradioactive hazardous waste. The present work examines 900 55-gallon drums containing paint solvent/waste which were to be shipped off plant for incineration. These drums contained up to three distinct phases of material with widely differing components. The materials were primarily mixed hydrocarbon solvents. A significant fraction of the material included paint solids, dust, and dirt. Additionally, many of the materials contained a measurable amount of water. All the samples contained the liquid organic material, but the other two phases were not always present. For these samples, neither the specific composition nor the contaminating isotopes could be assumed.

The radiometric methods used for these waste assays had to measure the total activity of waste samples in a relatively short turn around time. Sample preparation schemes such as fractional distillation for solvent cleanup, ashing of solid residue, digestion, solvent extraction, and either electroplating of a purified sample or pelletization of the original ash would have been simply too time consuming.

Methods that have been used for total low-level analysis of waste streams from other facilities also have such limitations. Traditional electroplating and alpha counting by semiconductor detector is a sufficiently sensitive method; however, this often requires lengthy chemical separations, and occasionally, sample ashing prior to plating of the sample. Gross plating of the original sample rarely proves satisfactory for such situations owing to the mixed character of the material and the time required. Because a large number of samples must be processed on a continuing basis, this was not considered a viable alternative.

Tritium was considered one of the most likely contaminants of liquid waste materials resulting from on-site, processes. On-line methods for tritium detection in organic matrices do exist,[2] however, because much of the waste also contained a significant fraction of solid material, neither the on-line method previously mentioned nor simple liquid scintillation of untreated samples could be relied upon.

The total radioactive concentrations of the various waste materials had to be measured; thus, analysis for individual isotopes was not required. Based on the SRS process history, the major radionuclides were expected to be alpha, beta, and gamma emitters, although minor contributions from electron capture and positron decay could not be ruled out. Primary manmade radionuclides could be fission products, tritium, transuranics including plutonium, and neutron activation products. Liquid scintillation is very efficient in detecting the charged particle emitters and thus yields the total activity for samples devoid of radionuclides that decay by electron capture. Some of the neutron activation products do decay by electron capture, and their activities are best appraised by gamma spectroscopy due to their weak charged-particle emissions. The percentage of activity due to electron capture was judged to be fairly low and therefore liquid scintillation was enlisted as the primary technique for determination of "total activity." High purity, germanium diode, gamma spectrometry (HPGe spectrometry) was used as a consistency check to ensure that assumptions concerning ec-gamma activity were correct. HPGe spectrometry of the samples used routine methods; however, liquid scintillation required some method development to overcome the various problems associated with these samples.

Liquid scintillation can provide a measure of the total activity of a sample. All ionizing radiation can be detected by liquid scintillation; thus all isotopes could be expected to yield some degree of signal in a liquid scintillation sample. Of course, the detection limit varies for counting time and isotope of interest. As an example, tritium is one of the lowest energy beta emitters and its detection limit on a modern beta liquid scintillation spectrometer is about 11 picoCuries per sample given that no quenching is present and that the sample is colorless.

HPGe spectrometry is ideal for assay of gamma emitting nuclides, especially for gamma energies above 100 keV. The detection limits of HPGe spectrometry depend on the isotope and the counting geometry. In the Savannah River Site/Environmental Technology Section (SRS/ETS) Underground Counting Facility (UCF) the absolute sensitivity for ^{60}C in soil/water may be as low as 0.007 mBq/cm^3, depending on sample configuration. This value is for a 24 hr counting time, and lower levels are attainable with longer counting periods. The samples must conform to a limited number of fixed geometries for the accurate use of HPGe spectrometry.

For many situations, low-level HPGe spectrometry is adequate for determination of total activity. In situations where the primary emission of the contaminate is a strong gamma-ray and the material can be made to conform to a known geometry, this technique is quite useful. Generally, little or no sample preparation is required for HPGe spectrometry. However, for appraising total activity, a significant amount of sample information is required prior to measurement. Also there must be a reasonable assurance that the primary contaminate is not tritium or some isotope such as [129]I which is not suitable for HPGe spectrometry.

ANALYSIS

Sample Preparation

Because of the large number of drums, it was decided to define an individual drum as a sampling unit and composite samples from these units were put in groups of $N = 2^n$ drums each. By using powers of two for the composite, a sample that was subdivided to trace a possible source of contamination (and minimize the amount of material that would have to be treated as low-level waste) could be subdivided down to a single barrel if necessary without the necessity of testing every single barrel in the original composite. This compositing allowed for much faster processing of the samples, but it did lead to increased background values as the dilution factor of N was now included in the detection limits. That made the acceptable maximum value for a single composite sample 2/N nCi/g. This smaller limit arose from the consideration that, in a single composite, there could be a single contaminated drum while all the other drums were clean. If N were too large, the analysis limit of 2/N would be too close to the detection limit of the instrument. Because of this, the maximum number of sampling units (55-gallon drums) in a single composite was determined by the background level of the instrument to be used. The number of sampling units in a single composite had to be selected so that 2/N nCi/g was higher than the background of the instrument. For most applications 16 or 20 was found to be a usable value for N. To create a composite sample, the drum contents were agitated, and then a long cylinder was used to retrieve a column of material extending from the top to the bottom of the drum. Uniform 25 mL aliquots were taken from each of these increments, and these subsamples were combined to give the analysis composite.

For the samples that consisted of multiple layers or phases of material, the components were separated into "homogeneous" samples for analysis; the entire sample could not be counted as a single entity or the heterogeneity would effect the results. After the phases were analyzed separately, the total activity would be found by factoring the relative masses of the phases into the summation of the individual activities. The relative masses of the phases were determined by measuring the volume (as a cylinder) of a single phase and using the empirically determined average density to calculate the mass of the phase in question.

For the organic liquid layer of the paint samples, quenching was dependent on the amount of the sample present in the cocktail. With the proper dilution of such a sample, it is possible to achieve an optimum count rate since both quenching and activity level would increase with concentration. Conceptually, this is shown in Figure 1. The resultant, optimum count rate, is the result of the increasing efficiency and the decreasing count rate with dilution. Assuming that the count rate (C), is the result of both the efficiency and the activity,

$$C = EA = Enm \tag{1}$$

where n is the specific activity and m is the mass.

$$\frac{\partial C}{\partial m} = En + \frac{\partial E}{\partial m}nm = 0 \tag{2}$$

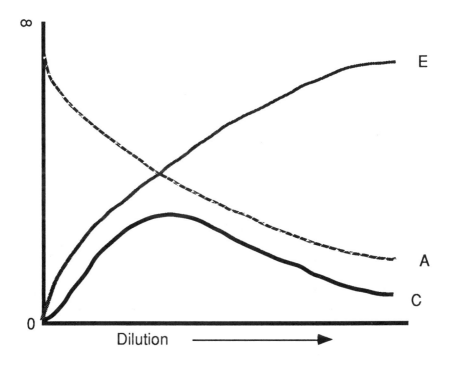

E: Efficiency   A: Activity

C: Observed Count Rate

Figure 1.   Graphic representation of dilution effects.

Substituting Equation 2 for the optimum mass we get,

$$m = m_0 = \frac{-E}{\partial E/\partial m} \tag{3}$$

To estimate an expression for E, it is assumed that the mass M of the sample in the cocktail reduces the scintillation light collection efficiency according to the Beer-Lambert law. A differential form of this law, as applied to the present work is

$$\frac{\partial E}{E} = -L\partial M \tag{4}$$

where L is a constant. This type of fractional decrease in efficiency with increasing sample mass is probably reasonable for a variety of quenching mechanisms; thus, Equation 4 was used for quenching in general as well as its integrated form

$$E = Eo^{-LM} \tag{5}$$

where Eo is a constant. From Equations 3 and 4 it is seen that Mo = 1/L. Thus Mo could be determined by solving Equation 5 for L, using Eo = 1 for the infinitely dilute (M ~ 0) case and E = 0.001 from tritium spike measurements. The resulting value for L is 2.24 gm^{-1}. Using this value for L a result of ~0.4 g was arrived at as the optimum sample size. With this as a starting point, a few empirical trials determined that the optimum dilution for the majority of the cases was 250 μL of organic material to a single cocktail. Each sample was combined with 20 mL of Opti-Fluor® scintillation cocktail and sufficient water to give a total volume of 23 mL.

Any aqueous phase was collected separately. This was achieved by reaching through the top organic layer with a disposable polyethylene pipette and transferring an aliquot of the aqueous material into a glass vial. The cocktail was then prepared out of this vial. The final sample preparation was identical to that for the organic layer with the exception that 1 to 1.5 mL of the sample was placed in the cocktail. In each case, the aqueous layer was less colored and exhibited less quenching than the organic layer.

Suspended solid materials presented the greatest problem. After examining several of the samples, it was discovered that most of the solids were chips of paint from 1 to 2 mm in "diameter." There were two primary concerns with this material: (1) most alpha events that occur in the chips themselves could be buried and never reach the scintillation cocktail, and (2) light resulting from alpha events in solution could be blocked or absorbed by the relatively large paint chips. Furthermore, the material was heterogeneous and not all particles were soluble in the Opti-Fluor® cocktail. Thus, it was necessary to reduce the size of the particles to the point where they would: (1) allow the majority of the alpha events to escape into the cocktail, (2) disperse homogeneously, and (3) remain suspended long enough to allow effective counting.

This dispersion and size reduction of the particles was accomplished by ultrasonically shattering the solids while they were immersed in a fairly viscous water/surfactant solution. For samples with a visible layer of solids, a polyethylene pipette was used to transfer 1.5 g of these solids to a tared glass vial. This material was diluted to a total volume of 10 mL with a 50/50 (v/v) mixture of Joy® detergent and deionized water. The vial was closed and immersed in an ultrasonic bath for a period of 15 min. One mL of this mixture was then used to prepare the liquid scintillation cocktail for evaluation of solids in the given sample. In previous applications at SRS laboratories, Joy® had been used for radiological cleanups and was found to harbor no abnormally high levels of activity. Also, blanks using tritium and Joy® showed no signs of quenching.

All samples were made up to a uniform volume of 23 mL. In all cases, both a normal sample and a spiked sample were prepared. For this work, the spike used was a dilution of a NBS standard tritiated water solution. A single spike consisted of 100 μL of tritiated water having a disintegration rate of 32,000 dpm/mL. All samples were counted for three to nine periods of ten minute durations. The number of count periods depended on the time for any luminescence or chemiluminescence of the samples to decay. All samples were counted on a Packard Tri-CARB 2000 CA/LL with the output divided into three windows, 0 to 2 keV, 2 to 18 keV, and 18 to 2000 keV.

Sample preparation for the HPGe counting was much simpler. This preparation consisted of merely weighing the total sample and measuring the thicknesses of the individual layers in the bottle. In order to assure that any material on the outside of the bottles would not contaminate the SRS ultra low-level counting facilities, each bottle was placed in a tight-fitting polyethylene bag which was taped closed. The mass and height of material in each bottle was incorporated into the efficiency calculation for the HPGe detection.

ANALYSIS

Activity Calculations

As previously stated, the purpose of this work was to generate upper limit values which could be used as the basis for a decision for off-site shipment of hazardous waste. To this end, the liquid scintillation calculations were designed to yield the tritium and total radioactive concentrations. A tritium window of 2.0 to 18.6 keV and a total activity window of 0 to 2000 keV were selected for the basic analysis. Each sample and corresponding tritium-spiked sample yielded a tritium concentration (A_t), and a psudo-total concentration (A_p). The value of A_p is larger that the actual total (A_a), because the tritium calibration spike is strongly quenched relative to other nuclides. Assuming that all other radionuclides are detected with an efficiency that is a factor of f greater than the tritium efficiency, the total activity concentration is given by:

$$A_a = A_t + \frac{(A_p - A_t)}{f} \tag{6}$$

In the present work, f is defined as the ratio of the ^{14}C and ^{3}H efficiencies[3] as shown in Figure 2. Here the plotted quench parameter (tSIE) is defined for each sample from its spectrum with an external ^{133}Ba source.[3] The magnitude of each efficiency curve is strongly correlated with the beta energy. Accordingly, the curve for ^{3}H betas, which range from 0 to 18 keV, is considerably lower than that for ^{14}C betas which range from 0 to 156 keV. Effectively, all other beta-emitting radionuclides have energies comparable to or greater than that for ^{14}C. Alpha particles are less efficient by a factor of 10 in converting their energies to scintillations; however, their higher energies ($\sim$5000 keV) result in scintillation responses that are well above those of the ^{14}C spectrum. Consequently, all radionuclides other than tritium are treated as having the same efficiency as ^{14}C; thus making the overall calculation somewhat conservative.

The individual region of interest calculations of A_t and A_p used the following formula:

$$A = \left(\frac{S}{M}\right) * \left(\frac{C - C_0}{C_S - C} - \frac{B - B_0}{B_S - B}\right) \tag{7}$$

where A = A_t or A_p for corresponding spectral window (dpm/g)
  S = Tritium spike activity (dpm/g)
  M = Mass of Sample (g)
  C/B = Sample/blank count rates
  C_0/B_0 = Sample/blank non-scintillant constant background rates
  C_S/B_S = Sample/blank tritium spike count rates

Calculation of Conservative Estimate and Best Estimate

For each composite sample, a conservative estimate (C_E) and a best estimate (B_E) were made for the total concentration of radioactivity in the entire sample. Both liquid and solid phases were treated and incorporated into a general formula:

$$E = F_l \times A_l + C_s \times F_s \times A_s \tag{8}$$

where E = Either C_E or B_E, as appropriate
  F_l = Liquid fraction of sample
  A_l = Activity for liquid fraction (nCi/g)
  C_s = Solids correction factor
  F_s = Solids suspension fraction of sample
  A_s = Activity of solids suspension of sample (nCi/g)

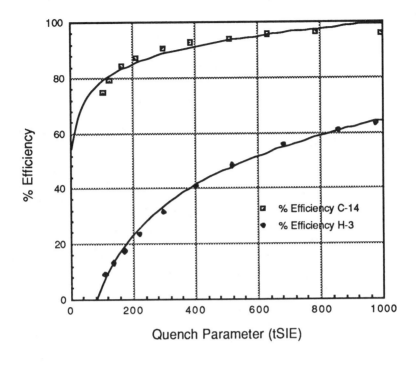

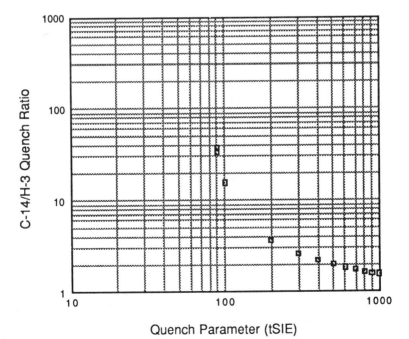

Figure 2. Individual isotope quench curves and ratio curve.

Here, only two phases are represented in the formula. In some cases, the formula was extended to handle an additional liquid phase (water).

The conservative estimate uses pseudo total A_p for activities A_l and A_s, and a conservative correction (C_s) of four. As discussed earlier, A_p would be the total activity if all the count rate were assumed to be tritium. Because tritium is detected least efficiently, A_p is noticeably greater than the actual total activity A_a. The solids correction factor of 4 corresponds to a 25% escape probability for 4.5 MeV alphas uniformly distributed within a sphere of density 1 g/cm^3 and a diameter of 200 μM. Actually, the escape probability for tritium betas would be even lower, but tritium is not expected to be trapped in these solids after disintegration in the ultrasonic bath. Most other betas would escape more readily than the alphas. The correction of 4 is also conservatively high as the actual diameters of the particles are at or below 100 μm, many alphas have higher energies than the values mentioned above, and no credit is taken for the distributional effects between solids and the suspending liquids.

The best estimate uses actual total A_a for activities A_l and A_s, and a correction factor of $C_s = 1$. The use of A_a is an obvious selection; however, arguments for choosing $C_s = 1$ need to be reviewed. Here a 53% escape probability is calculated for 5 MeV alphas uniformly distributed within a 100μm sphere of density 1. However, any contamination for paint solids is likely to be distributed on the surface of the spheres, because the ultrasonic breakup of the paint solids should cause fractures along regions which have been exposed to potential contamination sources. If all activity is on the surface of the 100μm spheres, the escape probability increases to 69%. The solids suspension contains >60% liquid; thus, if the alpha activity were distributed proportionately to the solid and liquid fraction, the effective alpha escape probability would be >88%. The estimate should be even higher because (1) it is unlikely that all the activity is due to alphas (which have the shortest range in solids), (2) the typical liquid/solids ratio of >100 in the scintillation cocktail favors activity in the liquid, (3) many of the particles are likely to be smaller than 100μm in diameter, and (4) any internal solids activities are not likely to be from SRS sources and thus should contribute only a minor component from natural radiation. Adoption of $C_s = 1$ assumes that, taken together, each of the four additional factors contribute an average increase of only 3%.

RESULTS

The results of the individual samples are graphically shown in Figure 3 and listed numerically in Table 1. All values contain the average of at least three trials for liquid scintillation and the activity detected by HPGe analysis. Not only are the conservative and best estimates shown, but two additional values, the highest likely and highest possible activities, are shown. In these two additional cases, the assumption was made that all potential activity came from a single sample in the composite (highest possible), or that the original

Liquid Scintillation Results

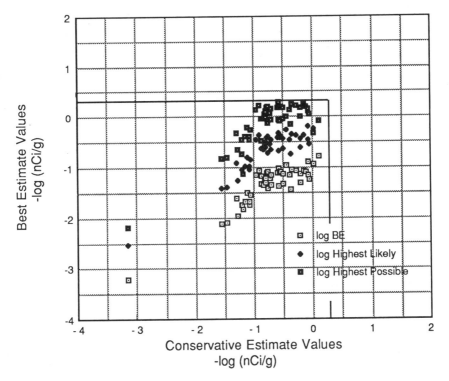

Figure 3. Plot of all composite and sub-composite samples (darker lines indicate limits for acceptance [log of 2 nCi/g]).

distribution of possible activity in the samples was based on a log-normal distribution with 99% confidence (highest likely).

The highest possible estimate is the product of the best estimate activity and the number of samples in the composite:

$$HPE = N \times B_E \tag{9}$$

This represented the highest possible activity for one single drum in the composite. Only in the cases where both this value and the conservative estimate were both less than 2 nCi/g were the samples deemed acceptable for off-site release.

The highest likely estimate used a factor of 5.01 in the place of the number of samples, N. This value results from a log-normal probabilistic distribution of the material among the drums which make up the composite.[5] These values are felt to be more realistic than the highest possible value above.

$$HLE = 5.01 \times B_E \tag{10}$$

Table 1. Numeric Results for all Composite and Sub-Composite Samples

Composite #	Samples	CE[a]	BE[a]	Highest Likely	Highest Possible
1	11	0.0007	0.0006	0.0030	0.0066
2	20	0.7600	0.0820	4.1080	1.6400
3	20	0.0280	0.0075	0.0376	0.1500
4	20	0.0510	0.0240	0.1202	0.4800
5	20	0.1070	0.0690	0.3457	1.3800
6	20	0.1250	0.0860	0.4309	1.7200
4	20	0.0340	0.0080	0.0401	0.1600
8	20	0.1580	0.0650	0.3257	1.3000
9	20	0.4370	0.0900	0.4509	1.8000
10	20	0.1520	0.0560	0.2806	1.1200
11	20	0.1340	0.0460	0.2305	0.9200
12	20	0.1590	0.0610	0.3056	1.2200
13	20	0.0620	0.0180	0.0902	0.3600
14	20	0.0830	0.0180	0.0902	0.3600
15	20	0.1660	0.0770	0.3858	1.5400
16	20	0.1750	0.0800	0.4008	1.6000
17	20	0.0860	0.0210	0.1052	0.4200
18	20	0.6230	0.0500	0.2505	1.0000
19	20	0.2930	0.0800	0.4008	1.6000
20	20	0.1750	0.0380	0.1904	0.7600
21	20	0.1880	0.0590	0.2956	1.1800
22	20	0.1960	0.0720	0.3607	1.4400
23	20	0.8430	0.0720	0.3607	1.4400
24	20	0.2010	0.0440	0.2204	0.8800
25	20	0.0870	0.0280	0.1403	0.5600
26	20	0.1440	0.0450	0.2254	0.9000
27	20	0.4160	0.0360	0.1804	0.7200
28	20	0.4160	0.0360	0.1804	0.7200
29	20	0.2650	0.0970	0.4860	1.9400
30	20	0.3180	0.0460	0.2305	0.9200
31	20	0.2620	0.0430	0.2154	0.8600
32A	10	0.0620	0.0180	0.0902	0.1800
32B	10	0.8160	0.1300	0.6513	1.3000
33	20	0.6930	0.0880	0.4409	1.7600
34	20	0.1600	0.0490	0.2455	0.9800
35A	10	0.2710	0.0820	0.4108	0.8200
35BA	5	0.1360	0.0670	0.3357	0.3350
35BB	5	1.2880	0.1670	0.8367	0.8350
36	20	0.3200	0.0750	0.3757	1.5000
37	20	0.0760	0.0310	0.1553	0.6200
38	20	0.2780	0.0770	0.3858	1.5400
39	20	0.6140	0.0750	0.3757	1.5000
40	9	0.3630	0.1110	0.5561	0.9990
41	20	0.8600	0.0570	0.2856	1.1400
42	20	0.2590	0.0760	0.3808	1.5200
43	20	0.5270	0.0840	0.4208	1.6800
44	20	0.4120	0.0700	0.3507	1.4000
45	20	0.0540	0.0110	0.0551	0.2200
46	4	1.0180	0.1180	0.5912	0.4720
47	20	0.0710	0.0210	0.1052	0.4200
48	5	0.0660	0.0150	0.0751	0.0750

[a]CE and BE stand for conservation estimate and best estimate respectively.

The grounds for acceptance of a sample are illustrated by the diagram in Figure 3. If both the conservative estimate (CE) and the highest possible estimate (HPE) were below the 2 nCi/g guideline, the sample was accepted as below DOT guidelines. If the sample fell outside the DOT guidelines on either of the two above parameters, the samples in the composite were subdivided to create two new samples ($N_{second} = N/2$) and two new composites made up for analysis. This was required for two composites and one subcomposite. In fact, none of the samples which were outside the acceptance limits for the first composite were ever traced to a single "contaminated" drum. This fact supports the argument that the samples, which did initially fail to meet criteria, did so because of mathematical multiplication of background levels.

CONCLUSIONS

The method of phase separation, sample processing, and liquid scintillation analysis proved effective for the analysis of mixed phase, hydrocarbon based waste samples. The described method was readily applicable to a variety of circumstances and was straightforward enough to be used by staff analysts after a minimum of training. The results were obtained in far shorter time (usually less than 2 hr for a single sample) than would be required for other typical methods of analysis. All samples analyzed by the methods presented above were found to have activities below the 2 nCi/g DOT guideline for shipment of non-nuclear waste.

The effects of quenching and luminescence deserve further attention in the future. Methods for luminescence approximation through graphic interpretation have been considered, as well as the use of chemical agents to decrease luminescence effects. Overall, the compositing scheme reduced the number of samples almost by a factor of 20.

The ultrasonic fracturing of the solid particles provides a method for rapid analysis of this particularly difficult material. In the future, a more effective approach will be developed. Trial use of a cellular disintegration probe, as opposed to a ultrasonic bath, is underway at Waste Management Technology. Also, other materials such as Triton X-100 would probably provide better suspension characteristics than Joy.®

REFERENCES

1. *Code of Federal Regulations*, 49CFR173.403(g), June 1986.
2. Gevirts, V.B., A.A. Palladiev, and T. Yu Pautova. *Soviet Radiochem.* 28 (4): 462–464 (1978).
3. *Operation Manual: Model 2000CA TRi-CARB Liquid Scintillation Analyzer: Publication No. 169-3023*, (Downer's Grove, IL: Packard Instrument Company, 1987), p. 5–79.
4. Winn, W.G., W.W. Bowman, and A.L. Boni. "Ultra-Clean Underground Counting

Facility for Low-Level Environmental Samples," *The Science of The Total Environment*, 69:107–144 (1988).
5. Bury, Karl V. *Statistical Models in Applied Sciences*, (New York: John Wiley and Sons, 1975), p. 295.

CHAPTER 36

Development of Aqueous Tritium Effluent Monitor

K.J. Hofstetter

ABSTRACT

A variety of techniques are being evaluated and tested in an attempt to develop a real-time monitor for low-level tritium in aqueous streams. One system being tested is a commercially available HPLC radioactivity monitor. This system uses crushed yttrium silicate as the scintillator and employs standard fast coincidence electronics to measure tritium. Laboratory tests of this unit indicate that the monitor can sense tritium at concentrations above 600 pCi/cc using a two minute counting interval. Pooling the count rate data over a longer interval, e.g., 24 hr results in a detection limit of ≈ 20 pCi/cc under constant background conditions. Unfortunately, the cells are easily plugged with debris even under laboratory conditions.

To overcome this problem, a prototype system using unclad fibers of plastic scintillator as the detection medium was designed and is being tested in the laboratory. Approximately 500 1-mm in diam fibers were assembled into a flow cell, with two 51-mm in diam photomultipliers (PMTs) coupled to the ends of the fiber bundles, to detect the scintillations. Fast coincidence, pulse shaping electronics are used to minimize the single photon and dark current backgrounds. The tritium counting efficiency, background, and sensitivity will be determined in the laboratory followed by field reliability testing.

The results of laboratory tests and a comparison of other types of scintillators (liquids, plastic beads and fibers, crushed inorganic, etc.) are presented. The sample preconditioning requirements (e.g., filtration, ion exchange, etc.) and known interferences (e.g., chemical and biological luminescence, natural radioactivity, etc.) for continuous monitoring of tritium in surface waters are discussed.

INTRODUCTION

Tritium is one of the most significant radioactive isotopes released to the environment by the nuclear industry. It is produced in light water reactors by ternary fission, in heavy water reactors by neutron capture in the moderator, in tritium production facilities at our defense facilities, in fallout from nuclear weapons testing, and as a byproduct of nuclear fusion development. The releases are predominantly in the form of tritiated water (HTO), its most active

421

biologically form. While tritium is produced by cosmic ray protons in the atmosphere, the manmade quantities dwarf the natural levels.[1,2]

The measurement of tritium is difficult due to its low energy beta emissions (maximum energy = 18.6 keV) and long half life (12.33 years).[3] Most monitoring methods for tritium have relied on grab sampling and laboratory analysis by liquid scintillation counting methods. The main interferences in the laboratory analysis are due to color and chemical quenching in the sample. These laboratory analysis techniques have been summarized in a variety of publications.[4] Real-time monitoring for tritium in the environment is not presently performed. Liquid scintillation techniques do not adapt readily to the measurement of tritium in flowing streams. A semicontinuous flow system has been tested using rapid mixing of the sample with the scintillation cocktail followed by injection into the counting chamber for a short count and then out to waste. The entire process is then repeated at short intervals to provide near real-time monitoring.[5]

The use of solid scintillators for tritiated water measurements in flowing streams has also been reported.[5] While these detectors are less sensitive than liquid scintillation techniques, they offer three main advantages: the use of solid scintillators permit continuous monitoring, are not affected by chemical and color quenching, and are less expensive to operate (the cost of liquid scintillation cocktail is high and disposal of the contaminated cocktail is difficult).

Background

Management at the Department of Energy Savannah River Site (SRS) requested the development of an on-line monitor to detect tritium in various aqueous effluent streams. The Environmental Technology Section of the Savannah River Laboratory is conducting laboratory and field testing of various systems; they are looking for environmental levels of tritium with response time suitable for corrective action. After reviewing various flow-through scintillation systems, counting cells loaded with solid scintillators were recommended over water/liquid scintillation cocktail mixtures since: no potentially mixed waste was generated, an improved response time was achieved, the sensitivity was sufficient, and the operation was much simpler.

The specific objective was to develop a monitoring system capable of detecting tritium at a concentration of 2000 pCi/cc within 2 min, detect changes at the 2 pCi/cc level over a 24 hr interval, and to measure concentrations within 10% at the normal reactor discharge levels over a 24 hr period. The continuous monitor was not necessarily intended to replace the existing tritium grab samplers, with subsequent laboratory analyses as a means of quantifying a release. The primary purpose was to detect an abnormal release and provide early warning for emergency response.

To efficiently detect tritium, one needs a detector with high surface area as the most probable beta energy is about 5.7 keV; this corresponds to an average range of only 6 μm in water. Most of the beta energy is lost in the water before reaching the scintillator, and relatively few photons are produced when there is an interaction. Maximizing the surface area increases the probability of a light producing event, while minimizing the detector volume reduces the background from cosmic rays and other beta emitters. Efficient light collection is mandatory.

Beads of plastic scintillator have been suggested as a tritium detector, but have large spacings relative to the short beta range. Only tritium in a thin sheath over the surface of the beads would be detected, resulting in a low counting efficiency and low sensitivity. Large sheets of plastic scintillator[6] and thin fibers coated with anthracene[7] have also been evaluated for tritium sensitivity. Crushed scintillators offer the highest surface to volume ratio and even with the decreased light collection efficiency, provides a sensitive tritium detection method.

EXPERIMENTAL

A commercially available tritium detection system was purchased for evaluation at SRS. The system was originally designed to detect tritium labeled compounds as they are eluted from a high-pressure liquid chromatography column.* The detector consisted of a Teflon measurement cell filled with crushed yttrium silicate solid scintillator interposed between two PMTs. Special pulse shaping and timing electronics are provided to minimize the background and maximize the counting efficiency for tritium. A computer-based data acquisition and analysis system is also provided as part of the system. Several solid scintillation materials are offered by the vendor for use as the tritium detector: cerium-activated lithium glass, calcium fluoride, yttrium glass, and yttrium silicate. The latter was chosen due to its high light output.[8] The inorganic scintillator was crushed and then suitable grain sizes were selected to optimize the counting efficiency while minimizing back pressure.

A recirculation system was set up in the laboratory to circulate aqueous solutions containing tritium and various contaminants. The detector system uses a HPLC pump capable of reaching system pressures of 100 psi at well-controlled flow rates. Due to the burst pressure of the Teflon cell, the system pressures were carefully monitored and did not exceed this value during the testing. Several cells of differing void volumes and scintillator loadings were evaluated. A simple diagram of the monitoring system is shown in Figure 1. Backgrounds were taken by circulating demineralized water through the system. The efficiency of the system was determined using demineralized water spiked with various quantities of tritium and by comparing the monitor output with the concentrations determined from aliquots (analyzed by standard liquid scintillation counting techniques in our laboratory). System stability and

---

*Distributed by Berthold Analytical, Nashua, NH 03063

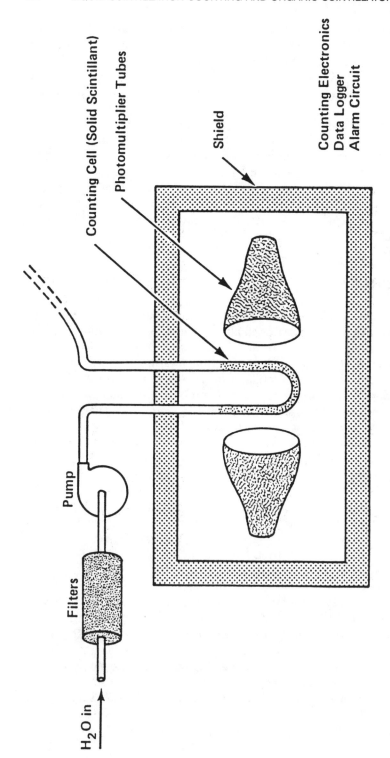

Figure 1. Simplified diagram of a continuous aqueous effluent tritium monitor using crushed scintillator.

Table 1. Performance Characteristics of Cells Containing Crushed Yttrium Silicate Inorganic Scintillator

Cell Void Volume (cc)	Background (cpm)	Efficiency (%)
0.4	12.5	1.77
0.5	14	0.4
1.0	18	0.35

reproducibility were determined by recirculating aqueous solutions through the system for extended periods (30 to 40 days).

The output of the tritium monitoring system was recorded using the computer-based data acquisition and analysis system. Two counting channels were set up on the system, the low-energy channel, set to encompass the tritium beta spectrum, and the high-energy channel, set to detect events above the tritium beta spectrum endpoint. Data were output at three different counting intervals, a short time interval (usually 10 min), an intermediate time interval (usually 10 hr), and a long time interval (usually 24 hr). The times were selected as typical for a field installation of the monitor, consistent with the overall goals of the project.

Three different counting cells were tested with three different cell configurations, each containing crushed yttrium silicate scintillator mesh sized to 35 microns. The largest cell had a void volume of 1 mL and was configured in a simple U-tube geometry using 7 mm OD × 1 mm wall tubing. The smallest cell was a compact spiral of 5 mm OD scintillator filled tubing and with a void volume of 0.4 mL. The intermediate cell was a single loop of 6 mm tubing with a void volume of 0.5 mL. In all cases, the cells were placed between the two photomultipliers and maintained at a fixed distance relative to each other. The cells were sealed from stray light and the entire assembly placed inside a lead background shield.

The results of the testing of these cells are summarized in Table 1. The backgrounds are the minimum attainable for each cell. It was observed that the background would increase monotonically over a period of several weeks. A cell wash procedure restored the background to its minimum value.

DISCUSSION

Sensitivity

The sensitivity for measuring tritium is a function of the efficiency, the background, and the counting time. The detection limits were calculated using the methodology given by Currie.[9] For completeness, the formalism used to evaluate the data are summarized below.

The minimum detectable concentration (pCi/cc) is defined as that concen-

Table 2. Minimum Detectable and Quantifiable Concentrations of Tritium for Cells Tested

Void (cc)	MDC (2 min)[a] (pCi/cc)	MDC (24 hr) (pCi/cc)	MQC (24 hr)[b] (pCi/cc)
0.4	600	20	60
0.5	2300	73	230
1.0	1450	48	150

[a]Minimum detectable concentration for 2 min count (see text).
[b]Minimum quantifiable concentration for 24 hr count (see text).

tration which has 95% probability of being above a threshold that is exceeded by only 5% of the background counts. It is given by:

$$MDC = DL / (V * t * \epsilon * 2.22)$$

where DL (counts) is the detection limit, V (cc) is the void volume of water in the active region of the cell, t (min) is the count time, ϵ (c/d) is the counting efficiency in the tritium channel, and 2.22 (dpm/pCi) is the rate conversion factor. The detection limit is given by:

$$DL = 2.71 + 3.29 * sqrt (B)$$

where the total background counts (B) is given by the product of the background count rate and the count time.

The minimum quantifiable concentration, as defined by Currie, is that concentration that will permit the quantification at a precision of 10% one sigma relative standard deviation as computed by the following:

$$MQC = QL / (V * t * \epsilon * 2.22)$$

where QL is the quantification limit (counts) as defined as

$$QL = 50 * [1 + sqrt[1 + (B / 25)]].$$

Using the data obtained from the cells tested in our laboratory and the above formalisms, the MDC and MQC for these cells for a short counting interval of 2 min and a long counting time of 24 hr can be calculated. The results are shown in Table 2. In all cases, it is obvious that the detection of a 2000 pCi/cc concentration is possible in a short period of time. At a 2 mL/min flow rate through the cells, careful engineering is required to minimize the holdup of the solution from the sampling point to the counting cell. It is also obvious that none of the cells meet the criteria for detecting or quantifying a release at the 2 to 10 pCi/cc concentration. A combination of a larger cell, with higher efficiency and lower backgrounds, is required. Figure 2 shows how the sensitivities could improve with reasonable advances in the cell and system design. A counting cell with a void volume of 3 mL a 7% counting efficiency, and a background of only 15 cpm would satisfy our requirements. The vendors claim that such improvements are possible, but to date, none have been attained.[10]

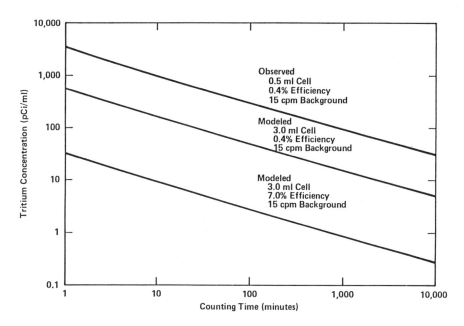

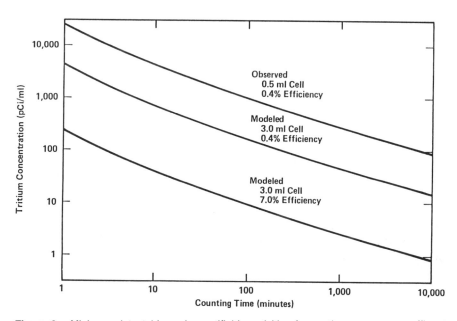

Figure 2. Minimum detectable and quantifiable activities for continuous aqueous effluent tritium monitor.

Stability

The variation of the efficiency or background of the detector will have a significant effect on the overall performance of the monitor. As stated previously, the background was observed to change by a factor of two over a several week period. This may be caused by the buildup of material on the surface of the scintillant or the presence of contaminants in the light-emitting water. Bioluminescence is a known interference with scintillation counting techniques.[11]

To test the long term performance in a laboratory environment, and validate the calculated sensitivities, a 3-week-long experiment was performed using the 0.4 mL cell. First the background was established by recirculating demineralized water through the system, periodically cleaning with NaOCl or dilute nitric acid. This was continued until a constant background was attained (12.5 ± 0.8 cpm). After adequate background stability had been reached, a small quantity of tritium was introduced into the water recirculation system, and the monitor response noted. The initial aliqout increased the concentration of tritium to about 140 pCi/cc; an aliquot was removed for laboratory analysis. The monitor count rate increased by about 20% in the tritium counting channel while the high-energy beta channel showed no change in count rate (see Figure 3 for the hourly observations). After a suitable recirculation time (at least 24 hr) another aliqout of tritium standard was spiked into the system, and the response was measured. This cycle was continued until the concentration in the recirculation system reached ≈ 1000 pCi/cc. At this point the monitor was recording a system count rate of twice the background in the tritium counting channel and no change in the gross beta-counting channel. By removing aliquots of each solution for laboratory analysis, the efficiency of the monitor at each concentration could be calculated (see Figure 4 for a plot of the efficiency as a function of concentration). It was found that over the concentration range tested, the efficiency (ϵ = 0.0177 ± 1σR.S.D.) appears to be independent of tritium concentration (correlation coefficient = –0.073).

At this point in the experiment, the system was flushed with repeated washings of fresh demineralized water, and the detector response returned to background suggesting no memory effects. To begin testing aqueous solutions typical of surface streams, a sample of secondary cooling water from one of the reactors was obtained for testing and analysis. The cooling water is taken from the streams in the SRS area, passed through the reactor primary heat exchanger, and then discharged into a cooling pond before it is released back into the surface streams. A composite sample of the water discharged from one of the reactors was circulated through the monitor in an attempt to determine the tritium concentration. The calculated tritium concentration, from the increase in count rate observed by the monitor, was 200 pCi/cc, while laboratory analysis of an aliqout showed a tritium concentration of only 50 pCi/cc. The sample was highly quenched. The monitor response to the reactor effluent is shown on the right hand side of Figure 3. It is noteworthy that the gross beta

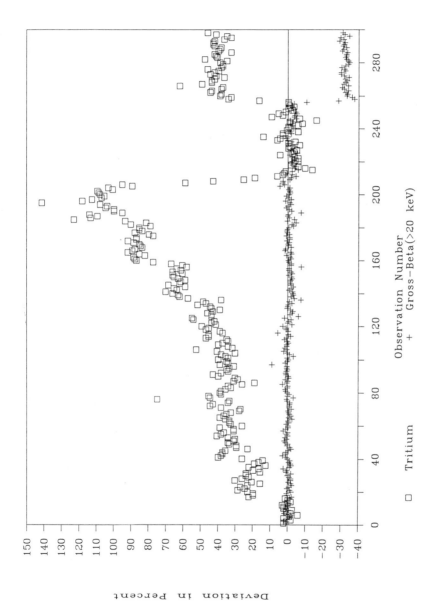

Figure 3. Monitor response to aqueous solutions containing tritium.

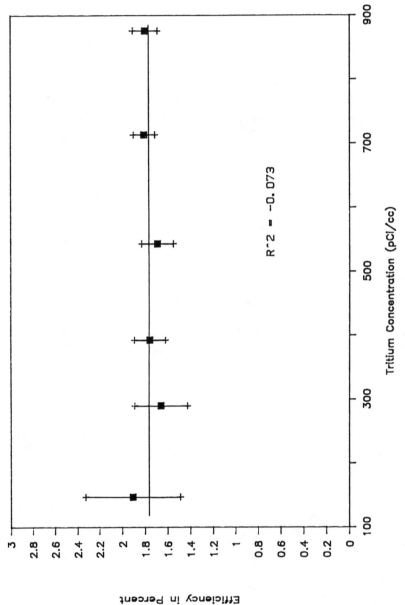

Figure 4. Variation of tritium counting efficiency with concentrations tested.

channel count rate decreased by ≈35% during the experiment with reactor effluent. During this run, the differential pressure rose from 20 psi (its equilibrium value during the demineralized water tests) to 100 psi after about 2 days of run time. The explanation of this phenomenon can be attributed to the microorganisms in the sample. Bioluminescence is occurring in the tritium channel, resulting in an apparent increase in count rate. Meanwhile the microorganisms are creating a slime on the scintillator which reduces the light output in the gross beta channel. Plugging of the scintillator is consistent with this speculation.

Care will have to be taken to precondition the solution prior to reaching the measurement cell. Sterilization of the sample stream using biocides, ultraviolet light, ultrafiltration, etc. is mandatory if solid scintillators are to be used in the monitor. Other conditioning steps that will be required include the filtration of debris to prevent cell plugging, the removal of organics (e.g., by activated charcoal) that might produce light, and the removal of inorganic impurities (e.g., by ion exchange). These technologies are available using standard swimming pool water purification processes.

DEVELOPMENT OF ALTERNATE SENSORS

Since the plugging of tritium measurement cells containing crushed inorganic scintillators was apparently weak environmental applications, development of alternate sensors is underway at SRS. To minimize the plugging problem, the spacing between the light detecting media should be made larger for less restriction to flow. To provide a sensitive tritium monitor, a large surface area with efficient light detection is required. Borrowing on the design of the anthracene coated fiber detector[7] and recent developments on the fabrication of plastic scintillator fibers, a flow cell was constructed with overall dimensions of 51-mm in diam by 51-mm in length, containing ≈500 fibers of 1-mm diam plastic scintillator* spaced on 2-mm centers. The ends of the fibers are cemented into a pair of light pipes which are, in turn, optically coupled to a pair of PMTs. The aqueous sample flows perpendicular to the fibers and the light is transmitted down the fibers to the PMTs. The high index of refraction of the plastic scintillator (1.58),[12] relative to that of typical aqueous solutions (1.3), should limit the loss of scattered light. The plastic scintillator is nearly transparent to the light produced by the tritium interactions (bulk light attenuation length = 250 cm[12]), so detectors containing long lengths of fibers could be used.

The fabrication of the detector is underway and will be tested in the laboratory when the construction is complete. Pulse shaping and fast coincidence electronics will be used to discriminate single photon events and dark current noise. The background of the cell should depend only on the amount of scintillator and the lead shielding from cosmic rays. A comparison of the

---

*Manufactured by Bicron Inc., Newberry, OH 44065

Table 3. Comparison of Measured and Expected Responses for Various Types of Solid Scintillator Detectors

Scintillator	Background (cpm)	Volume (cc)	Efficiency (c/d)	MDC (10 min) (pCi/cc)
Crushed	12.5	0.4	0.0177	251
Crushed	14	0.5	0.004	938
Crushed	18	1.0	0.0035	603
Plates (Ref. 6)	40	1.8[b]	0.006	286
Anthr. (Ref. 7)	30	0.28[b]	0.016	600
1mm fibers	42	1.0[b]	0.0104[c]	304
0.5mm fibers	18.5[a]	0.9[b]	0.0104[c]	228
0.25mm fibers	6.8[a]	0.65[b]	0.0104[c]	199
0.1mm fibers[d]	4.1[a]	0.029[b]	0.0104[c]	3551
0.25mm fibers[d]	4.3[a]	0.012[b]	0.0104[c]	8761
0.5mm fibers[d]	3.9[a]	0.0057[b]	0.0104[c]	17671
1.0mm fibers[d]	3.9[a]	0.0053[b]	0.0104[c]	19005

[a]Based on mass of scintillator relative to mass of plates and background observed in Ref. 6.
[b]Based on active surface area and 6 μm range of ^{3}H beta.
[c]Based on 65% light output of anthracene.
[d]Assumes 50% void volume in U-cell 8.5-cm long.

expected response to that of already tested systems is shown in Table 3. The basis for the comparison is the intrinsic light production of the scintillator, the surface area of the active material, and an estimate of the light collection efficiency.

The data presented in Table 3 show that the detectors using plastic scintillator fibers offer the best chance of reaching the lower detection sensitivities, even with the conservative background estimates. If modern pulse shaping and timing electronics and photomultipliers can improve the signal to background ratio, the MDC can be reduced even further. Additional shielding may also increase the sensitivity for making tritium measurements. All these enhancements will be evaluated by further laboratory and field testing.

REFERENCES

1. Jacobs, D.J. "Sources of Tritium and its Behavior Upon Release to the Environment," USAEC Report TID-24635, (1968).
2. Eisenbud, Merril. *Environmental Radioactivity*, 3rd Edition, (New York: Academic Press, 1987).
3. Lederer, C.M. and V.S. Shirley, Ed. *Table of Isotopes*, 7th Edition, (New York: John Wiley & Sons, 1978).
4. Crook, M.A., P. Johnson, and B. Scales, Ed. *Liquid Scintillation Counting*, Volume 2, (New York: Heyden & Son Inc., 1972).
5. Budnitz, R.J. "Tritium Instrumentation for Environmental and Occupational Monitoring—A Review," *Health Phys.*, 26:165 (1974).
6. Osborne, R.V. "Detector for Tritium in Water," *Nucl. Instrum. Meth.*, 77:170 (1970).

7. Moghissi, A.A., H.L. Kelly, C.R. Phillips, and J.E. Regnier. "Tritium Monitor Based on Scintillation," *Nucl. Instrum. Meth.*, 68:159 (1969).
8. Berthold Analytical, "Measuring Cells" Unpublished Results (1989).
9. Currie, L.A. "Limits for Qualitative and Quantitative Detection," *Anal. Chem.*, 40:586 (1968).
10. Reeve, D.R. "HPLC Radioactivity Monitors—Fact and Fiction," *Laboratory Practice,* 3 (1983).
11. Schram, Eric. "Bioluminescence Measurements: Fundamental Aspects, Analytical Applications and Prospects," in *Liquid Scintillation Counting—Recent Developments*, P.E. Stanley and B.A. Scoggins, Ed. (New York: Academic Press 1974), pp. 383–402.
12. Hurlbut, C.R. "Plastic Scintillators—A Survey," *Trans. Am. Nucl. Soc.*, 50:20 (1985).

CHAPTER 37

Rapid Determination of Pu Content on Filters and Smears Using Alpha Liquid Scintillation

P.G. Shaw

ABSTRACT

This chapter discusses a technique for rapidly determining plutonium content on filters and smears using alpha liquid scintillation. Filter and smear samples will be analyzed daily for plutonium (^{239}Pu) content during projected waste retrieval operations at the Radioactive Waste Management Complex (RWMC) of the Idaho National Engineering Laboratory. Daily monitoring will allow for trending of airborne and surface contamination. Present analysis techniques are time consuming, as both numerous naturally occurring isotopes, such as uranium and thorium daughters, and inert solids must be removed prior to counting to avoid interference with Pu detection. Alpha liquid scintillation (ALS) in conjunction with microwave digestion was investigated as a technique for rapid Pu analyses. Advantages offered by ALS are short turnaround time and field use with acceptable accuracy. A state of the art Photon Electron Rejecting Alpha Liquid Scintillation (PERALS) Spectrometer with pulse shape discrimination (PSD), and an oil filled photomultiplier tube counting chamber with 99.7% counting efficiency and 99.95% rejection of beta and gamma pulses, was used. Relatively clean filter samples could be directly counted in an all purpose scintillant, bis 2-ethylhexyl phosphoric acid (HDEHP), 4-biphenyl-6-phenylbenzoxazole (PBBO), toluene, and naphthalene. Laboratory preparation of soil samples and smears with high inert solids content was accomplished by dissolution of the sample in nitric and hydrofluoric acids using a microwave digestion system in teflon pressure vessels. The Pu in the dissolved sample was extracted into tertiary amine nitrate and counted in a HDEHP or 1-nonyldecylamine sulfate (NDAS) containing extractive scintillant. This method is applicable to the determination of total plutonium in air filters, smears, and soils. The minimum detectable activity (MDA) for direct counting of air filters is about 100 pCi/g (3.7 Bq/g) for an hour count. If the sample is dissolved and Pu extracted, activities near 1 pCi/g (0.037 Bq/g) can be seen with a 20 min count.

INTRODUCTION

Filter and smear samples will be regularly analyzed over the course of the day for ^{239}Pu content during projected waste retrieval operations at the Radioactive Waste Management Complex (RWMC) of the Idaho National Engineering Laboratory. Daily monitoring will allow for trending of airborne and surface contamination. Airborne contamination collected on filters and surface contamination collected on smears are the two normal forms of samples expected from the contamination monitoring system.

Alpha liquid scintillation (ALS) techniques developed with beta and originally used the same equipment as beta scintillation counting.[1-3] A specifically designed alpha liquid spectrometer (PERALS)[4] with 99.7% counting efficiency, 99.95% rejection of beta and gamma pulses, 5% (250 keV) resolution for energies in the Pu region of 5162 keV and 0.02 cpm electronic background was tested.

Relatively clean filter samples were directly counted in an all purpose scintillant, bis 2-ethylhexyl phosphoric acid (HDEHP), 4-biphenyl-6-phenylbenzoxazole (PBBO), toluene, and naphthalene. This formulation was found to be the optimum for ALS when compared to common beta cocktails.[3,4] Routine treatment for soil samples and smears with high inert solids content, however, requires laboratory dissolution and two extractions, first into a tertiary amine nitrate, and second into a HDEHP or 1-nonyldecylamine sulfate (NDAS) containing extractive scintillant.[4,5]

Special sample preparation techniques, such as microwave digestion, are important in ALS. Microwave digestion quickly reduces solid samples to an extractable form[6] within a nitrate system. Microwave digestion in specially designed, closed, raised-pressure, teflon vessels has been used for dissolution of a variety of materials and is currently being considered as an alternate EPA method for standard open vessel wet ashing techniques.[7] With the proper organic extractants, filters can be directly counted and soil samples extracted and counted by PERALS. This should be possible in a short amount of time (about 1 hr), in the field and with acceptable accuracy and peak discrimination against natural background components in the soil.

This chapter is a scoping study to test procedures needed in preparing samples, standards, and blanks for the PERALS system and gives preliminary results of filter and soil Pu analysis. Screening samples in less than an hour was tested rather than routine analytical analysis.

APPARATUS

Until recently adaptation of existing beta detectors was the only way to perform alpha liquid scintillation spectrometry.[1,2] However, these detector systems did not provide good spectrometric results. To obtain good results, we needed an alpha only detector with (1) improved counting chamber design with no air gap, (2) improved pulse shape discriminator (PSD) to separate alpha from beta and gamma pulse continuum, (3) multichannel analysis (MCA) rather than the pulse height analysis (PHA) counting of energy regions, (4) elimination of most quenching, (5) lower background through the rejection of afterpulses characteristic of cosmic radiation, and (6) improved efficiency through rejection of unwanted luminescence.

McDowell[4] has designed a workable ALS instrument called a Photon-Electron Rejecting Alpha Liquid Scintillation Spectrometer (PERALS). Currently an improved alpha liquid scintillation design is manufactured exclusively by Oak Ridge Detector Labs (ORDELA). The ORDELA PERALS Model 8200B detector uses pulse shape discrimination (PSD) and an oil filled photomultiplier tube counting chamber giving: 99.7% counting efficiency, 99.95% rejection of beta and gamma pulses, 5% (200 keV) resolution at energies in the Pu region (5162 keV), 0.02 cpm background.

A PERALS spectrum of ^{226}R and its daughters is overlaid on a typical one using silicon surface barrier alpha spectrometry (Figure 1). This comparison shows the radium, radon and polonium peaks resolved in both systems. The 250 keV for Pu resolution of PERALS, noted in the first peak when compared to about 20 keV in the surface barrier spectrum, does not allow resolution of the two ^{226}R peaks at 4.6 and 4.78; thus ^{241}Am, always found with ^{239}Pu, are also seen as one peak on the PERALS system.

Analysis equipment includes: the CEM 81D microwave digestion unit, assorted sizes of separatory funnels (30 to 500 mL), adjustable hot plates, heat lamps, a balance capable of weighing to 0.1 g, a dry argon sparging apparatus, (Figure 1), 10 × 75 mm culture test-tubes, cork stoppers, parafilm, and lambda pipettes. The ALS spectrometer was powered by a Canberra Model 3120 High Voltage Power Supply. Data is transferred through a Canberra Model 18075 A to D Converter to a Canberra MCA system 100 operating an IBM PC, for peak detection and analysis.

Four types of reagents are used in the procedure: (1) mineral acids for sample dissolution and organic stripping, (2) inorganic salts for oxidation state adjustment, (3) large organic amines or phosphates for sample extraction, and (4) scintillation grade organic reagents and fluors for cocktail preparations. High purity reagents decrease the probability of various unwanted reactions, minimize introduction of undesirable quenching species, and reduce background.

PROCEDURE

This section describes the ALS analytical scheme. The results give a preliminary performance evaluation of the ALS system, microwave digestion, and organic extraction as an analytical tool for both directly analyzed and digested/extracted filter and soil samples.

Certain air filters and smears may be counted directly in the extractive scintillator. Swipe or smear samples and lightly coated air filters must be relatively clean, contain mostly alpha activity, have low beta-gamma activity and have low inert solid content. If the sample contains a large amount of soil or matter, it will require digestion and extraction.

Sample dissolution was accomplished by wet ashing in nitric and hydrochloric acids using hydrofluoric acid to break down silicates. A microwave

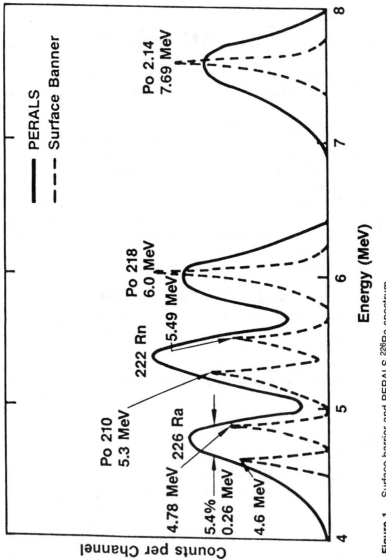

Figure 1. Surface barrier and PERALS ^{226}Ra spectrum.

digestion system with teflon pressure vessels contained the HF and reduced dissolution time.

The sample (soil, smear, or filter) is weighed, placed in a 120 mL high temperature teflon vessel, and a 7:3 mixture of aqua regia and hydrofluoric acid is added. The cap is tightened on the vessel to a prescribed torque. This allows the acid mixture to become pressurized to 120 psi. At this pressure, the temperature of the solution reaches 150°C and the HF remains in solution longer than in an open system. The microwave is operated for 5 min at full power (650 watts) then 10 min at 50% power.

Vessels are uncapped and 5 mL of nitric and 5 mL of 30% hydrogen peroxide are added to drive off chloride and fluoride ions and oxidize minute traces of organic matter. The sample is placed in a 30 mL beaker, and 3 mL aluminum nitrate is added to tie up free fluoride, prevent calcium fluoride precipitation, and adjust the ionic strength. Volume is reduced to 5 mL by evaporation to remove high acidity, chloride, and fluoride ions, giving a pure nitrate system. The solution should be about 1 M nitric acid and 3 M total nitrate.

The dissolved sample is converted to a suitable oxidation state for extraction into an an aqueous immiscible scintillator. Pu is primarily in the $+6$ state after nitric acid digestion. It is brought to $+4$ by reduction with ferrous sulfate or potassium metabisulfite (if the system contains appreciable iron). Any Pu $+3$ present is raised to $+4$ with sodium nitrite. After addition of these reagents the solution must be contacted immediately with the tertiary amine or Pu will disproportionate back into multiple oxidation states.

After ionic strength and oxidation state adjustments, Pu is extracted into the high molecular weight (>300) tertiary amine such as tri-octyl amine or, as used here, the proprietary formulation, Adogen-364. The amine is nitrated by contact with 0.7 M nitric acid before extraction. The purity of this amine is critical. Any primary or secondary amines present may keep the Pu from being extracted, bind it to the amine so it can not be stripped, or extract unwanted ions. Due to time constraints the purity of the amine could not be assayed and purifying procedures could not be undertaken. The main concern in the extraction step is the removal of Th, U, and any colorant such as iron. Removal of other elements in the extraction procedure is not as crucial in this method as the sample will not be plated out on a surface as a solid.

The acid solution is transferred to a 30 mL separatory funnel and shaken with the amine for several minutes by an automatic shaker. The Pu is stripped from the amine nitrate, which is highly quenched, and put into an organic with less quenching. It is stripped with a solution of 1 N sulfuric or perchloric acids and a small amount of an associated salt of each acid, lithium perchlorate or sodium sulfate. Salt provides a surface for Pu to adhere to upon evaporation and prevents Pu from plating on the container sidewall.

Acidity and volume of the stripped solution is reduced by heating. This also destroys any residual amine. The type of acid determines the final extractive scintillant. An extractive scintillant containing HDEHP was used for perchloric acid, NDAS for sulfuric. A diluent, 2-ethylhexanol, was to the trioctylamine nitrate (TANO$_3$) as an aid to stripping into 1 N sulfuric acid. The pH of the perchloric solution for HDEHP extraction should be greater than one, as the extractant is a weak acid. The pH for the sulfate system can be between zero and two. K$_2$S$_2$O$_8$ is added to ensure the Pu +4 for the perchloric system. Other than the extractants HDEHP or NDAS, the scintillant cocktails are formulated the same for either system with solvent, enhancer, and fluor.

A matrix standard soil was specially prepared to a 100 pCi/g concentration of Pu homogeneously distributed, and chemically and physically bound to the soil. Soil for the standard was obtained from the RWMC at an 8 ft depth near the location of the proposed retrieval effort. Soil was dried at 105°C in a laboratory oven and sieved through a 35 mesh screen (525 μm) followed by a 200 mesh screen (75 μm). Both fractions were weighed with more screened soil added until a total of 1 kg of screened soil was obtained.

The 1 kg of soil was then spiked by taking approximately 100 g from each fraction, wetting completely to a slurry consistency, and adding an accurately weighed aliquot of ^{239}Pu stock solution. These slurries were mixed thoroughly, dried, ground, resieved, and then mixed in a special dual cylinder mixer. Enough ^{239}Pu spike was added to each fraction so that the activity concentration was approximately 100 pCi/g. The final blended product had an activity concentration of 102.3 pCi/g. Sieving of the soil was done to enhance the particle size fraction that was less than 10 μm. This is the respirable fraction and should be the same size as found on most of the smears and filters to be analyzed.

An above background sample of Pu contaminated soil that had been environmentally aged and may have been "high fired" was obtained from the same source as most of the waste, Rocky Flats Plant, and used as another matrix standard. This soil has a high silica and low clay content, a more refractory (hard to dissolve) Pu oxide, and Pu that is more intimately bound to the soil. The RFP soil was obtained by RFP personnel, downwind from a former drum storage area. The Pu originated 20 years ago from leaking drums that contained contaminated cutting oil. The oil held a suspension of <3 μm Pu particles. This area was decontaminated in 1969 and covered with an asphalt pad.[9] The estimated ^{239}Pu concentration was 1000 pCi/g.

The RFP soil was dried and sieved to determine particle size distribution. The finest particle size range, less than 45 μm was used for most of the tests as this approximates air deposition or filter samples more closely than the bulk soil. A similar sized sample from the RWMC standard was also used for this reason. Both matrix standards, RWMC and RFP soil, were used to test the sample dissolution and extraction efficiencies and verify elimination of background interferences.

Matrix blanks were used to test the method, particularly the contribution of natural background alphas and reagent impurities to the peak. For a blank soil known to be Pu free, a subsurface soil sample from a basement excavation, presumably having no Pu fallout, was used. This soil was counted directly or was dissolved, extracted, and counted. Differences between the matrix standards and the matrix blank were used to determine the minimum detectable activity, as well as how the ALS system and chemistry is distinguishing Pu alpha from those alpha naturally in the soil.

The Pu and Am concentration of the standard and blank soil was determined by the Rocky Flats Plant, Golden, CO, UNC Geotech at Grand Junction, CO; and the INEL. They used dissolution and extraction, followed by conventional alpha spectrometry and whole sample counting of Am by gamma ray spectrometry. RFP and INEL used gamma ray spectrometry of [241]Am and assumed a 10:1 ratio of Am to Pu. UNC used total dissolution, organic extraction, ion exchange, precipitation, and alpha spec for Pu and Am. The results averaged 1010 pCi/g Pu and 110 pCi/g Am for the suspendable (less than 100 μm) portion of the soil. The blank had less than the detectable (about 0.1 pCi/g) for both isoptopes.

Following sample preparation, samples were purged of dissolved oxygen and dried before insertion in the oil filled counting chamber. In direct filter analysis, occluded air on the filter was also removed. A filter disc, smear, or portions there are folded and pushed with a lambda pipette into the counting test tube so it takes up no more space than the one mL volume scintillation solution. Air is removed by bubbling dry argon, saturated with toluene, through a lambda pipette while probing with the pipette to remove all bubbles from the test-tube. The paper becomes transparent and seems to disappear. A transfer lambda pipette tip was used as a sparging lance. Water in the gas is removed with molecular sieve and metallic sodium. Bubbling through toluene saturates the gas with the scintillant solvent so the sample volume remains constant.

The test tube is corked and sealed with parafilm, wiped clean, and placed in the sample holder oil bath. The light tight cap is replaced, the high voltage activated, and counting on the ALS spectrometer initiated. A directly prepared sample will count with near 100% efficiency if the alpha activity is on the surface of the fibers in a very thin layer. This method works best on samples with ultrafine particulates, such as air filter samples with only respirable size particulates. Sparged samples sealed with parafilm may last for several weeks, but slow evaporation of the scintillant still occurs. Only glass sealing and dark storage will ensure long term stability.

Two types of calibration are necessary in the ALS system: energy and amplitude. Energy calibration is done by extracting Pu from the standard aqueous solution into the same scintillant in the same manner as the sample. The peak energy of the sample should match that of the standard. Instrument and extraction efficiency are verified with check standards and spikes, assuming near 100% counting efficiency. When directly counting samples the calibration for both energy and amplitude is difficult, as color and chemical quenching shifts the spectrum, and inclusion of alphas in some particles lowers the efficiency. Results for direct counting of various types of blanks, prepared calibration standards, and standard spiked soil are given below.

RESULTS

This section discusses results of analysis of soils and filters on the PERALS system, both directly without treatment and following microwave digestion and extraction of the sample. Three experimental parameters will be reviewed as they apply to both direct analysis and analysis after dissolution and extraction: efficiency, resolution, and background. Optimum is 99.7% efficiency, 5% resolution, and <0.02 cpm background for the entire process (dissolution, extraction, and counting). Precision for these parameters, for direct and extracted soil standards, blanks, soil blanks, and soil spikes, for problems encountered, and for further work necessary are discussed.

Background includes electronic noise, cosmic rays, and chemical impurity contributions to the final count rate. Efficiency is the percentage of the analyte extracted and pulses successfully detected by the instrument and converted to the count rate. Resolution is the degree of energy separation of one group of pulses from another and the location on the energy scale of a specific peak. Precision is the stability of the instrument and reproducibility of the other parameters.

Resolution

Energy resolution, the separation of two alpha peaks by energy and the energy location on the spectrum, has both instrument and chemistry contributions. Resolution depends on the amount of light per pulse received; thus an efficient and stable scintillator and diffuse reflector are necessary. The characteristics of a PM tube and its physical relationship to a sample are critical to maximizing this parameter. The detector uses an oil-filled cavity, eliminating the air gap between sample and phototube. This prevents spectrum distortion caused by refractive index discontinuity and improves resolution.

Figure 1 shows a PERALS alpha spectrum for ^{226}R and its daughters overlaid with a surface barrier spectrum.[9] The radium, radon, and polonium peaks are resolved in both systems. The weak 4.6 MeV radium, however, is not resolved from the strong 4.78 MeV radium, illustrating the the practical limitations 5% resolution. The resolution of the major peak is about 20 keV in the surface barrier spectrum and 250 keV in the PERALS spectrum.

Resolution is needed to separate background nuclide activities from those activities of Pu in the soil. Energy stability is important in identifying the peak. The separation of the Pu peak from that of background Th depends on energy stability and resolution. A spectrum of a naturally occurring alpha emitter, Th, is shown with that of Pu in Figure 2 after one week of ingrowth. The separation of the single Pu peak (5.1 MeV) from the major Th (4.2 MeV) and daughter peaks (6.0 and 7.7 MeV) can be seen.

Table 1 gives resolution for prepared Pu standards. The full width at half maximum (FWHM) and the region of interest (ROI) width were both used as measurements of resolution. Most peaks had resolutions under 10%. We

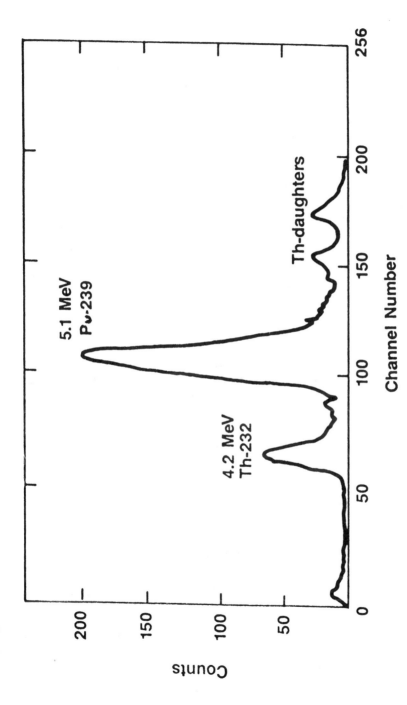

Figure 2.   Plutonium and thorium standard spectrum.

Table 1. Plutonium Peak Resolution on Prepared Standards

Standard PCI	FWHM KeV	Resolution Percent
0.1	290	5.2
0.7	502	9.7
5.3	546	10.6
6.0	490	9.5
74.4	557	10.8
110.2	638	12.4
142.9	585	11.3
Average	515	10
Standard Deviation	111	2.2

achieved a 5.6% resolution on a low-level sample (0.1 pCi) in the the HDEHP extractive scintillant. This is the scintillant used for Pu extraction in the perchloric system and direct filter counting. Peak location is also a factor in energy calibration and peak identification. The standard deviation (1 sigma) of peak energies (location) was less than 10% that of resolution was 2.2%.

The effect of the soil on resolution was tested by using direct soil laboratory spikes. Two soils were spiked with a standard Pu solution directly in the counting tube. The Pu peak was shifted by about 10% from that of the standard without soil and broadened by 30 to 50%. A slight increase in the beta continuum region channel 5–10 was also noted. Spike recovery was 90%.

Energy shifts and loss of resolution occur when adding soil to the scintillant in direct analysis. Resolution also changes with various soil types — blank soil, spiked RWMC soil, and RFP soil. The direct soil sample method often gives a highly colored sample, which gives an energy shift toward the beta continuum region, channel 1–10. This type of quenching may also decrease counting efficiency with the loss of resolution. Energy shifting and quenching is discussed in the background and precision sections.

Efficiency

Total efficiency is the sum of counting efficiency (the percentage of pulses successfully detected by the instrument) and the efficiency of dissolution and extraction, sometimes called chemical yield. Several types of samples were analyzed to measure these operational parameters and thus assess the quality of data achievable with the ALS system (instrument and chemistry). Efficiency of various samples and control standards in both the direct and extracted mode are given in the tables and discussion that follows. Problems and interferences that affect efficiency are given in the discussion of precision.

Table 2 gives combined counting and extraction efficiency for various soils both directly counted and digested-extracted before counting. At the current efficiency of about 20 to 25% in the direct mode, 100 pCi/g of Pu on RFP soil should be detectable on relatively clean filter samples. At this concentration (one tenth the RFP soil concentration 1000 pCi/g) and efficiency, the count

Table 2. ALS PERALS Analysis of RWMC Standard Soil, Rocky Flats Soil, Rocky Flats Soil on Filters

Soil Type	Method	Peak Centroid	Percent Efficiency	Percent Resolution
100 pCi/g Standard	Direct	94	45	14
1000 pCi/g RFP Soil	Direct	12	8.6	12
RFP Soil on Filter	Direct	19	28	28
Standard	Extraction	72	85	9
RFP Filter	Extraction	85	22	8

rate of 50 mg of sample is 2.2 dpm, or over ten times the background of 0.15 dpm.

The direct filter analysis had a higher efficiency than direct RFP soil analysis, 9 vs 28%. The suspension of the soil in the PMT viewing area and the limited settling of contents could account for this higher efficiency. The filter samples here were highly quenched, as evidenced by the peak shifting seen in Figures 3 and 4. Some of the RFP Pu was seen in the higher channel regions, with less quenching and count rates about 3 times the soil background. Count rates for RFP and blank soil (Tables 2, 4, and 5) indicate the Pu could be distinguished from the background activities present in non-Pu-containing blank soil in the direct analysis mode.

When the spiked soil standard was counted directly, the overall efficiency approached 45% (Table 2). The efficiency for the RFP soil with a much more refractory (hard to dissolve) form of Pu was only 9%. The count rate of 0.93 cpm (Table 5) for the spiked soil in the Pu region of interest (channels 90 to 110) is over 5 times (0.16 cpm) that of the blank soil. HDEHP is the most effective extractive scintillant for direct counting in actually leaching some of the Pu. The NDAS does not leach the Pu as well as the HDEHP and has a lower blank and standard spiked soil count rate.

RFP and standard soil samples that were digested and multiply extracted have the same type of relationship between spiked soil and RFP soil held in direct counting. Extraction of Pu from the RFP soil is more difficult than the RWMC standard soil. Efficiencies for the RWMC standard soil approached 85% and the RFP soil 22%.

Some RFP soil samples were digested and extracted directly into the extractive scintillator, HDEHP. Recovery efficiencies of 25% were achieved for single extractions, 22% for multiples giving a detectable peak in 1 hr of counting for 1 g of a 1 pCi/g sample. This is about 0.55 cpm, about 5 times greater than the background of 0.1 cpm (Table 1). The peak location on the average is somewhat lower for the standard soil extract than the RFP soil. The RFP soil had a higher counting uncertainty and gave a wider peak (Figure 5). Time did not permit full development in this area, but further work should bring this extraction process up to 99% efficiency.

Table 3 lists overall efficiencies for prepared standards. The extraction and counting efficiency of 95% approach the optimum 99.7% (counting wall

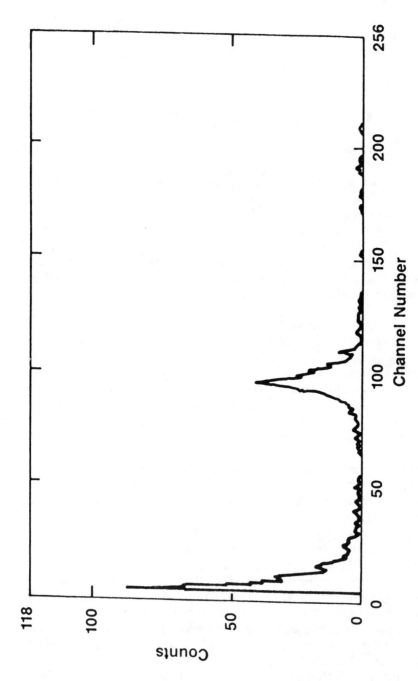

Figure 3. Direct Pu soil spectrum.

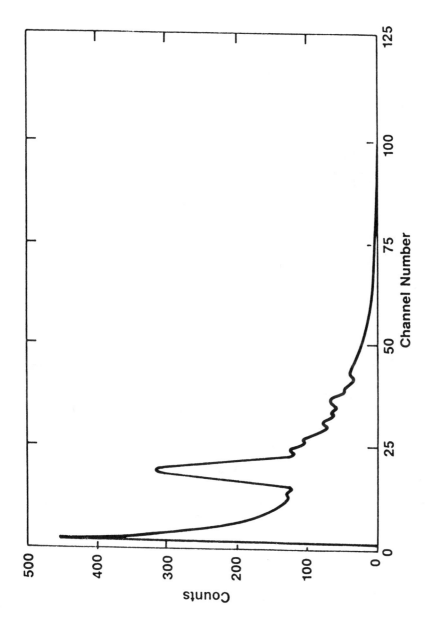

Figure 4. Direct RFP soil on filter spectrum.

Table 3. Plutonium Efficiency on Prepared Standards

Standard PCI	Efficiency Percent
0.1	98
5.3	92
74.0	93
88.0	95
100.9	89
142.9	101
177.0	95
Average	95
Standard Deviation	3.9

losses only). The main problem with efficiency was the extraction rather than the counting. Some standards were not in either the proper oxidation state or pH range for efficient extraction. The reproducibility of extraction was about 4%. Improvement in extraction efficiencies and consistent recoveries is desirable.

Background

There are two types of background radiations of concern in ALS, those from the beta and gamma emissions caused by naturally occurring substances in the soil such as ^{40}K, and those from other alpha emitters in the same region as Pu, such as the naturally occurring radionuclide daughters of the uranium and thorium chains. Total background is all counts in the region of interest (ROI) not from the desired element (Pu). This includes contributions from the reagents, scintillants, and other nuclides in the sample, including that of the instrument. Elimination of beta and gamma background is achieved by pulse shape discrimination (PSD). Decreasing alpha background requires clean

Table 4. Summary of Background

Background Type	Extractive Scintillant	Total DPM in ROI	ROI Centroid	Energy Shift Percent
Standard	HDEHP	1.69 ± 0.90	9.4	
Soil	HDEHP	1.34 ± 0.74	8.7	7
Method	HDEHP	1.58 ± 0.6	5.7	39
Standard	NDAS	0.92 ± 0.90	4.9	
Soil	NDAS	1.16 ± 0.71	4.7	4
Method	NDAS	0.71 ± 0.12	5.8	21
Standard	NDEHP	0.08 ± 0.04	105	
Soil	HDEHP	0.15 ± 0.02	92	12
Method	NDEHP	0.29 ± 0.21	81	23
Standard	NDAS	0.20 ± 0.07	75	
Soil	NDAS	0.04 ± 0.01	58	24
Method	NDAS	0.09 ± 0.07	55	28

Table 5. ALS Duplicate Analysis

Method	DPM in RO1	Relative Percent Deviation	Primary Peak Channel	RPD Peak Location
Standard	0.14 ± 0.02		108	
blank	0.15 ± 0.02	7	109	0.9
Soil blank	0.16 ± 0.10		62	
direct	0.18 ± 0.18	12	77	22
RFP soil	2.10 ± 0.38		13	
direct	1.14 ± 0.21	39	14	7
RFP filter	4.30 ± 0.30		9	
direct	3.63 ± 0.33	17	17	61
Standard soil	0.97 ± 0.12		98	
direct	0.73 ± 0.04	28	102	4
Standard	192.6 ± 9.4		103	
extraction	193.8 ± 12.8	0.6	92	11
RFP soil	83 ± 2.1		72	
extraction	88 ± 1.1	6	98	15
Standard soil	28 ± 0.84		82	
extraction	39 ± 0.4	31	83	0.6

reagents and a low radon working area. Currently the method blanks give a background 4 to 10 times higher than the optimum electronic background.

Various blanks (method, standard) were prepared for both extracted and directly analyzed samples. The background contributions can also indicate the efficiency of the scintillant and show energy shifting. The method blank is blank soil directly prepared in scintillant or digested and extracted into the scintillant. The standard blank is pure extractive scintillant.

Table 4 lists the average count rates for backgrounds in two different extractive scintillants. A typical background spectrum for the HDEHP extractive scintillator is shown in Figure 6. Peak location and width for two different regions of interest (ROI) are given, one region is the area of highest background near the beta continuum, containing possible thorium peaks, the other is further down field where Pu peaks should be located. The primary channel is the center of the peak where the largest number of counts are clustered. The peak width is given by the number of channels in the region of interest. The counts per channel gives the actual background that can be used for correction of sample peaks falling in those ROIs.

Background varies more at different regions of interest than with different types of blanks. This seems to indicate that the background contribution of natural alpha emitters in soil is negligible. Near the edge of the beta continuum (peak channel 5-9) the background is about 10 times higher, 0.9 to 1.7 dpm, than in the region of interest for Pu in a clean sample, 0.04 to 0.2 dpm. The HDEHP, which is the scintillant of choice for direct soil counting, has a higher background than the NDAS. The NDAS is perhaps easier to use for the final

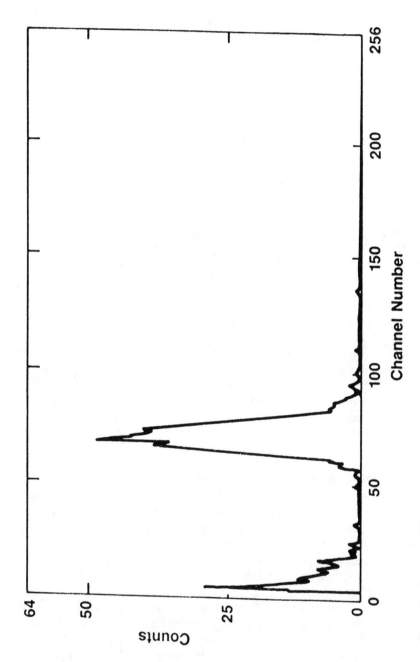

Figure 5. Extracted RFP soil spectra.

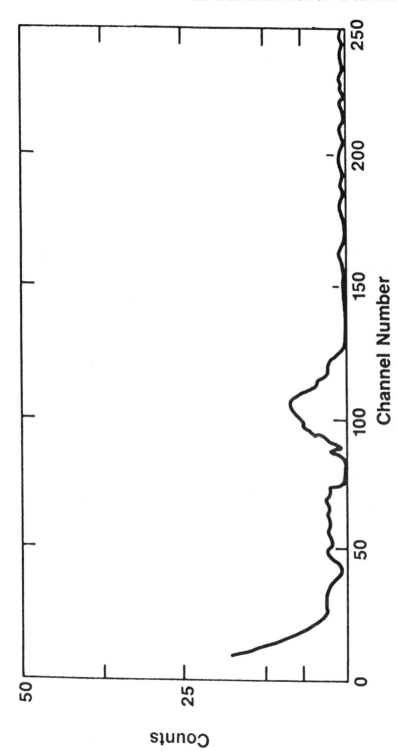

Figure 6.   Background spectrum.

secondary extraction than the HDEHP but does not partially extract Pu in the direct soil mode. Soil added to to the scintillant does not increase the background significantly but does shift the ROI.

From the background count rate of blank soil (Table 5, 0.16 dpm) the approximate detection limit of 100 pCi/g can be estimated. Assuming a detection limit of 5 (4.66 sigma) times background, the activity of the sample would have to be about 0.8 dpm. At 20% efficiency this is 4 dpm in the sample, or about 0.02 g of sample. The count rates for about 10 mg of the 100 pCi/g standard, 1000 pCi/g RFP soil, and soil on filter are also at least 5 times the background of the blank soil.

Precision

The precision associated with background, resolution, and efficiency for direct and extracted soil analysis can be seen in Tables 1, 3, and 4 with a summation in Table 5. Standard deviations (1 sigma) in Tables 1 and 3 give an idea of the stability of the instrument and standards and the reproducibility of the final extraction procedure.

The combined instrument and sample stability, as expressed by reproducible counts of the same standard or blank over 2 weeks time, is about 1.4%. Some degradation of standard was noted after 2 weeks. The extraction procedure is more of a factor in efficiency reproducibility than the instrument instability. Lack of temperature control in the basement lab must also be considered. The standard deviation of efficiency for multiple counts of different standards, covering a wide range of concentrations (0.1 to 200 pCi) on different days, was about 4%.

The peak location or energy stability varied by about 10% for the extraction and counting of standards that varied by over the three orders of magnitude in concentration. The percent standard deviation (1 sigma) in resolution, as expressed by FWHM, was 2.2%. The energy stability for a well sealed standard over time was about 2%.

Variations in extraction procedure are more of a factor in any energy shifts than the instrument instability. Energy shifts between soil types for both extracted and direct analysis can be seen in comparing ROI reproducibility in Table 5 and the spectra of the standard soil and the RFP soil on a filter Figures 3 and 4. The location of the peak for the spiked RWMC soil is not shifted noticeably from that of the standard, but there is a shift when comparing the direct RFP soil on filter sample. The primary peak activity for the RFP soil on filter was located in the 10–20 channel ROI, whereas the standard and standard soil peak is around channel 100.

Background stability can be seen when comparing the ROI of either extractant in the standard blank (scintillant only) to soil blank (soil and scintillant) in Tables 4 and 5. In HDEHP the shift is 7% from 9.4 to 8.7 in the low channels near the continuum and 12% from 105 to 92 in the Pu ROI (5100 keV). For

NDAS the shift is 4% in the higher background beta continuum region and 24% in the Pu ROI.

Considerable shifting of the peak is apparent when comparing direct soil and extracted soil analysis (Table 3). Rocky Flats soil and filters caused a greater energy shift than the directly prepared RWMC soil standard. The Pu on the standard soil is in a more extractable form than that of the aged RFP soil; thus some Pu was detected in true solution rather than from a soil particle. The clarity of the sample in direct analysis was critical and made a great difference in the peak location and the overall counting efficiency. After sparging, some samples settled more rapidly than others. Their clearing up gave rise to much of the variation between replicate counts.

A general idea of background stability can be seen in Tables 4 and 5 with the use of replicate background counts and uncertainties within a single count. Background can be influenced by instrument stability and chemical stability, and it is a factor in analysis precision. The uncertainties for the replicate blanks are the standard deviations (1 sigma) of multiple counts. The background spectrum lacked well defined peaks (Figure 6). The width of the ROIs for most of the background counts, the low count rates, and the energy shifting do not allow identification of specific contaminant isotope contributions to background activity. The counting uncertainties are higher than when activity is actually present (Table 5). High standard deviation of replicate runs and high blank count rates could have been from the poor location of the spectrometer in a known high-radon-background basement lab.

Digestion and a single extraction into the final scintillator gives most of the advantages not found in a directly counted sample, such as improved resolution and efficiency and background elimination, but it saves time by eliminating subsequent stripping and extraction steps. This was tried with several dissolution systems and extractive scintillators with some success. The primary problem was incomplete extraction and extraction of iron, making the solution highly colored. Getting the aqueous sample into the proper state, by adjusting pH and ionic strength without precipitating out calcium, was also a problem without the preliminary extraction. The time saved is significant and for some applications may be feasible, especially if some other organic extractant could be found. The ideal extractant needs to be selective for removing Pu from a nitrate or sulphate system, rejecting thorium, uranium, and iron, and not containing chloride, nitrate, or other quenchant groups. The tertiary amine nitrate now used is selective but is itself highly quenched due to the nitrate group.

Multiple extractions should give lower backgrounds, less energy shifting, and increased resolution. Possible interferences while extracting the aqueous solution from the digested sample into the organic amine or extractive scintillator, in order of importance are: (1) incomplete extraction of Pu, (2) extraction of unwanted ions, and (3) inability to strip the Pu from the first extractant into the aqueous solution. Some of the causes of these interferences are: incorrect Pu oxidation state, incorrect pH, insufficient ionic strength, impure extractants, and nonextractable Pu complexes in the aqueous solution. The spectral separation of Pu from an interfering alpha emitter such as thorium is shown in Figure 2. Thorium should not be a problem in direct analysis unless present in great amount and allowed to ingrow, as some thorium daughters will add to the Pu peak. Multiple extractions chemically remove thorium and uranium and eliminate alpha interference problems.

CONCLUSIONS

An ALS system called PERALS has been investigated for use in obtaining alpha radiation levels during the TRU waste retrieval. By directly analyzing lightly soiled filters, the ALS system can detect concentrations as low as 100 pCi/g Pu within 1 hr of receiving the sample. For heavily soiled filters or soil samples, a minimum detectable activity of 1 pCi/g can be obtained in 2 hr including the time to dissolve and extract the sample. Direct counting of filters gave efficiencies of about 20%. Alpha resolution of about 5% was sufficient to separate the Pu peak from the primary Th peaks.

Areas of future work include: improving extraction efficiency and reproducibility, determining detection limits in the presence of uranium and varying amounts of thorium, and decreasing the preparation time and complexities. Instrumental improvements such as better photomultiplier tubes, better light couplings, and better electronics are possible, though interferences in the instrumentation and counting area have been reduced to a considerable extent.

The largest gains can be made by improving the overall efficiency for the entire analytical scheme. Future work would involve spiking at various points during dissolution and extracting to pinpoint critical procedural parameters where losses are occurring. Noninstrumental interferences can hinder the separation of alpha and beta pulses and the resolution of alpha peaks. Chemistry improvements in the selection of fluors and scintillants, and elimination of quenching and interferences through organic extraction, are currently under investigation. The inherent advantages of near 100% counting efficiency make ALS a viable method for rapid Pu analysis and screening in conjunction with traditional analysis methods.

REFERENCES

1. Horrocks, D.L. *Applications of Liquid Scintillation Counting*, (New York: Academic Press, 1974), pp. 28–80.
2. Birks, J.B. *The Theory and Practice of Scintillation Counting*, (New York: MacMillan, 1964).
3. Polack, H. et al. *Advances in Scintillation Counting*, (Edmonton: University of Alberta Press, 1983). pp. 508.
4. McDowell, W.J. *Alpha Counting and Spectrometry using Liquid Scintillation Methods*, Nuclear Science Series, ORNL, NAS-NS-3116 (1986).

5. McDowell, W.J. and G.N. Case. "Methods for Radium, Uranium and Plutonium by Alpha Liquid Scintillation," ORNL TM 8531 (1982).

6. Matthes, S.A., R.F. Farrel, and A.J. Mackie. "A Microwave System for the Acid Dissolution of Metal and Mineral Samples," NTIS PB83–225391, U. S. Bureau of Mines Albany OR. (1983).

7. Joshi, B.M., L.C. Butler, and G. Lebanc. "Application of Microwave Oven to Digest Environmental Samples for Inorganic Analysis," Preliminary Investigation, Lockheed-EMSCOM, U. S. EPA, Las Vegas NV, (1988).

8. Langer, G. "Wind Resuspension of Trace Amounts of Plutonium Particles from Soil in a Semi-Arid Climate," *1st International Aerosol Conference, Minneapolis, Minn.* (1984) pp. 484–487.

9. Chanda, R.N. and R.A. Deal. *Catalogue of Semiconductor Alpha-Particle Spectra*, Idaho Nuclear Co., USAEC, IN-1261 (1970).

CHAPTER 38

Vagaries of Wipe Testing Data

Jill Eveloff, Howard Tisdale, and Ara Tahmassian

At the University of California at San Francisco, the common radioisotopes ^3H, ^{14}C, ^{35}S, and ^{32}P constitute over 80% of all radioisotope usage. An effective technique for determining contamination by these isotopes is wipe testing and liquid scintillation counting (LSC). The State of California requires all radioactive materials users at UCSF to maintain wipe test data of their laboratory to demonstrate contamination control. The Radiation Safety Department of the Office of Environmental Health and Safety at UCSF provides a special service to its research staff and radioactive materials users. This service performs wipe tests of their laboratories in accordance to their required monitoring frequency. This includes the analysis of the wipes taken.

Because of the large volume of wipe test analyses performed by our office, it is important to understand some of the vagaries associated with wipe testing data. This knowledge helps our office provide better service to the university community. The vagaries investigated include the following:

- effects on wipe test data of wiping two different surface materials
- effects on wipe test data using three different wipe media
- effects on wipe test data using varying amounts of cocktail
- effects on wipe test data of self-absorption in the wipe media
- frequent complications to tritium monitoring from fluorescence

INVESTIGATIONS OF SURFACE MATERIALS AND WIPE MEDIA

The efficiency of detecting low level contamination of ^3H, ^{35}S, and ^{32}P was tested using two lab bench surfaces, transite and wood. These materials were chosen because they are most common in laboratories at UCSF. Wipes of each isotope from each surface were made using three different wipe media; a two-inch diameter dry wipe, a two-inch diameter wet wipe, and a 0.25×4.0 inch piece of scotch tape. These wipes were conducted using controlled activities of 4000 cpm, per isotope, per wipe, and on controlled 3×3 inch surface areas. The results, % cpm removed by wipe, are shown in Table 1.

Table 1. % CPM Removed by Wipe

	Wood Surface			Transite Surface		
	^{3}H	^{35}S	^{32}P	^{3}H	^{35}S	^{32}P
Dry wipe	0.7	2.1	4.5	0.0	2.2	5.6
Wet wipe	0.2	9.3	18.7	2.8	37.3	33.9
Scotch tape	1.7	1.7	5.1	4.3	11.7	15.8

INVESTIGATIONS OF VARYNG COCKTAIL VOLUME AND SELF ABSORPTION-EFFECTS

The efficiency of detecting low level contamination of ^{3}H, ^{35}S, and ^{32}P was tested using different volumes of scintillation cocktail: 10 mL, 5 mL, and 2.5 mL. In addition, the of self-absorption effects of the isotope by the wipe media was investigated. Controlled amounts of isotope activities were placed directly on two-inch diam dry wipes and inserted in vials with specific volumes of scintillation cocktail. The results, % cpm recovery, are shown in Table 2.

FLUORESCENT COMPLICATIONS IN TRITIUM ANALYSIS

The detection of ^{3}H contamination was found to be complicated frequently by the interference of fluorescent compounds. With the wide variety of research done on the UCSF campus, there exists the probability of encountering chemical compounds with fluorescent components during typical wipe testing procedures. Fluorescence may frequently produce significant cpm in the ^{3}H counting channel (up to 18.6 keV). It is sometimes impossible to know the difference between fluorescence and ^{3}H contamination unless spectrum analysis is performed.

Most present day LSCs have spectral analysis capabilities; however, some have associated computer software which will generate graphic representations of the spectral analysis. These representations can show, most obviously, the differences between fluorescence and ^{3}H contamination. A typical ^{3}H spectrum is shown in Figure 1 and a typical fluorescence spectrum in Figure 2; however, fluorescence spectra, are not restricted to the type seen in Figure 2. The differences between the two spectra can be seen with close observation.

Table 2. % Recovery of CPM

ISOTOPE	ACTIVITY (CPM)	COCKTAIL VOLUME		
		10 mL	5 mL	2.5 mL
^{3}H	456	20.6	18.2	16.2
^{3}H	4748	19.7	16.1	13.1
^{35}S	766	56.8	63.7	65.9
^{35}S	6525	62.5	63.0	63.6
^{32}P	519	94.4	91.3	91.0
^{32}P	5098	99.5	100.0	100.0

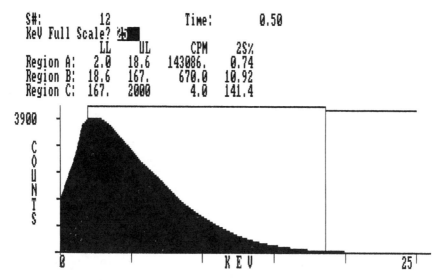

Figure 1. Typical ³H spectrum.

The differences include both the distribution of energies (i.e., up to Emax) and the average energy; therefore, it is always prudent to take one step further in ³H contamination analysis by evaluating whether the source of counts could be attributed to fluorescence.

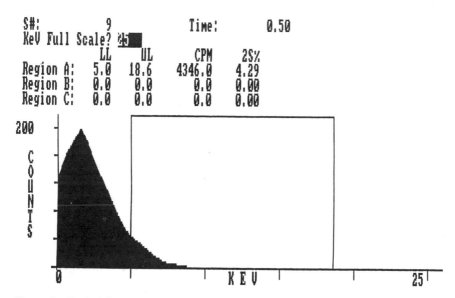

Figure 2. Typical flourescence spectrum.

RESULTS

A summary of the results of these experiments is as follows:

- Contamination can be more easily removed by wipe from transite as opposed to wood surfaces.
- Low level ^{3}H was difficult to detect under all conditions studied.
- Wet wipes provide the more efficient contamination removal.
- Scotch tape also increased the efficiency of contamination detection.
- Dry wipes produce up to 80% self absorption of ^{3}H samples, 40% self-absorption of ^{35}S samples, and very little self absorption of ^{32}P
- Very slight decreases in % recovery of cpm was seen in reduction of scintillation cocktail used per sample (The implications here for cost savings are incredible.)

The results of this somewhat shallow study produce thought trends that can be used in wipe test data manipulations and/or analysis. All samples used in this study were counted in a Packard LSC model 2000CA.

CHAPTER 39

The Determination of ^{234}Th in Water Column Studies by Liquid Scintillation Counting

R. Anderson, G.T. Cook, A.B. Mackenzie, and D.D. Harkness

INTRODUCTION

As a result of the pronounced differences in their geochemical properties, uranium and thorium exhibit markedly different solubilities in seawater under oxidizing conditions. Uranium is relatively soluble in seawater, with an average open ocean ^{238}U concentration of 0.04 Bq l^{-1} (3.2 μg^{-1}) and it exists as an anionic carbonate complex of the uranyl ion. In contrast, ^{232}Th is highly insoluble in seawater, with a concentration of about 2×10^{-7} Bq l^{-1} (6×10^{-5} μg l^{-1}) and it is highly susceptible to removal from solution by hydrolysis or by sorption on, and incorporation in, particulate material. Similarly, thorium isotopes produced *in situ* in seawater by decay of soluble parent radionuclides also have a very short residence time in solution. The resulting situations of radioactive disequilibrium which develop in the marine environment, eg., between ^{230}Th and ^{234}U, (members of the ^{238}U decay series shown in Figure 1), have long been recognized as a means of investigating the rates and mechanisms of a variety of oceanographic processes. For example, the systematics of ^{230}Th/^{234}U disequilibria have been used in the development of oceanic sediment chronologies,[1,2] investigation of uranium diagenetic chemistry and mobility in sediments,[3,4] and study of the efficiency of particulate scavenging of thorium and protactinium in different areas of the oceans.[5-7]

In recent years, however, it has become apparent that the short-half-life ^{234}Th (Figure 1) also has considerable potential for investigating the rates of a range of important marine processes; ^{234}Th/^{238}U disequilibrium has been used to investigate open ocean euphotic zone productivity rates,[8] nearshore scavenging processes,[9,10] and particle fluxes and reworking rates in deep ocean sediments.[11]

^{234}Th is therefore an extremely useful natural radio-tracer. There are, however, difficulties associated with its analysis; it is a weak β-emitter (E β_{MAX} = 190keV), it gives rise to a daughter nuclide, ^{234m}Pa, which is also a β^- emitter (E β_{MAX} = 2.33 MeV), and it has a half life of 1.17 minutes. ^{234m}Pa is generally

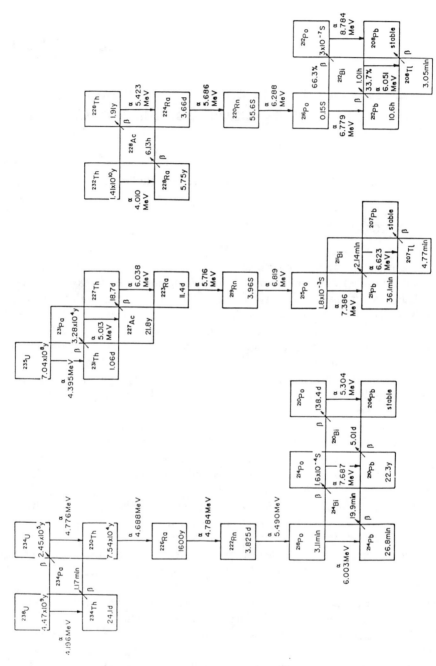

Figure 1. Natural decay series (major branching schemes).

present in secular equilibrium with ^{234}Th during its analysis. The conventional approach to the analysis of ^{234}Th concentrations in seawater involves spiking the filtered sample (typically of volume $\sim$ 40L) with ^{230}Th as a yield tracer, scavenging the thorium from solution by coprecipitation with ferric hydroxide, redissolution and extraction of the thorium by anion exchange separation, electrodeposition of the thorium on a planchette, and measurement of the combined ^{234}Th plus ^{234m}Pa activity using a G.M. or proportional counter. The ^{230}Th activity is subsequently determined by alpha spectroscopy using a surface barrier detector. The limitation on detection efficiency using these 2π counting systems is one of the main factors affecting the sensitivity of this analysis, and the method also requires intercalibration of the alpha and beta counting systems.

Gamma spectroscopy analysis of ^{234}Th using its gamma photon of energy (63.2 keV) offers an alternative method of determining ^{234}Th concentrations, but the intensity of 2.5% for this decay mode, in conjunction with the low efficiency of gamma photon detectors, gives rise to a relatively low sensitivity for this method (e.g. typical detection efficiency using a 130 cc Ge (Li) detector = 0.5%).

Modern liquid scintillation counters offer a highly attractive alternative for the analysis of ^{234}Th in that much higher detection efficiencies can be obtained and simpler chemical separation and source preparation techniques can be employed. The low-level analysis capability now routinely available with modern liquid scintillation counters provides suitably low backgrounds for the determination of ^{234}Th, and the provision of spectroscopy capability simplifies simultaneous counting of alpha and beta particles so that ^{234}Th and the ^{230}Th yield tracer can be analyzed in a single count. A description is provided below of the development of a liquid scintillation method for analysis of ^{234}Th in seawater. The work is being undertaken as part of the U.K. BOFS (Biogeochemical Ocean Flux Study) program which is associated with the international JGOFS (Joint Geochemical Ocean Flux Study) program.

Development of the method was performed using a Packard 2000CA/LL liquid scintillation counter. Comparison was made between the low level count mode which employs pulse shape analysis to discriminate true β events from background events[12] and the normal counting mode. As a result of this pulse shape discrimination, a small percentage of the β^- efficiency is lost, together with a proportionately higher percentage of the α events. The latter are well documented as having much broader pulse widths and therefore being more liable to discrimination. Despite this, it was envisaged that the achievable reductions in background would more than compensate for the loss of efficiency.

EXPERIMENTAL

Isolations of ^{230}Th and ^{234}Th

The ^{230}Th spike was prepared by anion exchange isolation of Th, from a mineralized uranium nodule in which the ^{238}U decay chain was in equilibrium, at least to ^{230}Th. The concentration of ^{232}Th was below the limit of detection as well. The ^{230}Th fraction was retained in dilute nitric acid with aluminium carrier. Aging for approximately 12 months ensures decay of ^{234}Th, ^{231}Th, ^{227}Th, and the corresponding decay products to a negligible level (Figure 1).

The uranium fraction was also retained and similar aging allows ^{234}Th to come to secular equilibrium with the ^{238}U. This provides a source of ^{234}Th free from ^{230}Th for method development. Following isolation of ^{234}Th from ^{238}U, a period of a few days should be allowed for decay of ^{231}Th ($t_{1/2} = 1.06$d) formed by decay of ^{235}U which was also present in the solution. The purities of both the ^{230}Th and the U/^{234}Th aged solutions were confirmed by electrodeposition and alpha spectroscopy. The work presented here was performed with spikes which were not completely aged, however, so an allowance was made for this in calculations.

Liquid Scintillation Counting of ^{230}Th and ^{234}Th

To obtain representative spectra, suitable aliquots of ^{234}Th, ^{230}Th, the combination of ^{234}Th and ^{230}Th, and background were prepared as follows:

- Aliquots were added to 7 mL Packard low potassium glass vials and gently taken to dryness.
- The Th was redissolved in 0.5 mL of 1 mol dm^{-3} HCl, and 5 g Packard Hionic Fluor was added.
- Each sample was counted on a Packard 2000 CA/LL liquid scintillation counter for 300 minutes, first in the low level mode and then in the normal mode.
- Spectra were saved for subsequent analysis using the Packard Spectragraph software package.

Following measurement of the ^{234}U activity, the ^{234}Th activity was calculated based on the ingrowth time since U/Th separation. To determine the counting efficiency of ^{234}Th (^{234m}Pa) the following steps were carried out: a known aliquot of ^{238}U/^{234}Th was spiked with the ^{230}Th yield tracer, and the Th was isolated again by anion exchange. The volume was reduced to 1 to 2 mL and quantitatively transferred to a 7 mL scintillation vial, taken to dryness, and prepared for counting as previously described. The same activity of ^{230}Th yield tracer was added directly to a further vial, taken to dryness, and similarly prepared.

Seawater analysis

Replicate 20 L seawater samples were collected from the River Clyde estuary. The ^{238}U concentration was determined by anion exchange separation and conventional alpha spectroscopy. The samples were filtered through 0.45 μm millipore membrane filters to remove particulate material, and the pH was lowered to 2 with HCl to maintain Th solubility. Three of these samples were then analyzed for ^{234}Th almost immediately, while a further four were stored to allow ^{234}Th ingrowth (23 to 25 days). Samples were first spiked with ^{230}Th and thereafter 500 mg of $FeCl_3$. $6H_2O$ was added to scavenge out Fe/U/Th etc. by pH adjustment to 9. To maximize recovery, the 20 L samples were stirred for approximately 2 hr and left overnight for precipitate settling. Th separation and vialing/scintillation counting was as previously described.

RESULTS

The spectra obtained from counting ^{234}Th, ^{230}Th, the combination of ^{234}Th and ^{230}Th, and a background, in both counting modes, are presented in Figure 2. Although there is a reasonable degree of spectral separation, it is not complete. There is some interference in the ^{230}Th region from the high energy ^{234m}Pa β^- emissions. Conversely, residual ^{234}Th in the ^{230}Th yield tracer brings about a small interference in the ^{234}Th region. Provided that the degree of sample quenching remains constant, the extent of interference will remain constant, and indeed this has been the case throughout the study. Using Spectragraph, windows of 0 to 80 keV and 100 to 220 keV were selected for ^{234}Th and ^{230}Th, respectively. For the ^{234}Th spectrum obtained under low-level conditions, a fully optimized window of 10 to 70 keV was determined using software developed at SURRC. The respective crossovers were then calculated in both counting modes and for both 0 to 80 and 10 to 70 keV counting windows. The ^{230}Th contribution into the ^{234}Th region will decrease with aging, i.e., as the residual ^{234}Th decays. The contribution of ^{234}Th/^{234m}Pa will remain a constant percentage of the net count rates in the 0 to 80 and 10 to 70 keV windows. As an example of this, under low-level conditions and using a 0 to 80 keV window, the contribution ^{234}Th/^{234m}Pa under the ^{230}Th peak was estimated at 16.5 ± 0.002%. The contribution, at the time of counting of ^{230}Th in the ^{234}Th region was 3.9 ± 0.93%. Using the low level option there is an approximate 10% loss in ^{234}Th counting efficiency and a 45% loss in ^{230}Th efficiency.

^{234}Th efficiencies were determined as follows: (1) gross counts in appropriate regions were obtained for all vials from stored spectra using Spectragraph. (2) Appropriate backgrounds were subtracted from each region and corrections for crossover interferences were made, thus yielding net ^{234}Th/^{234m}Pa and ^{230}Th count rates. (3) Yields were determined from the ratio of sample ^{230}Th count rates to those of ^{230}Th added directly to scintillation vials. (4) The count rates in the ^{234}Th/^{234m}Pa region were corrected for decay and the yield factor was applied. (5) The activity of ^{234}Th was calculated from the ^{238}U activity as 470.5 ± 7.6 dpm ml^{-1}. Since ^{234m}Pa is in equilibrium with ^{234}Th, the gross activity will be 941 dpm mL^{-1}. Table 1 presents triplicate values of the overall counting efficiency for ^{234}Th/^{234m}Pa in the two counting windows and in both counting modes together with yields representing the efficiency of the chemistry.

Table 2 indicates background count rates and E^2/B values for relevant counting windows and in both counting modes. The results show that E^2/B is maximized using a 10 to 70 keV window in the low level mode. However, this

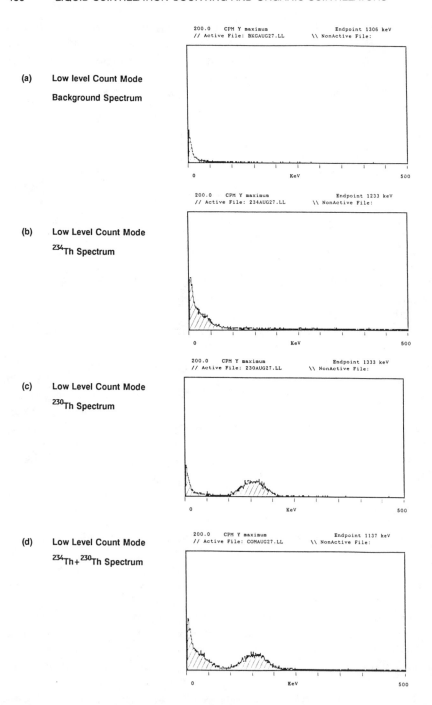

Figure 2a.   Low-level count mode spectra for background, ^{234}Th, ^{230}Th, and ^{234}Th + ^{230}Th.

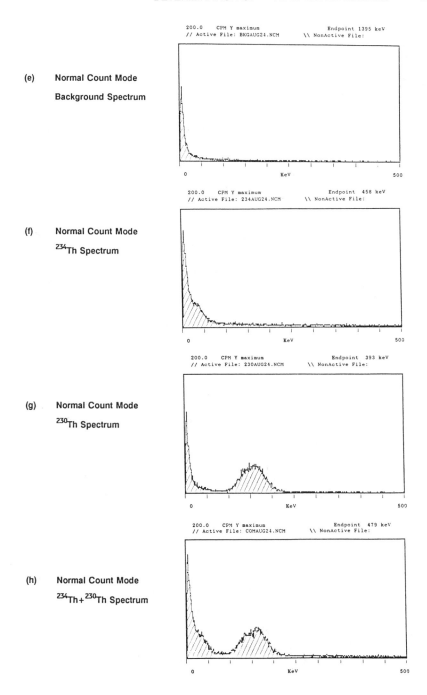

Figure 2b.   Normal count mode spectra for background, [234]Th, [230]Th, and [234]Th + [230]Th.

Table 1. ^{234}Th Chemical Yields and Counting Efficiencies Using Both the Low-Level and Normal Counting Modes and 0 to 80 and 10 to 70 keV Counting Windows

Replicate	Chemical Yield (%)	Normal Mode % Eff.		Low Level Mode % Eff.	
		0–80 keV	10–70 keV	0–80 keV	10–70 keV
1	80.0	136.9	100.3	115.5	85.0
2	79.1	138.7	102.0	118.2	85.7
3	73.0	137.8	101.3	120.0	88.2

Table 2. Background Count Rates and E^2/B Factors for ^{234}Th and Background Count Rates for ^{230}Th Using Both the Normal and Low-Level Counting Modes

	Normal Count Mode			Low Level Count Mode		
	^{234}Th (0–80 keV)	^{234}Th (10–70 keV)	^{230}Th (100–220 keV)	^{234}Th (0–80 keV)	^{234}Th (10–70 keV)	^{230}Th (100–220 keV)
Background (cpm)	16.75	6.73	2.27	6.46	2.77	0.93
E^2/B	1134	1522	—	2152	2689	—

improvement is partly negated at present by larger errors in the final net count rate calculations because the ^{230}Th spike is not fully aged. With a fully aged spike, overspiking with ^{230}Th, followed by counting in the low level mode, should be the optimum method.

Table 3 represents the measured ^{234}Th concentrations and the predicted values based upon the assumption of insignificant ^{234}Th activity initially present in these high suspended particulate content, shallow waters.[10] The results for the short term ingrowth of ^{234}Th are generally consistent with the predicted values, whereas systematically lower values of 80 to 90% of predicted values are observed for longer ingrowth periods. This is probably a result of partial scavenging during the longer ingrowth periods. This would not represent a problem in practical applications of the technique in which ^{234}Th separation is effected as quickly as possible after sample collection.

Table 3. Predicted and Measured ^{234}Th Concentrations in Replicate 20 L Estuarine Seawater Samples (Errors Quoted at the ±1 Sigma Level of Confidence)

Replicate Number	Ingrowth time (days)	Predicted ^{234}Th (dpm L^{-1})	Measured ^{234}Th (dpm L^{-1})			
			Normal Mode		Low Level Mode	
			0–80 keV	10–70 keV	0–80 keV	10–70 keV
1	2.0	0.11(0.01)	0.12(0.03)	0.15(0.03)	0.16(0.02)	0.16(0.02)
2	4.7	0.25(0.01)	0.25(0.03)	0.28(0.03)	0.26(0.02)	0.26(0.02)
3	4.7	0.25(0.01)	0.35(0.03)	0.36(0.03)	0.37(0.03)	0.36(0.03)
4	23.8	0.98(0.02)	0.87(0.02)	0.89(0.04)	0.90(0.03)	0.93(0.03)
5	23.8	0.98(0.02)	0.86(0.04)	0.88(0.04)	0.85(0.03)	0.84(0.03)
6	25.7	1.03(0.02)	0.82(0.03)	0.84(0.03)	0.81(0.03)	0.82(0.03)
7	25.7	1.03(0.02)	0.82(0.03)	0.88(0.03)	0.82(0.03)	0.83(0.03)

DISCUSSION AND CONCLUSIONS

The results to date indicate that ^{234}Th measurements from seawater samples are indeed feasible by this technique. The advantages of the method are that (1) the preparation chemistry is simpler, since no electrodeposition is required, (2) both isotopes may be measured in a single count on equipment employing an automatic sample changing facility, and (3) counting efficiencies are much higher for both isotopes of the order of 100% for ^{230}Th and 137% relative to ^{234}Th activity for combined ^{234}Th/^{234m}Pa. As a result of these, a higher through-put of samples can be achieved with greater precision. The possibility also therefore exists of carrying out the determinations on much smaller samples. This is a major advantage for seawater analysis, where sample size is often a limiting factor. Five litre samples should be well within the capabilities of this method.

In order to validate the method yield factors can and will be confirmed when all ^{234}Th has decayed. Where a fast turnaround time of results is not required, this should enable higher precision to be obtained. The possibility also exists of using scintillation counters employing simultaneous α/β separation and counting which should in theory virtually eliminate the need for crossover calculations.

ACKNOWLEDGEMENTS

This work is being undertaken as part of the Biogeochemical Ocean Flux Study Program for which the financial assistance of the U.K. Natural Environment Research Council under grant GST/02/302 is gratefully acknowledged.

REFERENCES

1. Turekian, K.K. and J.K. Cochran. "Determination of Marine Chronologies Using Natural Radionuclides," in *Chem. Oceanography*, vol 7, 2nd ed. J.P. Riley and R. Chester, Eds. (London: Academic Press, 1978).
2. Lalou, C. "Sediments and sedimentation processes," in *Uranium Series Disequilibrium. Applications to Environmental Problems*, M. Ivanovich and R.S. Harmon, Eds. (Oxford: Clarenden Press, 1982).
3. Colley, S. and J. Thomson. "Recurrent Uranium Relocations in Distal Turbidites Emplaced in Pelagic Conditions." *Geochim. Cosmochim. Acta*, 49:2339–2348 (1985).
4. Colley, S., J. Thomson, and J. Toole. "Uranium Relocations and Derivation of Quasi-isochrons for a Turbidite/Pelagic Sequence in the Northeast Atlantic," *Geochim. Cosmochim. Acta*, 53:1223–1234 (1989).
5. Anderson, R.F., M.P. Bacon, and P.G. Brewer. "Removal of ^{230}Th and ^{231}Pa from the Open Ocean," *Earth Planet. Sci. Lett.*, 62:7–23 (1983).
6. Anderson, R.F., M.P. Bacon, and P.G. Brewer. "Removal of ^{230}Th and ^{231}Pa at Ocean Margins," *Earth Planet. Sci. Lett.*, 66:73–90 (1983).

7. Shimmield, G.B., J.W. Murray, J. Thomson, M.P. Bacon., R.F. Anderson, and N.B. Price. "The Distribution and Behaviour of ^{230}Th and ^{231}Pa at an Ocean Margin, Baja, California, Mexico," *Geochim. Cosmochim. Acta*, 50:2499–2507 (1986).

8. Coale, K.H. and K.W. Bruland. "Oceanic Stratified Euphotic Zone as Elucidated by ^{234}Th:^{238}U Disequilibria," *Limnol. Oceanogr.*, 32:189–200 (1987).

9. Tanaka, N., Y. Tanaka, and S. Tsunogai. "Biological Effect on Removal of ^{234}Th, ^{210}Po and ^{210}Pb from Surface Water in Funka Bay, Japan," *Geochim. Cosmochim. Acta*, 47:1783–1790 (1983).

10. McKee, B.A., D.J. DeMaster and C.A. Nittrocier. "The Use of ^{234}U/^{238}U Disequilibrium to Examine the Fate of Particle Reactive Species in the Yangtze Continental Shelf," *Earth Planet. Sci. Lett.*, 68:431–442 (1984).

11. Aller, R.C. and D.J. DeMaster. "Estimates of Particle Flux and Reworking at the Deep Sea Floor Using ^{234}Th/^{238}U Disequilibrium," *Earth Planet. Sci. Lett.*, 67:308–318 (1984).

CHAPTER 40

Performance of Small Quartz Vials in a Low-Level, High Resolution Liquid Scintillation Spectrometer

Robert M. Kalin, James M. Devine, and Austin Long

ABSTRACT

To meet an increased demand for the radiocarbon dating of small samples via liquid scintillation counting, Haas[1] introduced a technique which employed square, optically flat, quartz vials, with optimal UV transmission characteristics, as counting solution containers. Benzene samples as small as 0.3 mL (240 mg carbon) were dated using these vials. With proper installation and handling, these vials were shown to provide excellent long-term stability and precision.

We now report results on small quartz vials used in a modern low-level LS counter, the LKB-Wallac 1220 Quantulus, in the underground counting laboratory at the University of Arizona. Counting solutions comprise sample benzene diluted by the addition of "dead" benzene to a counting volume of 0.3 mL, with approximately 0.95 weight % butyl-PBD scintillant. A low-background copper vial holder was designed to accommodate the vials; constant orientation and alignment during loading is maintained by the heavy weight of the holder. Samples and standards are counted for 1200 min (60 × 20 min count periods). Background cpm, based on running average, is 0.0454 ± 0.006 cpm at a counter efficiency of 67.4%. Figure of merit E^2/B = 100,130; signal/noise (net modern standard B cpm/background cpm) = 53.5. Maximum determinable age, based on the Stuiver and Polach[2] criterion, is 39,200 years before present. These performance characteristics compare favorably with data presented for 0.3 mL Teflon vials[3] and 3.0 mL glass vials.[4]

INTRODUCTION

In the past decade there has been an effort to push the limits of nuclear counting techniques. The introduction of improved liquid scintillation (LS) instruments[5-11] has reduced background noise dramatically. This decrease in noise enables the user to measure smaller samples with improved accuracy.

Small sample ^{14}C dating techniques use miniature gas proportional counting systems to measurement as little as 5 mg of carbon.[12-15] Haas[1] introduced the use of 0.3 mL quartz spectrophotometric vials for ^{14}C LS counting. Devine[16] reported a figure of merit value of 2477 when using these vials in an Intertechnique LS20 counter. The comparison of 0.3 ml quartz and teflon vials in various counters[3] shows that small sample counting has benefited greatly from developments in LS technology.

This chapter presents the results of small sample (0.3 mL of benzene) counting using quartz spectrophotometric vials in an LKB-Wallac 1220 Quantulus LS counter at the University of Arizona, Radiocarbon Dating Lab.

METHODS

Two quartz spectrophotometric vials of 1 mm path length, like those used by Haas,[1] were used (Figure 1). A special vial holder was manufactured from low-background, low-oxygen copper (Figure 2) such that the maximum area of sample can be seen by the photomultiplier tubes, and the vial holder blocks photomultiplier crosstalk in the rest of the sample chamber. The diameter and height of this vial holder is identical to that of a 20 mL glass LS vial.

Benzene samples are synthesized from Oxalic Acid I (OXI) primary standard and from Mississippian limestone background material. Spectrophotometric benzene is also used as background. Samples to be dated, yielding

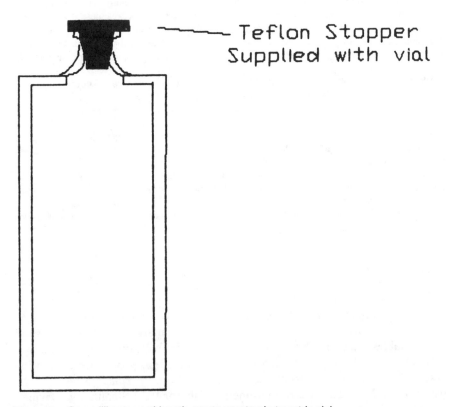

Figure 1. One millimeter pathlength quartz spectrophotometric vial.

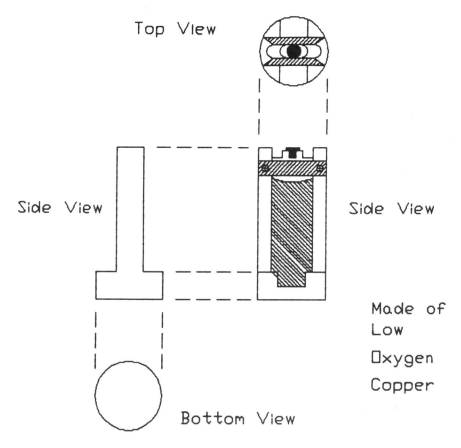

Figure 2. Low-background, low-oxygen copper vial holder.

benzene and weighing between 80 and 350 mg, are diluted to 350 mg carbon as benzene. Butyl-PBD, 0.95 wt-percent, is added to the benzene. The sample is introduced into the quartz vial via syringe, weighed, and placed into the LS counter, counted 60 × 20 min intervals, weighed after counting, and removed from the vial with a syringe. The quartz vials are flushed with spectrophotometric benzene repeatedly, then dried.

Time series data on OXI and background are compiled and averaged for age calculations.[2] Laboratory quality assurance–quality control (QA/QC) samples of well known age are analyzed as unknowns and used to check the validity of the time series data.

The energy spectrum of a synthesized OXI ^{14}C benzene and background (BKG) benzene samples for the LKB-Wallac 1220 Quantulus are shown in Figure 3. The energy windows for this study were set to maximize efficiency while avoiding the low energy background peak.

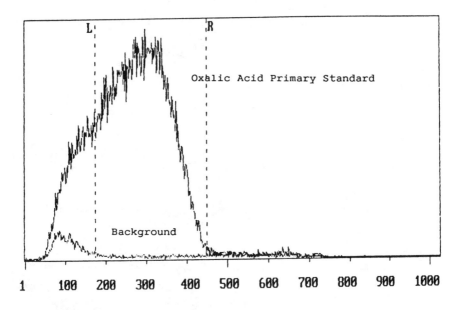

Channels 175 - 450   Efficiency = 61.22 %

Figure 3. Spectrum output from LKB-Wallac 1220 Quantulus of OXI primary ^{14}C standard and background benzene.

RESULTS AND DISCUSSION

Efficiency of photon measurement is related to the alignment of the quartz vial in the LS counter. The faces of the vial must be parallel to the faces of the photomultiplier tube. The vial holder was placed in the sample tray so that the optimal alignment occurs when the elevator lifts the sample into the chamber. The degree of rotation of the copper vial holder during elevator operation was determined by raising and lowering the vial holder. Some misalignment was found over 60 repeat events, but the weight of the copper vial holder produces no noticeable misalignment for one loading event. Therefore, no special alignment mechanism was manufactured, rather, samples are loaded into the chamber, and 60 × 20 min repeat counts are taken without unloading the sample vial.

Performance characteristics of the 0.3 mL quartz vials in the LKB-Wallac 1220 Quantulus counter (Table 1) compare favorably to those of the 0.3 mL Teflon vials,[3] and to 3 mL glass vials.[4] The difference in efficiency between the 0.3 mL Teflon vials and the 0.3 mL quartz vials is due to the exclusion of the low-energy end of the energy spectrum for this study. This area of low-energy background noise seen in the spectrum (Figure 3) is instrument dependent. A second LKB-Wallac 1220 Quantulus, which is next to this instrument, does not have this noise peak.[4] Differences in photomultiplier tube composition most

Table 1. Small Sample ^{14}C Results in 0.3 mL Quartz Vials, 0.3 mL Teflon Vials[3] and 3.0 mL Glass Vials[4]

	0.3 mL Quartz Vial	0.3 mL Teflon Vial	3.0 mL Glass Vials
^{14}C efficiency (E)	67.4%	80.8%	74.0%
Average background (B)	0.04537 CPM	0.05	0.49
Average modern[a] signal (N_0)	2.5286 CPM	2.66	27.5
Figure of merit E^2/B	100,130	130,570	11,090
Factor of merit (fM) N_0/B	11.9	11.9	39.1
S/N (modern/background)	53.46	53.2	55.7
Maximum determinable age	39,200[b]	43,230[c]	53,400
Error modern sample (yrs)	140	130	50

[a]Modern signal = 0.95 Oxalic Acid I primary standard.
[b]Sample Counted 1200 minutes.
[c]Sample Counted 3000 minutes.

likely leads to this variation among instruments (P. Makinen personal communication).

The distribution of measured events in each 20 min time period for OXI (Figure 4) and BKG (Figure 5) fit a Poisson distribution.

This technique was used for samples containing between 80 and 350 mg of carbon. The statistical uncertainty for samples less than 125 mg is usually

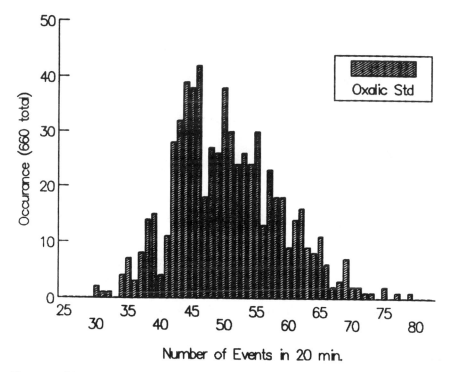

Figure 4. Distribution of the number of events measured during each 20 min time interval for OXI primary standard.

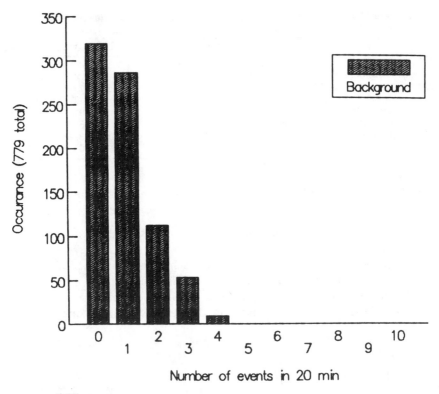

Figure 5. Distribution of the number of events measured during each 20 min time interval for BKG samples.

unacceptable (> 500 years), but for certain applications this information may be useful.

The results of three analyses of our in-lab QA/QC program, using the small quartz vials, is presented with the 1988 to 1989 results of the same sample for 3.0 mL vials from three different counters (Figure 6). These three samples were less than 350 mg carbon, thus they were diluted before measurement. At one standard deviation, the error bars overlap with the known age range of this sample.

The performance of small sample counting with the quartz vials makes LS counting an option for those who need radiocarbon dates on small samples but do not have access to, or funds for, AMS determinations. The disadvantage of small sample counting, compared to our routine determinations,[4] is the loss of some accuracy and a reduction of the age limit which we can attain.

Techniques of low-level counting, including small sample techniques, need not be limited to ^{14}C dating. Application of low-level counting to environmental testing and biomedical research would allow a decrease in dosimetry while maintaining a high signal to noise ratio. This could reduce costs of materials and may ease some of the problems of radioactive waste disposal.

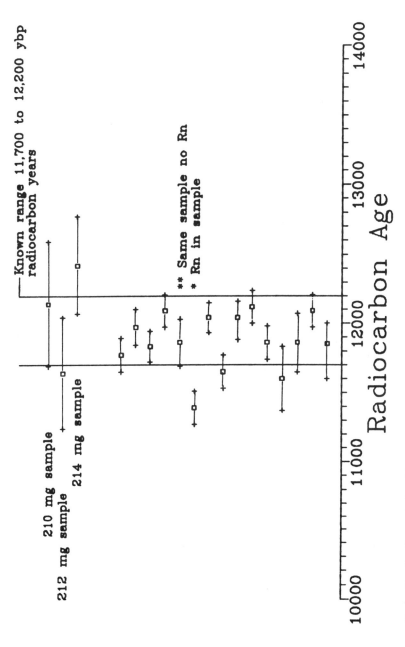

Figure 6. Small sample QA/QC results with the July 1988 to June 1989 QA/QC results of the Arizona Radiocarbon Laboratory.

CONCLUSION

The advent of new LS technology has allowed our lab to decrease the minimum sample size for ^{14}C dating to less than 300 mg of carbon. Special alignment mechanisms are not required to keep the vial in the optimal position, provided the vial is loaded only once into the chamber. The performance of the 0.3 mL quartz vials compares favorably with that of 0.3 mL Teflon vials and 3.0 mL glass vials. Signal and background events fit a Poisson distribution. Samples as small as 80 mg have been measured, but they lose accuracy. QA/QC samples determined with this method are within an acceptable range. Applications beyond ^{14}C dating of milligram size samples should be explored using techniques for small sample LS determinations.

REFERENCES

1. Haas, H. "Radiocarbon Dating," in *Proceedings of the 9th International ^{14}C Conference*, R. Berger and H. Suess, Eds. (1979), pp. 246–255.
2. Stuiver, M. and H.A. Polach. *Radiocarbon*, 19:355–363 (1977).
3. Polach, H.A., L. Kaihola, S. Robertson, and H. Haas. *Radiocarbon*, 30:153–155 (1988).
4. Kalin, R.M. and A. Long. in *Proceedings of the 13th International Radiocarbon Conference* A. Long and R. Kra, Eds. (Dubrovnic, Yugoslavia: Radiocarbon Vol. 31, 1989).
5. Noakes, J.E. and J.D. Spaulding. in *Liquid Scintillation Counting: Recent Applications and Developments Vol. 1, Physical Aspects*, C.T. Peng, D.L. Horrocks, and E.L. Alten, Eds. (New York: Academic Press, 1980), pp. 160–170.
6. Rajamae, R. and J.M. Punning. in *Proceedings of the 10th International Radiocarbon Conference*, M. Stuiver and R. Kra, Eds. (Bern: Radiocarbon 22, no. 2, 1980), pp. 435–441.
7. Kojola, H., H. Polach, J. Nurmi, T. Oikari, and E. Soini. in *Int. J. Appl. Radiat. Isotopes*, 30:452–454 (1984).
8. Polach, H.A., J. Nurmi, H. Kojola, and E. Soini. in *Advances in Scintillation Counting*, S.A. McQuarrie et al., Eds. (Edmonton: University of Alberta Press, 1984), pp. 420–441.
9. Oikari, T.H. Kojola, J. Nurmi, and L. Kaihola. in *Int. J. Appl. Radiat. Isotopes*, 38:9 (1987).
10. Noakes, J.E. and R.J. Valenta. in *Proceedings of the 13th International Radiocarbon Conference*. A. Long and R. Kra, Eds. (Dubrovnic, Yugoslavia: Radiocarbon Vol. 31, 1989), pp. 332–341.
11. Kessler, M. in *Proceedings of the International Workshop and Inter-Comparison of ^{14}C Laboratories*. A Long and R. Kra, Eds. (Glasgow: 1989).
12. Srdoc, D., B. Obelic, and N. Horvatincic. in *Proceedings of the 11th International Radiocarbon Conference*, M. Stuiver and R. Kra, Eds. (Seattle: Radiocarbon 25, no. 2, 1983), pp. 485–492.
13. Sheppard, J.C., J.F. Hopper, and Y. Welter. in *Proceedings of the 11th International Radiocarbon Conference*, M. Stuiver and R. Kra, Eds. (Seattle: Radiocarbon 25, no. 2, 1983), pp. 493–499.

14. Hut, G., J. Keyser, and S. Winjima. in *Proceedings of the 11th International Radiocarbon Conference*, M. Stuiver and R. Kra, Eds. (Seattle: Radiocarbon 25, no. 2, 1983), pp. 547–552.

15. Otlet, R.L., G. Huxtable, G.V. Evans, D.G. Humphreys, T.D. Short, and S.J. Conchie. in *Proceedings of the 11th International Radiocarbon Conference*, M. Stuiver and R. Kra, Eds. (Seattle: Radiocarbon 25, no. 2, 1983), pp. 565–575.

16. Devine, J.M. and H. Haas. *Radiocarbon*, 29,1:12–17 (1987).

CHAPTER 41

The Optimization of Scintillation Counters Employing Burst Counting Circuitry

G.T. Cook, R. Anderson, D.D. Harkness, and P. Naysmith

INTRODUCTION

Recognition of the versatility of liquid scintillation counters in environmental radioactivity, together with the advent of small gas proportional counters and the accelerator mass spectrometry technique (AMS) for small sample radiocarbon dating, has undoubtedly led to the development of a new generation of scintillation counters with much reduced background count rates and enhanced E^2/B values. This consequently enables much lower detection limits. The two instruments most commonly employed in U.K. laboratories are the LKB-Pharmacia Quantulus and the Packard 2000CA/LL (and subsequent models). The Quantulus has much enhanced passive shielding in addition to an active coincidence guard counter; the counter contains a mineral oil based liquid scintillant enabling the detection of cosmic and environmental gamma radiations. Further features to optimize performance include a user-set pulse amplitude comparator and high purity Teflon/copper counting vials, the latter enabling a 30% reduction in background compared to glass vials.[1,2] The Packard 2000CA/LL does not require an active or enhanced passive shielding. Its ability to reduce background count rates via burst counting circuitry enables the characteristics of sample scintillation pulses to be differentiated from those of background pulses through pulse shape/duration analysis.[3] Most background pulses have a number of afterpulses in addition to the fast prompt pulse common to all events. Although true β^- events, particularly at higher energies, may have a certain number of afterpulses, a good degree of discrimination is certainly achievable. The detector assembly incorporating the slow scintillating plastic, a feature of the Packard-2260XL, enhances this time resolved burst discrimination since its long decay constant increases the number of photons in the afterpulses of background radiation, therefore acting as a quasiactive guard.

One criticism of the Quantulus has been that a lack of uniformity in the

Teflon vial characteristics has meant characterizing each vial individually. This necessarily reduces the element of routineness obtainable from the more uniform glass vials, and hence caused their abandonment for ^{14}C dating at SURRC except with samples whose expected ages exceed 25,000 years. Here, the improvement in background can be critical in terms of age resolution. Nevertheless, for 2 mL (1.75 g) and 4 mL (3.5 g) geometries in 7 mL slimline glass vials, efficiencies of 65 and 67.6% and backgrounds of 0.44 and 0.70 cpm, respectively, are obtainable.

A criticism of the Packard 2000CA/LL has been that although background count rates are considerably reduced, there is also a considerable reduction in overall counting efficiency.[4] This implies that the burst counting circuit is not 100% effective and that a certain percentage of true β^- events are being discriminated out, presumably because their prompt pulse widths are sufficiently broad or they have an uncharacteristically large number of afterpulses. A logical extension to this argument is that by sharpening these pulses, much of this loss in efficiency could be regained.

The results presented here are a résumé of a considerable data set built up while assessing counters employing burst counting circuitry.

EXPERIMENTAL

This study can, in essence, be divided into three component stages. First, the influence of a range of scintillants, their concentrations, and their combinations were assessed to determine influences on counting efficiency. This was followed by an assessment of two modifications to the original system: (1) the detector assembly incorporating a slow scintillating plastic and (2) vial holders produced from the same material; studies were carried out on both ^{3}H and ^{14}C. Finally, for any radiocarbon dating laboratory the ultimate test must be the production of accurate age measurements. Naturally, the sample counting is only one aspect of the entire dating process; however, recent studies indicate that it is particularly important in terms of overall accuracy.[5]

This third section can be subdivided as follows:

1. A sample of benzene, synthesized from the peat used in stage 3 of the recent International Collaborative Study, was re-vialed into two 7 mL slimline glass vials. This sample had previously been dated using an "old technology" Packard 4530, with butyl-PBD/bis-MSB as the scintillant, giving an age of 3360 ± 50 years BP. The mean result for the 28 laboratories who dated this material was 3388 BP.[6]

2. One aspect which arose from the international study and which caused some considerable concern was that the NERC Laboratory dated two duplicate samples of marine shell material, each separated into inner and outer fractions, thus yielding four age measurements. When using old technology counters, all four results were well within the overall spread, however, when using the burst counting circuit, one of the four was enriched relative to

modern. In this instance, butyl-PBD alone (13 mg mL^{-1}) was the scintillant employed. To determine whether this effect was specific to the scintillant, the sample was re-vialed into three separate vials with appropriate dilutions and additions of toluene and bis-MSB to exactly duplicate the scintillant used in the University of Glasgow Laboratory.

Throughout the optimization studies involving different scintillant combinations and concentrations, standard 20 mL low potassium glass vials were employed; 4.5 g of a ^{14}C benzene standard (178.5 $\pm$ 2.0 dpm/g^{-1}) were used throughout. The essential criteria for optimization were (1) constancy of efficiency with variation in scintillant concentration as measured in a 0 to 156 keV window and (2) constancy of quenching as determined by t-SIE value. t-SIE is the spectral index of the transformed Compton spectrum of the external standard. Values can in theory range between 0 and 1000; a value of 1000 indicating no quenching. The 2000CA/LL employs a 20 μCi ^{133}Ba standard, the 2260XL an 8 μCi ^{133}Ba standard.

The optimization studies on the burst counting circuit, slow scintillant detector assembly, and vial holders were carried out in 7 mL glass vials using 2 g of a high activity ^{14}C standard (2340 $\pm$ 8 dpm/g^{-1} benzene) and 0.42 g of a butyl-PBD/bis-MSB (12 and 6 g/L^{-1} respectively) scintillant combination dissolved in toluene. The same geometry was also adopted for all dating studies. For ^{3}H assessments, tritiated water (410 dpm/g^{-1}) was used throughout.

Dating of the benzene sample synthesized from peat was carried out in the 2000CA/LL using the burst counting circuit, in the 2260XL using the burst counting circuit, and finally with the addition of the vial holders to the 2260XL system. Dating of the shell-derived sample was carried out on the 2000CA/LL.

RESULTS AND DISCUSSION

Table 1 indicates that with the burst counting circuit activated, open window counting efficiency decreases from 84.1 to 73.9% as the concentration of butyl-PBD increases from 2 to 14 mg/g^{-1} benzene. In parallel, t-SIE values increase to a plateau value at 10 mg/g^{-1} and above. With the burst circuit off, the same trend in t-SIE values is observed; however, efficiency remains constant between 4 to 20 mg/g^{-1} at approximately 93.5%. These features are found to be common to virtually all scintillants and scintillant combinations. That is:

1. t-SIE value trends are virtually identical whether the burst counter is on or off.
2. Throughout virtually all the concentration ranges, efficiency is constant at approximately 93.5% with the burst counter off, but variable, tending to a plateau value with the burst counter on.

Table 1. Effect to Butyl-PBD Concentration ^{14}C Spectral Stability and Counting Efficiency with and without the Burst Counter Circuit in Operation

Butyl-PBD (mg/g^{-1} of C$_6$H$_6$)	Eff. (%) BC on[a]	t-SIE$_b$ BC on	Eff. (%) BC off	t-SIE BC off	Difference in % Eff.
2	84.1	557	91.8	557	7.7
4	80.7	683	93.5	681	12.8
6	78.9	730	92.8	732	13.9
8	76.6	749	93.5	751	16.9
10	75.7	757	93.2	755	17.5
12	75.2	761	93.6	762	18.4
14	74.8	761	93.1	760	18.3
16	73.9	760	93.4	759	19.5
18	73.5	758	93.0	759	19.5
20	73.7	755	93.8	755	20.1

[a]BC = Burst counter.
[b]t-SIE = spectral index of hte transformed Compton spectrum of the external standard.

Table 2 represents the optima in efficiency for these plateau values using a range of scintillants and scintillant combinations. It should be noted that the highest efficiencies are always observed in the presence of bis-MSB, ie. bis-MSB alone, butyl-PBD/bis-MSB and PPO/bis-MSB, at about 89 to 90% compared with about 93.5% with the burst counter off. This suggests that bis-MSB differs in some respect from the other scintillants, a factor which undoubtedly merits further investigation in terms of the pulse shaping rather than the ultimate influence on efficiency.

Since the introduction of the Packard 2000CA/LL, several modifications have been introduced as previously described. Table 3 indicates the performance enhancements these can bring about. With the burst counting circuit off, i.e., used as a conventional liquid scintillation counter, optimum window efficiency is relatively poor (57.8%). This is due to the shape of the background spectrum. To maximize the E^2/B parameter, the lower discriminator had to be set to 18 keV. With the burst counter on, although open window efficiency is slightly reduced, optimum window efficiency is greatly enhanced (65.0%). This is due to a lower discriminator setting of 12.5 keV, enabling

Table 2. Optimum Concentrations of a Range of Scintillants/Scintillant Combinations for Achieving Maximum ^{14}C Efficiency with Stability of Quenching on the Packard 2000CA/LL

Scintillant (mg/g^{-1} of C$_6$H$_6$)	Eff. (%) (Burst circuit on)	Eff. (%) (Burst circuit off)	Difference in % Eff.
Butyl-PBD (16)	73.9	93.4	19.5
PPO (5.6)	83.6	93.5	9.9
PPO (5.6) + POPOP (0.7)	81.9	93.8	11.9
PPO (5.6) + Me$_2$ POPOP (0.7)	82.6	93.6	11.0
PPO (5.6) + bis-MSB (2.7)	89.6	93.7	4.1
Butyl-PBD (12) + POPOP (0.7)	78.7	93.2	14.5
Butyl-PBD (12) + Me$_2$ POPOP (0.7)	77.2	93.5	16.3
Butyl-PBD (12) + bis-MSB (4.0)	88.5	93.1	4.6
Bis-MSB (5.3)	88.5	93.0	4.5

Table 3. Progressive Optimization for ^{14}C of the Packard Low-Level Counting System, Employing Burst Counting Circuitry (Packard 2000CA/LL) and Using (1) a Detector Assembly with a Slow Scintillating Plastic (Packard 2260XL) and (2) Vial Holders Made from the Same Material (Scintillant = Butyl-PBD/Bis-MSB in Toluene)

Counting Conditions	Open window % Eff.	Opt. window % Eff.	Opt. Window Background (cpm)	E^2/B
2000CA burst circuit off	92.1	57.8	2.87	1166
2000CA burst circuit on	88.1	65.0	1.31	3226
2000CA burst circuit on + vial holders	87.5	66.0	0.85	5134
2260XL burst circuit on	86.9	70.2	0.94	5269
2260XL burst circuit on + vial holders	88.6	71.4	0.69	7383

E^2/B to be maximized while excluding all ^{3}H. A general feature of the modifications (slow scintillant detector assembly and vial holders) is that they enhance performance not only by an overall reduction in the background count rate, but by changing the background spectra shape and enabling reduced lower discriminator settings. E^2/B is enhanced from 3226 (Packard 2000CA/LL burst counter on) to 7383 (Packard 2260XL plus vial holder). Similar trends are also observed for ^{3}H (Table 4). With a standard 20 mL glass vial, using 10 mL of tritiated water and 10 mL Picofluor LLT, E^2/B is enhanced from 62 (2000CA/LL burst counter off) to 161 (2260XL). Similarly, with 3.5 mL tritiated water and 3.5 mL Picofluor LLT in a standard 7 mL slimline glass vial, E^2/B is enhanced from 155 to 400. The holders do not appear to enhance performance beyond that of the 2260XL. With the data currently available it is not possible to determine whether the differences between the 2260XL, 2260XL plus vial holders, and 2000CA/LL plus vial

Table 4. Progressive Optimisation for ^{3}H of the Packard Low-Level Counting System Employing Burst Counting Circuitry (Packard 2000CA/LL) and Using (1) a Detector Assembly with a Slow Scintillating Plastic (Packard 2260XL) and (2) Vial Holders Made from the Same Material

Counting Conditions	Open Window % Eff.	Opt. Window % Eff.	Opt. Window Background (cpm)	E^2/B
(a) 10 mL H$_2$O + 10 mL Picofluor LLT				
2000CA burst circuit off (glass vials)	25.1	19.7	6.30	62
2000CA burst circuit on (glass vials)	23.5	21.5	3.77	123
2260XL glass vials	25.1	23.2	3.33	161
2000CA burst circuit on (plastic vials)	22.6	18.5	1.93	178
260XL (plastic vials)	24.3	21.3	2.12	215
(b) 3.5 mL H$_2$O + 3.5 mL Picofluor LLT				
2000CA burst circuit off	26.9	19.7	2.51	155
2000CA burst circuit on	25.5	22.6	2.06	248
2000CA burst circuit on + vial holders	22.3	19.1	1.05	350
2260XL	25.5	23.5	1.38	400
2260XL + vial holders	23.7	22.0	1.32	367

Table 5. ^{14}C Age Determinations on Benzene Synthesized from Peat and Shell Material Used in the International Collaborative Study

	Counting conditions	Mean ^{14}C Age
(a)	Peat sample	
	2000CA/LL burst circuit on	3410 ± 40
	2260XL	3440 ± 60
	2260XL + vial holders	3410 ± 70
(b)	Shell sample	
	2000CA\LL burst circuit on	750 ± 70

holders are significant. Table 5 gives the age measurements made on the benzene synthesized from both the peat and shell samples. The peat results show excellent agreement with those of the international study regardless of the counting conditions employed. Results for the shell sample seem to confirm that the anomalous measurement on this sample is in some respect tied to the scintillant employed.

The mean age derived from the three replicate subsamples was calculated to be 750 ± 70, and considering the extra manipulations involved in re-vialing etc., this compares quite favorably with the mean result of 637 BP from 31 laboratories, and it is obviously significantly older than the previously derived modern age. It would appear that under certain circumstances a combination of vial plus contents and the butyl-PBD must enhance efficiency beyond the norm. The most obvious possibility would seem to be some impurity in the benzene, perhaps altering pulse shapes or variations in dissolved oxygen, which will influence the triplet state and reducing the delayed component.[2]

For ^{14}C dating with the 2260XL, the counting time of the external standard was increased from 15 to 60 sec to reduce the greater variability in t-SIE observed. Determining a t-SIE value in the 2260XL requires an additional step. Along with the Compton spectrum produced by the vial-content interaction, there is also a spectrum produced by the interaction of the γ radiations with the slow scintillating plastic in the detector assembly. The combination of these two spectra will not yield a true t-SIE value. To compensate for this, an additional spectrum must be measured, i.e., that of the external standard with an empty vial which, in effect, is brought about by interactions with the plastic. This second measurement subtracted from the first produces the "true" t-SIE value. A one minute counting time for the external standard plus sample reduces the variability in t-SIE to the same magnitude as found with the 2000CA/LL. Using the 2260XL with vial holders produces much greater variations in t-SIE which can not be overcome by increasing the external standard counting time, even to 4 min. To circumvent this problem, the following steps were taken. First, a uniform response of the vial holders has to be verified. This was carried out as follows.

1. Each vial holder was loaded with a vial containing 4.5 g of scintillation grade benzene and 0.95 g of butyl-PBD (12 g/L^{-1}) and bis-MSB (6 g/L^{-1}) in toluene.

Each vial was then counted for 5×100 min. The results were entirely consistent with a uniform background response.

2. A single high activity standard (2340 dpm/g^{-1} benzene) was counted sequentially in each of the vial holders (3×25 min counts). The vial holders with the lowest and highest count rates then underwent further counting (4×50 min). At the end of this counting period, open and optimum window counting efficiencies were identical (87.4 and 71.0%, respectively). It appears that despite the variability in t-SIE values, between successive counts from a single vial holder and values from different vial holders, the background and efficiency responses are essentially constant within the 20 individuals tested.

Having established this constancy of response, a quench curve was constructed using 16 vials of 2 g high activity benzene standard quenched to differing degrees with acetone. This was carried out first without the vial holders and subsequently with them. A second degree polynomial regression analysis of the former yielded an R^2 value of 99.8%, the latter yielded a value of 79.9%. Finally, the count rate using the vial holders was regressed against the t-SIE values without the vial holders yielding a value of 99.6%. The same system was then employed to determine the degree of quenching in samples, backgrounds, and modern reference standards, i.e., a large number (50) of very short (0.1 min) sample counts, with 1 min counts of the external standard, were undertaken without the vial holders. The samples were then counted using the vial holders for a minimum of 2000 min via the quasisimultaneous batch counting method as would normally be adopted.[7]

CONCLUSIONS

The results of these studies confirm that the burst counting circuit makes the efficiency of the Packard 2000CA/LL highly sensitive to scintillant concentration and type. Decreases of up to 20% in counting efficiency are observed compared to conventional counting. These can largely be overcome by careful manipulation of scintillant concentration and type. Best overall performance is always obtained in association with the secondary scintillant bis-MSB. Efficiencies approaching 90% are routinely observed, i.e., within approximately 4% of conventional levels. It is proposed that bis-MSB sharpens pulse widths thereby bringing many more true β^- events within the cut-off threshold of the burst counter. In some instances it has been observed that efficiency increases as the degree of quenching increases which is completely contrary to normal theory. This obviously reflects the complexity of the pulse shape analyses and must represent a complex balance of sufficient scintillant, self absorption, and the energy transfer process.

The use of the detector assembly incorporating both the slow scintillating plastic (2260XL) and the slow scintillating plastic vial holders enable significant enhancements in performance for ^{14}C. The burst counter and plastic detector assembly both enhance ^{3}H performance, although it is questionable whether the vial holders with the plastic detector assembly represents any further enhancement. With standard 20 mL plastic vials containing 10 mL tritiated water + 10 mL Picofluor LLT a limit of detection of 1.4 Bq/L is achievable for a 500 min count.

In terms of actual ^{14}C age measurements, the limited results so far indicate that accuracy is possible with burst counting circuitry and subsequent modifications to this system using butyl-PBD/bis-MSB as the scintillant. Many more measurements, however, are required to confirm this, particularly when the use of vial holders and the associated modifications of normal ^{14}C dating practices are envisaged. The use of butyl-PBD alone as a scintillant for ^{14}C dating obviously has associated problems, and until these can be identified and eliminated, its use should not be recommended.

REFERENCES

1. Kojola, H., H. Polach, J. Nurmi, T. Oikari, and E. Soini. "High Resolution Low-Level Liquid Scintillation β Spectrometer," *Int. J. Appl. Rad. Isotopes*, 35: 949–952 (1984).
2. Gupta, S.K. and H. Polach. *Radiocarbon Dating Practices at ANU-Handbook*, (Canberra: ANU, 1985).
3. Van Cauter, S. "Three Dimension Spectrum Analysis: A New Approach to Reduce Background of Liquid Scintillation Counters," *Packard Applications Bulletin*, No. 006 (1986).
4. Polach, H., G. Calf, D.D. Harkness, A. Hogg, L. Kaihola, and S. Robertson. "Performance of New Technology Liquid Scintillation Counters for ^{14}C Dating," *Nucl. Geophys.* 2(2):75–79 (1988).
5. Scott, E.M., T.C. Aitchison, D.D. Harkness, M.S. Baxter, and G.T. Cook. "An Interim Progress Report on Stages 1 and 2 of the International Collaborative Programme," *Radiocarbon*, 31 (1989).
6. Aitchison, T.C., E.M. Scott, D.D. Harkness, M.S. Baxter, and G.T. Cook. "Report on Stage 3 of the International Collaborative Programme," in *International Workshop on InterComparison of ^{14}C Laboratories*, 12–15 September, 1989, East Kilbride.
7. Stenhouse, M.J. and M.S. Baxter. "^{14}C Dating Reproducibility: Evidence from Routine Dating of Archaeological Samples," in *^{14}C and Archaeology*, Groningen, 1981 (1983), pp. 147–161.

CHAPTER 42

Statistical Considerations of Very Low Background Count Rates in Liquid Scintillation Spectrometry with Applications to Radiocarbon Dating

Robert M. Kalin, Larry Wright, James M. Devine, and Austin Long

ABSTRACT

The new generation of low-level liquid scintillation spectrophotometers have significantly reduced background count rates. These very low rates require non-Gaussian statistical treatment of data. An LKB Quantulus, employed for radiocarbon dating of samples of between 80 to 300 mg, received no events in 41% of the 780 20-min intervals on background benzene. The distribution of the background events is Poisson. For very small count rates, which are non-background (measurable ^{14}C content) benzene, Poisson statistical treatment of data is preferable to Gaussian treatment.

INTRODUCTION

This chapter is not meant to supercede previous discussions of counting statistics. The authors agree on strict adherence to the recommendations put forth by Stuiver and Polach[1] for use in the general case of radiocarbon date calculation. We would like to submit that very low background count rates may allow for the non-Gaussian treatment of data.

In the last decade, many advances in liquid scintillation (LS) technology have taken place. The incorporation of anticoincidence shielding reduced background count rates.[2-8] Recently, pulse shape discrimination techniques within active shielding[9] and novel energy discrimination — after pulse analysis techniques without active shielding[10,11] have decreased background count rate further.

The level of background noise in LS counters without these techniques can be approximated with Gaussian statistics. This assumes that the background events are stochastic, without a periodic noise from electronics or other source. As background events are reduced over two magnitudes of order, these events approach the predicted Poisson distribution for small numbers of stochastic events.

The statistical interpretation of ^{14}C data collected using decay counting has been discussed,[12-17] and for ^{14}C date calculations, Stuiver and Polach[1] published the guidelines which should be followed by all radiocarbon labs. Yet, when the number of measured events approaches zero, careful consideration of the distribution of the events is required.

The very low background count rates, and subsequent low sample count rates measured with LS counters, can be compared to the statistical analysis of microcontamination in ultraclean environments.[18]

VERY LOW BACKGROUND COUNT RATES

Measurements of very low background count rates (less than 0.1 cpm) have been measured both with gas proportional counting systems[17,19-21] and LS counting.[23,23]

Figure 1 shows background data collected using 0.3 mL vials in an LKB-Wallac 1220 Quantulus LS counter. These data compromise 13 different background samples counted for 60 to 20 min intervals, totaling 1200 min each. A chi-square test of this data confirms that they fit a Poisson distribution. The arithmetic mean of these data is 0.04539 cpm. The distribution of events from 11 separate 0.95 Oxalic acid ^{14}C standard samples (Figure 2), has 4 values which do not fit a Poisson distribution.

TREATMENT OF DATA COLLECTED

A Poisson distribution is quite asymmetrical when the number of counts is near zero, but it quickly approaches a Gaussian distribution by approximation as the count number increases. Tsoulfanidis states that the Gaussian distribution is almost identical with the Poisson distribution at a mean value of 25 counts/time.[24] Therefore, if an unknown were counted, such that an average 25 counts/time were measured, Gaussian statistics would be a close approximation.

Data from Otlet et al., on small gas proportional counting systems, reported background count rates ranging between 21 and 400 counts per day.[20] Increasing the time period is one method of treating the results as a Gaussian distribution. Important information on the stability and performance of an LS counter can be lost if repeat time series data is not collected.

There are many factors which can affect the fit of time series data to an expected Poisson distribution. Periodic external noise, i.e., electronic spikes, ^{222}Rn and ^{220}Rn contamination, and periodic internal noise will alter the fit of measured data. The signal and background will be a Poisson distribution if the LS counter is only recording random events.

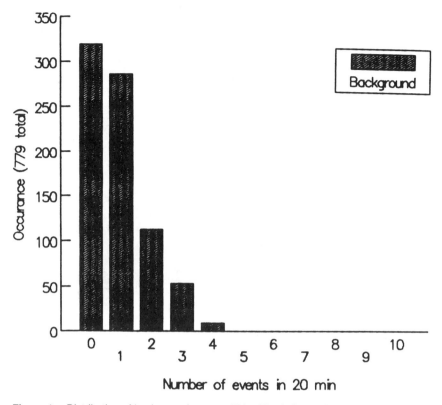

Figure 1. Distribution of background events, 780—20 min intervals.

The background data collected follow a Poisson distribution. There is not any evidence to suggest a nonrandom component of these data. The ^{14}C data have four values (of 660) which do not follow a Poisson distribution. Small amounts of Radon gas were detected in 2 of the 11 samples. We hypothesize that the non-Poisson values are the result of Radon contamination or an artifact of slight sample evaporation during the collection of data.

RADIOCARBON DATING AND POISSON STATISTICS

Questions arise when determining the age of a sample from a small number of events. How certain are the investigators that the events measured are from the original sample material and not from some source of contamination? An assessment of this can be obtained by analyzing background material which undergoes identical treatment along with samples.

A chi-square test of the time series background count rates can alert the investigator to Radon contamination or periodic noise. We also suspect that an analysis of the length of time between background events may be useful in

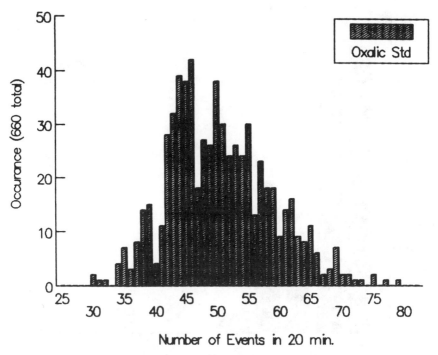

Figure 2.   Distribution of 0.95 oxalic acid ^{14}C standard, 660—20 min intervals.

determining a component of periodic noise, and we will investigate this possibility further.

The introduction of systematic errors by lab personnel can be determined by measuring repeat samples of known ^{14}C activity, such as lab standards or known values quality assurance — quality control samples.

The authors would like to suggest the use of Poisson distribution tables like that of Crow and Gardner for determining acceptance criterion when calculating radiocarbon ages.[21] For the background count rate in this study, an average of 55 events is expected during the 1200 min of counting. The two criteria of Stuiver and Polach for 55 events would calculate a "greater than" age using 17 events above background.[1] The Poisson tables of Crow and Gardener would calculate the "greater than" age, 95% confidence, using 16 events.[21] Although one event may be considered trivial, it represents 2% of the total events measured. Assuming that the laboratory researchers have very carefully studied all sources of error within the lab and those associated with the LS equipment, there is a probability that one event can be considered discrete. Rejecting that one event is a loss of information. As the background count rates approach zero, this discrepancy between Gaussian and Poisson distributions becomes a greater portion of the total signal.

It may be best to measure a sample for a period of time sufficient to collect enough events so that this discrepancy is negligible. But, this may not be cost efficient, or LS technology may advance to the point that the discrete nature of single events must be taken into account.

CONCLUSION

The background events measured follow a Poisson distribution. Four of the 660 events for the ^{14}C standard did not fall within the chi-square goodness of fit criteria. These events are hypothesized to be radon contamination or artifacts of evaporation during counting. For very low background count rates, the goodness of fit to a Poisson distribution should be calculated, and an attempt to identify any nonrandom component of the background should be made. Each lab should have a degree of confidence for samples prepared in the lab based on reproduceablility of backgrounds and known activity samples. And finally, by using Poisson statistics, slightly more information may be obtained than when using the recommendations of Stuiver and Polach (1977).[1]

REFERENCES

1. Stuiver, M. and H.A. Polach. *Radiocarbon*. 25, 2:458-492 (1977).
2. Peitig, von F. and H.W. Scharpenseel. *Atompraxis*. 7:1-3 (1964).
3. Punning, J.M. and R. Rajarmae in *Low-Radioactivity Measurements and Applications*. P. Povinec and S. Usacev, Eds. (Bratislava: Slov Pedagog Nakladatelstov, 1975), pp. 169-171.
4. Allesio, M., L. Allegri, F. Bella, and S. Improta. *Nuclear Instruments and Methods*. 137:537-543.
5. Noakes, J.E. in *Liquid Scintillation Counting, Vol. 4*, M.A. Crook and P. Johnson, Ed. (London: Heyden and Sons, 1977), pp. 189-206.
6. Broda, R. and T. Radoszewski. *International Conference on Low Radioactivities* P. Povinec, Ed. (Bratislav: VEDA, 1982), pp. 189-206.
7. Jaing, H., S. Lu, S. Fu, W. Zhang, T. Zhang, Y. Ye, M. Li, P. Fu, C. Peng, and P. Jaing, in *Advances in Scintillation Counting*, S.A. McQuarrie et al., Eds. (Edmonton: University of Alberta Press, 1983), pp. 478-493.
8. Kojola, H., H. Polach, J. Nurmi, T. Oikari, and E. Soini. *Int. J. Appl. Radiat. Isotopes*. 35:949-952 (1984).
9. Oikari, T., H. Kojola, J. Nurmi, and L. Kaihola. *Int. J. Appl. Radiat. Isotopes*. 38:9 (1987).
10. Noakes, J.E. and R. Valenta. *Proceedings of the 13th International Radiocarbon Conference*, A. Long and R. Kra, Eds. (Dubrovnic: Radiocarbon 31, 1989), in press.
11. Kessler, M. *Proceedings of the International Workshiop on Inter-Comparison of 14C Laboratories*, A. Long and R. Kra, Eds. (1989), in preparation.
12. Currie, L.A. *Anal. Letters*, 4:777-784 (1971).
13. Currie, L.A. *International Conference on Radiocarbon Dating, 8th Proceeding*, T.A. Rafter and T. Grant-Taylor, Eds. (Wellington: Royal Society of New Zealand, 1973) pp. 598-611.

14. Pazdur, M.F. *Int. J. Appl. Isotopes.* 20:179–184 (1976).
15. Walanus, A. and M.F. Pazdur. *Radiocarbon.* 22, 4:1021–1027 (1980).
16. Sheppard, J.C., J.F. Hopper, and Y. Welter. *Radiocarbon*, 25,2:493–500. (1983).
17. Currie, L.A., R.W. Gerlach, G.A. Klouda, F.C. Ruegg, and G.B. Tompkins. *Proceedings of the 11th International Conference*, M. Stuiver and R, Kra, Eds. (Seattle: Radiocarbon, 24,2 1983), pp. 553–564.
18. Bzik, T.J. *Microcontamination*, May 1986, pp. 59–99.
19. Loosi, H.H., M. Heimann, and H. Oeschger. *Radiocarbon*, 22,2:461–469.
20. Otlet, R.L., G. Huxtable, G.V. Evans, D.G. Humphreys, T.D. Short, and S.J. Conchie. *Radiocarbon* 25,2:565–575.
21. Crow, E.L. and R.S. Gardener. *Biometrika*, 46:441–453 (1964).
22. Polach, H.A., L. Kaihola, S. Robertson, and H. Haas. *Radiocarbon*, 30,2:153–155 (1988).
23. Kalin, R.M., J.M. Devine, and A. Long, personal observation.
24. Tsoulfanidis, N. In *Measurement and Detection of Radiation*. (Washington, D.C.: Hemisphere Publishing Corp., 1983), p. 42.

CHAPTER 43

Liquid Scintillation Counting Performance Using Glass Vials in the Wallac 1220 Quantulus™

Lauri Kaihola

ABSTRACT

Low-potassium glass vials can obtain reduced background count rates for beta particles in liquid scintillation counting with the aid of pulse amplitude comparisons and pulse shape analysis; they also retain relatively high counting efficiencies. Adding secondary fluor to the solvent, to shorten the pulses from sample decay events, helps the effectively separate these from the slow glass fluorescences.

The power of this method in improving counting performance was examined for:

(1) ^{14}C in benzene in normal and low background environments using the Wallac low-level liquid scintillation counter, 1220 Quantulus™, with an active anticoincidence guard detector. The primary fluor was 15 mg/mL butyl-PBD, and the secondary fluor was 1 mg/mL bis-MSB,

(2) ^{3}H in aqueous solution with commercial cocktails.

In a normal environment, using Wheaton 20 mL low-^{40}K vials filled with 5 mL of benzene, the best ^{14}C figures of merit were achieved in a window of 65% counting efficiency. It gave a 0.78 cpm background count rate using low bias with the pulse shape analyzer (PSA) alone and a 0.51 cpm background count rate using the pulse amplitude comparator (PAC) in conjunction with PSA. The corresponding figures in a low background environment were 0.57 and 0.27 cpm.

Tritium background count rates were 2.8 and 2.9 cpm at 27 and 22% counting efficiency using Quickzint™ 400 and Optiphase Hisafe™, respectively, with PSA + PAC.

INTRODUCTION

Using glass vials in low-level liquid scintillation counting offers some advantages over plastic and teflon vials. The glass is inert and easy to clean (cleanliness can be checked by visual inspection). It is dimensionally stable and impermeable to aromatic and volatile solvents such as the benzene used in ^{14}C dating. For long term low-level counting, the accompanying plastic caps can be replaced with metal caps lined with indium gaskets to prevent vapor loss of the volatile sample.[1] Teflon vials were specially designed for counting ^{14}C using benzene and still reach the best performance, but they require very careful calibration and cleaning.[2] The search for an ideal vial material, in terms of low-level counting performance, is in progress; quartz might be the one.[3,4,5] However, the cost of inherently low radioactivity synthetic quartz is very high. The ability to use standard low-^{40}K glass vials would be a good choice for low-level counting if the background can be reduced to acceptable levels at a relatively high counting efficiency.

METHOD

Glass vials exhibit unquenchable and quenchable background radiation components. The unquenchable background is caused by 1.3 MeV beta particles emitted in the ^{40}K decays in the glass and by high energy particles interacting with the vial and PMTs (approximately 10 cpm in the Quantulus™; this is seen with empty glass vials as well). Further, fluorescence events are also created in the glass, mainly in the low energy region of the spectrum. These pulses are slower than the Cerenkov radiation created within the glass. This part of the background severly interferes with ^{3}H beta spectrum. The quenchable, mainly high-energy background, is caused by photons from ^{40}K decays in the vial and PMTs; the result is a Compton continuum in separable from a beta particle spectrum. Beta particles from ^{40}K decays and alpha particles (U,Th) from the inner surface of the vial reach the cocktail, contributing to the high energy background.[6]

Bias Threshold

The Quantulus has a user controlled bias threshold which can be set either low or high. The acceptance levels of the pulse amplitudes are tested for both PMT signals individually, before the coincidence summing of pulse heights.[7] At low bias, maximum ^{3}H detection efficiency is retained, while high bias effectively rejects cross talk, Cerenkov, chemiluminescence, and low energy pulses (e.g., part of the ^{40}K spectrum in glass and most of the ^{3}H spectrum). It is very effective in reducing background for ^{14}C, but it cannot be used for this purpose with low energy radiations.

Pulse Amplitude Comparator

The Wallac Quantulus electronics contain a user adjustable pulse amplitude comparator (PAC) that can to reject those events in the PMT tubes which vary more in their amplitudes than specified by their selected PAC setting.[8,9] PAC is designed to discriminate differently between low energy (^{3}H and others) and high energy (e.g., ^{14}C) pulses; the lower the energy of the ionizing event, the milder the PAC effect. The ^{14}C background cross talk events at low bias are more likely to be rejected by PAC due to their greater pulse amplitude disparity than the sample decay events. In this way one is also able to improve the ^{14}C S/N ratio without significant efficiency losses.

Pulse Shape Analysis

The pulse shape analyzer in the Quantulus is an analog device which integrates the tail of each pulse.[10] This integral is then compared with the total pulse integral to produce an amplitude independent parameter that relates to the pulse shape. Typically a true beta pulse decays exponentially in a few nanoseconds, while an alpha pulse decays non-exponentially in a few hundred nanoseconds. The application of this pulse shape parameter is user selectable and variable. Pulse shape analysis (PSA) is primarily intended for alpha/beta particle spectrum separation. It can also be used for background reduction, provided that the pulse shape of the sample and background signals differ.

Glass Vial Backround Reduction

In ^{14}C counting of benzene, butyl-PBD is widely accepted as the fluor due to its high efficiency and quench resistance in the presence of impurities in the sample solute.[2] It is, however, a quite slow fluor requiring the addition of a secondary fluor, e.g., bis-MSB, to improve pulse shape contrast between the fluorescent glass vial backround events and the fast ^{14}C sample pulses. The secondary fluor shifts the emission peak wavelength. This is essential to matching the peak transmission of the guard in the new Packard low-level counters, where it aids in achieving higher counting efficiency at lower background.[11,12,13]

The scheme was tested in an ultralow background liquid scintillation spectrometer, the Quantulus, inside and outside our low-level laboratory at Wallac. The Quantulus is fitted with a cosmic guard counter whose performance is totally independent of sample characteristics. There cannot be any beta particle energy transfer out of the actual sample spectrum into the guard. Therefore user selectable PSA enable free discrimination of long fluorescent pulses from the fast sample events in the glass, thus reducing the background without losing any efficiency through erroneous interpretation of their origin. Both the accepted and rejected pulse spectra can be analyzed and recorded to evaluate PSA and PAC performance at different settings and under different cocktail and environmental conditions.

RESULTS

Radiocarbon

Standard 20 mL low-^{40}K glass vials (Wheaton) were used for the experiments at low-bias settings. The PSA settings were scanned from 1 to 25 in steps of 5,

Table 1. ^{14}C Efficiency and Background Variations, Using 20 mL Low-^{40}K Glass Vials with a 5 mL Benzene Sample, in a Normal and Low Radiation Environment with PSA and PSA + PAC, at low bias

Mode Efficiency ^{14}C %			PSA B cpm B reduction %		PSA + PAC B cpm B reduction %	
			N	LR	N	LR
77%	B	= 1.34		1.00	0.78	0.49
	E^2/B	= 4400		5900	7600	12100
	B_{red}	= 46%		59%	72%	82%
65%	B	= 0.78		0.57	0.51	0.27
	E^2/B	= 5400		7400	8450	15650
	B_{red}	= 31%		38%	56%	67%

Note: B = background cpm
B_{red} = background reduction % in the selected window with respect to no PSA or PSA + PAC electronic discrimination
N = normal environment
LR = low radiation environment

and the PAC was scanned from 180 to 250 in steps of 10. These setting ranges have perceptible effects on the counting performance in the cocktail, 15 mg/mL butyl-PBD and 1 mg/mL bis-MSB. The sample 5 mL volume was placed in an unmasked vial. Spectroscopic grade benzene was used as the background sample. ^{14}C labeled fatty acid was used as the reference standard. Performance figures are given at optimum windows and at balance point, with the highest figure of merit (E^2/B) for the specified vial type, size, and sample volume (Table 1). Typically, a reduction of 31 to 72% in the background is observed at 65 and 77% efficiency windows when only PSA or PAC and PSA are activated, respectively, in a normal environment. The background is reduced by some 38 to 82% from the original when PAC also is activated in a low radiation environment. There is some 40% reduction in backround count rate when the instrument is taken from a normal environment into the low-level laboratory. More significantly, however, when using PAC + PSA, there is a background reduction in low-^{40}K glass vials of 82% at a counting efficiency of 77%, and a 67% reduction at 65% efficiency.

Tritium

The ^{3}H beta emission in aqueous solutions and glass fluorescence pulse amplitude spectra are very similar and overlap. The highest figure of merit is therefore achieved in the widest window. Water samples were tested as a mixture of 8 mL H$_2$O and 12 mL scintillation cocktail, Quickszint™ 400 from Zinsser and Optiphase Hisafe™ 3 from Pharmacia-Wallac. In a low-level radiation environment the background count rate in a full window was 2.8 and 2.9 cpm respectively, at 27 and 22% counting efficiencies, using PSA at maximum effect and PAC at 255. In a normal environment the increase is marginal, on the order of 0.5 cpm, leading to background count rates less than 3 cpm at efficiencies up to 26% for aqueous solutions.

In Teflon vials the best reported performance gave a background of 0.42 cpm at 27.9% with Quickszint 400.[14] In a normal environmental gamma flux the background would be below 0.8 cpm. In plastic vials the background is marginally higher (0.05 cpm). Modern biogradeable cocktails do not cause problems with Teflon and plastic vials, therefore there is no need to run low background water samples in glass vials; plastic vials are equally cheap and give better performance. Counting in Teflon and plastic vials in the Quantulus allows direct measurement of most environmental ^{3}H samples without enrichment.

DISCUSSION AND CONCLUSIONS

The use of standard low-^{40}K glass vials is of merit, and it is recommended for low-level ^{14}C determinations when maximum resolution of weak signals (close to background) is not essential. This is the case with close to Modern samples, while for very old samples, Teflon retains its merit due to very high counting efficiencies and ultralow background count rates. Application of PAC and PSA at high bias does not improve the performance of Teflon vials. In glass vials background reductions are achieved at the expense of some 10 to 20% loss in counting efficiency.

Optimum performance is not critical in all applications. Moreover, it depends on environment, vial, sample purity (lack of quench), cocktail selection, and concentration. The inherent stability of the Quantulus ensures reproducible performance under optimal counting parameter settings.

Performance in the ^{3}H energy region is affected by the residual radioactivity of the low-^{40}K glass, hence true, low background counting is not possible. Plastic and Teflon vials make use of the full power of low-level liquid scintillation spectrometers.

ACKNOWLEDGEMENTS

Hannu Kojola, Wallac Oy, Turku, and Henry Polach, ANU, Canberra, critically read this text. Their comments were appreciated.

REFERENCES

1. Pearson, G. University of Belfast (personal communication).
2. Polach, H.A., J. Gower, H. Kojola, and A. Heinonen. "An Ideal Vial and Cocktail for Low-Level Scintillation Counting," in *Proceedings, of the International Conference on Advances in Scintillation Counting* (Banff, Canada: University of Alberta Press, 1983), pp. 508-525.
3. Hogg, A., H. Polach, S. Robertson, and J. Noakes. "Applications of High Purity Synthetic Quartz Vials to Liquid Scintillation Low-Level ^{14}C Counting of Benzene," paper presented at this conference.

4. Devine, J.M., R.M. Kalin, and A. Long. "Performance of Small Quartz Vials in a Low-Level, High Resolution Liquid Scintillation Spectrometer," paper presented at this conference.

5. Haas, H. "Low Level Scintillation Counting with a Wallac Quantulus: Established Optimal Parameter Settings," paper presented at this conference.

6. Kaihola, L. and T. Oikari. "Some Factors Affecting Alpha Particle Detection in Liquid Scintillation Spectrometry," paper presented at this conference.

7. Polach, H.A., J. Nurmi, H. Kojola, and E. Soini. "Electronic Optimization of Scintillation Counters for Detection of Low-Level ^{3}H and ^{14}C,"in *Proceedings, of the International Conference on Advances in Scintillation Counting* (Banff, Canada: Univesity of Alberta Press, 1983), pp. 420–441.

8. Laney, B.H. "Electronic Rejection of Optical Crosstalk in a Twin Phototube Scintillation Counter," in *Organic Scintillator and Liquid Scintillation Counting*, (New York: Academic Press, Inc., 1971), pp. 991–1003.

9. Soini, E. "Rejection of Optical Cross-Talk in Photomultiplier Tubes in Liquid Scintillation Counters," *Wallac Report*, (Turku, Finland: Wallac Oy, 1975), p. 9.

10. Oikari, T., H. Kojola, J. Nurmi, and L. Kaihola. "Simultaneous Counting of Low Alpha-and Beta-Particle Activities with Liquid-Scintillation Spectrometry and Pulse-Shape Analysis," *Appl. Radiat. Isot.* 38(10):875–878 (1987).

11. Cook, G.T., D.D. Harkness, and R. Anderson. "Performance of the Packard 2000 CA/LL and 2250 CA/XL Liquid Scintillation Counters for ^{14}C Dating," paper presented at the 13th International Conference on Radiocarbon Dating, Dubrovnik, Yugoslavia, June 20–25, 1988.

12. Harkness, D.D. and B. Miller. "Recent Developments in LS Counting: Too far, too fast," paper presented at the International Workshop on Inter-Comparison of ^{14}C Laboratories, East Kilbride, Scotland, September 12–15, 1989.

13. Polach, H., G. Calf, D. Harkness, A. Hogg, L. Kaihola, and S. Robertson. "Performance of New Technology Liquid Scintillation Counters for ^{14}C Dating," *Nuc. Geophysics*, 2(2):75–79 (1988).

14. Schönhofer, F. and E. Henrich. "Trace Analysis of Radionuclides by Liquid Scintillation Counting," Report UBA-STS-85-02, Vienna, 1985, 25 pp.

CHAPTER 44

Time-Resolved Liquid Scintillation Counting

Norbert Roessler, Robert J. Valenta, and Staf van Cauter

ABSTRACT

A comparison is made between standard, two-tube coincidence liquid scintillation counting and the newly developed technique of time resolved-LSC (TR-LSC). In conventional LSC, coincidence requirements are fulfilled only by the fluorescence from excited primary singlets. Any delayed component that results from triplet-triplet annihilation is ignored by the circuit. TR-LSC, however, makes use of slow decaying pulse components. Certain scintillators, such as calcium fluoride, and scintillating glasses have much longer fluorescence decay times. Others are characterized by a prompt scintillation pulse followed by after pulses. These different pulse characteristics are analyzed by TR-LSC and used as discrimination criteria to validate scintillation counts.

TR-LSC recognizes and discriminates background in low-level LS counting. It uses the long pulse duration due to the interaction of cosmic rays with the glass of the photomultiplier tubes, the scintillation vials,and any other material surrounding the sample. This results in a two to fourfold improvement of the figure of merit (E^2/B) as compared to conventional LSC. The issues of cocktail composition and quench correction are addressed.

TR-LSC can also be used in dedicated liquid scintillation counters. The long pulse duration is used as a criterion to accept counts while discriminating pulses from thermionic emission. This allows single photomultiplier detection with efficiencies and backgrounds comparable to two-tube coincidence counting. The effect of cocktail and sample composition on performance will be correlated with lifetime data.

INTRODUCTION

Two-tube coincidence counting has remained the state of the art in liquid scintillation counter design. Instruments and cocktails were optimized for this configuration because it gives the best overall counting performance. Cocktails using scintillators with short lifetimes allow for maximum tube noise reduction using coincidence circuits with short coincidence resolving times. High efficiency is maintained because short lifetime scintillators release all of their energy within a few nanoseconds and give the maximum possible pulse height at the photomultiplier anode. When cocktails containing long lifetime scintillators are used, the photons emitted during the tail end of the scintillation pulse arrive too late and are ignored by the coincidence counting circuit.

This leads to an appreciable loss in efficiency. In specialized applications, such as radioactivity flow monitors based on heterogeneous counting, solid scintillators (e.g., calcium fluoride) and doped glasses are used. These have much longer luminescence decay times than dissolved organic fluors. Consequently, coincidence resolving times in the microsecond range must be used to obtain an acceptable efficiency. This in turn increases the background due to tube noise significantly. Time resolved LSC[1] makes use of the fact that scintillators do not give off their light energy instantaneously.

TIME RESOLVED COUNTING

The scintillation pulse originating from a scintillator is a burst of photons lasting from a few nanoseconds to several hundred microseconds. In Figure 1 the photoluminescence decay curve of a typical cocktail is shown in schematic form. This is not the pulse shape of a single decay, rather it is a function describing the probability of a photon being emitted as a result of that decay. Immediately after the decay, we have the prompt pulse or fast component. In liquids this time lasts from 2 to 8 nsec and represents the fact that most scintillation energy is emitted as direct fluorescence from excited singlet states of the secondary scintillator. The delayed pulse or slow component is attributed to the delayed fluorescence emission from a process called triplet-triplet annihilation. In this process, two scintillator molecules in the electronically excited triplet state collide to form one excited singlet state from which fluorescence occurs.

In Figure 2, the average pulse shape due to a high energy decay is shown for a conventional scintillation cocktail and a solid scintillating particle. The relative amount of light originating from prompt singlet emission and the delayed

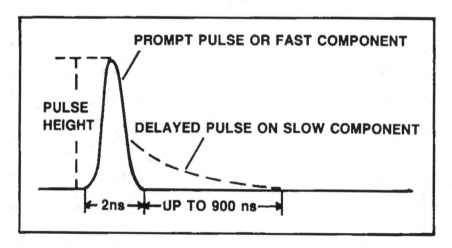

Figure 1. Typical cocktail photoluminescence decay curve.

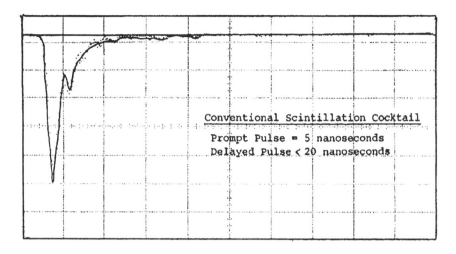

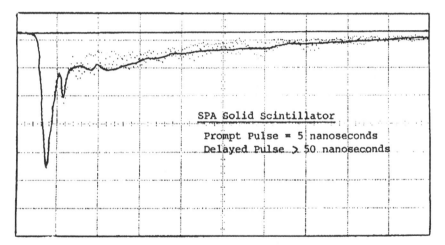

Figure 2. Averaged pulse shape due to high energy decay: conventional cocktail (top), solid scintillator.

component depends on the specific ionization of the sample. Since beta parti-cles, cosmic rays, and Compton electrons have lower specific ionization than alpha particles, the concentration of triplets formed in the track is lower. As a result, triplet-triplet annihilation is less likely to occur and the delayed compo-nent is diminished. This is the basis for alpha particle discrimination using pulse height analysis.

Figure 3 is a schematic which emphasizes the fact that a typical afterpulse pattern consists of individual photoelectrons. The prompt pulse is usually much larger than the afterpulses since it contains a large number of photons emitted in a time too short for the dynode circuit of the photomultiplier to resolve. The photons causing afterpulses, due to the slow component, are

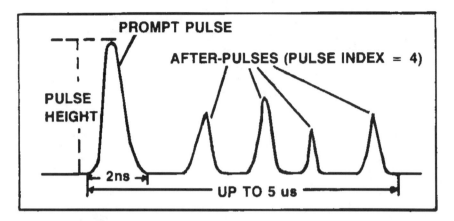

Figure 3. Typical pulse pattern due to beta decay.

fewer in number and spread out over time so as to be distinguishable as single photoelectrons.

The triplet concentration in the track does not depend solely on the specific ionization; however, it also depends on the composition of the cocktail. The concentration of oxygen in a cocktail exposed to air, for instance, is high enough to scavenge most triplets and practically eliminate afterpulses due to the delayed component. In deoxygenated solutions, triplets survive to recombine and cause afterpulsing.

Solid scintillators, such as glass, exhibit significant afterpulsing for a different reason. Oxygen quenching and other diffusional mechanisms do not occur in solids so the triplet states formed from the nuclear decay are not quenched, but they can emit light with their characteristically long lifetime. This results in afterpulsing due to an "unquenchable" delayed component.

LOW-LEVEL COUNTING

Time resolved techniques can be used in low-level counting to recognize background pulses from the natural radioactivity of the glass vial and the envelope of the PMT. The most likely number of afterpulses for a given energy is greater for glass than for liquid scintillation cocktails; thus, when a cosmic particle passes through the system, it can be discriminated against by counting the number of afterpulses—the burst counting technique. In this technique, each coincidence opens a burst counting window which counts the number of afterpulses occurring to about five microseconds after the event. The total number of afterpulses is defined as the pulse index.

Using the pulse index it is possible to create a 3-D spectral plot containing time resolved information on the delayed component (see Figure 4). We see that a background sample gives of an appreciable number of afterpulses at the low energy end of the spectrum. An unquenched ^3H sample (see Figure 5) gives

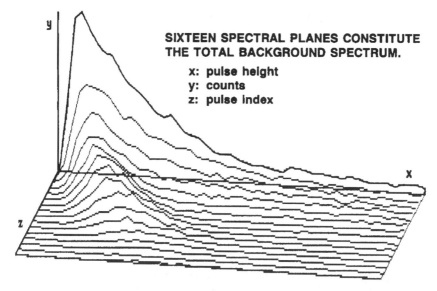

Figure 4. Three-dimensional plot of pulse height spectrum of background sample. The pulse index is the third dimension.

off few afterpulses and only at the high energy end of the spectrum. An air quenched ^{3}H sample (see Figure 6) gives off almost no afterpulses.

By accepting only counts with a low pulse index, a spectrum free from glass

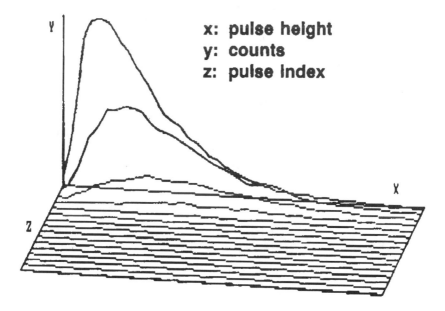

Figure 5. Three-dimensional spectrum of an unquenched ^{3}H sample.

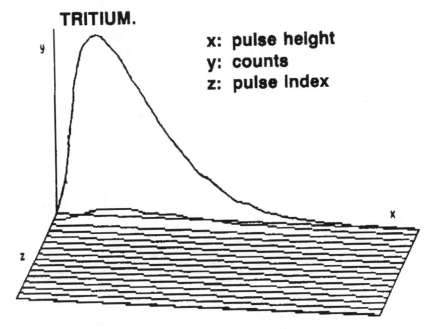

Figure 6. Three-dimensional spectrum of an air quenched ^{3}H sample.

scintillation counts is obtained. This background reduction comes at the price of a reduction in efficiency, since there is a significant probability that sample events will also result in a pulse index, leading to rejection of the count. This is especially true for deoxygenated samples and long lifetime scintillators which have a significant delayed component. Nevertheless, the effect of using the pulse index on low-level performance can be dramatic (see Table 1) because a large part of the background is due to glass scintillations. When air quenched samples are analyzed, the figure of merit (E^2/B) can be improved significantly, because the reduction of efficiency due to the time resolved circuit is overshadowed by the background reduction.

The data in Table 1 contains the results of a ^{14}C benzene sample prepared with a benzene synthesizer. A small glass vial was used to analyze 3.5 mL sample volume. On optimization, the counting region for this sample was

Table 1. Effect of Pulse Index Discrimination on Background, Efficiency, and E^2/B for ^{14}C-Benzene

Degree of Pulse Index Discrimination	^{14}C Efficieny (%)	Background (CPM)	E^2/B (Figure of Merit)
None	83.45	9.67	720.15
Normal	81.87	7.07	948.05
High sensitivity	78.50	4.74	1300.05
Low level	70.70	1.38	3560.00

Note: 3.5 mL Benzene with 4 g/L in small glass vial, O_2 quenched, 10 to 100 keV.

Table 2. Effect of Pulse Index Discrimination on Background, Efficiency and E^2/B for ^3H

Degree of Pulse Index Discrimination	^3H Efficieny (%)	Background (CPM)	E^2/B (Figure of Merit)
None	26.50	18.45	38.06
Normal	26.24	12.75	54.08
High sensitivity	24.68	9.25	65.85
Low level	22.59	3.33	153.25

Note: 10 mL InstaGel and 10 mL H_2O in large glass vial, 0.5 to 5.0 keV, O_2 quenched.

found to be 10 to 102 keV. As pulse discrimination is applied, the efficiency drops from 84% to 70%, but the background drops from 9.67 to 1.38 counts. This results in an almost fivefold increase in figure of merit.

Similar results were obtained on a tritiated water sample (see Table 2). The sample consisted of 10 mL tritiated water mixed with 10 mL of Insta-Gel in a large glass vial. The optimal counting region was determined to be 0.5 to 5 keV. While the efficiency is reduced only 15%, the background is reduced to 18% of its original value. This results in a fourfold increase in the figure of merit.

ANTI-COINCIDENCE GUARD SHIELDING[2]

The time resolved technique can be used to implement anticoincidence guard shielding in a low-level counter. For this application the normal reflector is replaced by a guard shield that has a scintillating plastic with a long lifetime. For small volume samples in small glass vials, a special adapter fitting into large vial cassettes can fulfill the same function. When cosmic or environmental radiation excite the slow fluor in the plastic, they produce afterpulses that lead to a high pulse index and a rejection of the count.

In Table 3, data showing the effect of using various degrees of pulse rejection are displayed for a ^{14}C benzene sample in a small vial. The scintillator used was 6 g/L PPO and 0.2 g/L POPOP. We see that the time resolved counting coupled with the slow fluor coincidence shielding raises the figure of merit from 1167 to 9520. The data for tritium (see Table 4) also shows an improvement, from 272 to 1745. Here, however, the combination of vial holder with guard shield does not give a significant advantage over either used alone.

Table 3. ^{14}C Benzene in PPO (6 g/L) and POPOP (0.2 g/L)

Degree of Pulse Index Discrimination	Sample	^3C Efficiency (%)	Background (CPM)	E^2/B (Figure of Merit)
None	Sample only	64.99	3.62	1167
Maximum	Sample only	54.05	0.76	3844
Maximum	+ vial holder	60.99	0.63	5904
Maximum	+ guard elevator	66.20	0.51	8593
Maximum	+ vial holder & guard detector	63.98	0.43	9520

Table 4. ^{3}H Benzene in PPO (6 g/L) and POPOP (0.2 g/L)

Degree of Pulse Index Discrimination	Sample	^{3}H Efficiency (%)	Background (CPM)	E^2/B (Figure of Merit)
None	Sample only	54.71	10.99	272
Maximum	Sample only	50.94	3.05	851
Maximum	+ vial holder	49.38	1.37	1778
Maximum	+ guard holder	38.44	0.86	1718
Maximum	+ vial holder & guard detector	37.13	0.79	1745

Table 5 contains results from a study of large volume tritiated water samples in Insta-Gel. The use of the guard detector significantly improves the figure of merit for counting environmental water samples.

SINGLE TUBE COUNTING[3]

Time resolved LSC can also be applied to single tube counting of high activity samples. In this case, the pulse index alone is used as a time resolved coincidence criterion. Only those pulses followed by one or more afterpulses are accepted. The physical basis for the operation of the circuit is shown in Figure 7. We see that the typical pulse due to PMT noise has a short width, determined exclusively by the characteristics of the dynode string. The Ultima Gold pulse, on the other hand, has a more complex shape even with a distinguishable afterpulse.

A burst count circuit was modified to test this concept. In two pulse-mode, the circuit counts those events which result in two distinguishable pulses within the coincidence resolving time. In three-pulse mode, a triple coincidence is required.

The data in Figure 8 was obtained with this modified burst count circuit. The fact that efficiency varies with the cocktail used can be attributed to variations in the proportion of the delayed component. The lower efficiency in the three-pulse mode is due to the fact that the emission has decayed by the time the circuit has recovered from the first and second pulse. The samples were also counted in a regular 2000CA to show the performance of the cocktails. Because the geometry of the breadboard is different from the production

Table 5. ^{3}H Water Analysis in Large Glass Vial (8 mL sample, 12 mL Pico-Fluor LLT)

Degree of Pulse Index Discrimination	Sample	^{3}H Efficiency (%)	Background (CPM)	E^2/B (Figure of Merit)
None	Sample only	22.85	6.12	5460
Minimum	Sample only	25.39	6.63	6223
Maximum	Sample only	25.63	3.87	10863
Maximum	+ guard holder	23.77	2.29	15791

Note: 8 mL sample, 12 mL Pico-Fluor LLT.

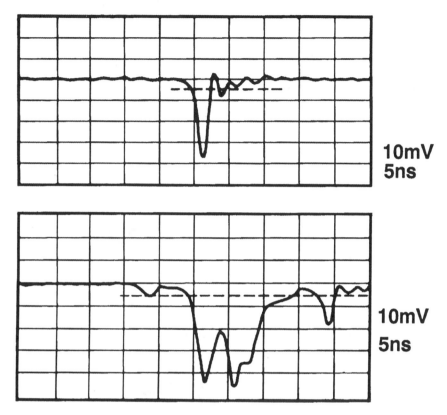

10mV
5ns

10mV
5ns

Figure 7.   Oscilloscope traces of typical pulses: top) PMT dark noise, bottom) ^{3}H in Ultima Gold.

counter, it is difficult to assess the reason for the lower efficiency of the single tube circuit.

CONCLUSION

The results presented above demonstrate that the addition of digital time resolved techniques to liquid scintillation counting has already resulted in technical innovations that advance the state of the art. Further research into this area, combined with a better understanding of the mechanisms involved in creating the delayed component of the scintillation pulse in solid and liquid phase detectors, promises to widen the field of applications for soft beta counting.

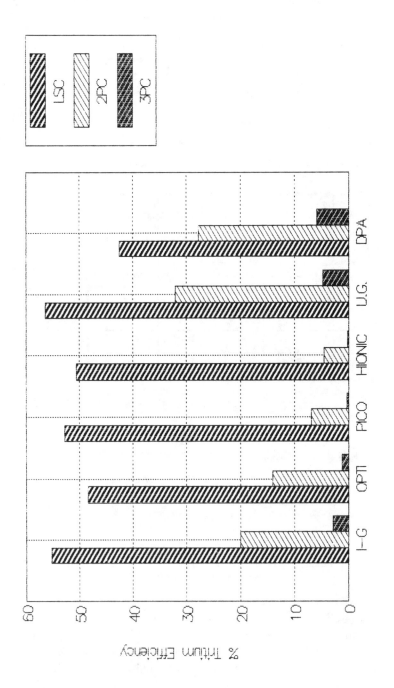

Figure 8. Tritium efficiency for different cocktail formulations. LSC: liquid scintillation counter, 2PC: single tube counter with time resolved double coincidence, 3PC: single tube counter with time resolved triple coincidence.

REFERENCES

1. Valenta, Robert J. "Reduced Background Scintillation Counting," U.S. Patent 4,651,006 March 17, 1987.
2. Valenta, Robert J. and John E. Noakes. "Improved Liquid Scintillation System with Active Guard Shield," U.S. Patent 4,833,326 S/N 07/167,407, March 14, 1988.
3. Valenta, Robert J. "Method and Apparatus for Measuring Radioactive Decay," U.S. Patent 4,528,450, July 9, 1985.

CHAPTER 45

Comparison of Various Anticoincident Shields in Liquid Scintillation Counters

Jiang Han-ying, Lu Shao-wan, Zhang Ting-kui, Zhang Wen-xin, and Wang Shu-xian

ABSTRACT

In low level liquid scintillation counters, most of the background can be eliminated by an anticoincident detector. NaI(Tl) crystal is well-known as an effective shielding detector in gamma ray spectrometers or in liquid scintillation counters. The plastic scintillator and liquid scintillator have also been used in commercial instruments, but all of these detectors are large and expensive. Four types of anticoincident detector have been tested in our laboratory.

As a result of the experiments, G-M tubes, made of stainless steel or glass, are very effective in reducing the background from cosmic rays, in particular for ^{14}C. Background in ^{14}C measuring channel is lower than 1 cpm, with the efficiency over 75% for 5 mL benzene samples. So, G-M tubes provide a cheap, compact, and effective anticoincident shield.

Bismuth germanate(BGO) crystal, in pieces as an anticoincident detector, is only about 2 kg. The effect is as well as NaI(Tl) crystal which weighs about 14 kg. Furthermore, different from NaI(Tl), BGO is not hygroscopic and damagable.

The effect of the plastic scintillator as an anticoincident detector is between the effect of solid crystals and G-M tubes.

The background spectrum for all four types of detector and the analysis are given and analyzed.

INTRODUCTION

As is known, an anticoincident shielding can effectively decrease the background of the main detector. J. E. Noakes[1] first introduced NaI(Tl) crystal as an anticoincident shielding detector (AD) to the low background liquid scintillation counter. Then plastic scintillator and liquid scintillation solution were introduced to the liquid scintillation counter and good results have been obtained. Since the '80s, counting tubes and multiwire proportional counters have been tested,[2-4] but people have not got the anticipated results. Therefore low background liquid scintillation counters are always associated with being large, heavy, and expensive.

A lot of experiments about various AD for liquid scintillation counters have been completed in our laboratory during the past few years. They are NaI(Tl) crystal, plastic scintillator, bismuth germanate(BGO), and common or special G-M counter. From the results, we can see that even when the common G-M counter is used as an AD, better results can be obtained. G-M counting tubes provide the cheapest AD. The γ ray efficiency of BGO crystal is higher than NaI(Tl) crystal, so a better result can be expected.

INSTRUMENT AND METHOD

Installations and Instruments of Liquid Scintillation Counters with AD

1. First, thirty-two MC-6 G-M counting tubes surrounding the main detector are arranged as a rectangle (Figure 1a). The matter shielding is 5-cm-thick lead. Second, an annular multiwire G-M halogen counting tube (Figure 1b) is specially designed (technology and manufacture completed by Beijing Nuclear Instrument Factory). Different from common commercial counting tubes, the shell is made of stainless steel. Because the product rate is too low and the cost is too much, it will not be used in commercial, low background liquid scintillation counters. The matter shielding is 5- or 10-cm-thick lead. Third, a kind of stainless steel cylindrical G-M counting tube is manufactured with shape and function similar to the common G-M γ counting tube. Fifteen counting tubes are arranged between two metal cylinders. The main detector is surrounded by the inner cylinder; the outer cylinder is surrounded by 10-cm-thick lead (Figure 1c). Fourth, the main detector is in 5-cm-thick lead annular and encircled by twenty-six common commercial glass counting tubes. Another 5-cm-thick lead annular encircles it (Figure 1d). The glass is replaced by stainless steel in order to study the effect of ^{40}K in glass on the background.
2. The anticoincident annular of plastic scintillator consists of seven detectors. Each of them includes a ϕ 76 $\times$ 150 mm plastic scintillator cylinder and a GDB-52 PMT. The distance between the center of each detector and the main detector is 105 mm. The seven detectors are arranged in various types as shown in Figure 2. The anticoincident effect of various arrangements are studied. In some experiments, only a few of the seven plastic scintillator detectors are operated.
3. The NaI(Tl) crystal annular (200 mm OD, 80 mm I.D., and 150 mm in length) is not purified for potassium. Six GDB-52 PMTs are used for collecting the photons from the NaI(Tl) crystal. This NaI(Tl) crystal is used in a DYS-1[5] low-level liquid scintillation counter.
4. The BGO AD consists of several pieces of BGO and a pair of GDB-52L PMTs. The weight of the crystals is about 2 kg. A 100 $\times$ 100 $\times$ 100 mm cubic case is made to contain these crystals. There is a ϕ 60 mm hollow in the middle of the case for placing the main detector. Please note that 2, 3, and 4 all have a 10-cm-thick lead shielding.

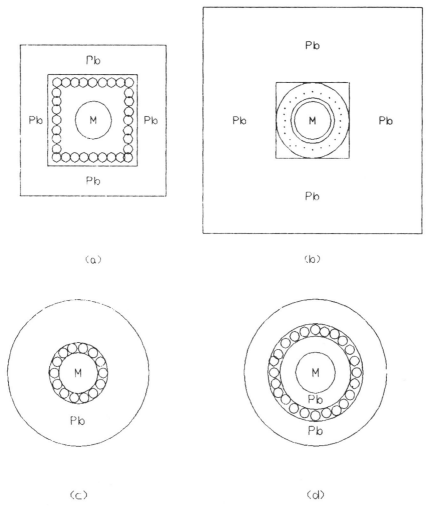

Figure 1. Four types of counting tube AD.

Electronics

Most of the electronics in these experiments are the same as the DYS-1. Some experiments are completed with a model DYS-84II automatic, low level liquid scintillation counter which has a capacity of twenty-four samples, is controlled by a Zijin II microcomputer, and has a software multichannel analyzer.

Standard Sample and Blank Sample

All of these samples are contained in quartz vials. The OD of the vial is about 27.5 mm. Different height vials are made for different volumes of samples. The

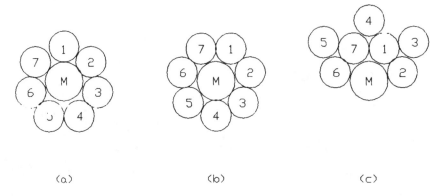

Figure 2. The arrangement of plastic scintillator detectors.

5 mL ^{14}C sample contains 5 mL benzene and 30 mg B-PBD. The 10 mL and 20 mL tritium water samples contain 60% emulsion scintillation solution (7 g PPO + 0.5 g POPOP/l toluene:triton X – 100 = 2:1) and 40% water. 20 mL tritium standard contains 20 mL benzene and 0.1 g PPO + 0.001 g POPOP. Standard samples are made up by standard solution. Blank samples are free from the radioactivity.

Main detector

The main detector consists of two EMI 9635QB PMTs and a type DYS sample chamber which can hold a 5 to 20 mL sample vial. The PMT is selected from many PMTs. Unfortunately, all of these experiments do not use the same pair of PMTs. E. Experimental condition

Experimental Condition

Counts per minute in total energy channel were measured for 20 mL benzene sample, in the operation points for 20 mL ^{3}H benzene sample, 10 mL tritium water sample, and 5 mL ^{14}C sample. When the AD operates, the counting rate of blank samples is abbreviated as AB. If the AD does not operate, the counting rate of blank samples is abbreviated as CB. The energy spectra of 20 mL blank and standard samples are measured. When there is no sample in the sample chamber, the measured cpm is named as EAC and ECC separately, corresponding to operation or non operation of AD.

RESULTS

Table 1 shows the results from G-M counting tube as AD in various types and conditions. It is shown that whatever the G-M counting tube is, commercial or specially made, all of them can effectively decrease the background, particularly for the ^{14}C channel. For 5 mL ^{14}C sample, when the efficiency is

Table 1. The Effect of AD of Counting Tubes

1a

The type of AD	Operating State						A.E. (1)
	No			Yes			
	E (%)	B (cpm)	E^2/B	E (%)	B (cpm)	E^2/B	
Annular tube	77.0	3.81 ± 0.06	1560	76.8	1.32 ± 0.03	4470	65%
32 MC-6 tube	81.9	4.15 ± 0.16	1616	81.1	2.02 ± 0.04	3260	51%

Note: 5 cm-thick lead

1b

Type	5 mL ^{14}C Operating State					
	No			Yes		
	E (%)	B (cpm)	E^2/B	E (%)	B (cpm)	E^2/B
Figure 1(d)	76.5	3.42 ± 0.18	1711	76.5	0.88 ± 0.09	6650
Figure 1(c)	76.7	2.91 ± 0.05	2022	76.7	0.73 ± 0.03	8059

1c

Sample		Operating State						A.E. (1)
		No			Yes			
		E (%)	B (cpm)	E^2/B	E (%)	B (cpm)	E^2/B	
20 mL	^{3}H	54.3	4.70 ± 0.07	600	53.7	2.59 ± 0.05	1100	45%
	^{14}C	79.3	6.10 ± 0.09	1030	79.3	2.56 ± 0.04	2450	58%
	total		24.89 ± 0.17			7.83 ± 0.10		68%
10 mL	^{3}H$_2$0	33.5	3.27 ± 0.05	340	33.4	1.97 ± 0.04	560	40%
5 mL	^{14}C	77.6	3.28 ± 0.05	1830	77.5	0.83 ± 0.03	7250	75%

Note: (1).A.E.: the efficiency of anticoincident.

more than 75%, the background can decrease below 1 cpm. In the ^{14}C channel, although 5cm-thick lead is used to absorb ^{40}K, which comes from the shell of counting tubes, the background is still a bit higher, but it is clear that in general, the result is acceptable. The anticoincident efficiency for tritium is approximately 40%.

Table 2 shows the results from the plastic scintillator AD in different conditions. When all seven detectors are operated (Table 2[A1]), the anticoincident effect is the best. For 20 mL benzene sample, the anticoincident efficiency for blank samples in the total energy channel is about 75%. For empty chambers it is about 87%. The EAC is 2.08 ± 0.03 cpm and the AB is 5.99 ± 0.07 cpm. When only four detectors are operated (Table 2 [B1]), the EAC is 4.18 cpm and the AB is 8.70 cpm. If the same number of detectors operate and there is a gap in the AD ring (Table 2 [A2 and B2, A3 and B3]), with one is on the top of the main detector, and the other below it, the results illustrate that there is no obvious difference between the two conditions. It is proved that the contribution of cosmic radiation is isotroic; the larger the gap, the higher the background.

The results of NaI(Tl) and BGO crystal detectors are shown in Table 3. In

Table 2. The Comparison of Resuts in Various Arrangements of 7 Plastic Scintillator Detectors

Figure 3		AD Operation (Yes or No)		Background (cpm)	cpm of Empty Chamber
A	1	1,2,3, 4,5,6,	Yes	5.99 ± 0.07	2.08 ± 0.03
		7	No	23.76 ± 0.14	15.06 ± 0.08
	2	2,3,4 5,6,7	Yes	7.32 ± 0.05	2.90 ± 0.03
			No	23.75 ± 0.09	14.81 ± 0.07
	3	1,2,3 6,7	Yes	8.35 ± 0.07	3.96 ± 0.04
			No	23.64 ± 0.11	14.98 ± 0.08
B	1	1,2 6,7	Yes	8.70 ± 0.06	4.18 ± 0.1
			No	23.78 ± 0.10	14.87 ± 0.09
	2	1,2,3 5,6,7	Yes	7.16 ± 0.05	2.92 ± 0.05
			No	23.72 ± 0.10	14.82 ± 0.12
	3	2,3,4 5,6	Yes	8.50 ± 0.07	3.81 ± 0.04
			No	23.92 ± 0.11	15.0 ± 0.07
C		1,2,3 4,5,6	Yes	7.52 ± 0.06	3.08 ± 0.05
		7	No	23.82 ± 0.10	14.78 ± 0.10

order to make a comparison, the main experiment results of G-M counting tube and plastic scintillator are also shown in Table 3.

Figure 3 shows all spectra of four anticoincident liquid scintillation counters. There are spectra of anticoincident and coincident background, spectra of empty chamber, and spectra of tritium and ^{14}C.

Table 3. Comparison of Four Types of ADs

		B Total (cpm)	20 mL ^{3}H			10 mL ^{3}H$_2$O			5 mL ^{14}C		
			E (%)	B (cpm)	E^2/B	E (%)	B (cpm)	E^2/B	E (%)	B (cpm)	E^2/B
A	Y	6.31	53.4	2.16	1320	29.7	1.23	720	80.0	0.54	11852
	N	31.56	53.8	6.03	480	30.0	3.14	290	80.1	4.09	1569
	AE	80%		64%			61%			87%	
B	Y	5.92	50.6	1.83	1400	28.5	1.07	760	74.5	0.40	13870
	N	23.66	51.4	4.15	640	29.4	2.38	360	74.8	2.98	1880
	AE	75%		56%			55%			87%	
C	Y	6.01	48.2	1.73	1343	28.5	1.31	626	76.2	0.55	10557
	N	24.17	48.4	4.13	567	28.7	2.46	335	76.4	2.43	2402
	AE	75%		58%			47%			77%	
D	Y	6.95	52.8	2.32	1200	34.1	1.77	655	74.6	0.74	7520
	N	23.96	52.9	4.04	690	34.2	2.84	410	74.6	3.39	1641
	AE	71%		42%			37%			78%	

Note: A:NaI(T1); B:BGO; C:Plastic scintillator; D:G-M counting tube
Y:operating AD; N:not operating AD;
AE:anticoincident efficiency.

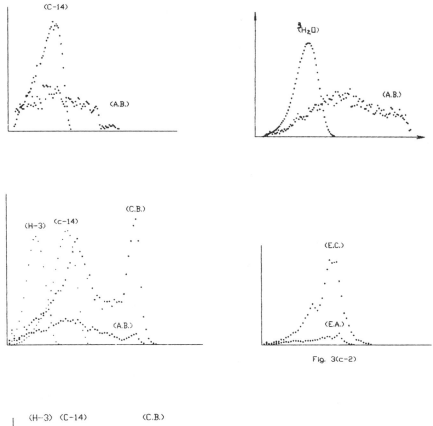

Figure 3.   (a). The spectra of Table 3a. (b). The spectra of Table 3b. (c-1). The spectra of Table 3c. (c-2). The spectra of Table 3c. (d). The spectra of Table 3d.

DISCUSSION

From Table 3, except for NaI(Tl) crystal, the other three types of anticoincident detectors all have approximately the same CB, about 24 cpm in the total energy channel. That is to say, these ADs, as matter shielding, not evidently decrease background. For NaI(Tl) crystal, this background is more than 30 cpm. The reason is explained elsewhere.[6] Except for G-M counting tubes, the other types of anticoincident detectors all have almost the same anticoincident background in total energy channel, about 6 cpm. For G-M counting tubes, this background is 7 cpm or so, slightly more than others. From the background spectra, it is shown that μ meson peak is basically elimited. Although the γ efficiency is very low, the main detector is not very sensitive for γ-rays, and the heavy shielding absorbs most environmental radiation. This is the reason why the G-M counting tube can get better effects. The requirement for electronics is very simple and the cost is less than others. In particular, it can be made longer without the price rising too much. So if large vials are used it can provide the cheapest large anticoincident shielding. We think if the longer G-M counting tube is combined with the BGO, perhaps the best result can be obtained.

The plastic scintillator anticoincident detector gives a better result, particularly for measuring tritium. The advantage is clear when comparing with a G-M counting tube, but it is similar to NaI(Tl) crystal, in that more PMTs and complex electronics are necessary and the volume is large. Therefore the matter shielding is larger than others.

The NaI(Tl) and BGO crystals all have high absorption for γ-ray and good data are attained if using one of them as an anticoincident detector. The BGO crystal can more effectively absorb γ-ray and cosmic radiation than NaI(Tl), and it is not hygroscopic and damageable. It is expensive, but only 2 kg are used. The real cost is largely lower than NaI(Tl) because NaI(Tl) crystal weighs 14.5 kg; therefore, a more compact and effective AD can be made of BGO. The disadvantage of BGO is the high cost, therefore large crystals are too expensive to use.

REFERENCE

1. Noakes, J.E., M.P. Neary, and J.D. Spaulding. "Tritium Measurements with a New Liquid Scintillation Counter," *Nucl. Instruments Methods*, 109:177 (1973).
2. Iwakura, T. et al. "Behaviour of Tritium in the Environment," IAEA-sm-232:63 (1979).
3. Rajamae, R., et al. *Radiocarbon*, 22(2):435 (1980).
4. 1220 Quantulus A multi-parameter low level LSC system from LKB-Wallac.
5. Jiang Hanying et al. "Model DYS Low-Level Liquid Scintillation Counter," in *Advances in Scintillation Counting*, (Edmonton, Canada: University of Alberta, Edmonton, 1983), pp.478–493.
6. Jiang Hanying and Lu Shao-wan. "The Effect of Altitute on the Background of Low-Level Liquid Scintillation Counter," in *An International Conference on New Trends in Liquid Scintillation Counting and Organic Scintillators* (1989).

CHAPTER 46

The Effect of Altitude on the Background of Low-Level Liquid Scintillation Counter

Jiang Han-ying and Lu Shao-wan

ABSTRACT

The background counting rate increases more than 100% in Xining over Beijing for the same low-level liquid scintillation counter with NaI(Tl) anticoincident shield. Another experimental counter with bismuth germanate(BGO) anticoincident shield has been transported to Xining, and a series of experiments have been performed. The source of the increasing background is analyzed and the methods of decreasing the background are provided.

The altitude of Xining is 2000 meters higher than Beijing, so the earth surface intensity and the constituent of cosmic rays is different in the two cities. Based on our experiments, it seems that the increasing background rate in Xining is not only due to the increased cosmic ray intensity, but also due to the high atomic number of shielding material, e.g., NaI(Tl), BGO, and lead. It is possible that most of the increasing background comes from the interaction between cosmic rays and shielding material. Appropriate selection of shielding material can weaken the effect of the interaction and decrease the background.

INTRODUCTION

The first low-background liquid scintillation counter with an anticoincident shielding was developed by J.E. Noakes.[1] Following a growing application and requirement of the radioisotopes, such as ^{14}C and tritium, in the fields of geohydrology, geography, geology, archaeology, and others, many types of low-level liquid scintillation counters[2] with various anticoincident shielding have been rapidly developed. Most of them are used at elevations in the hundreds of meters. It is interesting to find what will happen when an instrument is operated at an elevation of more than one thousand meters. In 1985, a model DYS-3 low-background liquid scintillation counter with a NaI(Tl) crystal as an anticoincident shielding was transported to the Institute of Qinghai Salt-Lake. We found that the background in Xining increased more than twofold. We completed a series of experiments to find out how environmental radiation affects background. From the experimental results, it is thought that the increasing background in Xining is mainly due to the interaction between cosmic rays and shielding materials (lead and NaI(Tl)). If the shielding material is appropriately readjusted, the background can be obviously decreased.

INSTRUMENTS

The Model DYS-3 automatic low-level liquid scintillation counter was made at the Institute of Biophysics, Academia Sinica, Beijing. A sketch of the detector is shown in Figure 1. An annular NaI(Tl) crystal(< 200 mm OD, 80 mm ID and 150 mm in length) with a ϕ 50 mm is used as an anticoincident detector. The sample is elevated from the changing machinery to the sample chamber through the hollow. The matter shielding is 10-cm-thick lead. The changing of sample, data acquisition, and processing are automatically completed by an APPLE II microcomputer.

An experimental installation with bismuth germanate(BGO) crystal is the anticoincident shielding. Several pieces of BGO are in a $100 \times 100 \times 100$ mm cubic case. A ϕ 60 mm hollow in the middle of the cuboid is for placing the sample chamber and PMTs and a ϕ 30 mm hollow of the case over the sample

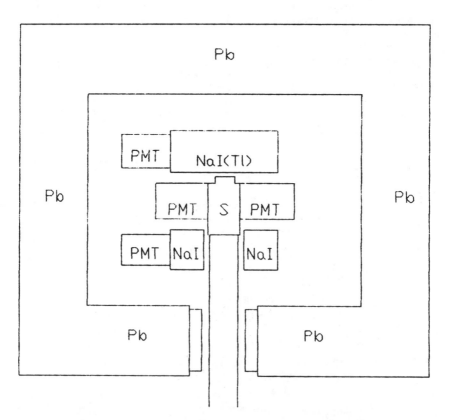

Figure 1. DYS-3 detector made at the Institute of Biophysics, Academia Sinica.

chamber is for changing the sample. The weight of the BGO crystal is about 2 kg. Two windows (ϕ 50 mm) on the surface of the cuboid allow for two PMTs to collect the photons from the BGO (Figure 2). The matter shielding is 10-cm-thick lead.

An experimental installation with plastic scintillator is an anticoincident shielding. Most of it is the same as the DYS-1,[3] except for the anticoincident shielding. It consists of seven detectors in annular. Each one has a ϕ 76×150 mm cylindrical plastic scintillator and a PMT.

EXPERIMENT AND RESULT

The blank sample is 20 mL benzene scintillation cocktails (5 g PPO + 0.05 g POPOP/l benzene) in a 20 mL quartz vial. The vial is sealed by 914 adhesive agent. All of the counters are operated at high voltage for tritium measuring. The background counts per minute are measured in total energy channel. If

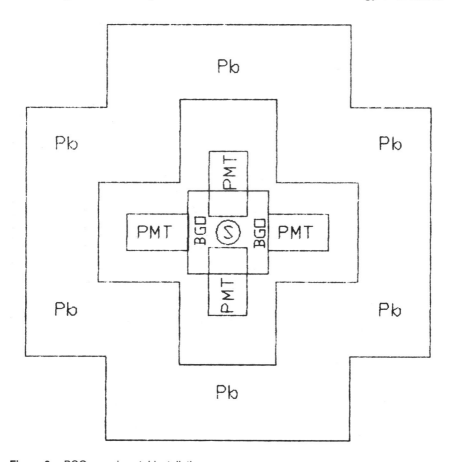

Figure 2.   BGO experimental installation.

Table 1. The Comparison of the Background in Beijing and Xining

| | DYS-3 | | | | BGO Installation | | | |
	CB	AB	EB	EA	CB	AB	EC	EA
Beijing	30.19	5.64	18.07	1.44	24.40	5.39	15.63	1.32
Xining	71.47	13.75	29.95	1.75	55.88	17.11		

Note: E.C.: empty sample chamber C.B.; E.A.: empty sample chamber A.B.

the anticoincident detector does not operate, the background measured is named by coincident background (CB). If the anticoincident detector operates, the background measured is named by anticoincident background (AB). When there is no sample in the sample chamber, the counting rate is named by empty chamber background (separately ECB and EAB).

The experiments are done separately in Beijing (the elevation is about 100 meters, 40° N) and in Xining (the elevation is about 2300 meters, 36.5° N). The background and the empty chamber counting rate are measured first in Beijing and then in Xining. The results (Table 1) show the AB of two counters all increase about threefold.

The shielding experiments are completed in a BGO installation in Xining. All of the experimental results are listed in Table 2. We find, if the lead shielding is increased from 10 cm to 20 cm, the background increases. The CB increases from 55.88 cpm to 59.44 cpm and the AB increases from 17.11 cpm to 19.49 cpm. If we append the lead with plastic material (polystyrenee) and water, the CB decreases to 50.37 cpm and AB decreases to 14.20 cpm. The 10 cm lead chamber is enlarged to test the radon effects on the background. There

Table 2. The Data of the Shielding Experiments in Two Places

| Place | | Xining | | Beijing | |
Shielding Condition		CB	AB	CB	AB
BGO installation					
A	Only 10 cm lead under the detector	295.5	171.7	300.8	163.8
B	Only 20 cm lead under the detector	338.9	198.6		
C	Entire 10 cm lead shielding	55.88	17.11	24.40	5.39
D	C + appending 10 cm lead on top	57.07	17.68		
E	C + appending 10 cm lead around	59.44	19.49		
F	C + many polystyrene pellets fills the empty space of the lead chamber and 30 cm thick polystyrene on top of and 601 water around the lead chamber	50.37	14.20		
G	Enlarged 10 cm lead chamber	54.90	17.93		
H	G + 30 kg. paraffin wax and polystyrene fills the empty of the lead chamber	44.77	10.57		
I	H + 60 L water + 60 kg. paraffin wax around and on the lead chamber	44.65	9.67		
DYS-3					
	10 cm lead shielding	71.47	13.75	30.19	5.64
	The empty of the lead chamber are filled with 30 kg. of paraffin wax	58.00	9.41		

Table 3. The Estimated Intensity of the Cosmic Rays at 50° N

	Sea Level	2000 meters Elevation	800 g/cm² Atmosphere
Total intensity (particles /sec.cm², in all directions	0.020	0.035	
Hard constituent (particles /sec.cm², in all directions)	0.013	0.018	
Soft constituent (particles /sec.cm², in all directions)	0.007	0.017	
E:electron and positron with energy more than 10 MeV	1×10^{-3}/cm².sec.Ω		6×10^{-3}/cm².sec.Ω
μ:μ mesons with various energy	8×10^{-3}/cm².sec.Ω		1.5×10^{-2}/cm².sec.Ω
P: proton with energy more than 400 MeV	3×10^{-5}/cm².sec.Ω		2×10^{-4}/cm².sec.Ω

are 10 cm empty spaces above and under the detector. The background obviously does not change, but if the empty spaces of the lead chamber are filled by 30 kg paraffin wax, the background decreases. The CB decreases from 55.88 cpm to 44.77 cpm; the AB decreases from 17.11 cpm to 10.57 cpm.

If the empty space of the lead chamber of DYS-3 is filled with paraffin wax, the CB of DYS-3 decreases from 71.47 cpm to 58.00 cpm; the AB of DYS-3 decreases from 13.75 cpm to 9.41 cpm.

DISCUSSION

Comparing the empty chamber background in two places (Table 1), we find the CB of the empty chamber of DYS-3 in Xining is 1.66 times higher than that in Beijing, but the AB of the empty chamber in Xining is only 1.22 times higher than in Beijing. Table 3 shows that the total intensity of various energy μ mesons is 1.88 times higher at the elevation with 800 g/cm atmosphere (or elevation of 2000 meters or so) than at sea level. According to our past experiments, most contributions of the μ mesons to the background can be eliminated by any type of anticoincident detector; thus it is suggested that the background of the empty chamber in coincident counters at sea level is mainly from the μ mesons of cosmic rays. The increasing background of the blank sample in Xining mainly does not come from the increasing μ mesons.

In Beijing, most of the low-level liquid scintillation counters with different anticoincident detectors all have a CB value of about 24 cpm. The counter with NaI(Tl) crystal as an anticoincident detector, however, has a CB of about 30 cpm; it is 6 cpm higher than the others. In the past we thought this came from the ⁴⁰K in NaI(Tl). Now, from Table 2, under the same experimental condition, the CB of DYS-3 is 15.6 cpm higher than the CB of BGO installation in Xining. Obviously, it can not be assigned the ⁴⁰K of the NaI(Tl). Maybe it comes from the interaction between NaI(Tl) crystal and some high energy

Table 4. The Anticoincident Effect of Two Detectors for Neutron

	Coincident Counting Rate for Neutron (cpm)	Anticoincident Countingrate for Neutron (cpm)	The Anticoincident Effect for Neutron
BGO installation	214.2 ± 1.5	158.1 ± 1.3	19.5%
Plastic scintillator installation	186.9 ± 2.7	81.6 ± 1.4	53.3%

particles in cosmic radiation; new radiation is created, and it increases the background.

In Xining, the greater the lead thickness, the greater the background. In Beijing, the background decreases if the lead thickens.[2] When the lead is only placed under the detector and there is no lead above and around the detector, the background is obviously no different in two places, but the difference rapidly increases as the thickness of the lead increases. From this, we think the increasing background is partly from the interaction between the lead and high-energy cosmic radiation. When a particular cosmic ray shoots the lead, it produces heavy radiation.

If the empty space between the detector and lead is filled with paraffin wax, the background decreases. We feel the interaction between cosmic radiation and lead produces a nuclear reaction, and a lot of neutrons are produced. The difference between CB and AB, after and before being filled with paraffin wax for DYS-3, is respectively 13.47 cpm and 4.34 cpm. It means the effect of anticoindence of NaI(Tl) for this part of the background is about 68%. For BGO installation, the difference of the CB and AB after and before being filled with paraffin wax is 10.13 cpm and 7.37 cpm, respectively; the effect of anticoincidence is only about 27.2%. In Beijing, 5×10^5/sec Am-Be neutron (in the state for transportation) is placed out of the anticoincident counters of BGO and plastic scintillator, and the counting rate is measured with the same condition as above. The results are shown in Table 4. From Table 4, the anticoincident effect for BGO and plastic scintillator is 20% or so separately and more than 50% combined. We think it probably depends on the volume, the atomic number and geometric condition of the anticoincident detector, and also the energy spectrum of the neutron and the interaction between neutron and matter. According to the experimental results, the increasing background can be decreased by paraffin wax, and there is a difference of anticoincident effect between two counters in Xining. We consider that the increasing background is due to the neutrons produced by the lead interacting with the cosmic radiation. The other part of the increasing background for DYS-3 is difficult to decrease. It comes from the interaction between the NaI(Tl) and cosmic radiation.

To sum up, plastic scintillators have better anticoincident effect. When interacting with cosmic radiation, it does not produce much second radiation; it does not lead to increased background for liquid scintillation counters as

NaI(Tl) crystal does. In a high elevation area, as an anticoincident shielding, it is better than NaI(Tl). A complex matter shielding of lead and paraffin wax for a liquid scintillation counter is necessary.

REFERENCES

1. Noakes, J.E., M.P. Neary, and J.D. Spaulding, J.D. "Tritium Measurement with a New Liquid Scintillation Counter," *Nuc. Instrum. Methods*, 109:177 (1973).
2. Yang Shouli, Jiang Peidong, and Lin Han. "The Advance and Application in Liquid Scintillation Measurement Technique," p.88 (1987).
3. Jiang Hanying et al. "Model DYS Low-Level Liquid Scintillation Counter," in *Advance in Scintillation Counting* (S.A. McQuarrie, et al.: (Ed. Edmonton, Canada: University of Alberta, 1983).

CHAPTER 47

Calculational Method for the Resolution of ^{90}Sr and ^{89}Sr Counts from Cerenkov and Liquid Scintillation Counting*

Thomas L. Rucker

INTRODUCTION

Environmental samples taken around reactor facilities potentially contain both 50.4 day half-life strontium-89 (^{89}Sr) and 29 year half-life strontium-90 (^{90}Sr). The health risk associated with exposure to ^{90}Sr is considered to be a factor of 20 greater than that associated with ^{89}Sr due to its longer half-life and its energetic daughter yttrium-90 (^{90}Y). It is therefore often important to distinguish between these two isotopes when monitoring for health and environment protection. Most common methods for determining both ^{89}Sr and ^{90}Sr are based on time allowed ingrowth of ^{90}Y after the total strontium is chemically purified and counted. This usually requires from 7 to 14 days to allow sufficient ^{90}Y ingrowth for adequate statistics.

A method has been developed that determines both ^{89}Sr and ^{90}Sr without requiring an ingrowth period for ^{90}Y; therefore, the results can be obtained much sooner. The method is similar to that developed more than a decade ago by Randolph[1] which uses sequential Cerenkov and liquid scintillation counting after chemical purification. A new calculational method has been employed, however, which corrects for the fraction of the Cerenkov counts due to ^{90}Sr and the fraction of both counts due to the early ingrowth of ^{90}Y. The method does not rely on the spectral resolution of ^{89}Sr, ^{90}Sr, and ^{90}Y through the use of channel ratios, as has previously been required with the combined liquid scintillation of a Cerenkov technique. Therefore, higher detection efficiencies are obtained which result in better sensitivity.

*Research performed under the auspices of the U.S. Department of Energy under Contract DE–AC09–76SR00001 with E. I. DuPont De Nemours & Co. and Subcontract AX–715305 with Science Applications International Corporation.

Table 1. Nuclide Characteristics

Nuclide	^{89}Sr	^{90}Sr	^{90}Y
Half-life	50 d	29 Y	64 h
Max. beta energy (keV)	1491	546	2284
Avg. beta energy (keV)	583	196	934
Rel. health risk	1	20	—
M.P.C. (pCi/L) drinking water	80	8	—

THEORY

Cerenkov radiation is produced when a charged particle passes through a transparent medium at a velocity greater than the speed of light in that medium. A threshold condition must be met which is defined by

$$vn/c = 1, \tag{1}$$

where v is the velocity of the particle, c is the speed of light, and n is the refractive index of the transparent medium. Furthermore, the Cerenkov photon yield increases proportionally as the energy of the particle increases above the threshold.

The nuclear decay characteristics for ^{89}Sr, ^{90}Sr, and ^{90}Y are shown in Table 1. The combined Cerenkov and liquid scintillation counting technique takes advantage of the large difference in average beta energies of ^{89}Sr and ^{90}Sr to discriminate between the two after the ^{90}Y is separated chemically. The threshold energy for the beta particle stimulation of Cerenkov radiation in water is 263 keV; therefore, in a modern liquid scintillation counter, Cerenkov counting provides over 40% efficiency for the 583 keV average beta energy ^{89}Sr, while providing less than 1% counting efficiency for the 196 keV average beta energy ^{90}Sr. The total strontium activity is determined by liquid scintillation counting, which provides over 90% counting efficiency for both isotopes. The combined information obtained from these two counts yields the activity of each isotope.

OPERATIONAL METHOD

Chemical Separation

A standard procedure, similar to that used in EPA Method 905.0,[2] is used to chemically purify the sample. Calcium and magnesium are removed by precipitation of the strontium as the nitrate. Barium and radium are removed by a chromate precipitation. Yttrium and other impurities are removed with hydroxide scavenging. A final strontium precipitation is made as the carbonate

Table 2. Experimental Efficiencies and Backgrounds

Nuclide	Cerenkov Eff. (%)	Liquid Scint Eff. (%)
^{89}Sr	40	95
^{90}Sr	1	91
^{90}Y	61	95
	(cpm)	(cpm)
Background	15	62

in a tared liquid scintillation vial. The precipitate is washed, dried, and weighed to calculate the recovery of the standardized strontium carrier. The average chemical recovery is approximately 80%.

Counting

The precipitate is dissolved in 5 mL of 0.3 M HCl, and the aqueous solution is counted on a liquid scintillation counter without scintillator for 20 min for Cerenkov radiation, primarily from ^{89}Sr. Seventeen mL of "Atomlight" liquid scintillator is then added and mixed with the sample. The solution is again counted for 20 min on the liquid scintillation counter for the measurement of both ^{89}Sr and ^{90}Sr.

Randolph used optimized discriminator channel settings to provide energy windows for each count. He then used the count ratios in each widow to calculate factors for correcting the fraction of the Cerenkov counts due to ^{90}Sr and the fraction of both counts due to the early ingrowth of ^{90}Y. However, due to the severe energy spectrum overlap of each of the isotopes, the optimized windows only contain a fraction of the total counts from the isotope of interest. Therefore, the sensitivity is reduced.

In the present experiment, the novel calculational method is used to correct for the fraction of the Cerenkov counts due to ^{90}Sr and the fraction of both counts due to the early ingrowth of ^{90}Y. The discriminators were set for each count such that the energy window contained the entire peak of the isotope or isotopes of interest, thus improving the sensitivity of the method. Beckman LS-9800 liquid scintillation counters were used in this study. The experimental efficiencies for each isotope and the backgrounds obtained for each counting method are shown in Table 2.

CALCULATIONAL METHOD

The Cerenkov counts are primarily from ^{89}Sr, with a small contribution from ^{90}Sr and the early ingrowth of ^{90}Y. The net Cerenkov count rate (cpm), C_1, is given by

$$C_1 = {}^{89}Sr\ cpm_c + {}^{90}Sr\ cpm_c + {}^{90}Y\ cpm_c, \tag{2}$$

where cpm_c represents the Cerenkov counts per minute obtained from each nuclide. The liquid scintillation counts are due primarily to both ^{89}Sr and ^{90}Sr, with a small contribution from the early ingrowth of ^{90}Y. The net liquid scintillation count rate (cpm), C_2, is given by

$$C_2 = {}^{89}Sr\ cpm_l + {}^{90}Sr\ cpm_l + {}^{90}Y\ cpm_l, \tag{3}$$

where cpm_l represents the liquid scintillation counts per minute obtained from each nuclide. The count rate due to early ingrowth of ^{90}Y can be defined in terms of the ^{90}Sr count rate by knowing the elapsed time from chemical separation of ^{90}Y to counting. Equations 2 and 3 then become

$$C_1 = {}^{89}Sr\ cpm_c + ({}^{90}Sr\ cpm_c)\ (A_1) \tag{4}$$

and

$$C_2 = {}^{89}Sr\ cpm_l + ({}^{90}Sr\ cpm_l)\ (A_2), \tag{5}$$

where

$$A_1 = 1 + (D_1)(1 - e^{-\lambda t_c}) \tag{6}$$

and

$$A_2 = 1 + (D_2)(1 - e^{-\lambda t_l}), \tag{7}$$

where D_1 is the ^{90}Y Cerenkov efficiency to ^{90}Sr Cerenkov efficiency ratio, t_c is the time from chemical separation to Cerenkov counting, D_2 is the ^{90}Y liquid scintillation efficiency to ^{90}Sr liquid scintillation efficiency ratio, t_l is the time from chemical separation to liquid scintillation counting, and λ is the ^{90}Y decay constant.

The Cerenkov and liquid scintillation count rates for each nuclide are related by their counting efficiency ratio of each for the two counting methods such that

$$^{89}Sr\ cpm_c = ({}^{89}Sr\ cpm_l)\ (B_1) \tag{8}$$

and

$$^{90}Sr\ cpm_c = ({}^{90}Sr\ cpm_l)\ (B_2), \tag{9}$$

where B_1 is the ^{89}Sr Cerenkov efficiency to ^{89}Sr liquid scintillation efficiency ratio and B_2 is the ^{90}Sr Cerenkov efficiency to ^{90}Sr liquid scintillation efficiency ratio. By replacing these terms in Equation 4,

$$C_1 = ({}^{89}Sr\ cpm_l)\ (B_1) + ({}^{90}Sr\ cpm_l)\ (B_2)\ (A_1). \tag{10}$$

If the ratios of counting efficiencies are determined by counting pure standards, Equations 5 and 10 contain two unknowns and can be solved simultaneously to yield

Table 3. Results from Measuring Mixtures of Isotopes (Lower Level)

Sample/ Nuclide	Known (pCi/L)	Measured (pCi/L)		2 Sigma CL	Ratio
1 ^{89}Sr	120	130	+/−	11	1.08
1 ^{90}Sr	0	0.7	+/−	5	—
2 ^{89}Sr	120	125	+/−	15	1.04
2 ^{90}Sr	220	234	+/−	11	1.06
3 ^{89}Sr	240	219	+/−	16	0.91
3 ^{90}Sr	220	259	+/−	13	1.18
4 ^{89}Sr	120	130	+/−	19	1.08
4 ^{90}Sr	440	491	+/−	16	1.11
5 ^{89}Sr	0	8	+/−	12	—
5 ^{90}Sr	220	260	+/−	10	1.18

$$^{90}\text{Sr cpm}_1 = [(B_1)(C_2) - C_1] / [(B_1)(A_2) - (B_2)(A_1)] \tag{11}$$

and

$$^{89}\text{Sr cpm}_1 = C_2 - (A_2)(^{90}\text{Sr cpm}_1). \tag{12}$$

The activity concentration of each nuclide can then be determined by Equations 11 and 12 and by

$$^{90}\text{Sr (pCi/L)} = (^{90}\text{Sr cpm}_1) / [(E_1)(V)(R)(2.22)] \tag{13}$$

and

$$^{89}\text{Sr (pCi/L)} = (^{89}\text{Sr cpm}_1) / [(E_2)(V)(R)(2.22)] \tag{14}$$

where E_1 = ^{90}Sr liquid scintillation efficiency,
E_2 = ^{89}Sr liquid scintillation efficiency,
V = Sample Volume (L), and
R = Chemical Recovery.

RESULTS

Known mixtures of ^{89}Sr and ^{90}Sr, ranging to 100% of each, have been measured by this technique at both low and moderate activity levels. The results are shown in Tables 3 and 4. While it appears from the lower level data that a slight positive bias may exist for ^{90}Sr, it is likely that the apparent bias is due to counting uncertainty since the bias disappears at the higher activity level. The error due to uncertainty in correction for the early ingrowth of ^{90}Y is minimized by counting the samples as soon as possible after separation.

Since this method requires two counts, the uncertainty must be considered when estimating the total measurement due to counting statistics. An estimate of the standard deviation, s, due to counting uncertainty can be made by

Table 4. Results from Measuring Mixtures of Isotopes (Higher Level)

Sample/ Nuclide	Known (pCi/L)	Measured (pCi/L)		2 Sigma CL	Ratio
6 ^{89}Sr	2044	2042	+/−	47	1.00
6 ^{90}Sr	0	2	+/−	41	—
7 ^{89}Sr	2044	2041	+/−	53	1.00
7 ^{90}Sr	2199	2045	+/−	44	0.93
8 ^{89}Sr	4088	3938	+/−	71	0.98
8 ^{90}Sr	2199	2145	+/−	60	0.98
9 ^{89}Sr	2044	2074	+/−	66	1.01
9 ^{90}Sr	4398	4371	+/−	52	0.99
10 ^{89}Sr	0	38	+/−	36	—
10 ^{90}Sr	2199	2177	+/−	26	0.99

$$^{90}\text{Sr cpm, s} = \{[(B_1)^2(G_2/T_2 + F_2/T_2) + G_1/T_1 + F_1/T_1] \\ /[(B_1)(A_2) - (B_2)(A_1)]^2\}^{1/2} \tag{15}$$

and

$$^{89}\text{Sr cpm, s} = [(G_2/T_2 + F_2/T_2) + (A_2)^2(^{90}\text{Sr cpm, s})^2]^{1/2}, \tag{16}$$

where G_1 = gross Cerenkov count rate (cpm),
G_2 = gross liquid scintillation count rate (cpm),
F_1 = background Cerenkov count rate (cpm),
F_2 = background liquid scintillation count rate (cpm),
T_1 = Cerenkov counting time (min), and
T_2 = liquid scintillation counting time (min)

when the background and sample counting times are equal. Similarly, the lower limit of detection (LLD), considering only counting statistics, can be estimated by

$$\text{LLD (cpm)} = 2.71/T_2 + 4.65(\text{sb}) \tag{17}$$

where sb is the standard deviation of the blank as defined by

$$^{90}\text{Sr sb} = \{[(B_1)^2(F_2/T_2) + F_1/T_1] / \\ [(B_1)(A_2) - (B_2)(A_1)]^2\}^{1/2} \tag{18}$$

and

$$^{89}\text{Sr sb} = [F_2/T_2 + (A_2)^2(^{90}\text{Sr sb})]^{1/2} \tag{19}$$

The liquid scintillation counters have a higher background than the low-background gas-proportional counters typically used for counting environmental strontium samples. This results in reduced sensitivity. However, the higher efficiencies obtained with liquid scintillation counters allow lower limits of detection which are only a factor of 2 to 3 higher than those obtained using

Table 5. Calculated Lower Limits of Detection

Cerenkov and Liquid Scintillation Count
^{90}Sr = 12.8 cpm = 7.9 pCi/L
^{89}Sr = 15.9 cpm = 9.4 pCi/L

Traditional 14-Day Two-Count Method
^{90}Sr = 1.4 cpm = 2.4 pCi/L
^{89}Sr = 2.0 cpm = 3.6 pCi/L

Note: All LLDs assume 20 min counts on 1 L samples and 80% chemical recovery.

the same count time on gas-proportional counters. The lower limits of detection obtained in this study are shown in Table 5 with the LLDs obtained with the traditional 14-day two-count method (assuming 1-L samples counted for 20 min and 80% chemical recovery for both). If low-background liquid scintillation counters are used, the lower limits of detection may be even lower.

CONCLUSION

The described method provides a useful way to analyze both ^{89}Sr and ^{90}Sr without a 14 day wait for ^{90}Y ingrowth. The use of the derived equations provides a correction for early ^{90}Y ingrowth and the ^{90}Sr Cerenkov counts, without the use of energy window ratios. Better sensitivity has been provided than has previously been possible with this technique.

The sensitivity of this method is adequate for monitoring plant effluents and many environmental level samples. Lower limits of detection, below the maximum contaminant concentration allowed for drinking water, can be achieved when a 1-L sample and a 20 min count time are used. If monitoring for lower levels of contamination are necessary, a larger sample or a longer count time may be used.

REFERENCES

1. Randolph, R.B. "Determination of Strontium-90 and Strontium-89 by Cerenkov and Liquid-Scintillation Counting," *Int. J. Appl. Radiat. Iso.*, 26:9–16 (1975).
2. Krieger, H.L. and E.L. Whittaker. "Prescribed Procedures for Measurement of Radioactivity in Drinking Water," EPA-600/4-80-032 (1980), pp. 58–74.

CHAPTER 48

Determination of ^{222}Rn and ^{226}Ra in Drinking Water by Low-Level Liquid Scintillation Counting—Surveys in Austria and Arizona

Franz Schönhofer, J. Matthew Barnet, and John McKlveen

ABSTRACT

One important source to man for both ^{222}Rn and ^{226}Ra is drinking water. ^{226}Ra may be ingested and ^{222}Rn is liberated by household activities; thus adding to the radon concentration already present in indoor air.

Very simple yet specific methods, avoiding any chemical separation, are presented for the determination of both radionuclides, using a commercially available ultra low-level liquid scintillation counter. The lower limit of detection is 40 mBq/L (1.1 pCi/L)—based on 500 min count and 3 σ of the background. These methods have been applied for surveys both in Vienna and Arizona, the results of which are presented.

HEALTH HAZARDS OF ^{226}RA AND ^{222}RN

Most countries have strict regulations about the maximum permissible concentration (MPC) of ^{226}Ra in drinking water (^{226}Ra is a bone seeker). MPCs range in different countries from 0.11 to 1.85 Bq/L. In Austria it is 0.122 Bq/L.

^{222}Rn is regarded as hazardous, since it is abundant in all indoor air in enhanced concentrations. It emanates from ground or building material, and it is liberated from water by household activities like cooking, washing, toilet flushing, etc. It is estimated[1] that tap water containing 100 kBq/m^3 gives off an additional effective dose equivalent to about 0.4 to 0.7 mSv/a from the liberated radon and its daughters. An air change rate of 0.5 changes per hour is assumed. Modern isolation techniques used for energy conservation have altered the air change rate by approximately 0.1 in most cases, enhancing radon concentration in houses. The above mentioned additional dose corresponds to the average additional dose the Austrian population received in the first year after Chernobyl. (Austria was one of the most contaminated European countries outside the U.S.S.R.)

No MPCs for ^{222}Rn are known in any country, but in some countries, for instance Sweden, national health boards give recommendations.[2] Below 100 Bq/L no restrictions or counter-measures are considered necessary. Between 100 and 1000 Bq/L cases have to be considered individually with regard to other sources of radon and ventilation. If water contains more than 1000 Bq/L it is likely that the concentration of radon daughters in the air exceeds 400 Bq/ m^3 — which is the action level — and remedial measures should be undertaken.

It is expected that the U.S. Environmental Protection Agency will set recommendation levels for radon in water soon, and they will probably be lower than the Swedish ones at present. In principle, it is possible to apply the recommendation of 4 pCi/L air (0.15 Bq/L), taking the liberation from water into account. Doses from ingestion of radon are negligible compared to the lung doses from inhalation.

Determination of ^{222}Rn and ^{226}Ra in water is therefore of great importance in health physics.

DETERMINATION OF ^{222}RN AND ^{226}RA IN WATER

^{222}Ra is highly soluble in toluene and other organic solvents frequently used in cocktails for LSC. Since α-particles are counted with approximately 100% efficiency, LSC is nowadays widely used for measurement of radon.[3-5]

One simple method is to mix water with a gel-forming scintillation cocktail and wait about three hours until equilibrium with the daughters is established. Normally ^{222}Rn is present in such high concentrations that any contribution of other naturally occurring radionuclides is negligible. Figure 1 shows the spectrum of an actual water sample, 8 mL of which were mixed with 12 mL of Quickszint 400 (obtained from Zinsser Analytic, Frankfurt). The vial used was a PTFE coated polyethylene vial (from Zinsser). The left peak is the unresolved sum of the peaks of ^{222}Rn (5.49 MeV) and ^{218}Po (6.00 MeV), the right, well-resolved one is due to ^{214}Po (7.69 MeV). The resolution is estimated to be about 300 to 400 keV. The x-axis shows the logarithmic pulse height and not the energy. Efficiency is between 300 and 400%, depending on vial and cocktail. It is determined with a ^{226}Ra standard treated the same way and taking the 100% efficiency of the Ra-α-particles into account.

Another method uses radon extraction from the water phase into a water-immiscible, mineral oil-based scintillation cocktail.[5] The cocktail employed in this research was PSS-007H from NEN. Ten mL of water is mixed with 10 mL cocktail, shaken vigorously, and counted after about three hours. In this time, equilibrium is established, the sample is cooled, and the phases are separated. The resolution is better (Figure 2) than in the case of the gel-forming cocktail. The efficiency is equally high and is determined with a ^{226}Ra standard. ^{226}Ra is not extracted into the organic phase.

^{226}Ra is usually enriched and isolated chemically. Measurement is frequently done with LSC.[6-7] We have been able to develop a method which completely

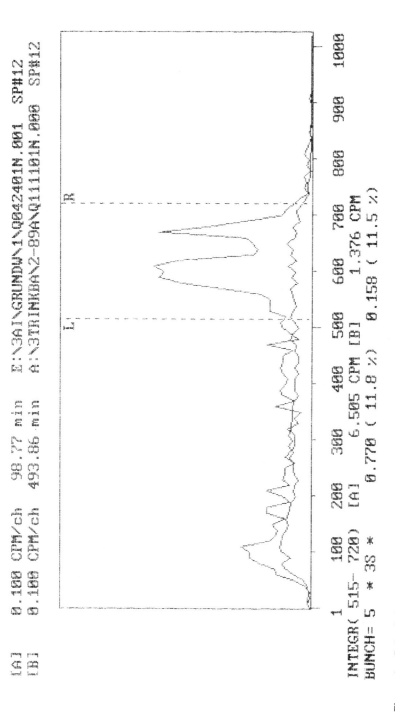

Figure 1. Pulse height spectrum of a water sample containing 222Rn, 8 mL mixed with 12 mL Quick szint 400, and PTFE coated polyethylene vial.

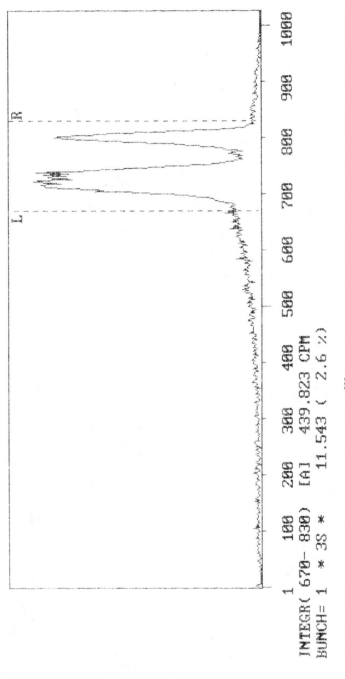

Figure 2. Pulse height spectrum of a water sample containing ^{222}Rn, radon extracted from 10 mL water into 10 mL of NEN cocktail, and PTFE coated polyethylene vial.

avoids the chemical isolation by using the ultra low-level LS counter "Quantulus" (Pharmacia — Wallac, Turku, Finland). This instrument exhibits extremely low background due to heavy passive and active shielding consisting of an anticoincidence unit based on a liquid scintillator. Moreover activity of the selected construction material is very low. Pulse height spectra are recorded automatically and can be used for control of possible interferences. Tests with standards of [226]Ra showed that the above mentioned methods for [222]Rn (simple mixing of water with the above mentioned cocktails) were well applicable for [226]Ra and they gave a lower limit of detection of about 37 to 48 mBq/L (1.0 to 1.3 pCi/L) (based on 3σ of the background and 500 min counting time), which is well below the maximum permissible concentration in Austria of 0.122 Bq/L (3.3 pCi/L).

The mixture of water with the gel-forming cocktail is not recommended for [226]Ra, since in such low concentration ranges heavy interferences from other naturally occurring radionuclides, especially [40]K, may occur. Quench effects from the water may also cause heavy interference. The extraction with the mineral oil–based cocktail, however, is specific for [226]Ra. After ingrowth of [222]Rn from [226]Ra (or decay of excess and unsupported [222]Rn) only [222]Rn in equilibrium with [226]Ra is extracted. [220]Rn and [219]Rn would not interfere because each has a very short half-life.

In practice, water was analyzed for [222]Rn immediately after it received the sample by the mineral oil cocktail method. Then it was stored for the appropriate time to allow for decay of unsupported [222]Rn. The same sample was then measured again for [226]Ra, thus the manual labor involved for determination of both radionuclides was simply the pipetting of 10 mL of sample and 10 mL of the cocktail — not to forget the shaking. For [222]Rn the measurement time is usually 60 min. For [226]Ra 500 min is was chosen, which means automatic determination of three samples per day, also during weekends and holidays. The high costs of our instruments have been compensated for by long time savings of personnel and labor.

RESULTS

The afore described methods were applied for surveillance purposes in Austria, and the method for [222]Rn has been used to survey risk area in Carefree/Cave Creek Basin in Arizona.

In Carefree, 26 wells were tested. The frequency of the concentrations found can be seen in Figure 3. Eleven wells (42%) show concentrations above the 100 Bq/L, which the Swedish authorities regard as the concentration below which no countermeasures need to be considered. They result in an additional effective dose equivalent of 0.4 to 0.7 mSv/a, which is much higher than the impact of nuclear power plants on people living in the surrounding area.

Concerning [222]Rn in Austria, the results from a risk area in the northeastern part of the country are presented. After a nationwide survey, work was con-

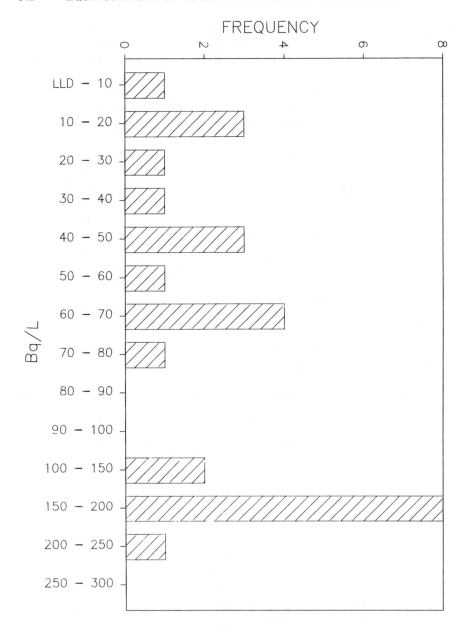

Figure 3. Distribution of concentrations of [222]Rn in water, Carefree/Cave Creek Basin, Arizona.

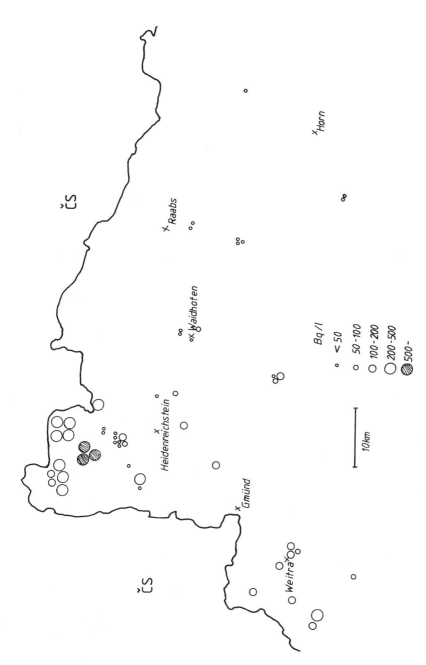

Figure 4.   Geographical distribution of ^{222}Rn in water in a risk area in northeastern Austria.

centrated on an area where the ground consists partly of granite and partly of metamorphicum (mostly from sediments). The regional distribution of ^{222}Rn concentrations in household water is shown in Figure 4. Clearly, small areas with extreme concentrations can be distinguished from areas with rather high concentrations. The former ones coincide well with granite areas and the latter ones with metamorphicum.

In spite of the extreme values for ^{222}Rn, even in the risk area, only a few samples exceeded the lower limit of detection of 40 mBq/L for ^{226}Ra. By far, the highest value observed was 490 mBq/L, followed by 180 mBq/L.

CONCLUSIONS

The simple methods for both ^{222}Rn and ^{226}Ra determination have made it possible to survey a large number of samples within a short time. The results show that both Arizona and Austria have risk areas for high ^{222}Rn concentrations in water, and analysis has to be extended to other risk areas. The concentrations found are significant health concerns, and after limits are defined, countermeasures have to be undertaken.

Regarding the instrumentation, it is attempted to test the α-β-discrimination unit of "Quantulus" thoroughly for its application in ^{226}Ra determination. Since the background for α emitters can be reduced practically to zero, it is expected that a much lower LLD for ^{226}Ra can be achieved. Careful investigations of vials and cocktails will be necessary.

ACKNOWLEDGEMENT

The authors thank Christian Tschapka and Jeffrey Brown for performance of measurements and evaluation of data used in this chapter, Robert Mecl, Franz Oesterreicher, Kurt Hierath (water department of the Federal Institute), and Oswald Ruttner (BBSUA, Vienna) for collecting and providing most of the Austrian water samples.

The research in Arizona was financially supported by the Arizona Department of Environmental Quality, which is gratefully acknowledged.

REFERENCES

1. *Naturally Occurring Radiation in the Nordic Countries—Recommendations*, (Reykjavik: The Radiation Protection Institutes in Denmark, Finland, Iceland, Norway, and Sweden, 1986).
2. Kulich, J., H. Möre, and G.A. Swedjemark. *Radon och radium i hushållsvatten, (Radon and Radium in Household Water), Swedish Radiation Protection Institute, Report SSI-rapport 88-11, 1988 (in Swedish).*
3. Parks, N.J. and K.K. Tsuboi. "Emulsion Scintillation Counting of Radium and Radon," *Inst. J. Appl. Radiat. Isotopes*, 29:77 (1978).

4. Asikainen, M. and H. Kahlos. "State of Disequilibrium Between ^{238}U, ^{234}U, ^{226}Ra, and ^{222}Rn in Groundwater from Bedrock," *Geochimica et Cosmochinica Acta*, 45:201 (1981).
5. Pritchard, H.M. and T.S. Gesell. "Radon-222 in Municipal Water Supplies in the Central United States," *Health Phys.* 45:991 (1983).
6. Prichard, H.M., T.F. Gesell and C.R. Meyer. "Liquid Scintillation Analyses from Radium-226 and Radon-222 in Potable Waters," in *Liquid Scintillation Counting, Recent Applications and Developments, Vol. I, Physical Aspects.* (New York: Academic Press Inc. 1980).
7. Cooper, M.B. and M.J. Wilks. *An Analytical Method for ^{226}Ra in Environmental Samples by the Use of Liquid Scintillation Counting*, (Yallambie, Victoria: Australian Radiation Laboratory, ARL/TR040, 1981).

CHAPTER 49

A New Simplified Determination Method for Environmental ⁹⁰Sr by Ultra Low-Level Liquid Scintillation Counting

Franz Schönhofer, Manfred Friedrich, and Karl Buchtela

ABSTRACT

⁹⁰Sr is a bone seeker and when in equilibrium with its short-lived, hard-beta-emitting daughter ⁹⁰Y, it may cause damage to the bone marrow. There is considerable interest in determining ⁹⁰Sr in a wide variety of environmental material, particularly food.

Normally, time-consuming chemical separation steps for isolation and decontamination from interfering radionuclides are necessary. Using a commercially available ultra low-level liquid scintillation counter, the required time for analysis could be reduced to a minimum. Extremely low LLDs allow for very small samples, and the entire separations can be performed in centrifugation tubes. The LLD achieved depends on the measurement method (Cerenkov or LS counting), and is between 50 and 90 fCi/sample (1.7 to 3.0 mBq/sample), based on 3 σ of the background and 500 min counting time. The method has been tested with certified reference materials, and some results from analysis of food contaminated by Chernobyl fallout will be given.

INTRODUCTION

⁹⁰Sr is of considerable interest in radiation protection. It is produced by nuclear fission, and its fission yield is approximately the same as for ¹³¹I or ¹³⁷Cs (5.7%). ⁹⁰Sr, originating from atmospheric bomb tests, has been spread all over the world and found its way to man through the food chain; there is direct deposition on plants and root uptake from the soil. The main source is milk. Transfer from soil to plants is normally about four times higher for ⁹⁰Sr as compared to ¹³⁷Cs; transfer in the cow to milk is lower. ⁹⁰Sr is incorporated into bone tissue and therefore has an extremely long biological half-life. ⁹⁰Sr is a pure β-emitter with a maximum energy of 0.546 MeV and a half-life of 28.5 years. Its daughter ⁹⁰Y is a pure β-emitter as well, with a maximum energy of 2.27 MeV and a half-life of 64.1 hours, so it is very fast in equilibrium with ⁹⁰Sr. The high energetic β-particles may damage bone marrow. Since ⁹⁰Sr is a

highly radiotoxic nuclide maximum permissible levels in food are very low, and sensitive methods are needed for its determination.

Since the ban of atmospheric nuclear weapons testing in 1962 the ^{90}Sr concentration in food has decreased considerably. In the case of the nuclear accident at Chernobyl, ^{90}Sr was emitted to a much lower fraction than ^{137}Cs because it is less volatile. In Western Europe the total ^{90}Sr deposition was much lower than the one originating from the atmospheric bomb tests.[1] Nevertheless measurement of ^{90}Sr is necessary both for radiation protection purposes and public interest. It has been shown for Austria, by measurements with a conventional method, that in 1986 the additional intake of ^{90}Sr was equal to the intake of remaining ^{90}Sr from atmospheric bomb tests.[2]

This paper will deal only with ^{90}Sr, in the absence of short lived fission products, and ^{89}Sr (half life 50.5 d). Analysis of ^{90}Sr and ^{89}Sr will be the next step in development of fast measurement methods for the case of a nuclear accident.

STRATEGY FOR THE DEVELOPMENT OF THE METHOD

The traditional "nitrate method"[3] has some serious disadvantages, the most important is handling the large amounts of fuming nitric acid for repeated precipitations. It was therefore attempted to avoid or at least reduce the number of purification steps with fuming nitric acid.

Liquid scintillation counting has been used for ^{90}Sr determinations since at least 1965 (an overview is given in Dehos[4]). The determination of low levels of ^{90}Sr has been considered by Salonen,[5] who also gave a first comparison between LSC and gas flow counters. In LSC the sample has a 4π geometry, a very high efficiency compared to gas flow counters.

Before choosing a separation method it was necessary to determine the lower limits of detection of the liquid scintillation counters for ^{90}Sr.

The equipment used was the commercially available ultra low-level liquid scintillation counter "Quantulus" (Pharmacia—Wallac, Turku, Finland). This instrument exhibits extremely low background due to heavy passive and an active shielding.

The concentration of ^{90}Sr can be determined by several methods. In samples where ^{90}Y is in equilibrium with ^{90}Sr, the former can be extracted by, e.g., HDEHP, and measured either by Cerenkov counting or after mixture with a suitable scintillation cocktail. (This method was first applied to cheese, but interference by an unidentified radionuclide with a half-life of approximately 6 hours was found.) ^{90}Sr can be measured directly after separation and mixture with a suitable cocktail and it can be remeasured after ingrowth of ^{90}Y for control. ^{90}Y also can be measured in the final solution after ingrowth by Cerenkov counting followed by scintillation counting. Data from Cerenkov counting can be obtained only after some waiting time, but then possible interferences from lower energy beta emitters and gamma emitters can be excluded.

First tests were conducted at the low-level laboratory ("LOLA") of Wallac OY in Turku, Finland. On one hand the environment shows high background radiation due to granite rock, on the other hand the laboratory has heavy concrete shielding. For all measurements, PTFE coated polyethylene vials (Zinsser, Frankfurt) were used. Cerenkov measurements were done in 10 ml 0.5 M hydrochloric acid; in LSC measurements 4 ml of 0.5 M hydrochloric acid solution were mixed with 6 ml Quickszint 400 (Zinsser).

Figure 1 shows the obtained pulse height spectra from Cerenkov counting both for background and standard. The window was optimized for highest figure of merit, using standard software for Quantulus. Efficiency was 69.25%, background 0.753 cpm. In Figure 2 ^{90}Sr and ^{90}Y are in equilibrium and mixed with cocktail. The window is adjusted for the sum of the radionuclides, giving an efficiency of 184.57% and a background of 1.988 cpm. If the window is adjusted for the part of the ^{90}Y spectrum above the maximum energy of ^{90}Sr (Figure 3), then an efficiency of 58.12% and a background of 0.603 cpm results.

Table 1 is a summary of the results obtained in terms of figure of merit (E^2/B) and lower limit of detection both for counting times of 100 and 500 min. The LLD is based on 3 σ of the background.

DETERMINATION OF ^{90}SR

The extremely low LLDs achievable with our equipment makes it possible to use considerably lower amounts of material than in conventional methods. In the conventional nitrate method described, 20 g milk or cheese ash are used. We have used for our analysis between 1 and 2 g of ash. The following procedure was tested and is now used in routine work:

> The ash of dairy products is dissolved in dilute nitric acid; carrier solutions of strontium, barium, iron, and chromate are added. The resulting volume in the first step is about 120 mL. Upon addition of ammonium acetate and adjustment to pH 5, barium chromate, iron phosphate, and basic iron acetates are precipitated and centrifugated. Phosphate is thus removed from the sample. Radium is also removed and it can be recovered and isolated from the precipitation. After precipitation with ammonium carbonate, work can be continued in small volumes of 10 to 20 mL. After dissolution in nitric acid, hydrogen peroxide, and yttrium carrier, the hydroxides of chromium and yttrium are removed. The final solution can be either acidified and measured directly or have one carbonate precipitation step added.

The use of fuming nitric acid is avoided completely. The chemical recovery, which is determined by atomic absorption spectroscopy, is in the range of 70%, which is regarded as sufficient compared with the ease of the procedure. In three days 20 samples of cheese ash have been processed. The limiting and

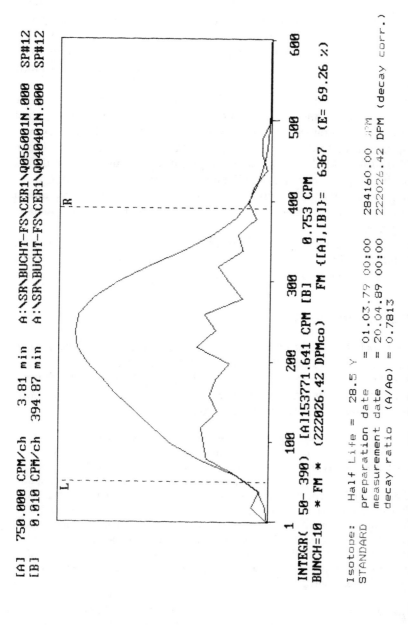

Figure 1. Pulse height spectra of ^{90}Sr standard and background, and Cerenkov counting in 0.5 M hydrochloric acid, with window optimized.

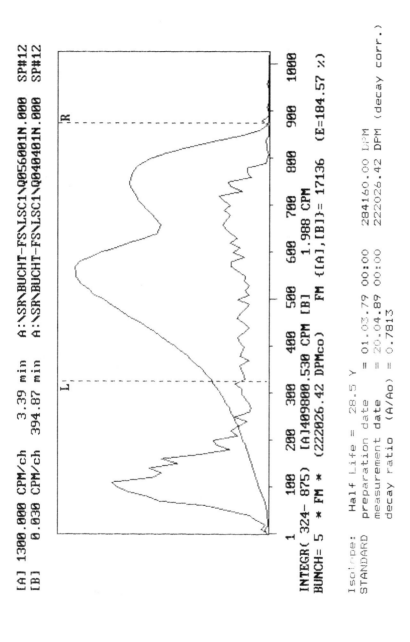

[A] 1300.000 CPM/ch   3.39 min   A:\SR\BUCHT-FS\LSC1\Q056001N.000   SP#12
[B]   0.030 CPM/ch   394.87 min   A:\SR\BUCHT-FS\LSC1\Q040401N.000   SP#12

INTEGR(324- 875)   [A]409800.530 CPM   [B]   1.988 CPM
BUNCH= 5   * FM *   (222026.42 DPMco)   FM ([A],[B])= 17136   (E=184.57 %)

Isotope: Half Life = 28.5 Y
STANDARD   preparation date = 01.03.79 00:00   284160.00 DPM
             measurement date = 20.04.89 00:00   222026.42 DFM (decay corr.)
             decay ratio (A/Ao) = 0.7813

Figure 2. Pulse height spectra of ^{90}Sr standard and background, and mixture with Quickszint 400, with window optimized for sum of ^{90}Sr and ^{90}Y.

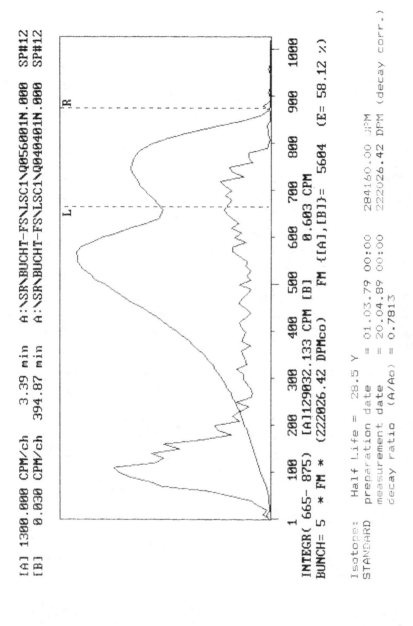

Figure 3. Pulse height spectra of ^{90}Sr standard and background, and mixture with Quickszint 400, with window optimized for ^{90}Y alone.

Table 1. Performance of Quantulus for ^{90}Sr for Different Conditions

Measurement Condition	Figure of Merit	LLD (mBq) (100 min)	LLD (mBq) (500 min)
Cerenkov counting	6,367	6.3	2.8
LS counting, open window	17,136	3.8	1.7
LS counting, ^{90}Y window	5,604	6.7	3.0

most time consuming step is ashing the samples. In the case of fresh milk, ion exchangers for concentrating strontium are planned, thus avoiding ashing.

RESULTS

The liquid scintillation method has been applied to a large variety of food samples collected after the Chernobyl accident. They are predominantly measured for ^{137}Cs and ^{131}I. Figure 4 shows the results for ^{137}Cs in emmental cheese from Salzburg—a highly contaminated area. Since ^{137}Cs is concentrated in whey, its activity concentration in cheese is low compared with milk. Figure 5 shows the activity concentration of ^{90}Sr for some of the cheese samples. Not all of the samples which had been measured for ^{137}Cs were available for ^{90}Sr analysis, so the increased speed of activity concentration in milk cannot be concluded. As in the case of ^{137}Cs a slow decrease can be seen about two weeks after the contamination occurred. The activity of directly deposited ^{90}Sr per kg grass ingested by the animals decreased due to dilution by plant growth. Samples from March 1987 showed concentrations of ^{90}Sr in cheese between 1.5 and 3.6 Bq/kg fresh weight (40 to 97 pCi/kg).

Concerning meat, another important source, concentrations were low—from below LLD to 0.1 Bq/kg freshweight (2.7 pCi/kg), with some exceptions in highly contaminated areas of up to 0.4 Bq/kg (2.5 pCi/kg).

The analyses of many samples collected after Chernobyl is going on and a report on the impact of the nuclear accident regarding ^{90}Sr is under preparation.

CONCLUSION

A method has been developed to analyze environmental samples and especially milk products for ^{90}Sr without using fuming nitric acid. Due to the extremely low LLDs of the used liquid scintillation counter, a very small sample amount can be used, and the whole separation procedure can be carried out in centrifuge tubes. The chemical yield is about 70% and it is determined by AAS. Twenty cheese ashes were processed within three days. The only limiting step is the ashing of the samples. The method saves time, chemi-

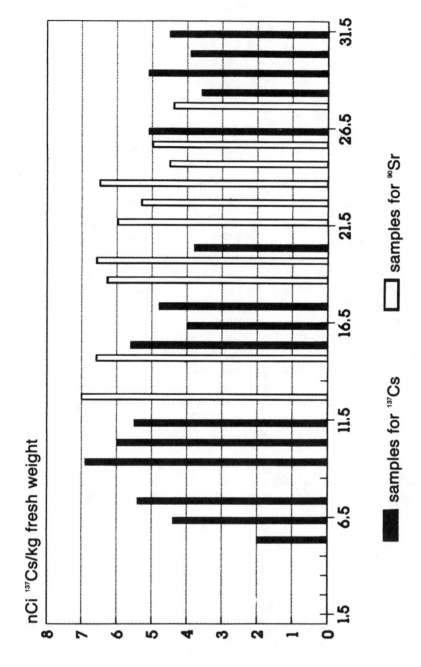

Figure 4. Contamination of emmental cheese form Salzburg with ¹³⁷Cs following the nuclear accident at Chernobyl. Values are given for fresh weight.

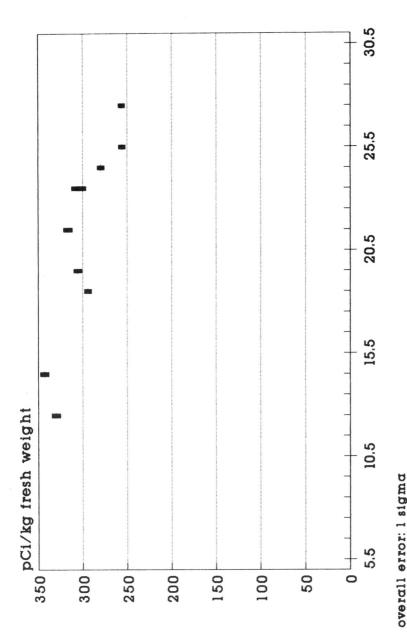

Figure 5. Contamination of some emmental cheese samples from Salzburg with ^{90}Sr. Values are given for fresh weight.

cals, and equipment, and it is accurate as was proved with certified reference material.

ACKNOWLEDGEMENT

The authors thank Ing. E. Findeis for saving a lot of precious samples during the hectic days of Chernobyl and for his skillfull ashing and sample preparation. Thanks to Dr. A. Graf for her help in obtaining cheese samples from the desired time and area, as well as to all institutions which provided the samples.

REFERENCES

1. Schönhofer, F., W. Ecker, H. Hojesky, W. Junger, K. Kienzl, H. Nowak, A. Riss, P. Vychytil, and J. Zechner. *Chernobyl and Its Impact on Austria*, Austrian Ministry of Health and Environmental Protection, Nov. 1986 (in German).
2. Mück, K.S. Striet, F. Steger, and K. Mayr. *The Contribution of ^{90}Sr to the Ingestion dose after the reactor accident in Chernobyl*, Austrian Research Center Seibersdorf, OEFZS−4452, ST−157/88, June 1988.
3. *Methods of Radiochemical Analysis*. (Geneva: World Health Organization, 1966), p. 52 ff.
4. Dehos, R. *Determination of ^{90}Sr in Human Bones, Tissues and Excretion*. STH-Berichte 13/1979, (Federal Republic of Germany: Bundes gesundheitsamt [Federal Health Autority], 1979), in German.
5. Salomen, L. "Determination of ^{90}Sr and ^{89}Sr in Environmental Samples by Liquid Scintillation Counting," paper presented at the International Symposium on Scintillation Counting and Related Techniques, Bath, England, September 13-16, 1977.

CHAPTER 50

Mixed Waste—A Review from a Generator's Perspective

John Hsu and Jeanne K. Krieger

The production of radioactive labeled compounds generate a variety of waste streams for research applications. Aggressive programs of segregation, compaction, and process improvements have reduced volume. From 1982 to 1988, the volume of waste generated and prepared for disposal by the manufacture of labeled chemicals declined by nearly 70%.

LOW LEVEL WASTE REDUCTION STRATEGIES

The DuPont NEN low level waste is segregated into six categories to facilitate handling the volume reduction. A variety of management techniques and physical methods have been developed to minimize the waste volume designated for shipment.

For example, jugs and containers comprise 10 to 12% of the prepared waste. Freezing the plastic ware with liquid nitrogen and pulverizing it while in a friable state reduces this waste stream. Aqueous waste, generated from the operation of stack scrubbers and chemical processing, is adsorbed in cement, in a ratio of 33 gal of water per 55 gal drum. Laboratory trash, i.e., gloves and lab coats are compacted at a pressure of 19,000 psi by a commercially available press to achieve a 6- to 15-fold reduction in volume. These physical methods of volume reduction are based on a management program to create awareness and good practices among the workforce. Feedback on waste generation within the working group and rewards for novel solutions create the incentive to participate in waste reduction programs.

In addition to the waste designated for shallow land burial, waste containing short-lived nuclides, ^{32}P, ^{35}S, and ^{125}I, is held for decay. Nearly 60% of the total waste volume generated from the manufacture of labeled chemicals is held for decay.

The most effective radioactive waste reduction strategy is recycling, which for our waste stream means isotopic separation. Approximately 20kCi per year of tritium gas are purified and returned to the process.

Recently, we embarked on a program of crushing scintillation vials for bulk disposal of scintillation fluid. Spent scintillation cocktail is disposed in 55 gal containers at a cost of approximately $10/gal of liquid, in contrast to the $167/gal incurred when vials were merely layered with adsorbent in drums for disposal. Vial crushing is achieved with a VYLEATER, manufactured by S & G Enterprises, Milwaukee, WI.

MIXED WASTE IN THE MANUFACTURE OF LABELED CHEMICALS

Mixed waste, low-level waste with hazardous waste properties, comprises approximately 25% of the waste barrel volume generated in the manufacture of labeled chemicals. Currently, there is no site licensed to dispose of this waste form. Mixed waste, originating from the manufacture of labeled compounds, is organic solvents contaminated with tritium and ^{14}C. In the case of tritium, the radioactivity is often incorporated directly into the solvent by chemical exchange and does not exist as part of a chemically distinct species. In contrast, carbon-containing waste is usually constituted of low concentrations of radioactive solute in cold hazardous carrier. This waste stream is the inevitable by-product of organic syntheses conducted with radioactive compounds. Mixed waste originates from the solvents used to mediate the reactions and as effluent from chromatographic purifications. Even the product itself, i.e., labeled benzene or toluene is classified as a mixed waste.

This particular waste stream, hazardous organics chemically indistinguishable from the radioactive component, is unique to the manufacture and application of labeled chemicals. No other process produces low level mixed waste of this composition and chemical form. Preventive steps are constantly being taken to reduce the volume of this waste stream, but the mixed waste generated from the synthesis of tritium and ^{14}C compounds cannot be completely eliminated. Minimization can be achieved by segregation and reduction at the source. Where possible nonhazardous solvents are used. Analysis of the chemical composition of the waste stream indicates that 30% is aqueous and can be segregated in the laboratory for disposal adsorbed on cement. Process changes influence the generation of mixed waste. Higher yields, improved product stability, and alternate purification procedures impact the synthesis frequency and concomitantly the solvent volume used. Despite minimization programs, mixed waste is an inevitable result of the manufacture of labeled chemicals. An approved disposal or treatment plan is mandatory.

ALTERNATIVE PROCESSES

In addition to the standard waste reduction methodologies of minimization and recycling, innovative approaches to render the mixed waste nonhazardous have been considered. These approaches are based on the principle of converting the waste stream to radioactive carbon dioxide and water, chemical substances that are nonhazardous and can be collected and disposed of in accord with regulatory guidelines in an environmentally responsible manner. Among the techniques that have been considered are microwave pyrolysis, both conventional and catalytic combustion, supercritical fluid oxidation, and electrochemical destruction. These techniques require minimal external fuel and can be run as closed systems, facilitating the required containment of the radioactive effluent.

None of these treatment protocols has been developed for disposal of radioactive waste. Considerable engineering and development effort is required to verify the technical efficacy of these alternatives. Unfortunately, successful development of these techniques from an engineering viewpoint is only half the battle. Evaluation of these "delisting" techniques, involves more than simple technical issues. Processing of mixed waste by any of these techniques may well require sanction by both the EPA and the public. An institution seeking to implement any of these technical processes may be required to obtain a RCRA Class B permit, a procedure that because of its political nature will be costly, slow, and characterized by low probability of success. Compounding the engineering uncertainty with a political uncertainty limits the successful outcome of a responsible technical solution. Through minimization and recycling programs, the labeled chemical industry has demonstrated a commitment to environmentally sound waste practices. The EPA and their equivalent state agencies are frustrating this pursuit by their restrictive definition of waste.

RECOMMENDATIONS

Dual regulation of mixed waste by both the NRC and the EPA leads to redundancies and technical inconsistencies. Technical inconsistencies arise in the definitions of site design specifications accepted by the two agencies. The EPA favors a dual liner and leachate collection system. The NRC discourages use of trench lines. Agency policies also conflict on waste package inspection.

The effects of NRC regulation, which controls the potential radiological hazard, provide the necessary protections to the environment and public health and safety. Safe radiological management practices extend to waste management. A system of proper management does exist.

The complexity of licensing a facility to meet requirements of two agencies has resulted in having no facility licensed for disposal of mixed low-level waste. Interim storage at institutional facilities is creating a practice of "second-best" waste management. Continued accumulation of mixed waste is making an eventual solution much more complex. The inability to agree upon licensing criteria is frustrating the process of siting low level waste compacts in accord with the federal Low-Level Radioactive Waste Policy Act. The judicious resolution of this conflict is for the NRC with the agreement of state agencies, to adopt sole jurisdiction for the disposal of mixed waste.

Ultimate resolution of the mixed waste issue lies in a technical, not a political solution. The labeled chemicals industry needs support in identifying and implementing innovative technical schemes for rendering mixed waste nonhazardous. Let us regard treatment of radioactive organic waste as a logical extension of the manufacturing process, not as a waste treatment requiring additional licensing. Development and implementation of environmentally sound, technical programs cannot be encumbered by political barriers.

CHAPTER 51

Disposal of Scintillation Cocktails

John W. McCormick

The advent of recent regulations restricting land disposal of certain types of liquid scintillation cocktails has sent the users searching for alternative disposal methods. These alternatives have been in the form of incineration and even drain disposal.

Normally, cocktail users have a mixed (radioactive and hazardous) waste when they generate spent cocktails in vial or bulk form. Most users do not have an on-site incinerator which is permitted to burn hazardous or radioactive wastes, so they send material off site for processing. Users of the new "drain disposable" cocktails may not be able to pour these cocktails down the drain due to state and local regulations.

For the past several years, the method that uses some of the best available treatment technologies is the process of crushing the vials, rinsing them, and then using bulk cocktails and the fluids generated by the crushing and rinsing operation as fuel in a concrete aggregate kiln. Normal incineration burns the vials or liquids and does not recover any beneficial use of the material.

Some users have attempted in-house processing, but have usually been unable to rinse crushed vials, or burn the fluid in their own incinerator. The burning was not for reuse, but for destruction and therefore needed a permit.

Currently three facilities in the country are permitted by their regulatory agencies to take mixed, radioactive or hazardous cocktails, in vial or bulk form, and process them. One of these uses fuel recovery as a disposal method while another uses deep well injection.

TECHNOLOGY

Scintillation Cocktails as Fuel

Using wastes as fuel in kilns has proven both environmentally and economically superior to incineration, and it is the primary alternate since landfilling is prohibitive. Organic solvents, which are blended to provide a high BTU fuel,

561

meet the needs of kilns and replace the fuel oil currently used in making concrete aggregate.

The EPA has found that cement kilns demonstrate high destruction efficiencies and low risk. Economically, since cement kilns are using the waste to make a product, the cost is a fraction of incineration and the long term liability is removed.

Most scintillation cocktails, including "drain disposable cocktails", contain a blend of organic solvents which provide a highly rated BTU fuel. Cement kilns are found in every state, and some are able to take hazardous waste as fuel.

CLEANING VIALS

Since scintillation vials usually contain radioactive and hazardous wastes (mixed wastes), the vials can not be emptied and then discarded. Incinerating the vials would produce a waste; a way to clean the vials had to be found. After doing some research, it was found that rinsing the vials with a liquid that could strip the mixed wastes off the vials turned out to be the best method we could find. Opening each vial, even with automation, proved to be costly. Crushing the vials, rinsing them, and then shredding them turned out to be the most efficient method.

Currently, some commercial processors use a large custom unit to separate the vials from the absorbents and then crush them into small strips. These shreddings are rinsed three times. The fluids from the vials and the rinsate are burned as fuel. The crushed vials are clean and, as such, nonhazardous; they are sent for disposal in an industrial landfill (Figure 1). The typical process flow of bulk liquids is shown in Figure 2.

PERMITS AND LICENSE

The State of Florida is the regulating body for both radioactive and hazardous materials. Through its separate regulating bodies the state issued processors the necessary authorization to perform the work described.

The radioactive license, issued by the Department of Health and Rehabilitative Services (HRS), allowed the facility to receive, process, and send to a kiln 29 different isotopes contained in liquid scintillation type fluids. HRS, like the NRC, was the only regulating body for mixed wastes. Now, the handling, storage, or treatment of mixed wastes is regulated by the EPA and/or a state agency.

The Department of Environmental Regulation, the Florida hazardous regulatory agency, issues permits for the storage of hazardous wastes. This permit allows the procesor to hold and bulk a number of hazardous wastes, including scintillation cocktails, for shipment to a kiln or other treatment facility.

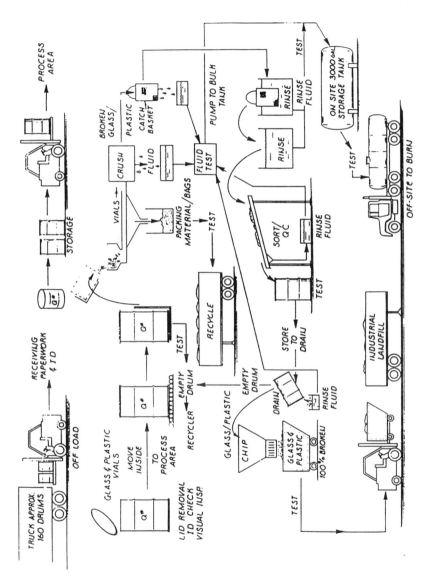

Figure 1. Typical process flow of glass/plastic vials.

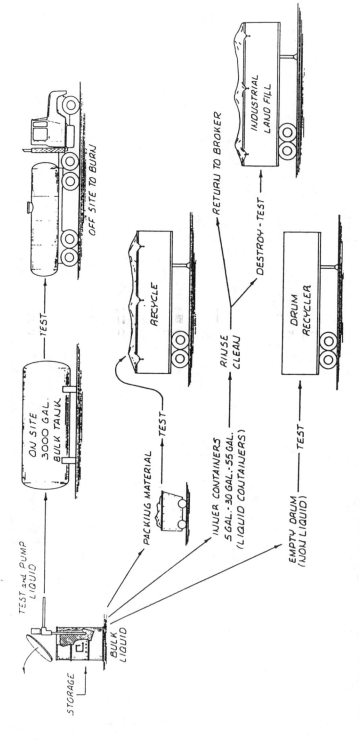

Figure 2. Typical process flow of bulk liquids.

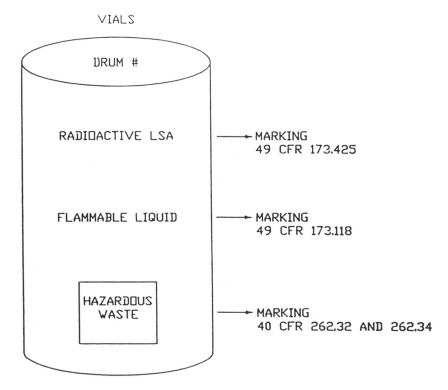

VIALS

DRUM #

RADIOACTIVE LSA ——→ MARKING
49 CFR 173.425

FLAMMABLE LIQUID ——→ MARKING
49 CFR 173.118

HAZARDOUS
WASTE ——→ MARKING
40 CFR 262.32 AND 262.34

Figure 3a.   Drum markings for vials.

QUALITY ASSURANCE PROGRAM

Every shipment sent off site to the kiln for use as a fuel needs to be checked to be sure it complies to the kiln specifications, and more importantly, complies to license limits. In order to ensure that every shipment is in compliance, development of a Q.A. program is necessary.

Samples from bulk containers, the fluid and the rinsate from vial containers should be analyzed on an individual and generator batch basis, respectively. These samples should be analyzed for the presence and concentration of known and unknown isotopes. A liquid scintillation counter can be used for ^{3}H, ^{14}C and beta, a single channel analyzer can be used for gammas, and a multi-channel analyzer is used for unknowns.

Any material found to be above our license limits should not be processed and we return them to the generator. In the case of already processes vials, the fluids and rinsate above limits can be returned to the generator. The vials have been cleaned, and they are disposed of properly.

Material within limits can then be added to the bulk tank in small amounts (50 to 100 gal), and then before making a shipment, the bulk tank should be analyzed.

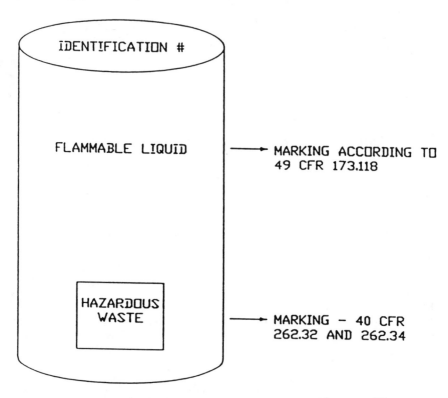

Figure 3b. Drum markings for vials containing a concentration of ^{3}H and/or ^{14}C not greater than 0.05 μCi/mL of media or containing radioactive matter except for ^{3}H and/or ^{14}C with a concentration less than 0.002 μCi/mL.

Using this method, we have never had a problem with the bulk tank. Anytime any material has been over the limit, or contained any unlicensed isotope, the generator of the waste was easily identified. Mixing generators and sampling at the end would prevent this identification.

REGULATORY BACKGROUND

Generators of waste material, including scintillation cocktail users, have a responsibility under federal regulations to identify and dispose of the waste materials they produce correctly. Since cocktail users often work with radioisotopes and fluids which may be hazardous, the following regulations would apply: EPA 40 CFR 260–265, NRC 10 CFR, or agreement state regulations, and DOT 49 CFR 171–173. Figure 3 a–d shows the drum markings for vial and bulk liquids.

Cocktails containing toluene, xylene, and benzene are considered hazardous

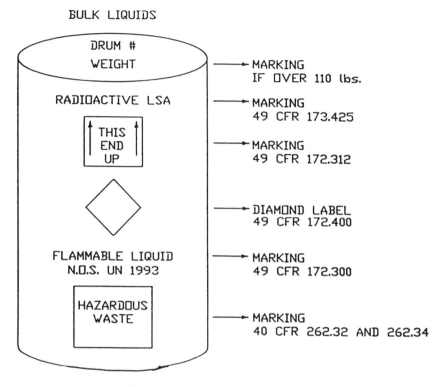

Figure 3c.   Labeling and marking of hazardous waste bulk liquids.

waste. Hazardous wastes can only be stored on a generator for ninety days (some exceptions allow up to 270 days) and need to be sent to a permitted facility for disposal (Table 1).

Many facilities who have been handling materials on site, are now being told by the EPA and state regulatory agencies that they lack the proper permits. Some facilities are able to secure the proper permits, while others are being required to ship off site.

Cocktails usually contain a radioisotope in tracer amounts. NRC regulations exempt for disposal liquid scintillation cocktails containing ^{3}H and/or ^{14}C at a concentration at or below 0.05 μCi/mL, provided the cocktails are disposed of according to any other applicable regulations. Nonscintillation cocktails with ^{3}H and/or ^{14}C below 0.05 μCi/mL are not exempt from the NRC regulation.

All cocktails containing radioisotopes, with the exception of the exempt ^{3}H and ^{14}C vials must be disposed of as a radioactive material at a licensed facility. If the cocktails are also a hazardous waste, the cocktail is a mixed waste and must be disposed of in accordance with both its hazardous and radioactive content.

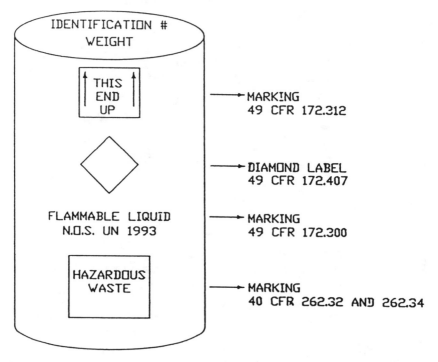

Figure 3d. Hazardous waste bulk liquids can contain liquids having a U.S. EPA hazardous waste number of F001, F002, F003, F005, or D001 defined in 40 CFR 261.

The Department of Transportation exempts isotopic concentrations below 0.002 μCi/mL as well as ^{3}H and/or ^{14}C less than 0.05 μCi/mL from being transported as a radioactive material. The DOT exemption may be misleading; the exemption is only for transportation. Licensed radioactive material is to be shipped from one licensee (the generator) to another licensee (the processor/disposer).

Waste considered hazardous is usually manifested on a hazardous waste manifest (Figure 4), and radwaste or hazardous waste on a HWM and radioactive shipment manifest (Figure 5). The exact shipment requirements can be found within the previously mentioned regulations or in a copy of processor facility instructions.

Table 1. BTU Ratings

Cocktail Type	BTU/gal
Xylene	39,000
Toluene	137,000
Benzene	135,000
Drain Disposable	34,000

UNIFORM HAZARDOUS WASTE MANIFEST

1. Generator's US EPA ID No. | Manifest Document No. | 2. Page 1 of | Information in the shaded areas is not required by Federal law.

GENERATOR

3. Generator's Name and Mailing Address
4. Generator's Phone ( )
A. State Manifest Document Number
B. State Generator's ID
5. Transporter 1 Company Name
6. US EPA ID Number
C. State Transporter's ID
D. Transporter's Phone
7. Transporter 2 Company Name
8. US EPA ID Number
E. State Transporter's ID
F. Transporter's Phone
9. Designated Facility Name and Site Address
10. US EPA ID Number
G. State Facility's ID
H. Facility's Phone

11. US DOT Description (*Including Proper Shipping Name, Hazard Class, and ID Number*) HM	12. Containers No.	Type	13. Total Quantity	14. Unit Wt/Vol	I. Waste No.
a.					
b.					
c.					
d.					

J. Additional Descriptions for Materials Listed Above
K. Handling Codes for Wastes Listed Above

15. Special Handling Instructions and Additional Information

16. **GENERATOR'S CERTIFICATION:** I hereby declare that the contents of this consignment are fully and accurately described above by proper shipping name and are classified, packed, marked, and labeled, and are in all respects in proper condition for transport by highway according to applicable international and national governmental regulations.

If I am a large quantity generator, I certify that I have a program in place to reduce the volume and toxicity of waste generated to the degree I have determined to be economically practicable and that I have selected the practicable method of treatment, storage, or disposal currently available to me which minimizes the present and future threat to human health and the environment; **OR**, if I am a small quantity generator, I have made a good faith effort to minimize my waste generation and select the best waste management method that is available to me and that I can afford.

Printed/Typed Name | Signature | Month Day Year

TRANSPORTER

17. Transporter 1 Acknowledgement of Receipt of Materials
Printed/Typed Name | Signature | Month Day Year

18. Transporter 2 Acknowledgement of Receipt of Materials
Printed/Typed Name | Signature | Month Day Year

FACILITY

19. Discrepancy Indication Space

20. Facility Owner or Operator: Certification of receipt of hazardous materials covered by this manifest except as noted in Item 19.
Printed/Typed Name | Signature | Month Day Year

ORIGINAL — RETURN TO GENERATOR

Figure 4. Uniform hazardous waste manifest.

CONCLUSION

The process described here is a viable alternative to the usual incineration of liquid scintillation cocktails. It is a working solution to the current generator's problem of mixed waste, and it provides fuel for working, permitted kilns. It is

RADIOACTIVE WASTE SHIPMENT & DISPOSAL MANIFEST

PAGE ______ OF ______ SHIPMENT NO ______

SHIP TO ____ FROM ____
QUADREX HPS, INC.
1940 N. W. 67th PLACE
GAINESVILLE, FL 32606
904/373-6066

LICENSE NO. ____________

GENERATOR NO. ____________

SHIP TO ________ FROM ________
NAME ____________
ADDRESS ____________
____
PHONE ____________
CONTACT ____________
LICENSE NO. ____________

CARRIER INFORMATION:
CARRIER NAME: ____________
TRACTOR LIC. NO. ________ TRAILER LIC. NO. ________
OPEN ____ CLOSED ____ PLACARDED: YES ___ NO ___
EXCLUSIVE USE: YES ___ NO ___ SEALED: YES ___ NO ___
SHIPPING WT. ________ CUBIC FEET ________
DRIVER'S SIGNATURE ____________

TOTAL FOR EACH CLASS: # OF PACKAGES	WEIGHT (POUNDS)	PROPER SHIPPING NAME & HAZARD CLASS (PER 49 CFR 172 101)	ID NUMBER
		Radioactive material, empty packages	UN2908
		Radioactive material, fissile, N.O.S.-radioactive material	UN2918
		Radioactive material low specific activity N.O.S.-radioactive material	UN2912
		Radioactive material, N.O.S.-radioactive material	UN2982
		Radioactive material, limited quantity, N.O.S.-radioactive material	UN2910
		Radioactive material, speci[illegible] N.O.S.-radioactive material	UN2974
		Radioactive material, instruments and articles-radioactive material	UN2911
		Thorium nitrate-radioactive material	UN2976
		Uranyl acetate (RQ-5000/2270)-radioactive material	NA9106
		Uranyl nitrate, solid (RQ-5000/2270)-radioactive material	UN2981

(9) ITEM NO.	(10) CONTAINER TYPE	(11) Container Volume (Cu. Ft.)	(12) Container weight (pounds)	(13) Physical form (see Note # 1)	(14) Waste description (see Note # 2)	(15) Solidification or absorbent media (see Note # 3)	(16) Chemical form/ Chelating agent (10 CFR 20.311)	Check below if < 0.1 % chelating agent by weight	(17) Radionuclide (10 CFR 20.311)	(18) Activity □ Curies □ Millicuries	(19) D.O.T. Sub-Type (A1, A2 LSA, etc.)	(20) Special Nuclear Material (Grams)	(21) Source Material (Kilograms)	(22) Waste Form Class 10 CFR 61.55	(23) Stability Class (See Note # 4) 10 CFR 61.56	Radiation Levels MR/HR (24) Container Surface	(25) Container At 1 Meter	(26) Transport Index	(27) Fissile Class	(28) D.O.T. Label/ Classification Label 49 CFR 173.444 10 CFR 61.57
																				Radioactive—
																				Radioactive—
																				Radioactive—
																				Radioactive—
																				Radioactive—
																				Radioactive—
																				Radioactive—

(7) SHIPMENT TOTALS

VOLUME CU. FT.	TOTAL # OF PACKAGES	SOURCE MATERIAL KGS	SPECIAL NUCLEAR MATERIAL (GRAMS) U-233	U-235	PLUTONIUM	TOTAL

SUPPLIES DELIVERED:

ACTIVITY

ACTIVITY TOTALS:	TRITIUM	C-14	Tc-99	I-129	ALL ISOTOPES
□ Curies					
□ Millicuries (10 CFR 20.311)					

Note # 1 - Physical Form Code
1. Solid
2. Liquid
3. Gaseous
4. Animal Carcasses
5. Scintillation Vials
6. Other

Note # 2 - Waste Description Codes
1. Vials
2. Dry Solid
3. Solidified Liquid
4. Biological
7. Filter Media
8. Dewatered Resins
9. Solidified Resins
10. Absorbed Aqueous Liquid
11. Absorbed Scintillation Liquid
*99. Other

Note # 3 - Solidified or Absorbent Media Codes
1. Diatomaceous Earth
2. Speedi Dry
3. Celatom (m.P78)
4. Floor Dry/Super Fine
5. Hi Dri
6. Florco or Florco X
7. Instant Dri
8. Safe-T-Sorb
9. Oil Dri (Safe n Dri)
10. Zonolite, Grades 2, 3, 4
11. Dow Media
12. Cement
13. Asphalt
14. Delaware Custom Media
15. Envirostone
99. Other

Note # 4 - NRC Stability Class Code
S - Stable
N - Not Stable

"OTHER" AS A DESCRIPTION MUST INCLUDE A WRITTEN DESCRIPTION ATTACHED TO THIS MANIFEST.

THIS IS TO CERTIFY THAT THE MATERIALS NAMED IN THIS MANIFEST ARE PROPERLY CLASSIFIED, DESCRIBED, PACKAGED, MARKED AND LABELED AND ARE IN PROPER CONDITION FOR TRANSPORTATION ACCORDING TO APPLICABLE REGULATIONS OF THE DEPARTMENT OF TRANSPORTATION AND ARE IN COMPLIANCE WITH ALL REQUIREMENTS APPLICABLE AT THE DESIGNATED DISPOSAL SITE, AND THAT THE MATERIALS ARE CLASSIFIED AND DESCRIBED IN ACCORDANCE WITH THE REQUIREMENTS OF 10 CFR, PART 61 OR EQUIVALENT STATE REGULATIONS

____________ AUTHORIZED SIGNATURE ____________ TITLE

DRIVER'S SIGNATURE ____________

Figure 5. Radioactive waste shipment and disposal manifest.

lower in cost than deep well injection, and it terminates the generator liability. For all these reasons, the method of crushing the vials, rinsing them clean, and burning the resulting fluids becomes the obvious choice for quick and complete disposal.

CHAPTER 52

Liquid Scintillation Waste

Ara Tahmassian, Jill Eveloff, and Howard Tisdale

Scintillation fluids are commonly used in many institutions for detecting low levels of low-energy beta emitters. This is particularly evident in bio-medical research institutions where radionuclides such as ^3H, ^{14}C, and ^{32}P constitute over 80% of all radioisotope usage. Once the samples have been analyzed and the results verified, they are considered waste and disposed of appropriately. At first glance this appears to be a relatively simple process, however, recently it has become increasingly more difficult.

Until a few years ago, the scintillation fluid waste containing radioisotopes was classified as either: 1. "De minimus"/"exempt quantity", or 2. "regulated" waste.

De minimus or exempt quantity waste was defined as "any liquid scintillation fluid with less than 0.05 $\mu Ci/mL$ of ^3H or ^{14}C." This waste was exempt from regulations governing radionuclides and was, therefore, allowed to be disposed of as hazardous chemical waste in appropriate landfills.

The regulated waste, on the other hand, was that which contained any of the radioisotopes not meeting the definition of the De minimus category. This was disposed of as "low-level" radioactive waste.

This convienient disposal method was halted abruptly on July 3, 1986, when the U.S. Environmental Protection Agency (EPA) issued a notice in the Federal Register regarding the applicability of the Resource Conservation and Recovery Act (RCRA) to the management of radioactive "mixed" waste.

Mixed waste is subject to the U.S. Nuclear Regulatory Commission (NRC) because of the presence of source, special, or byproduct material. In addition, it is subject to EPA regulations because it is classified as "hazardous" under RCRA regulations.

This joint NRC-EPA jurisdiction has, in effect, stopped the disposal of mixed waste in any of the three existing low-level radioactive waste disposal sites (Beatty, Nevada; Hanford, Washington; and Barnwell, South Carolina) in the U.S. This is a direct result of having to meet the conflicting requirements of both the NRC and the EPA. See Table 1 for differences in NRC and EPA

Table 1. Differences in NRC 10CFR Part 61 Regulations and EPA Requirements Under RCRA

	NRC	EPA
Land disposal	ALARA	Prohibited unless can prove no migration
Performance objective	Sited, designed, used, operated, and closed	None (No migration)
Siting	Dry areas	Out of floor plane or withstand 100 year floor. 1/4 mile from fault
Trenches	Unlined	Double plastic-lined
Leachate collection system	No	Yes
Manifest	Each shipment	Each shipment
Sample inspection	Periodic	Incoming
Container structural stability	High integrity	None
Liquids	Absorbents (limit 1% liquid) or solidified	Lab pack
Treatment	Yes	Recommends
Storage	Allows for decay of short-lived isotopes, extended storage	Three months
Monitor	Radioactivity only	Toxic pollution
Post closure institutional control	Up to 100 years	30 years
State and federal ownership	Yes	No
Financial requirements for corrective action	Under development	Extensive and detailed

Source: "Mixed Waste Producers Face Mixed Management Rules," Environmental Management News, Vol. 3, No. 3, April 1988.

regulations. Until November of 1985, mixed waste was routinely and safely disposed of at all three sites. After that time, it became necessary for the site operators to obtain a RCRA Part B permit in order to dispose of the mixed waste. To date, no waste disposal facility has been granted such a permit.

The mixed waste disposal problems are having a serious, adverse effect on the research activities of the generators. These generators are not allowed to use storage and decay methods to convert the mixed waste into typical hazardous chemical waste prior to disposal. If a generator stores the mixed waste for more than 90 days, then it must have a federal RCRA permit to be a treatment/storage/disposal facility (TSD permit). If the generator was approved as a RCRA facility, it would be subject to significant additional regulatory requirements, as well as legal risks.

There are scintillation fluids being marketed presently which, under the current EPA regulations, are classified as nonhazardous and even biodegradable. These fluids have a potential to remedy some of the immediate problems caused by the mixed waste issue. The long term solution to the mixed waste disposal issue, however, lies with the return of full jurisdiction to the NRC. This will in no way compromise public health and safety; however it will demonstrate appropriate exertion of regulatory authority.

In addition, it is strongly recommended that major research efforts be devoted to volume reduction techniques for mixed waste.

CHAPTER 53

Liquid Scintillation Counting
of Radon and its Daughters

Daniel Perlman

Measurements of airborne concentrations of radon and its daughter products are based upon the detection of those ionizations associated with its radioactive decay. One common method of detection employs ion chamber continuous monitors, which are portable, self-contained instruments that can accommodate a scintillation material to measure alpha particles that are emitted in a small collection chamber. Though these measurements can be completed in the order of 10 min, they represent radon concentrations at a specific time and do not necessarily reflect average concentrations measured over longer periods of time.

Etched-track detectors are another means of detecting radon concentrations. These are plastic detectors which are placed within the areas to be measured and then returned to a laboratory. There the tracks, produced in the plastic by the passage by alpha particles, are measured and correlated in ^{222}Rn concentration. These devices, however, are difficult to standardize and calibrate to ambient radon levels.[1]

Charcoal canisters are the principal monitoring devices currently being used. Diffusion canisters containing known amounts of charcoal adsorb ^{222}Rn over a period of approximately one day to one week. These canisters are then sealed and returned to a laboratory where they are counted.[2] Planchette gamma ray counters, typically used to count these detectors, are relatively inefficient, and much of the ^{222}Rn decay signal is not detected.

Liquid scintillation counting has determined the environmental concentration of gaseous radon and its daughters.[3-5] Since this is the only analytical method capable of counting the alpha and beta emissions of radon and its daughter products (Figure 1), it is theoretically the most sensitive procedure available. It was not until an inexpensive and portable means of radon adsorption was developed, however, that liquid scintillation could be considered a practical alternative to those methods previously mentioned.[6] Though it is two or three orders of magnitude more sensitive than planchette counting, liquid

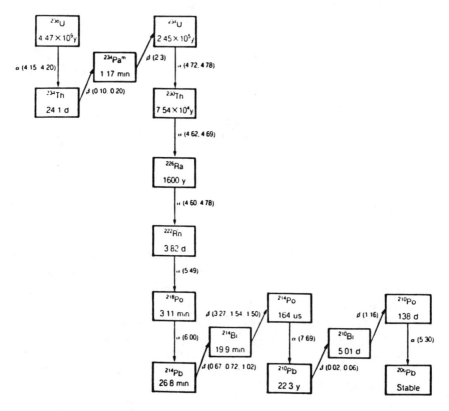

Figure 1. Radon and its daughter products.

scintillation counting inherently involves the generation of liquid scintillation waste.

Optimal extraction of ^{222}Rn and detection of its decay products has been attained in low-molecular-weight alkyl benzenes such as toluene and xylene. Prichard and Marien[6] discovered that radon could be extracted from charcoal by toluene and thereafter counted in conventional toluene based scintillators. Incubation of radon, bearing activated charcoal in xylene or toluene scintillation solutions, resulted in extraction equilibrium and radon daughter ingrowth within 3 to 12 hours. This depended on the size and quantity of charcoal granules and the temperature of extraction. These solvents, however, are associated with relatively high inhalation toxicity and environmental hazard.

Scintillation solvents that have been developed more recently, such as phenylxylylethane and its derivatives, alkyl and dialkylnaphthalene derivatives such as diisopropylnaphthalene, linear alkylbenzenes, substituted biphenyls and other high-molecular-weight aromatic molecules, have been tested for their ability to extract radon from charcoal. Unfortunately, some of these solvents, though biodegradable and much less toxic than toulene or xylene, have only exhibited as little as 20% the radon extraction efficiencies of toluene.

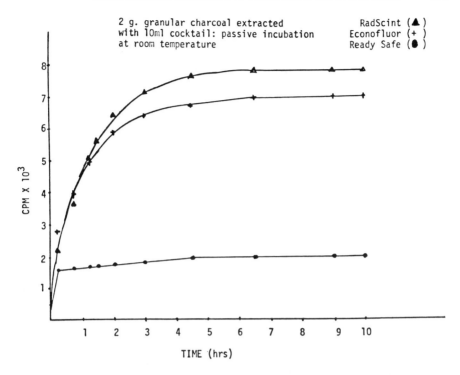

Figure 2.   Optimal extraction time of RadScint, Econofluor, and Ready Safe.

A major factor in charcoal detector application is the adsorption of water onto available charcoal binding sites. Increases in relative humidity at the test site will increase moisture pickup by the charcoal, and thus reduce the number of sites available to adsorb radon. Recently it has been reported that in diffusion-limited passive adsorption charcoal detectors desiccant, used in sufficient proportion to charcoal, prevents moisture accumulation in the charcoal.[7]

Experiments were performed to evaluate the efficiency of several commercial scintillation solutions to extract radon from charcoal.

Two grams of dry activated charcoal were placed in conventional 20 mL borosilicate glass scintillation vials. Since varying amounts of water can affect counting and extraction efficiencies, vapor was adsorbed to the dry charcoal at levels of 7.5% (w/w). This represented a relative humidity of approximately 40%, and at 17.4% (w/w), represented a relative humidity of approximately 70%. Vials were exposed in a controlled radon chamber for 96 hr. The vials were removed, filled with 10 mL of liquid scintillation each to be evaluated, allowed to incubate for 8 hr. and then counted. The results of the evaluation are presented in Table 1.

Optimal extraction time was also evaluated for several scintillators (Figure 2). High-molecular-weight solvents with surfactant had a tendency to plateau

Table 1.

	Material	High Humidity Charcoal ≈ 70% Relative Humidity		Low Humidity Charcoal ≈ 40% Relative Humidity	
		Net CPM	Relative Efficiency %	Net CPM	Relative Efficiency %
CONVENTIONAL SCINTILLATORS	Toluene (5 g/L PPO + 0.05 g/L POPOP)	980	67	1470	100
	Econofluor (DuPont)	968	66	1426	97
	ACS (Amersham)	1011	69	1435	98
	Instafluor (Packard)	953	65	1427	97
	Ready Organic (Beckman)	713	49	1036	70
LOW HAZARD BIODEGRADABLE SCINTILLATORS	Optifluor (Packard)	279	19	294	20
	Ready Safe (Beckman)	204	14	212	14
	Ecolume (ICN)	234	16	270	18
	RadScint (National Diagnostics)	1034	70	1471	100

quickly, but at a relatively low total extraction efficiency (Ready Safe). Conventional solvent based scintillators (Econofluor) generally require 6 to 8 hr to reach equilibrium. Modified high-molecular-weight scintillators, however, equilibrate at approximately the same time as conventional scintillators, 6 to 8 hr but at slightly higher total extraction efficiency.

Clearly, the advantages of low cost, high sensitivity, greater extraction efficiency, and greater reproducibility render liquid scintillation counting the preferred method for radon detection. Problems associated with the disposal of scintillation waste and the occupational exposure hazards of conventional toluene and xylene scintillators can be eliminated by employing the less hazardous, biodegradable solvents that have been modified to improve the extraction efficiencies of radon from charcoal. Scintillation solutions, such as National Diagnostics' RadScint™, have been designed for radon detection with these modified solvents. RadScint enables laboratories to optimize counting and extraction efficiencies of radon samples, while also reducing occupational exposure and hazardous waste generation.

REFERENCES

1. Alter, H.W. and R.L. Fleischer. "Passive Integrating Radon Monitor for Environmental Monitoring," *Health Physics*, 40:693 (1981).
2. Cohen, B.L. and E.S. Cohen. "Theory and Practice of Radon Monitoring With Charcoal Adsorption," *Health Physics*, 45:501 (1983).
3. Assaf, G. and J.R. Gat. "Direct Determination of Short Lived Radon Daughter Products on Air Filters By Liquid Scintillation Counting Using a Delayed Coincidence Technique," *Nucl. Instrum. Methods*, 49:29–37 (1967).
4. Kurosawa, R. "Determination of Concentration of Radon Daughters by Liquid Scintillation Technique," *Waseda Daigaku Rikogaku Kenkyusho Kokoku*, 51:1–11 (1971).
5. Horiuchi, K. and Y. Murakami. "A New Method For the Determination of Radon In Soil Air By the 'Open Vial' and Integral Counting With a Liquid Scintillation Counter," *J. Radioanal. Chem.*, 80(1–2):153–163 (1983).
6. Prichard, H.M. and K. Marien. "Desorption of Radon From Activated Carbon Into A Liquid Scintillator," *Anal. Chem.*, 55(1):155–157 (1983).
7. Perlman, D. Inventor. "Method of and Passive Apparatus for Detecting Radon," Brandeis University, Assignee. U.S. Patent 4812648. March 14, 1989.

CHAPTER 54

History and Present Status of Liquid Scintillation Counting in China

Shou-li Yang

INITIAL STAGES OF LSC IN CHINA

Studies of the liquid scintillation counting (LSC) technique began in 1958 at the Institute of Atomic Energy (IAE), Chinese Academy of Sciences (CAS, also know as Academia Sinica). This work was carried out with an imported liquid scintillation counter with single photomultiplier (PM). At the same time, Professor Shih-Chen Wang, Ph.D., of the Institute of Radiation Medicine (IRM), Chinese Academy of Medical Sciences (CAMS), was preparing to translate *Isotopic Tracers in Biology* (Martin D. Kamen, New York: Academic Press, 1957). While the translation was in progress, he read the revised and enlarged edition of the book which was published in December 1959 and noted immediately the increased number of entries in the appendix on the application of liquid scintillation counting to the major nuclides used in biology. Realizing the importance of LSC in bio-medical research, he not only purchased an Ecko single tube LS counter, but also took on (in 1962) two postgraduate students, Zong-qin Xia and Han Lin, and allowed them to study cholesterol and thyroid hormone metabolism, using LSC for detection of ^3H and ^{14}C.

In 1965, I also went to the Institute and studied extensively the methodology of LSC. Since then, due to extensive interchange of scientific personnel within the country, LSC methods have been spread throughout China. The scientists who worked in the Institute of Radiation Medicine not only have made a great contribution to the development of LSC applications in China, but have also become well-known internationally in the field of nuclear medicine. Professor Shih-Chen Wang held the post of the first president of the Chinese Society of Nuclear Medicine (CSNM) and is now the academic standing member of CAS, the highest honor in Chinese academic circles.

The design and manufacture in China of liquid scintillation counters came later. In 1968, the scientists of IAE, CAS, and Xi'an Nuclear Instrument Factory cooperated in the design of a liquid scintillation counter with two photomultipliers and two single channel analyzers. This counter was called

Model FJ-353 double channel liquid scintillation counter. Only a few were produced in 1970, but actual availability of the counter did not occur until "the National Training Course in Isotopic Techniques and Their Biomedical Applications" in 1973, where it was shown as a demonstration instrument. In 1972, Pei-dong Jiang, Han-ying Jiang, et al., the Institute of Biophysics, CAS, and engineers of the Instrument Factory of Beijing University cooperated to begin the design of the first automatic liquid scintillation system controlled by computer. This instrument, called model YS-1 automatic liquid scintillation spectrometer, was completed in 1976. Then, personnel at Xi'an Nuclear Instrument Factory, Institute of Biophysics, Shanghai Institute of Nuclear Research (SINR), and Beijing Nuclear Instrument Factory (BNIF) designed and successfully produced a variety of LS counters; these became the prototypes for the counters presently made in China.

PROFFESIONAL COMMITTEES

Along with the active development, design, and manufacture of liquid scintillation counters, Chinese specialists have held many meetings for appraising the quality of LS instruments and discussing scintillation methodology. In August of 1979, LS specialists belonging to more than forty units came together in Beijing and drafted a document proposing a national conference on LSC and an LSC committee. The document was submitted to the Preparatory Committee of the Chinese Society of Biophysics (CSBP). At present, the Chinese Professional Committee of Liquid Scintillation Counting (CPCLSC) reports to the Chinese Society of Nuclear Electronics and Detection Technology (CSNEDT) and to the CSNM.

The CPCLSC consists of the ten following members:

1. Han-ying Jiang, Institute of Biophysics, CAS,
2. Pei-dong Jiang, Institute of Biophysics, CAS, general head,
3. Bao-Sheng Liang, BNIF,
4. Zhang-zhi Wang, SINR,
5. Xi-fan Wu, Xi'an Nuclear Instrument Factory,
6. Zhong-qin Xia, Shanghai Second Medical University,
7. Jing-cheng Xiao, IAE, Chinese Academy of Agriculture,
8. Li-ping Yang, an Nuclear Instrument Factory, in charge of contact with CSNEDT,
9. Shou-li Yang, Institute of Clinical Medical Sciences, China-Japan Friendship Hospital, in charge of contact with CSNM,
10. Huai-xin Zhang, IAE, CAS.

Shao-wan Lu, Institute of Biophysics, CAS, is the secretary of the committee.

In addition, the Radiocarbon Dating Professional Group that reports to the Chinese Quaternary Research Committee is also involved in developing LS technology. The head of the group is Shi-hua Chou, Institute of Archaeology, Chinese Academy of Social Sciences.

Table 1. Four National Conferences on LSC Counting

	1st	2nd	3rd	4th
Dates	—	—	Nov. 12–16 1986	Nov. 24–27 1988
Location	—	—	Wuxi	Xi'an
Delegates	—	—	90	118
No. of papers	109	71	98	67
Supported society	CSBP & CSNEDT	CSNEDT	CSNEDT	CSNEDT
Financial support	XNIF, etc.	BNIF, LKB, XNIF, ect.	Beckman, HMIF, etc.	Aloka, Beckman LKB, Packard, XNIF, etc.

NATIONAL CONFERENCES

The CPC LSC has organized four national conferences (one every two years). These are listed in Table 1. The delegate composition of the conferences indicates that more than 50% of the delegates were with various institutes in the fields of biology and medicine. The importance of instrument factories making LSC counters lies not only in the production, but also in the development of new products and financial support of academic exchanges. Table 2 contains a classification of papers in the four national conferences according to their contents. Obviously, sample preparation and its applications, instrument quench correction, and instruments, electronics, and vials are the three main topics studied by Chinese LSC specialists.

I should like to note that LSC specialist Professor C.T. Peng, attended the first national conference at the invitation of the Professional Committee and gave a lecture about distinguishing multi-nuclides in a sample.

The Radiocarbon Dating Professional Group organized four national conferences on ^{14}C dating from 1982 to 1988. The number of delegates and papers in these conferences was fewer than the national conferences on LSC, but the topics discussed were more focused.

BOOKS ON LSC

In 1973, under the direction of the Ministry of Public Health of the People's Republic of China, the Institute of Radiation Medicine, CAMS organized an advanced isotopic technique training course in which the theory and practice of LSC were taught and demonstrated. Drs. Zhong-qin Xia, Shou-li Yang, Pei-dong Jiang, and Ni-na Jiang taught cocktail and sample preparation, quench correction, instrumentation methodology, liquid scintillation counter systems, and absolute activity measurement. The lecture notes were later published as a book, entitled *Isotope Technique and Their Biomedical Applications*, (Beijing: Science Press, 1977 and 1979). Because all the contributors in

Table 2. Classification of the Papers in the Four National Conferences on LSC

Categories	1st %	1st No.	2nd %	2nd No.	3rd %	3rd No.	4th %	4th No.	Total %	Total No.
Instrument, electronics, vial	18.3%	20	12.7%	9	15.3%	15	17.9%	12	16.2%	56
Instrumental methodology, quenching correction	19.3%	21	22.5%	16	19.4%	19	22.4%	15	20.6%	71
Scintillator, cocktail	6.4%	7	9.9%	7	10.2%	10	10.4%	7	9.0%	31
Sample preparation, and its applications	20.1%	22	23.9%	17	23.5%	23	23.9%	16	22.6%	78
Chem-(bio-) luminescence, B,r,Cerenkov counting	11.9%	13	16.9%	12	9.2%	9	6.0%	4	11.0%	38
Low-level counting	10.1%	11	5.6%	4	11.2%	11	6.0%	4	8.7%	30
Absolute activity counting	9.2%	10	2.8%	2	5.1%	5	9.0%	6	6.7%	23
Others	4.6%	5	5.6%	4	6.1%	6	4.5%	3	5.2%	18
Total		109		71		98		67		345

the book are prominent in Chinese nuclear medicine, the book is widely distributed and well known. It won an award from the National Science Congress in 1979 for excellence in both technical depth and clarity of presentation. At present, the second edition of this book is in press (edited by Shih-Chen Wang, Han Lin, and Chien Chou, Science Press, Beijing).

Eight other books on LSC have been published in China; two of them are written by Chinese scientists, one is translated and edited, and the other five are translations. Among these, two books are widely circulated and well received by the readers: *Advances and Applications in Liquid Scintillation Counting*, (edited by Shou-li Yang, Pei-dong Jiang, and Han Lin, Science Press, Beijing, 1987) and *Liquid Scintillation Counting and Its Application to Biology* (translated and edited by the LSC group of Institute of Biophysics, CAS, Science Press, Beijing, 1979). The former is in the second edition while the latter is taken mainly from *Applications of Liquid Scintillation Counting* by D. L. Horrocks (New York: Academic Press, 1974) and *Biological Applications of Liquid Scintillation* by V. Kobayashi and D. V. Maudsley (New York: Academic Press, 1974).

In addition, the Radiocarbon Dating Professional Group has edited *Proceedings of the First National 14C Dating Conference* (Beijing: Science Press,

Table 3. Main Journals Published Papers of LSC

Translated English Name	Chinese Spelling Name[a]	Publishing Period	Sponsored Society or Unit	Start Year
Nuclear Electronics and Detection Technology	HEDIANZIXUE YU TANCE JISHU	bimonthly	CSNEDT	1981
Chinese Journal of Nuclear Medicine	ZHONGHUA HEYIXUE ZAZHI	quarterly	CSNM	1981
Science and Technology of Atomic Energy	YUANZINENG KEXUE YU JISHU	bimonthly	CIAE	1964
Nuclear Techniques	HEJISHU	monthly	CNS	1978
Nuclear Science and Engineering	HEKEXUE YU GONG-CHENG	quarterly	CNS	1981
Nuclear Science and Techniques (English)		quarterly	CNS	1990
Radiation Protection	FUSHE FANGHU	bimonthly	CSRP	1981
Chinese Journal of Radiological Medicine and Protection	ZHONGHUA FANGSHE YIXUE YU FANGHU	bimonthly	CSRM	1981
Progress in Biochemistry and Biophysics	SHENGWU-HUAXUE YU SHENGWUWULI JINZHAN	bimonthly	IBP, CAS	1974
Archaeology	KAOGU	monthly	IA, CASS	1959

[a]Some important international abstracts quoted these names.

1984), in which there are 35 papers, including 11 on the methodology of liquid scintillation counting.

PROFESSIONAL JOURNALS

The papers and reviews on liquid scintillation counting are published mainly in professional journals sponsored by the societies of nuclear science (Table 3). They are also disseminated in the journals of other professional societies or learned units. The starting publication dates of the journals listed were mostly in the early 1980s. Before then, the majority of the research papers on LSC were published in informal learned journals, for example, *Radiation Medicine*, *Nuclear Protection*, and *Nuclear Instrument and Methodology*. Although these journals have stopped publication, they have carried many papers and reviews about LSC, nuclear medicine, nuclear instruments, and methodology, and therefore played an important part in information exchange during the early period. In fact, *Radiation Medicine* and *Nuclear Protection* established foundations for the publications *Chinese Journal of Radiological Medicine and Protection* and *Radiation Protection*, respectively, in 1981.

It must be emphasized that the *Nuclear Science and Techniques*, which is a new professional journal in English, will start as a publication of the Chinese Nuclear Society in 1990. This event shows the increasing growth of information exchanges between China and nuclear scientists around the world.

INTERNATIONAL ACADEMIC EXCHANGES

The first international academic exchange in LSC took place in 1979 at Beijing Normal University, when Dr. C.T. Peng, School of Pharmacy, University of California, gave a series of penetrating lectures on LSC and labeled isotope techniques. The training courses lasted one month and the number present from Beijing and other provinces was about two hundred; it was an exceptional occasion. Since then, Dr. Peng has visited China many times and participated in a variety of publications and discussions.

Dr. J.E. Noakes, Center for Applied Isotope Studies, University of Georgia, U.S.A., has visited China at the invitation of the Institute of Biophysics, CAS, after the conference on "Advances in Scintillation Counting" in 1983. Not long after, Dr. H. Polach, Radiocarbon Dating Research Laboratory, Australian National University, visited the Chinese Institute of Geology. They not only gave special reports but also exchanged low-level counting and instrumentation techniques with the Chinese LSC specialists.

Japanese LSC specialist Dr. H. Ishikawa visited the China-Japan Friendship Hospital, BNIF, and the Chinese Academy of Metrologic Sciences in 1988. He gave an excellent report about the efficiency tracing technique (ETT).

We hope that more and more LSC scientists could visit China. At the same time, we also hope that Chinese LSC specialists could be more extensively engaged in the LSC research being carried out in many distinguished foreign laboratories.

INSTRUMENTS

The Xi'an Nuclear Instrument Factory is the major manufacturer of LSC instruments and produces more than 80% of the LSC instruments made in China. Xi-fan Wu was the leader of this factory and Li-ping Yang served as the main LSC electronic specialist. They made great contributions to early LSC instrument manufacture in China. At present, there are about twenty LS counter models made in China; these are detailed in Tables 4 and 5.

In the 1980s, significant LSC interests in China included the development of (1) routine LS counters with multi-channel analyzer capability and software

Table 4. LSC Instruments Made in China

Model	Appeared Year	Features	Designer & Manufacturer
FJ–353	1970	Double channels, 3 discriminators, sample capacity 25, semi-automatic, GDB-52 tube, E (%) of ^{3}H > 40%, background < 40 cpm	IAE, XNIF
FJ–2101 (G)	1979	NIM system, automatic print, others the same as JF–353	XNIF
YS–1	1976	Controlled by model J30 date processor, 3 + 2 channel system, 100 samples, ^{137}Cs external standard	IBP
YS–2–1	1979	Automatic measurement and date processing, 3 + 2 system, 200 samples, ^{137}Cs external standard	IBP
YS–2–2	1979	No date processing function, rest is the same as YS–2–1	IBP
YS–3	1977	Single sample, manual, 2 + 2 channels	IBP
YS–A	1980	60 samples, manual, 2 + 2 channels	TMIF, IBP
YS–B	1980	200 samples, automatic, 2 + 2 channels	TMIF, IBP
YSJ–76	1976	3 channels, ^{137}Cs external standard, 4 counting models, automatic print, 200 samples	SINR, SSMU
FJ–2100	1979	The main features are the same as YSJ–76	SINR, SNIF
YSJ–78 (YSJ–1)	1978	An improved model of YSJ–76	SINR
FJ–2105	1982	Controlled by Z80 microcomputer, 3 + 2 system, 200 samples	XNIF
FJ–1907	1982	Double channels, 24 samples, manual	BNIF
FJ–2107G	1985	NIM standard, Apple II desk computer, others are the same as JF-2105.	XNIF
FJ–2107	1985	No computer, others are the same as JF-2107G	XNIF
FJ–2108	1986	3 + 2 system, cell grown index measurement, sample chamber with constant temperature and shaking	XNIF
FJ–2112	1986	Apple II desk computer for control and data processing, 300 standard vials or 480 minivials, 3 + 2 system	XNIF
FJ–2115	1987	Controlled by single plate computer rack type conveyer, 10 samples	XNIF
FT-1913	1986	TRS–80 microcomputer, CRT display 2048 channels	BNIF

Table 5. Low-level LSC Instruments Made in China

Model	Appeared Year	Features	Designer & Manufacturer
DYS-1	1980	Single sample, manual, NaI anti-coincidence shield (20 × 150 mm, hole 80 mm), sample volume: 2–100 ml, for 5 mL ^{14}C-benzene, E = 80%, B = 0, 54 cpm, E^2/B > 10,000, detection limit for ^{3}H in 50 mL water (add 50 mL Instagel cocktail) is 1 Bq/L. water (t = 30 min, E = 68.3%)	IBP
DYS-2	1982	No anticoincidence, sample volume: 5, 10, 20 mL, 10 samples, automatic, 5 mL ^{14}C-benzene E > 70%, B < 3 cpm [glass vial]; FOM[a] > 1700 (glass vial), > 2000 (Teflon vial); detection limit for ^{3}H: 40 T.U.	IBP
DYS-3	1981	5 samples, sample volume: 5–100 mL NaI anticoincidence, controlled by Apple II computer	IBP
FH-1915	1982	NaI anticoincidence shield, TP-80 mirocomputer controlled, ^{14}C: E = 76.3%; B = 0.4 cpm; FOM = 14550; ^{3}H: E = 52.6%, B = 1.11 cpm, FOM = 2492	BNIF
FH-1916	1983	An improved model of DYS-2,	IBP,BNIF
FH-1935	1985	New soft multichannels, TRS 80 microcomputer low-level counting software packages	BNIF
DYS-84	1984	Anticoincidence with cyclic counting tube or matrix of counting tube, 24 samples, controlled by Zijin-II microcomputer	IBP
DYS-86	1986	With soft multichannels, the rest is the same as DYS-84	IBP
DYS-88	1988	Anticoincidence with BGO crystal, the rest is the same as DYS-84	IBP

[a]FOM: figure of merit

sophistication, (2) low-level background liquid scintillation counters for environmental protection and radiocarbon dating, such as Models DYS-86 and FH-1935, (3) economical, simple and convenient LS counters for use in basic hospitals and small research units, such as FJ-2115, and (4) counters that can simultaneously measure both beta- and gamma-emitting nuclides.

The anticoincidence shield method in low-level LS counters is another subject of increased attention in China. Han-ying Jiang, Shao-wan Lu, and Ting-kui Zhang, Institute of Biophysics, reported BGO crystals as an anticoincidence shield and compared it with other anticoincidence methods. Some authors have studied the cross talk and cross talk discriminator and applied them to the products of the LS counter.

ORGANIC SCINTILLATORS, COCKTAILS, AND SAMPLE PREPARATION

In the 1970s, Nankai University, Shanghai Institute of Pharmacology, and Shanghai No.1 Reagent Works, etc., first produced PPO. Since then, Nankai University has studied butyl-PBD and BZOB. Presently, Shanghai No.1 Reagent Works has produced more than twenty scintillators, such as p-TP, PPO, PBD, butyl-PBD, BZOB, POPOP, DM-POPOP, BBO, DPA, DPH, bis-MSB, etc.

In 1982, Chun-liang Xu prepared a new scintillator, p-TTP [2,5-di(p-tolyl)1,3,4-oxadiazole], an analog of m-TTD [2,5-di(o-tolyl)1,3,4-oxadiazole]. Its luminescence efficiency is the same as butyl-PBD and its optimum concentration is 0.4 to 0.8%. In 1984, Zhen-gue Yang, Shanghai No.1 Reagent Works, synthesized Triton X-100 for LSC. Before then OP-115, an analog of Triton X-100, had been used as an emulsifier, but its background is higher and counting efficiency is lower than Triton X-100.

The counting of substances supported on filter paper or millipore filters is a simple method of sample preparation. Therefore, it has been widely used in the assay of bio-medical samples in China. In the 1960s, Sheng-li Huang et al., used this technique to measure ^{14}C in digested animal tissue. This method was popular up to the late 1970s, when glass filter paper appeared in the Chinese market. The metabolic experiments of a variety of important drugs in that period, such as the male contraceptive drug, gossypol, were also completed using this method. We also analyzed the energy spectrum of ^3H sample supported on millipore filter and described a LSC method for ^{125}I on the supporters. At present, the supports being used are glass filter papers and millipore filters. About half of the total bio-medical samples in China are assayed via this technique.

RADIOCARBON DATING

In the 1970s, the radiocarbon laboratory in China completed the conversion of dating methods from gas phase techniques to LSC. The most famous radiocarbon dating laboratories are parts of the Institute of Archaeology, Chinese Academy of Social Sciences (CASS), Department of Archaeology, Beijing University; Institute of Geology, CAS, etc.. They have made great contributions to Chinese prehistory and archaeology and are widely recognized internationally.

ACKNOWLEDGEMENTS

I wish to thank the conference organizers Drs. H.H. Ross, J.E. Noakes, and J.D. Spaulding for the financial assistance to attend this conference and Prof. C.T. Peng for his encouragement. I also wish to thank Miss Isabel Li for her valuable assistance in polishing up my English and lastly Mr. Ming Hu for typing the manuscript.*

---

*Note: This manuscript has undergone extensive revision and condensation in order to meet the style requirements and space limitations of the proceedings. We sincerely hope that we have not changed the substance or connotation of any information intended to be conveyed by the author. Eds.

CHAPTER 55

LSC Standardization of ^{32}P in Inorganic and Organic Samples by the Efficiency Tracing Method

L. Rodriguez, J.M. Los Arcos, and A. Grau

INTRODUCTION

The wide field of applications of ^{32}P in Medicine, Agriculture, and Biochemistry requires the highest achievable accuracy in its standardization.

Since ^{32}P is a pure beta emitter, liquid scintillation counting is the most appropiate method of measurement. Nevertheless, this method presents several difficulties, the wall adsorption in glass vials and the instability of phosphoric compounds in toluene-based scintillators, that very often make the detection efficiency unpredictable.

In order to overcome the adsorption problem, the commonly used procedures have recourse to a previous siliconing treatment of the vial walls[1] or to the addition of a variable amount of inactive carrier that in some cases does not produce the expected effect.[2]

On the other hand, the low solubility of phospor-derived inorganic compounds in liquid scintillators can be overcome by synthesizing a ^{32}P labeled organic compound. This allows sample preparation showing a high stability for toluene-based and other scintillation cocktails.[3]

In this chapter, a systematic study of the behavior of $H_3{}^{32}PO_4$ aqueous samples was first made; either the vial walls were treated with silicon, or carrier was added in variable amounts. Secondly, in view of the results obtained with inorganic samples, a triphenyl-phosphine oxide compound labeled with ^{32}P has been synthesized in an attempt to obtain better samples when incorporated into organic scintillators.

Finally, both the organic and inorganic stable samples have been used to standardize the ^{32}P by the efficiency tracing method, with an experimental calibration curve of ^{3}H efficiency obtained from a set of quenched standards.

PREPARATION OF INORGANIC SAMPLES

All the inorganic samples have been prepared from a carrier-free, aqueous solution of H_3 $^{32}PO_4$. These samples have been measured in INSTAGEL, since a very high water content can be incorporated without any problem. The volume of scintillator has been 10 mL to minimize the wall effect of high energy electrons.

Four different kinds of samples have been prepared and studied.

1. Type A samples were obtained by direct addition of $H_3^{32}PO_4$ into the liquid scintillator with no other process involved.
2. For the type B samples, the vial was previously inmersed in silicone oil for 24 hr at room temperature before adding the $H_3^{32}PO_4$.
3. Type C samples were prepared with six different amounts of carrier, by adding 6, 12, 24, 36, 48, and 60 μg of inactive H_3PO_4 to the scintillator volume before the active solution was incorporated.
4. Finally, for type D samples, the carrier adsorption was compelled by exposing the vial to an inactive solution of H_3PO_4, with 85% concentration, diluted with the same volume of water. After a 24 hr exposure at room temperature, the solution was removed and the vial was filled with the scintillator and the active solution.

PREPARATION OF ORGANIC SAMPLES

After a structural analysis of several organic compounds that could be labeled with ^{32}P and a previous work with tributyl-phosphate labeled samples,[3] the triphenyl-phosphine oxide was selected due to its low oxygen contents and ability to reduce wall adsorption.[4] The triphenyl phosphine oxide was obtained in two steps. In the first step, the $^{32}POCl_3$ was synthesized from the H_3 $^{32}PO_4$, by the reaction[5]:

$$H_3^{32}PO_4 + H_3\,PO_4 + 4PCl_5 + H_2O \longrightarrow 5\,^{32}POCl_3 + 5\,HCl \qquad (1)$$

This process was carried out by mixing 10 μCi of $H_3^{32}PO_4$, 0.08 mmol of H_3PO_4 and 13 mmol of H_2O in a two-necked flask.

One neck was connected to a micro dropping funnel and the other one was equipped with a drying tube. A dry ice-acetone bath was used to cool the flask. In the funnel was placed 12.9 mmol of phosphorus pentachloride. The reaction started by adding small amounts of phosphorus pentachloride and allowing the mixture to thaw slowly, until the reaction became controlled without freezing. At that point, the remainder of the phosphorus pentachloride was rapidly added.

Once the reaction mixture was at room temperature, it was refluxed for 15 min and then immersed in a dry ice-acetone bath. Warm water circulated through the condenser to help distilling from the con denser into the flask. The phosphoryl-^{32}P chloride was obtained by cooling the flask at $-5°C$ and distillating under vacuum. The radiochemical yield was about 90%.

In the second step, the triphenyl phosphine oxide was produced through the reaction[6]

$$C_6H_5MgBr + {}^{32}POCl_3 \longrightarrow (C_6H_5)_3\,{}^{32}PO + MgBrCl \qquad (2)$$

The 36 mmol of the Grignard agent were diluted to 100 mL of absolute ether, then 12 mmol of the phosphoryl ^{32}P chloride were slowly added into the flask. The flask was cooled in a dry ice-acetone bath. Once the reaction mixture warmed to room temperature, it was heated at $45°C$. The white precipitate was hydrolyzed, extracted with absolute ether, washed with water, and distilled under vacuum. The radiochemical yield attained 45%. The solid triphenyl phosphine oxide labeled with ^{32}P was diluted in benzene and then dispensed to the measurement vials.

The ^{32}P labeled samples obtained in this way have been measured with three different scintillators: a toluene-based solution composed of 1 L of toluene, 5 g of PPO, and 0.3 g of dimethyl POPOP and the commercial cocktails HISAFE II and INSTAGEL. Vials containing 10 mL of INSTAGEL and 15 mL of HISAFE II and toluenic scintillator were used.

EQUIPMENT AND CHEMICALS

The chemicals used for the preparation of organic and inorganic samples were PCl_5, H_3PO_4 (85%), Na_2SO_4, C_6H_5MgBr of analytical grade, an aqueous solution of $H_3{}^{32}PO_4$ containing about 40 MBq/mL and an n-hexadecane-^{3}H standard solution around 50 kBq/mL. Cl_4C was used as a quenching agent.

Glass vials with an internal diameter of 2.5 cm and very low Potassium contents were used in all the measurements and the scintillator volumes were dispersed by means of Brand instruments, having a calibration uncertainty lower than 1%.

Gilson micropipettes with an uncertainty calibration lower than 1% were used to incorporate radioactive samples into the scintillator volume.

The measurement system was liquid scintillation counting equipment from LKB, model Rackbeta 1219 Spectral. It has a 10 μCi source of ^{226}Ra as external standard source for the quench determination.

RESULTS OF SAMPLE PREPARATION

The liquid scintillation spectrometer checked the stability of the samples. A 24 hr period of thermal conditioning inside the counter has been always respected before starting the final measurements.

Stability of inorganic samples

The time stability of type A samples was studied for 17 days. Figure 1 shows the time evolution of the counting rate in the ^{32}P window from channel 531 to channel 830. This helped avoid the possible interference of the ^{33}P contaminant commonly accompanying the ^{32}P. Radioactive decay correction has been applied to take into account the ^{32}P half-life of 14.29 days. A strong counting loss was observed in these samples. Initially, it amounted to more than 5% a day and seemed to reach the maximum degradation in 15 days; it showed a total count loss of more than 35%.

The unstability is also evident in Figure 2, where the spectra were measured at 0, 6, and 17 days after the preparation. The spectra have been normalized in the figure to the highest count rate so that the radioactive decay of ^{32}P is compensated and the degradating effect of wall adsorption becomes clear.

Type B samples were measured for 12 days. The count rate is shown in Figure 1, and the spectra taken at 0, 6, and 12 days appear on Figure 2. The effect of the wall treatment is not clear. The initial counting loss was even greater than for type A samples, but after five days, it seemed to moderate with strong, nonstatistical fluctuations. The spectral degradation also changed, corresponding to the wall treatment, but the sample still is clearly unstable.

The behavior of type C samples was studied for 12 days. Figure 3 shows the count rate evolution of samples containing 6, 24, and 60 μg of carrier. The stability is good enough for values greater than 24 μg with no systematic trend, and it is good enough for statistical fluctuations with 0.2% standard deviation in agreement with the experimental measurements. The 6 μg sample shows a strong counting loss of 0.8% a day. The spectral evolution appears in Figure 4, where the degradation is only evident for the 6 μg carrier sample.

The type D sample also reveals a good stability for 20 days, better than 0.06% a day, even though it has a greater ^{33}P contribution, due to its late preparation date, as can be appreciated in Figures 5 to 6.

Stability of Organic Samples

The behavior of the organic samples has been investigated in three different scintillators, in toluene and HISAFE II for 44 days and in INSTAGEL for 12 days. Counting rates refer to the same ^{32}P window set for inorganic samples, and they have been corrected for ^{32}P radioactive decay.

Figures 7 and 8 show the time evolution of the count rate and the spectra measured at 0, 20, and 44 days for the toluene and the HISAFE II scintillators. The stability in ISTAGEL along the 12 day term and the spectra at 0, 6, and 12 days are shown in the same figures.

In consequence, the diphenyl phosphine oxide sample is stable enough in all three scintillators with no significant trend in the half-life corrected count rate values. The fluctuations observed are within the statistical limits, with stan-

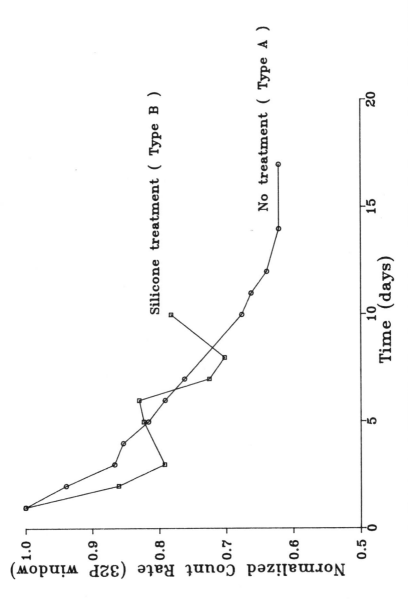

Figure 1. Time evolution of H$_3$^{32}PO$_4$ samples in INSTAGEL. O: Type A, no treated samples; □: Type B, silicone treated walls. The count rates are half-life corrected and then normalized to the value in the first day.

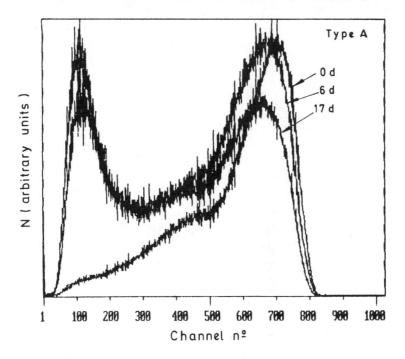

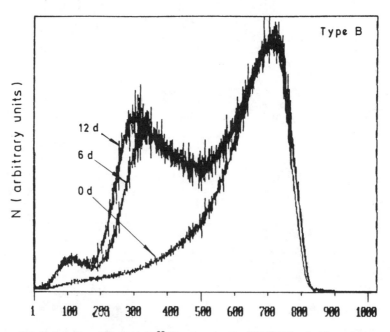

Figure 2. Spectral evolution of $H_3{}^{32}PO_4$ samples in INSTAGEL. A: Type A, no treated samples; B: Type B, silicone treated walls.

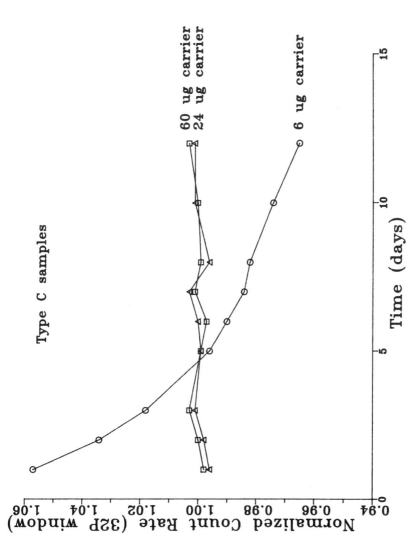

Figure 3. Time stability of H$_3$^{32}PO$_4$, type C, samples in INSTAGEL. 0: 6 μg carrier; □: 24 μg carrier; △: 60 μg carrier. The count rates are half-life corrected and then normalized to the mean value in the measurement period.

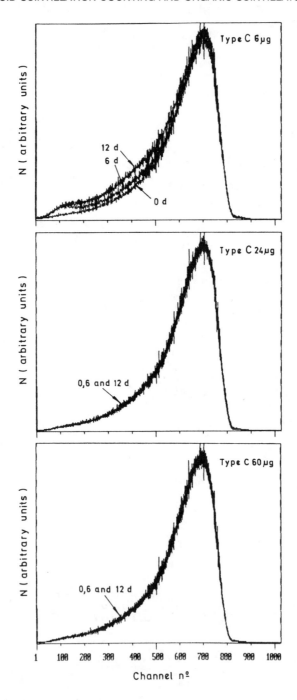

Figure 4.   Spectral evolution of $H_3{}^{32}PO_4$, Type C, samples in INSTAGEL, for 6, 24, and 60 μg of carrier.

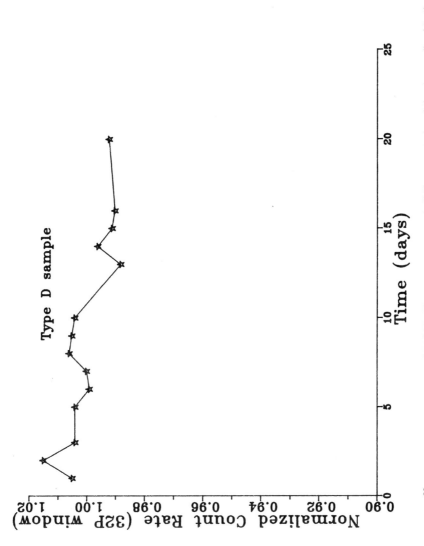

Figure 5. Time stability of $H_3{}^{32}PO_4$, type D, samples in INSTAGEL. The count rates are half-life corrected and then normalized to the mean value in the measurement period.

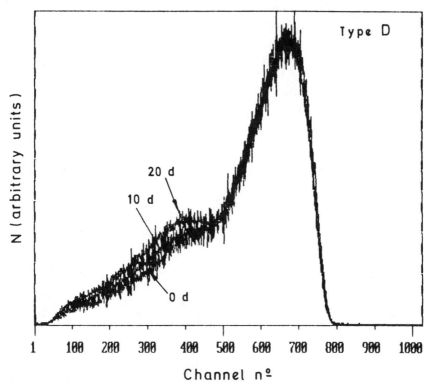

Figure 6. Spectral stability of $H_3{}^{32}PO_4$, type D, samples in INSTAGEL.

dard deviations of 0.4% in toluene, 0.3% in HISAFE II, and 0.15% in INSTAGEL, according to the total number of counts stored. The spectral changes in the three scintillators are due to ^{33}P interference that is clearly observed in the long term samples.

STANDARDIZATION OF ^{32}P BY THE EFFICIENCY TRACING METHOD

The organic sample in toluene as well as the type D inorganic sample in INSTAGEL have both been used to standardize the two ^{32}P solutions by the efficiency tracing method.[7] Both samples were calibrated within 48 hr after preparation, so that the inorganic sample 0.06% unstability did not affect the final results.

The activity concentration of each radioactive solution can be obtained from the expression

$$A = \frac{N}{\epsilon\,V} \tag{3}$$

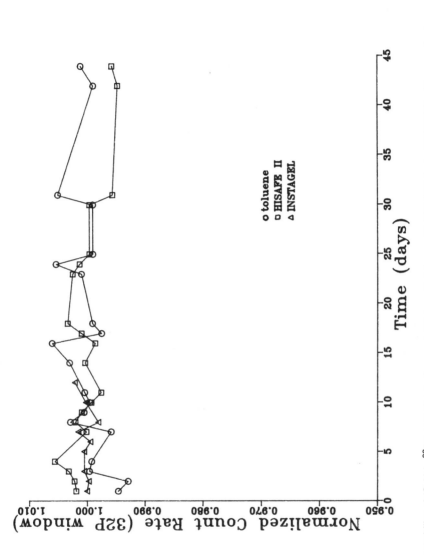

Figure 7. Time stability of $(C_6H_5)_3$ ^{32}PO samples in three scintillators. O: toluene; □: HISAFE II; △: INSTAGEL. The count rates are half-life corrected and then normalized to the mean value in the measurements period.

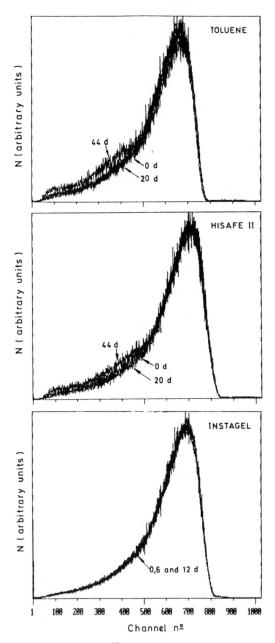

Figure 8.   Spectral stability of $(C_6H5)3*$ $^{32}P0$ samples in toluene, HISAFE II, and INSTAGEL.

where V is the volume of the active solution, N is the total count rate of the ^{32}P spectrum, and ϵ is the detection efficiency.

The radioactive solution has been dispensed with micropipettes GILSON, in amounts of 100 or 200 μL, depending on the freshness of the radioactive solution.

The number of total counts in each spectrum has been kept greater than 900,000, so that the statistical uncertainty is better than 0.1%.

The total counting efficiency ϵ has been determined by the efficiency tracing method[7] which needs an experimental curve of a set of tritium standards and the experimental measurement of [32]P.

The samples are affected by the ^{33}P contamination to a different extent depending on the time elapsed before measurement. The partial count rates N_{32} and N_{33} of ^{32}P and ^{33}P have been determined by a least-squares fitting procedure to the decay function

$$N(T) = N_{32}e^{-0.693t/T_{32}} + N_{33}e^{-0.693t/T_{33}} \qquad (4)$$

where $T_{32} = 14.29$ days and $T_{33} = 25.4$ days are the respective half-life values. Figure 9 shows the fit for the organic sample in toluene.

The counting efficiency was computed with the program EFFY[8] using the well known β-decay scheme of Figure 10 for [32]P.[9]

Both ^{32}P samples have been measured at different degrees of quench using increasing amounts of Cl_4C to check the consistency of the procedure. The measurements were made for figures of merit in the range 1.5 to 3.0 in INSTA-GEL and 1.5 to 4.5 in toluene, representing a ^{3}H equivalent efficiency between 45% and 15%.

Figure 11 shows the experimental and computed curves of efficiency vs figure of merit for the $H_3{}^{32}PO_4$ sample in INSTAGEL and the $(C_6H_5)_3{}^{32}PO$ sample in toluene. The maximum discrepancy is 0.8% for the inorganic sample and 0.5% for the organic sample.

An evaluation of the uncertainty factors of the ^{32}P computed efficiency is shown in Table 1, according to the recommendations of the Comité International des Poids et Mesures.[10] The total composed uncertainty is 0.525%, which agrees very well with the experimental determinations. The activity concentration of each sample was computed by the Equation (3) and its total uncertainty for both samples is shown in Table 2.

CONCLUSIONS

Several kinds of samples to calibrate ^{32}P by liquid scintillation counting have been analyzed.

Inorganic samples of phosphoric acid, $H_3{}^{32}PO_4$, having more than 24 μg carrier in 10 mL of scintillator are stable for more than 12 days with no counting loss observed. Samples whose vial walls have been previously saturated with carrier are also stable enough for 20 days, showing count losses lower than 0.06% a day.

Organic samples of triphenyl-phosphine-oxide, $(C_6H_5)_3{}^{32}PO$, synthesized in

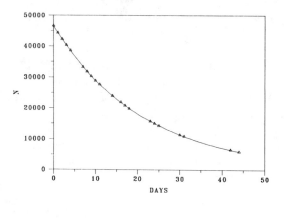

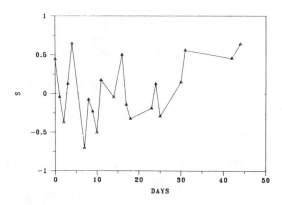

Figure 9. Total count rate fit and residuals for $(C_6H_5)_3$ ^{32}PO sample in toluene. N,S are stated in cpm and standard deviations, respectively.

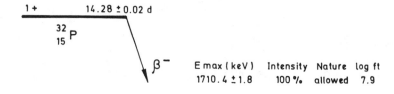

Figure 10. Decay scheme data of ^{32}P.

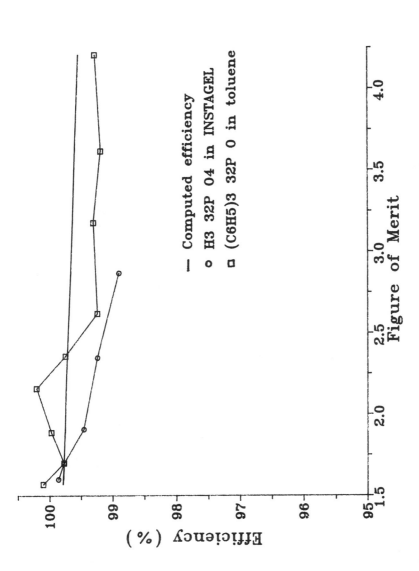

Figure 11. Counting efficiencies for H_3 $^{32}PO_4$ samples in INSTAGEL and $(C_6H_5)_3$ ^{32}PO samples in toluene. 0: inorganic sample; △: organic sample; -: computed by the efficiency tracing method.

Table 1. Contribution of the Diverse Uncertainty Factors (1 Standard Deviation) of the ^{32}P Efficiency Computed by the Efficiency Tracing Method

Factor	Original Uncertainty	Resulting Uncertainty on ϵ (^{32}P)
^{3}H standard calibration	3%	0.010%
Quench determination for ^{3}H standard	1%	0.003%
Counting (^{3}H)	0.3%	0.001%
Non detection probability (^{3}H, figure of merit 1–15)	2–7%	<0.100%
Volumetry	1%	0.325%
Quench determination for ^{32}P sample	1%	<0.0001%
Nuclear constants	—	0.300%
Phototube asymmetry	—	<0.250%
Ionization quench	—	0.100%
Composed uncertainty		0.525%

Table 2. Estimated Uncertainty Factors (1 Standard Deviation) Asociated to the Activity Concentration

Factor	Uncertainty
^{32}P computed efficiency	0.525%
Volumetry	1.0%
Counting	0.1%
Composed Uncertainty	1.38%

the laboratory show a very good long term stability, with no systematic trend, for 44 days in three different scintillators: toluene, HISAFE II, and INSTAGEL.

Both kind of samples have been calibrated by the efficiency tracing method. The discrepancies between experimental and computed values of the efficiencies are lower than 0.8% for inorganic samples and 0.5% for organic samples. Efficiency tracing allowed the ^{32}P solutions to standardize at a total composed uncertainty of 1.4%.

REFERENCES

1. Petroff C.P., P. Nair, and D. Turner. "The Use of Siliconized Glass Vials in Preventing Wall Adsorption of Some Inorganic Radioactive Compounds in Liquid Scintillation Counting," *Int. J. Appl. Radint. Isot.* 15:491 (1964).
2. Tykva, R. "Stability of Liquid Scintillation Counting for to Sample Sorption to the Counting Vial," *Int. J. Appl. Radint. Isot.* 26:495 (1975).
3. Rodríguez, L., A. Grau, J.M. Los Arcos, and C. Suráez. "Preparación y Calibración por Centelleo Líquido de una Muestra de ^{32}P," CIEMAT Report 623 (1988).
4. Wigfield, D. and V. Srinivasan. "Liquid Scintillation Counting and Sample

Adsorption. Structural Factors in Organic Molecules Controlling Likelihood of Adsorption," *Int. J. Appl. Radint. Isot.* 25:473 (1974).

5. Murray, A. and L. Willians. *Organic Syntheses with Isotopes*, Part II (New York: Interscience Publishers Inc., 1958) p. 1904.

6. Grignard, V. and J. Savard. "Sur les derivés Magnesiens de la Dichlorotry-phenylphosphine et sur les Pentaphosphines," *Comptes-Rendus* 139:674 (1904).

7. Grau, A. and E. García-Toraño. "Evaluation of Counting Efficiency in Liquid Scintillation Counting of Pure β-ray Emitters," *Int. J. Appl. Radint. Isot.* 33:249 (1982).

8. E. García-Toraño and A. Grau. "EFFY, a Program to Calculate the Counting Efficiency of β-particles in Liquid Scintillation," *Comp. Phys. Comm.* 23:385 (1981).

9. Lagoutine, F., N. Coursol, and J. Legrand. *Table de Radionucléides*, (Giff sur Ivette, France: Laboratoire de Métrologie des Rayonnements Ionisants, 1983).

10. Kaarls, R. "Rapport du Groupe de Travail sur l-Expression des Incertitudes," Procès Verbaux du Comité International des Poids et Mesures, 49:A1–A12 (1981).

CHAPTER 56

LSC Standardization of Multigamma Electron-Capture Radionuclides by the Efficiency Tracing Method

J.M. Los Arcos, A. Grau and E. Garcia-Toraño

INTRODUCTION

The basic problem in the LS standardization of radionuclide decay by electron capture (EC) processes is the accurate determination of the counting efficiency.[1-3] It must take into account all the energy absorbed in the scintillator by the different radiations emitted after the EC decay.

For pure EC nuclides it has been recently described an efficiency tracing method[4-5] that predicts the LSC efficiency as a function of a figure of merit. It characterizes the scintillator-equipment system as a whole and does not depend on the radionuclide to be measured.

This method[6] leans against an experimental quench curve of ³H standards and a model based on the X-rays and Auger-electrons emission. This includes the nonlinear response due to the X-ray escape and ionization quenching.

The efficiency evaluation is a little more difficult[7-8] when the electron capture leaves the daughter nuclide in an excited level, which later may decay by a single or double gamma transition in coincidence, like it occurs in ^{54}Mn, ^{85}Sr, ^{125}I, ^{57}Co, and ^{92}Nb. For these nuclides, photon-scintillator interaction and eventually conversion electrons and the subsequent atomic rearrangement, must be considered in order to obtain an accurate value for the detection efficiency. The difficulty is even greater for radionuclides where pure EC is in competition with single or double gamma transitions, like ^{88}Y or ^{65}Zn.

Nevertheless, in all these cases the efficiency evaluation is straight forward because the total number of intermediate rearrangement pathways is kept to a reasonable value and several codes have been developed for toluene-based scintillators.

This is no longer valid when the daughter nuclide shows a complex gamma spectrum involving more than three excited levels, because the number of intermediate rearrangement steps increases very quickly.

An additional difficulty for EC nuclides arises from the photon-scintillator interaction. While most β–γ emitters are not very sensitive to the scintillator because of the intrinsically high β efficiency, EC nuclides show a rather significant efficiency dependence on the chemical composition of the scintillator, its intrinsic efficiency is usually lower and the Compton interactions greatly modify the spectrum of detected electrons. As a consequence, a more general evaluation model is needed to obtain realistic values in practical situations. Some special cases require complex decay schemes or nontoluene-based scintillation cocktails for sample preparation.

This chapter presents a new systematic procedure to overcome these problems. First, an in depth description of the models and its limitations will be developed. Then, the results for several scintillators will be discussed and compared to experimental measurements for [133]Ba in INSTAGEL dioxane-naphtalene scintillators.

COUNTING EFFICIENCY FOR SIMPLE EC NUCLIDES

In order to derive the partial efficiencies for some simple decay schemes it will be assummed that the liquid scintillation counting system is composed of two phototubes with the same gain, working in coincidence.

The counting efficiency can then be written as a function of the figure of merit as a sum of contributions for each atomic rearrangement:

$$[1 - \exp(-\sum_\lambda E_\lambda Q(E_\lambda)/2M)]^2 \tag{1}$$

where E_λ = the energy of any radiation absorbed in the scintillator after a single EC process,

$Q(E_\lambda)$ = the ionization quench correction,

$E_\lambda \cdot Q(E_\lambda)$ = the effective energy,

M = the figure of merit.

The function $Q(E)$ can be evaluated with the Birks' semiempirical formula[9]

$$Q(E) = \frac{1}{E} \int_0^E \frac{dE}{1 + kB \frac{dE}{dx}} \tag{2}$$

A good aproximation to equation 2, computed with recent values of dE/dX[10,11] and taking $KB = 0.0075$ cm/MeV, is given by a rational function:

$$Q(E) = \frac{a_1 + a_2 \log E + a_3 \log^2 E}{1 + a_4 \log E + a_5 \log^2 E} \tag{3}$$

where   a1 = 0.357478
         a2 = 0.459577
         a3 = 0.159905
         a4 = 0.0977557
         a7 = 0.215882

Pure EC Nuclides

Assuming a K- and L-shell EC model and neglecting L-subshells, there are 21 different rearrangement pathways; the probabilities and effective energies are given in a previous work.[5,6] An additional M-shell capture, with negligible effective energy, can be kept to combine EC with other subsequent processes in complex decay schemes.

Therefore, for pure EC nuclides, the total counting efficiency is:

$$\epsilon_o (M) = \sum_{\eta}^{22} W_n \left(1 - \exp(-E_n^{EC}/2M)\right)^2 \qquad (4)$$

where W_n is the probability of the n-th rearrangement pathway and E_n^{EC} is the effective energy of that pathway.

EC Nuclides with a Single Gamma Transition in Coincidence

The procedure to compute the efficiency can be easily extended to EC nuclides followed by a single gamma transition in coincidence, like ^{54}Mn or ^{51}Cr.

In this case, three alternative processes can occur depending on whether the photon emitted escapes, is detected in the liquid scintillator, or transformed into a conversion electron ejected from an atomic shell.

Therefore, the efficiency can be written as:

$$\epsilon_1 (M) = \sum_{l=1}^{3} \psi_{1l}\epsilon_{1l}(M) \qquad (5)$$

where ψ_{1l} = the probability for the process mL,
         ϵ_{1l} = the partial efficiency for the process mL,
         l = 1, 2, 3, indicate the EC-photon escape, EC-photon detection, and EC-internal conversion processes, respectively.

The probabilities for each process are:

$$\begin{aligned}\psi_{11} &= P_g \, [1 - P_{in}] \\ \psi_{12} &= P_g \, P_{in} \\ \psi_{13} &= [1 - P_g]\end{aligned} \qquad (6)$$

where P_g is the probability of photon emission in the gamma transition, and P_{in} is the photon-scintillator interaction probability.

Since the photon escape can be treated as a pure electron capture, its partial efficiency ϵ_{11} is evaluated, as was indicated previously in this section.

The photon-scintillator interaction gives rise to an electron absorption spectrum $N_c(E_c)$ which includes the photoelectric peak as well as the Compton electrons distributions so that the efficiency is:

$$\epsilon_{12}(M) = \int_0^{E_g} N_c(E_c) \sum_i^{22} W_i \left\{ 1 - \exp\left[- (E_i^{EC} + E_c Q(E_c) / 2M \right] \right\}^2 dE_c \quad (7)$$

where E_g is the energy of the gamma transition.

The internal conversion process is similar to the EC process because both create a primary vacancy in an atomic shell that leads to the subsequent X-ray and Auger-electron emission. The difference lies in that for IC, instead of being absorbed by the nucleus like it happens in EC, an orbital electron is ejected away and contributes to the detection efficiency.

Neglecting IC in shells higher than M, there are 22 different atomic rearrangement pathways, and their weights are similar to those of the EC process, substituting the K- , L- conversion probabilities CK and CL for the capture probabilities PK, PL. The energies are also similar, adding the effective energy of conversion electrons. The partial efficiency can be written:

$$\epsilon_{13}(M) = \sum_\gamma^{22} V_j \sum_i^{22} W_i \left\{ 1 - \exp\left[-(E_i^{EC} + E_j^{IC})/ 2M \right] \right\}^2 \quad (8)$$

where V_j, E_j^{IC} are the probability and the effective energy of the j-th pathway in the IC process.

EC Nuclides with a Double Gamma Transition in Coincidence

When the EC decay is in coincidence with a double gamma cascade, there are nine possible rearrangement pathways according to the three different possibilities for each single gamma transition.

The detection efficiency can be written as:

$$\epsilon_2(M) = \sum_{l=1}^3 \sum_{n=1}^3 \psi_{21n} \psi_{21n} (M) \quad (9)$$

where

$$
\begin{aligned}
\psi_{211} = \ & P_g(32) \, [1 - P_{in}(32)] \, P_g(21) \, [1 - P_{in}(21)] \\
& P_g(32) \, [1 - P_{in}(32)] \, P_g(21) \, P_{in}(21) \\
& P_g(32) \, [1 - P_{in}(32)] \, [1 - P_g(21)] \\
& P_g(32) \, P_{in}(32) \, P_g(21) \, [1 - P_{in}(21)] \quad\quad (10) \\
& P_g(32) \, P_{in}(32) \, P_g(21) \, P_{in}(21) \\
& P_g(32) \, P_{in}(32) \, [1 - P_g(21)] \\
& [1 - P_g(32)] \, P_g(21) \, [1 - P_{in}(21)] \\
& [1 - P_g(32)] \, P_g(21) \, P_{in}(21) \\
\psi_{211} = \ & [1 - P_g(32)] \, [1 - P_g(21)]
\end{aligned}
$$

where

$P_g(st)$ = the probability of photon emission for the gamma transition st,

$P_{in}(st)$ = the total probability of interaction for a photon emitted in the gamma transition st,

st = the gamma transition between levels s and t.

The respective efficiency for an electron capture-internal conversion and photon interaction pathway is given by:

$$
\epsilon_{rst}(M) = \int_0^{E_g(st)} N_c \sum_{j=1}^{22} V_j \sum_{i=1}^{22} W_i \left\{ 1 - \exp\left[-\left(E_i^{EC} + E_j^{IC} + E_c Q(E_c)\right) / 2M \right] \right\}^2 \quad (11)
$$

Similar expressions hold for other combinations of processes.

Three Level EC Nuclides

Any EC nuclide involving no more than three levels in the daughter nuclide, can be accommodated in the general decay scheme of Figure 1. Eventually, four different alternative paths can be followed after an EC occurs. Accordingly, the total counting efficiency can be written as:

$$
\epsilon_T(M) = \sum_{l=1}^{4} \psi_l \epsilon_l (M) \quad (12)
$$

where ψ_l = the probability of the l-th path,

$\epsilon_l(M)$ = the partial efficiency for the l-th path,

M = the figure of merit.

All four paths begin with an EC decay, the first one being a pure process that ends at the ground level. In the second and the third path, the EC is

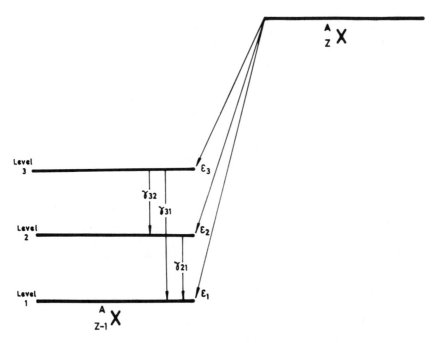

Figure 1. General decay scheme for 3 level EC radionuclides.

followed by a single gamma transition while in the fourth one it is accompanied by a double gamma cascade in coincidence. The probability of each path is:

$$
\begin{aligned}
\psi_1 &= P_{EC}(1) \\
\psi_2 &= P_{EC}(2) \\
\psi_3 &= P_{EC}(3)\, I_\gamma(31) \\
\psi_4 &= P_{EC}(3)\, I_\gamma(32)
\end{aligned}
\tag{13}
$$

where $P_{EC}(m)$ is the EC branching ratio to the m-th level in the daughter nuclide and $I_\gamma(mn)$ is the relative intensity of the gamma transition connecting the levels m-n in the daughter nuclide.

COUNTING EFFICIENCY FOR MULTIGAMMA EC NUCLIDES

Pathway Description

A direct evaluation of partial efficiency expressions for each alternative pathway leading to the ground level turns out to be impractical for EC nuclides involving more than three excited levels; the number of gamma cascade steps increases rapidly, as can be appreciated in Figure 2.

Instead, a different approach is preferred. One that sequentially simulates

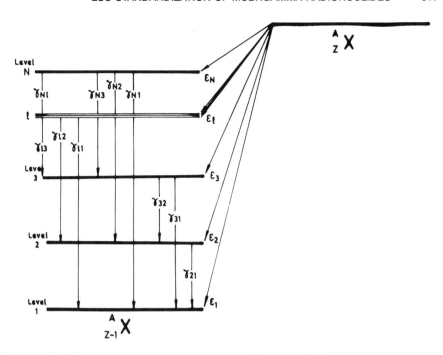

Figure 2.   General decay scheme for multigamma EC radionuclides.

each pathway down to the ground level and systematically combines the different transition modes for each intermediate step, photon detection, photon escape, or internal conversion.

If a "route" is defined as a set of gamma transition steps following the EC process, the number of different routes starting at the n-th excited level is:

$$r = 2^{n-2} \qquad (14)$$

Eventually, selection rules for gamma transitions apply and not all the routes are allowed.

After the starting level of a route has been fed by the EC, the number of different pathways associated to the j-th route is given by

$$P_j = 3^{S_j} \qquad (15)$$

where Sj is the number of steps which the route is made of.

A simple numbering algorithm for the k-th pathway allows all the possible routes and all different pathways in a route to be generated.

Efficiency Evaluation

Once the different modes in the k-th pathway starting at the n level have been stated, the total effective energy can be evaluated by adding the effective energies of the S_j modes to the effective energy EECn of the EC process. The pathway probability is obtained as a product of the EC branching ratio and the probabilities of the S_j modes, so the partial efficiency for that pathway is:

$$\epsilon_{njk}(M) = P(n) \prod_{\substack{EC \\ l=1}}^{S_j} P_{jkl} \left\{ 1 - \exp\left[-\frac{1}{2M}\left(E_n^{EC} + \sum_{l=1}^{S_j}E'_{njkl}\right) \right] \right\}^2 \qquad (16)$$

where E'_{njkl} and P_{jkl} are the effective energy and probability of the l-th step in the K-th pathway of the j-th route starting at the n-th excited level, and have identical expressions to Equations 10 and 11.

The partial efficiencies are accumulated as a function of the figure of merit during the pathway generation so that at the end, the total counting efficiency is available for each figure of merit previously stated.

RESULTS AND DISCUSSION

A computer code, CEGAN, has been written to carry out all the efficiency computations described in the previous section for any nuclide included in the general decay scheme of Figure 2 . It allows specification of the vial geometry and the chemical composition of scintillator.

These parameters concern the Monte Carlo computation of the photon interactions and the subsequent Compton electron distribution.[12]

A single database file for atomic and nuclear parameters makes it easy to specify any nuclide of interest in terms of capture probabilities, branching ratios, allowed gamma transitions, internal conversion coefficients, electron binding energies, fluorescence yields, and X-ray and Auger electron energies and intensities.

The standardization procedure has been tested with ^{133}Ba, which has the decay scheme[13] shown in Figure 3. ^{133}BaCl$_2$ samples were incorporated into 10 mL of INSTAGEL and dioxane-naphtalene, and a set of quenched samples was measured in an LS counter, LKB 1219 RackBeta spectral. A set of ^{3}H standards was used to obtain the quench calibration curve.

Figure 4 shows the efficiency values vs the figure of merit for both scintillators. The discrepancy between experimental and computed values is lower than 0.8%, with figures of merit in the range 1.5 to 4 for dioxane and 1.0 to 3.0 for INSTAGEL. An important prediction of the model is the efficiency dependence (several per cent) on the scintillator being used. The differences between some commonly used scintillators can be seen in Figure 5.

The pair INSTAGEL/HISAFE II as well as the PCS/READYSOLV can not be resolved from the point of view of its efficiency. Also, there is a very small difference between the two pairs.

Finally, it has to be noted that the efficiency estimations depend strongly on the uncertainty of the nuclear constants, especially the K-shell electron capture probability which transmits almost all uncertainty to the efficiency value.

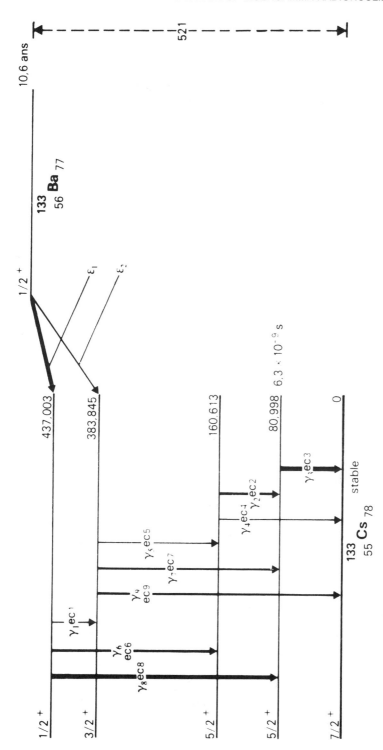

Figure 3.   Rearrangement pathways in ^{133}Ba. F: photon detection; G: photon escape; C: internal conversion.

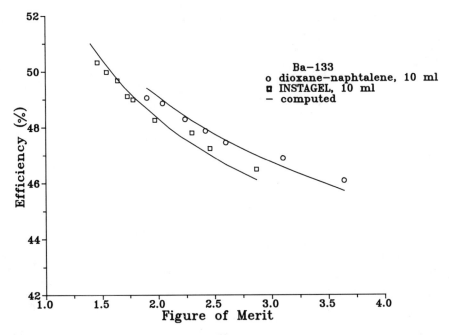

Figure 4.   Efficiency vs figure of merit for ^{133}Ba.

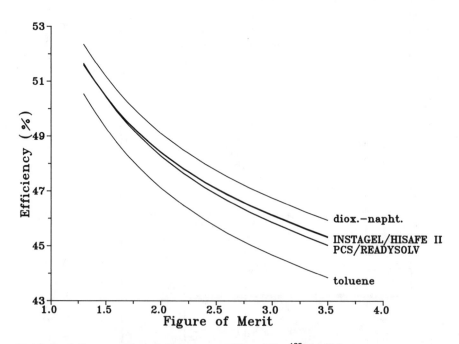

Figure 5.   Influence of the scintillator composition on the ^{133}Ba efficiency.

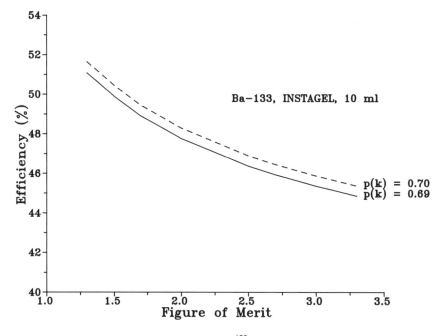

Figure 6. Influence of nuclear constants on the ^{133}Ba efficiency in INSTAGEL. -: Pk = 0.69; —: pk = 0.70.

Figure 6 shows the influence of an apparently small variation of the K-shell EC probability on the efficiency values in INSTAGEL, as computed by the code.

CONCLUSIONS

The LSC standardization of EC nuclides by the efficiency tracing method has been extended to complex decay schemes with an arbitrary number of coincident gamma emissions. It was done by means of an efficiencies model which takes into account the X-ray and Auger electrons, as well as the interaction of photons with the scintillator, eventually the conversion electrons emitted.

A computer code that performs the pathway evaluations has been prepared and applied to the decay of ^{133}Ba. It shows a good agreement, within 0.8% with experimental measurements and in the range of figures of merit 1 to 4.

A warning should be made against the unconcerned use of nuclear constant values that could transmit big errors on the efficiency evaluation. Also some predictions of the model about the behaviour of several commonly used scintillators have been analyzed.

REFERENCES

1. Steyn, J. and A.S.M. De Jesus. "Coincidence Measurements in Nuclides Decaying by Electron Capture," National Physical Research Laboratory Report NP-15912 (Pretoria, South Africa: Council for Scientific and Industrial Research, 1966).
2. Steyn, J. "Internal Liquid Scintillation Counting Applied to the Absolute Desintegration-Rate Measurement of Electron Capture Nuclides," in *Standardization of Radionuclides* IAEA/STI/PUB/139 (Vienna, Austria: International Atomic Energy Agency, 1967) p. 35
3. Vatin, R. "The use of Liquid Scintillators in Radionuclide Metrology," BIPM Monograph 3 (1980).
4. Grau, A. "Calculo de la Eficiencia de Detecci n de L Quidos Centelleadores. Nucleidos de Captura Electr Nica," JEN Report 521 (Madrid, Spain: Junta de Energ a Nuclear, 1982).
5. Grau, A. "Counting of Efficiency for Electron-Capturing Nuclides in Liquid Scintillator Solutions," *Int. J. Appl. Radiat. Isot.*, 33:371 (1982).
6. Los Arcos, J.M., A. Grau, and A. Fernandez. "VIASKL: A Computer Program to Evaluate the LSC Efficiency and Its Associated Uncertainty for K-L-Atomic Shell Electron-Capture Nuclides," *Comp. Phys. Comm.*, 44:209 (1987).
7. Grau, A. and A. Fernandez, "Counting Efficiency for Electron-Capture Nuclides in Liquid Scintillation Detectors. Single Coincidence Gamma-Ray," Anales F s. A 81:160 (1985).
8. Grau, A., E. García-Toraño and J.M. Los Arcos. "Free Parameter Codes to Compute the Counting Efficiency in Liquid Scintillators," *Trans. Am. Nucl. Soc.*, 55:55(1987).
9. Birks, J.B. *The Theory and Practice of Scintillation Counting*, (Oxford: Pergamon Press, 1965), p. 185.
10. Seltzer, S.M. and M.J. Berger. "Evaluation of the Collision Stopping Power of Elements and Compounds for Electrons and Positrons," *Int. J. Appl. Radiat. Isot.*, 33:1189(1982).
11. Seltzer, S.M. and M.J. Berger. "Procedure for Calculating the Radiation Stopping Power for Electrons," *Int. J. Appl. Radiat. Isot.*, 33:1219 (1982).
12. García-Toraño, E. and A. Grau. "EFYGA. A Monte Carlo Program to Compute the Interaction Probability and the Counting Efficiency of Gamma Rays in Liquid Scintillators, *Comp. Phys. Comm.*, 47:341 (1987).
13. Lagoutine, F., N. Coursol, and J. Legrand. *Table des Radionuclides,* (Giff sur Ivette, France: Laboratoire de M trologie des Rayonnements Ionisants, 1983).

CHAPTER 57

ANSI Standards for Liquid Scintillation Counting

Yutaka Kobayashi

The history of ANSI standards has been previously chronicled by Roger Ferris;[1] therefore, only a brief sketch of ANSI committees will be given here. The acronym, ANSI, stands for the American National Standards Institute based in New York. ANSI publishes industrial standards for instruments and relevant technology associated with their use. Once a standard is published, it remains in force for 5 years, after which it may either be reaffirmed for another 5 years or allowed to expire. These ANSI committees are composed of unpaid volunteers representing instrument manufacturers, relevant suppliers, government agencies, and knowledgeable users of the technology.

ANSI committee N42 was formed in 1975 for the expressed purpose of writing standard methods of calibration of nuclear detection instruments. A subcommittee, N42.2, was formed and delegated to do the actual writing of the standards. The chairman of the writing committee for N42.15, "Performance Verification of Liquid Scintillation Counting Systems,"[2] was Roger Ferris. N42.15 was first published in 1980 and reaffirmed in 1984. Yutaka Kobayashi succeeded Ferris for the revision of N42.15 which is currently under review. Kobayashi is also the writing committee chairman for N42.16, "Specifications for Sealed Radioactive Check Sources Used in Liquid Scintillation Counters."[3] N42.16 was published in 1986. The purpose of both of these standards was to provide users with concise and simple directions for calibrating their liquid scintillation counters. It was the hope of the committee that instrument manufacturers would incorporate and encourage the use of these standards, and in the same vein, regulatory agencies would require these or similar procedures to be adopted into the quality assurance programs of those using liquid scintillation counters under jurisdiction.

The objective of the instrument standard was to provide simple tests to verify the satisfactory performance of the instrument. It was decided early to exclude any discussion of quench correction methods. This was in keeping with our objective of maintaining the length of the standard to a few pages. We focused on only three parameters for instrument evaluation: performance, stability, and background. It was the consensus of the committee that tritium, generally the weakest radionuclide most commonly counted, would be a desirable check source with which to assess performance. If the instrument performed well with a tritium check source, measuring the more energetic radionuclides should not be a problem; however, instrument performance can be assessed using a check source containing any other radionuclide which may be more pertinent to a particular laboratory. The requirements are that the check source used be chemically stable and that the sample container dimensions conform to those specified in the International Electro-technical Commission (IEC) Standard Number 582.[4] The IEC standard 582 specifies the dimensional limits for glass and plastic vials for both the standard and miniature sizes used in liquid scintillation counters. This standard is international and was written for the benefit of instrument manufacturers in Europe and the United States. It should be noted that, today, only the standard for the large vials, the 20 mL size, is followed. The miniature vials counted today vary in size from microvials which hold only a few hundred μL to those which hold 7 mL.

To evaluate performance, commercially produced check sources usually sold as "unquenched standards" are recommended. They are usually sold as sets consisting of tritium, ^{14}C, and background check sources prepared in an oxygen-free atmosphere. If a check source with energy greater than ^{14}C is required, a ^{36}Cl check source (beta emission with an E_{max} of 712 keV) may be purchased. The advantage of using a commercial unquenched standard set is that they are usually traceable to the National Institute of Standards and Technology (NIT&S) and can be used to evaluate counting efficiency within known statistical limits. The question of establishing traceability for regulatory purposes is an ongoing problem in this field. To evaluate performance, the sample chemistry is not critical so long as the check source is stable over a period of time and the same check source is used daily. It is essential to use the same check source each time to eliminate any systematic error which may be introduced by another check source containing the same radionuclide. The ANSI standard recommends that each instrument be assigned its own set of check sources. To monitor stability, it is essential that the daily readings of the check sources be recorded. A daily inspection of this record will be a good indicator of the stability of the instrument. An example of a useful log sheet is appended to this standard.

In Table 1, the Table of Contents for ANSI Standard N42.15, "Performance Verification of Liquid Scintillation Counting Systems," is shown. The essence of both the introduction and scope has already been discussed above.

The definitions given are concise, simple, and accurate. They were not written to be encyclopedic; and thus do not meet with everyone's approval. Our objective was to make the definitions easy to understand and appropriate for this standard. In this section, we have also defined standard sources as those which are directly traceable to NIT&S and those which are supplied by other sources which certify traceability to NIS&T. The standard specifies four different types of check sources which can be used to conduct the various

Table 1. Table of Contents of ANSI N42.15, "Performance Verification of Liquid-Scintillation Counting Systems"

instrument performance tests. Only the flame-sealed check sources of known activity can be used in all tests.

Section 5, Operations and Tests, describes the frequency of testing, tests for counting efficiency, determination of background, and statistical methods for assessing instrument stability. Although this standard uses a tritium check source for determining counting efficiency, any check source more appropriate for a given laboratory may be used. The standard recommends the monitoring of instrument performance following installation, after instrument service, replacement of a check source, or after any other circumstance which may influence performance. The key recommendation is daily monitoring and data recording in an instrument log.

The last section, Precautions, discusses various factors which may influence the counting data. They include correcting for short-lived radionuclide decay, the instrument resolution time with respect to sample activity, chemiluminescence, and photoluminescence. These and other topics with which the user should be familiar are included in this section.

The Table of Contents shown here is for the current standard. A revised version of this standard is currently out for ballot. The revision was necessary because the current standard does not mention the fact that all new liquid scintillation counters use the equivalent of a multichannel analyzer instead of the classic pulse height analyzer. The new standard also includes a more detailed revision of precautions relative to photomultiplier tube performance.

A complimentary standard, ANSI N42.16, "Specifications for Sealed Radioactive Check Sources Used in Liquid Scintillation Counters," has been published. The Table of Contents is reproduced in Table 2. This standard

Table 2. Table of Contents: ANSI N42.16, "Specifications for Sealed Radioactive Check Sources Used in Liquid-Scintillation Counters"

covers only tritium, ^{14}C, and background check sources. It specifies the range of radioactivity to be contained in the two radioactive check sources and the source for both to be radio-labeled toluene. In summary, for the check sources, the solvent specified is analytical-grade (or better) toluene, the primary scintillator is diphenyloxazole (PPO), the secondary scintillator is one of three listed in the standard, and the check source is purged with an inert gas before sealing. The glass container must meet the IEC standard 582 previously discussed, and the final volume of the sample contained in the check source shall be 15 mL + /- 0.2 mL. The useful life of the check source is limited to 5 years from the time of sealing.

In conclusion, it should be stressed that these standards have been written for the benefit of all users. The ANSI committee welcomes suggestions for any other standard which may be useful to users of liquid scintillation counters.*

REFERENCES

1. Ferris, R. "ANSI Standards for L.S. Counters," in *Liquid Scintillation Counting: Recent Applications and Development*, C.T. Peng, D.L. Horrocks, and E.L. Alpens, Eds. (New York: Academic Press, 1980) Vol.1, p. 241.

---

*Copies of any ANSI standard may be obtained by writing:
IEEE, Inc
345 East 47th Street
New York, New York 10017

2. American National Standard. *Performance Verification of Liquid-Scintillation Counting Systems*, ANSI N42.15–1980(R84) (New York: IEEE,Inc).
3. American National Standard. *Specifications for Sealed Radioactive Check Sources Used in Liquid Scintillation Counters*, ANSI N42.16–1986 (New York: IEEE,Inc).
4. Dimensions of vials for liquid scintillation counting, IEC Pub No 582, Technical Committee No. 45: International Electrotechnical Commission,(IEC), Nuclear Instrumentation, Geneva, Switzerland. 1977.

CHAPTER 58

A General Method for Determining Sample Efficiencies in Liquid Scintillation Counting

Willard G. Winn

ABSTRACT

Many modern liquid scintillation counters use two coincident photomultiplier (PM) tubes to reduce background noise. These counters detect most charged particle emitting nuclides with an efficiency near 1; however, they are noticeably less efficient for detecting lower energy emissions from nuclides such as tritium, because each decay produces so few photons that any coincident detection is less probable. For quenched samples, the photon production is further reduced, yielding a corresponding contraction of the observed energy spectra. The probability (or efficiency) for detecting a charged particle increases from 0 to 1 as its photon production (or its energy) increases from 0 to higher values. A model for this probability was developed and tested against experimental measurements.

A trinomial probability distribution modeled the efficiency as a function of observed spectral energy. The three probabilities of the trinomial include two for photon detection by the two PM tubes and one for nondetection. For a TRI-CARB 2000 CA/LL liquid scintillation analyzer, reasonable efficiency models resulted when using the trinomial formalism as a basis for empirically selecting the efficiencies. Using measurements of low-energy Auger and conversion electrons of quenched ^{57}Co standards as an experimental guide, four efficiency curves were modeled and tested. Scintillation spectra of ^{3}H and ^{14}C quenched standards were convoluted with the efficiency curves to test each model. The dpm predicted by the best model agreed with that known for the standards, provided that cpm/dpm > 0.25. The method may be extended down to cpm/dpm ≈ 0.1 by incorporating a "bootstrap" correction technique.

INTRODUCTION

At the Savannah River Site (SRS), waste that is classified as nonradioactive hazardous liquid waste must be certified as having a total activity less than the DOT requirement of 2 nCi/g before it can be shipped off plant for disposal.[1] Reliable methods for determining the total activity of samples, therefore, have received much attention at SRS.

Liquid scintillation counting is capable of detecting all types of nuclear decays, making it an attractive waste appraiser. It is particularly suited for detecting nuclides that decay by α and $\beta^{\pm}$ emission. Some nuclides decay by electron capture (EC), but even these emit detectable Auger electrons and X-rays. Also, any accompanying τ-ray and conversion electron (ce) emissions may be detected. Unfortunately, the detection efficiency for the various decay modes can differ substantially, and many can have efficiencies well below 100%.

Over the years, various techniques have evolved for establishing liquid scintillation counting efficiencies. Examples of these include the channels ratio method,[2] efficiency tracing,[3-5] extrapolation,[6] internal/external quench standards,[2,7,8] and direct modeling of the efficiencies per basic detection processes.[9-14] These techniques have primarily been developed for β-emitters, but some can be extended to other types of radionuclide emissions. In particular, direct modeling promises to be useful for all emissions of interest. It assumes that the energy spectrum is known. This allows efficiency modeling based on the number of photons generated in the liquid scintillation cocktail and their fraction detected by the photomultiplier tubes. By contrast, in waste samples, a priori knowledge of the true energy spectrum cannot be assumed; the detected spectrum is distorted by an energy-dependent efficiency that results from the photon generation and detection probabilities. Nevertheless, the detected spectrum does include information for inferring the nature of the true spectrum and its detection efficiency.

The present work explores a method for using the detected energy spectrum to infer the efficiency for a given sample. The method is developed using a TRI-CARB 2000 CA/LL scintillation counter which incorporates two coincident PM tubes to reduce background noise. Such a design is very popular in modern scintillation counters; however, the probability for detecting a pair of coincident photons from a nuclear decay event varies from 0 to 1 depending on the total number of photons produced. Because the spectral energy corresponds to the number of coincident photons detected, each energy of a spectrum should correlate with a probability for detection. It is clear that this detection probability should be close to 0 for the lowest spectral energies and should approach 1 as the energy increases. The present study develops and tests models for this probability.

THEORY

Conceptual Approach

Each decay event detected within a liquid scintillation spectrum has a probability associated with its detection, and its energy response is proportional to the number of photons detected. The probability of coincident photon detection increases with the number of photons emitted per decay event; thus, the detection probability increases as the spectral energy increases, as shown in Figure 1. For a typical sample, the detected energies are distributed over a spectral range. The probability for the number detected at each energy must be

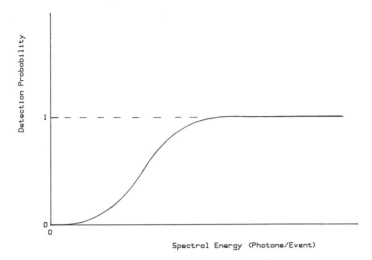

Figure 1.   Figure 1. General trend of probability for detecting a pair of coincident photons produced by a nuclear decay event.

considered. Using these probabilities to perform an efficiency correction for counts collected at each point in the spectrum, the integral of these results should yield the total decays represented by the spectrum.

An example of the above concept is inherent in typical quench curves for ^{3}H and ^{14}C, as shown in Figure 2a.[15] The spectra of quenched samples are contracted to lower energies. Their efficiencies are lower because fewer photons are produced per decay event. The plotted quench parameter tSIE is proportional to the energy range of spectra, calculated from the Compton spectrum of an external ^{133}Ba source.[8,15] A tSIE of 1000 corresponds to the unquenched case. As the spectra are contracted to lower energies (more quenching), the attendant dropoff in efficiency is consistent with the type of detection probability illustrated in Figure 1.

Because all β-spectra have approximately the same shape when scaled to their maximum endpoint energies E_β, it is interesting to combine the ^{3}H and ^{14}C quench curves into a single plot, as shown in Figure 2b. In this plot, the efficiencies are plotted against tkeV $\equiv$ tSIE/1000 $\times$ E_β, where E_β = 18.6 keV for the ^{3}H data and E_β = 156.5 keV for ^{14}C. For tSIE = 1000, the defined tkeV corresponds to the true endpoint of the unquenched spectrum. For lower tSIE, the tkeV approximates the endpoints of the contracted β-spectra, which, in turn, approximate real unquenched spectra with these endpoint energies. The tkeV approximation yields a discrepancy between the ^{3}H and ^{14}C values in the overlap region at 15 to 20 keV; this is probably caused by resolution broadening of the E_β endpoint in the observed spectrum, which is discussed later in this chapter. Indeed, the discrepancy disappears if the tkeV values are corrected for resolution broadening, as shown in Figure 2c. The overall trend of these data readily illustrates the influence of a probability like that in Figure 1.

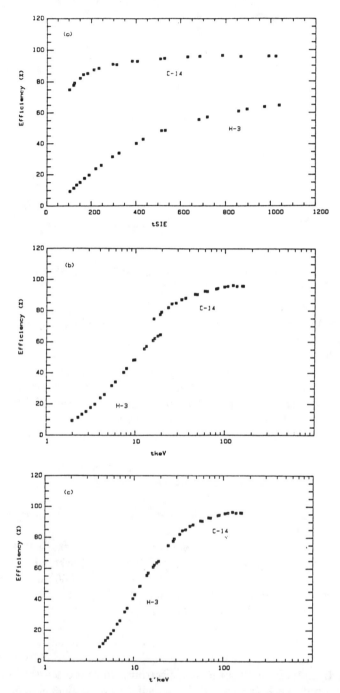

Figure 2. Typical efficiency data for TRI-CARB 2000 CA liquid scintillation analyzer: (a) efficiency vs tSIE quench parameter, (b) efficiency vs tkeV $\equiv$ tSIE/1000 $\times$ E_β and (c) efficiency vs t'keV = tkeV corrected for resolution broadening.

Mathematical Basis

A perfectly ideal liquid scintillation detector would count a sample spectrum $R(E)$ with 100% efficiency, and its energy scale would correspond to the total photons produced per decay event. Integrating $R(E)$ over its range yields the sample decay rate, r_d. The detected sample spectrum $S(E^|)$ does not have 100% efficiency, and its energy scale corresponds only to different fractions of the total photons per event. Thus, integration of $S(E^|)$ over its range yields an observed count rate, r_c, that is less than, r_d. These two rates define the sample counting efficiency as

$$\epsilon = \frac{r_c}{r_d} = \frac{\int S(E^|) \, dE^|}{\int R(E) \, dE} \tag{1}$$

Here, as in all subsequent discussion, the energy integrals are over their respective spectral ranges (0 to max) for the sample.

The present approach explores the relation between the observed $S(E^|)$ and the ideal $R(E)$ spectra, so that ϵ may be determined without using known standards. Proper usage of standards can be cumbersome for samples containing multiple α, β, τ, ce, X-ray, and/or EC/Auger emitting radionuclides. The 100% detection $R(E)$ spectrum is transformed to $S(E^|)$ by the scintillation detection processes, viz

$$S(E^|) \, dE^| = \int R(E) \, dE \, p(E^||E) \, dE^| \tag{2}$$
$$S(E^|) = \int R(E) \, p(E^||E) \, dE$$

where $p(E^||E) \, dE^|$ is the probability that E total photons per decay yield $E^|$ detected photons in the range $dE^|$. Assuming that an inverse transform exists for Equation 2, $R(E)$ is solved as

$$R(E) = \int S(E^|) \, f(E|E^|) \, dE^| \tag{3}$$

where $f(E|E^|)$ is the inverse transformation kernel. Equations 2 and 3 require that

$$
\begin{aligned}
R(E) &= \int [\int R(E^*)p(E^||E^*)dE^* \,] \, f(E|D^|) \, dE^| \\
&= \int R(E^*) \, [\int p(E^||E^*)f(E|E^|)dE^| \,] \, dE^* \\
&= \int R(E^*) \, \delta(E^* - E) \, dE^* \\
&= R(E)
\end{aligned}
\tag{4}
$$

where $\delta(E^* - E)$ is the Dirac delta function, and E^* is a dummy integration variable. This delta function defines $f(E|E^|)$ in relation to $p(E^1|E^*)$ as

$$\int p(E^||E^*) \, f(E|E^|) \, dE^| = \delta(E^* - E) \tag{5}$$

Equation 5 is obtained directly from Equation 4, and solving it for $f(E|E^|)$ requires that $p(E^|E^*)$ be defined. A useful example is

$$p(E^|E^*) = \delta(E^|/c - E^*) \, p(E^*)$$
$$f(E|E^|) = \delta(cE - E^|) \, f(E^|) \tag{6}$$

whereby each $E^|$ of $S(E^|)$ corresponds to a single E or $R(E)$, and vice versa, per $E^| = cE$. To illustrate that the Equation 6 relations constitute a solution of Equation 5, they are substituted directly into Equations 2 and 3, to yield

$$S(E^|) = \int R(E^*) \, \delta \, (E^|/c - E^*) p(E^*) dE^* = p(E^|/c) \, R(E^|/c)$$

$$R(E) = \int S(E^|) \, \delta \, (cE - E^|) f(E^|) dE^| = f(cE) \, S(cE) \tag{7}$$

Substituting $E^| = cE$ in the above expressions and then multiplying yields the relationship between $p(E)$ and $f(E)$. The specific relations are

$$S(cE) \, R(E) = p(E) \, R(E) \, f(cE) \, S(cE)$$

$$p(E) \, f(cE) = 1 \text{ or } p(E) = 1 \, / \, f(cE) \tag{8}$$

Using Equations 6 and 8 in Equation 5 yields

$$\begin{aligned}
\int p(E^|E^*)f(E|E^|)dE^| &= \int \delta(E^|/c - E^*) \, p \, (E^*) \, \delta \, (cE - E^|) \, f \, (E^|)dE^| \\
&= \delta(E - E^*) \, p(E^*) \, f(cE) \\
&= \delta(E - E^*) \, p(E) \, f(cE) \\
&= \delta(E - E^*)
\end{aligned} \tag{9}$$

which completes the proof. The above example is somewhat ideal, because any number of coincident photons ($E^|$), which is less than the emitted total (E), may be detected. Yet, for sufficiently narrow distributions of $E^|$, the δ-function approximation with $E^| = cE$ can be reasonable. This approximation is tested in the present study, where the $R(E)$ and $S(E^|)$ are related by $p(E)$ per Equations 7 and 8.

Model Approximation

The number of photons produced per scintillation event ultimately defines the detection efficiency for the event. The TRI-CARB 2000 CA/LL spectra are calibrated so that about 10 photons are produced for each keV for unquenched β-emissions.[15] For α emissions, photon production per keV is lower by an order of magnitude;[15] however, their monoenergetic α-energies are well above 4000 keV, producing numerous photons and efficiencies of almost 100%. For quenched samples, the photon production is lower, as reflected by their contracted spectra.

The two PM tubes detect only a fraction of the photons that are emitted in an event. These PM tubes detect the photons with probability p^1 and p^2, and the probability for non-detection is p_0. For N emitted photons, the trinomial distribution gives the probability for detecting n_1 and n_2 photons on the PM tubes and missing detection of $n_0 = N - n_1 - n_2$. This trinomial distribution is described by

$$P(n_0, n_1, n_2 | N) = \frac{N!}{n_0!\, n_1!\, n_2!}\; p_0^{n_0}\, p_1^{n_1}\, p_2^{n_2}$$

$$_{all}\Sigma\; P(n_0, n_1, n_2 | N) = (p_0 + p_1 + p_2)^N = 1 \tag{10}$$

where the summation "all" refers to all possible n_0, n_1, n_2 combinations. Suppose that at least n coincidences for the PM tubes are required for event detection. The probability for detecting n or more coincidences is

$$P(\geq n | N) = \sum_{\geq n} P(n_0, n_1, n_2 | N) \tag{11}$$

where the "$\geq n$" means that both n_1 and n_2 are at least as large as n for terms included in the summation. The $P(\geq n | N)$ is essentially the same as the $p(E^l | E)$ discussed in the preceding section, where E is proportional to N emitted photons and E^l represents the distribution of detected photons governed by the $\geq n$ range. With the δ-function approximation $P(\geq n | N)$ becomes $p(E)$. Equations 10 and 11 were used to generate the $P(\geq n | N) = p(E)$ plotted in Figure 3, and these illustrate that the detection probability is very low for small N and monotonically increases toward an asymptotic value of 1. An energy scale based on 10 photons per keV is added for reference to E; the corresponding average E^l would be expected to be $(p_1 + p_2) \times E$; however, the energy scale of the TRI-CARB 2000 CA/LL is calibrated so that $c = 1$, yielding $E^l = E$. Thus, the $p(E)$ and $p(E^l)$ are identical in the present study.

It should be emphasized that the above $p(E)$ models all use the δ-function approximation that was discussed in the preceding section. Such models are reasonable for these preliminary tests, and should act as a guide to future improvements with more refined models. In fact, a good case can be made for the present trinomial distribution as a model by examining the distribution of the $(n_1 + n_2)$ detected photons, which represents the E^l distribution. The average $<n_1 + n_2>$ and its variance σ^{-2} are calculated using a binomial representation of the trinomial distribution, which is described as

$$1 = (p_0 + p_1 + p_2)^N = \sum_{all} - \frac{N!}{(n_1 + n_2)!\, n_0!} - (p_1 + p_2)^{n_1 + n_2}\, p_0^{n_0}$$
$$<n_1 + n_2> = (p_1 + p_2)\, N \tag{12}$$
$$\sigma^2 = (p_1 + p_2)\,(p_0)\, N = <n_1 + n_2>\, p_0$$

A more accurate estimate would sum over "$\geq n$" as opposed to "all"; however, the above approximation is sufficient for the present study. Because $n_1 + n_2$ is proportional to E^l and N is proportional to E, these relations imply that $c =$

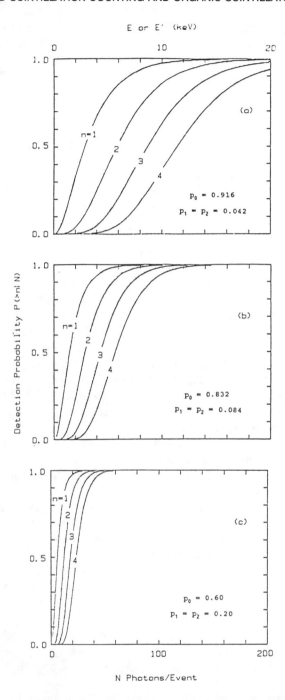

Figure 3. Detection probability p($\geq$n|N) vs photons/event N, for threshold detection of n coincident photons. The present study appraised the p($\geq$1|N) of Figures a and b, and the p($\geq$2|N) of Figure c.

$E^|/E = <n_1 + n_2>/N = p_1 + p_2$; however, as mentioned earlier, the TRI-CARB 2000 CA/LL is calibrated so that $c = 1$. Accordingly, Figure 3 includes scaled horizontal axes N, E, and $E^|$. Because the σ fluctuations about $< n_1 + n_2 >$ are proportional to $N^{1/2}$, their range is narrowest for the lowest $E^|$, where the inverse transformation factor $f(E^|) = 1/p(E)$ is largest. For larger $E^|$, its broader range is not as crucial, because the transformation factor is relatively constant with a value near 1.

MEASUREMENTS

Two types of measurements were performed. The first used conversion electrons and Auger electrons of ^{57}Co to help identify reasonable p(E) from calculations as shown in Figure 3. The second series of measurements accumulated ^{3}H and ^{14}C scintillation spectra to test the p(E) for correcting the total count rate of the observed spectrum, $S(E^|)$, to the total decay rate of the real spectrum, R(E).

A nominal 1.5 μCi/mL ^{57}Co aqueous solution, dissolved as $CoCl_2$, was provided by Isotopes Products Laboratories. A set of 20 quenched standards was prepared from this solution, in which 0, 50, . . . , and 950 μL of acetone were added to base cocktails, each composed of 18 mL of Optifluor, 2 mL distilled water, and 50 μL of the ^{57}Co solution. Standard 25 mL polyethylene counting vials were used. The TRI-CARB 2000CA/LL counted each sample for 10 min and stored its spectrum on floppy disk. The disk spectra were transformed for display analysis on an IBM/XT computer programmed with an EG&G Ortec ADCAM multichannel emulator. Representative spectra are shown in Figure 4.

Measurements for ^{3}H and ^{14}C were performed similarly. Some of the solutions were commercial standards from Packard Instrument Company; these were both quenched and unquenched and contained in nominal 25 mL glass vials. Other samples were prepared on site and used the 25 mL polyethylene vials. In particular, 500 μL of an NBS standard solution (NBS SRM-4927) with 33,720 dpm tritium was added to 2 mL of distilled water and 18 mL Optifluor for a base cocktail. Ten of these base solutions were counted, and then they were quenched by adding 50, 100, . ., and 500 μL of acetone and recounted. Following this 500 μL of acetone was added to each of these samples to yield a set quenched with 550, 600, . . . , and 1000 μL of acetone. The spectrum for each sample was counted for 10 min and stored on floppy disk.

RESULTS

Selection of p(E) from ^{57}Co Studies

The three spectra of Figure 4 illustrate how quenching contracts the observed spectrum $S(E^|)$ to lower energies. The Figure also illustrates the effect

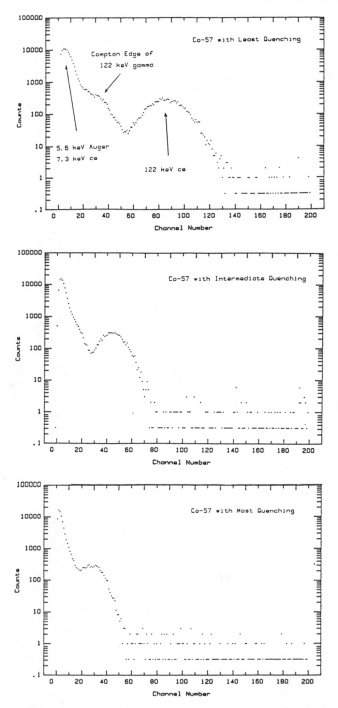

Figure 4. Liquid scintillation spectra of ⁵⁷Co with (a) least, (b) intermediate, and (c) most quenching. The dominant spectral components are identified.

of p(E) on the lower end of these spectra. In particular, because S(E¹) = p(E)R(E) per Equation 7, a larger fraction of the real spectrum R(E) is eliminated from S(E¹) as the spectra are contracted to lower energies.

Three peak regions appear in the spectra, which are all identifiable as ⁵⁷Co emissions.[16] The highest peak includes K- and L-shell conversion electron (ce-K and ce-L) emissions of 115 and 122 keV associated with the 122 keV τ-ray and similar emissions of 129 and 136 keV associated with the 137 τ-ray. The 122 ce dominates this peak. A contribution for the photoelectrons produced by these gammas is also included in this highest energy peak. The lowest energy peak is due to ce- and τ-emissions associated with the 14.4 keV transition and also Auger and X-ray emissions associated with the EC decay. The 7.3 keV ce-K is the dominant emission for the 14.4 keV transition, and a comparably intense 5.6 keV Auger emission dominates the Auger/X-ray cases. Emissions contributing to the lower peak effectively range from 5.6 to 14.4 keV. (Other Auger emissions of less than 1 keV were either quenched beyond detection or detected per summing with the larger peaks). The middle peak region is not well pronounced and includes Compton electrons produced by the 122 and 136 keV gammas, corresponding to Compton edges of 39 and 47 keV. The Compton edge of the 122 keV gamma is dominant.

The ⁵⁷Co spectral data were used as a guide for selecting the p(E). Analyses of the 20 quenched spectra yielded peak centroids and areas. Figure 5 plots the centroid ratio of the highest and lowest peaks against the centroid of the highest peak. The ratio is about 18 and constant, if the quenching does not move the highest peak below channel 55. As quenching contracts the peaks to

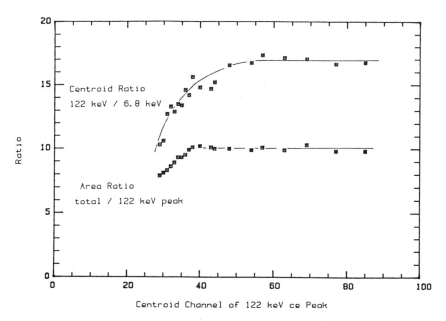

Figure 5. Centroid ratio and area ratio, measured for quenched ⁵⁷Co samples.

lower channels, the true centroid of the lowest peak is lost due to its low p(E), and an artificially high centroid is determined. Because this drop in ratio begins as the highest peak falls below channel 55, the p(E) begins to affect the lowest peak when it drops below channel $55/18 = 3.1$. The ratio of total spectral counts to highest peak area provides a less sensitive check, as Figure 5 illustrates. Its departure from constancy is just observable when the highest peak centroid drops below channel 40. Overall, these tests imply that the true departure is above channel 3. All trinomial predictions of Figure 3 were selected to be consistent with this condition.

As a further guide in selecting p(E), the centroid and FWHM of the 122 keV peak in Figure 4a were examined in terms of Equation 12. Specifically, a $<n_1 + n_2>$ corresponding to centroid channel 85 and its σ of FWHM/2.355 = 13.6 yields

$$<n_1 + n_2>^2/\sigma^2 = (p_1 + p_2) N / p_0 = (85/13.6)^2 = 39.1$$

Using the 10 photons/keV and 0.5 keV/channel calibration information,[15] $N = 85 \times 10 \times 0.5 = 425$, and this yields $p_1 = p_2 = 0.042$ and $p_0 = 0.916$. These values are comparable to values used in efficiency models. Using a minimum of $n = 1$ photon coincidences, Equation 11 yields the p(E) of Figure 3a for this case, which is identified as Model #1 p(E).

The Model #1 p(E) must have $E \geq 5$ keV (channel 10) before it begins to approach 1, and this is considerably higher than the $E \geq 1.5$ keV (channel 3) deduced from the [57]Co studies; thus, other models of p(E) that lie between these extremes were also examined. Model #2 rises twice as fast as Model #1, whereby $E \geq 2.5$ keV as p(E) begins to approach 1. Model #3 is a modification of Model #2, in which p(E) is reduced for the lower E values, in an attempt to provide a better empirical model. Finally, Model #4 requires a minimum of $n = 2$ coincidences, and the p_0 and $p_1 = p_2$ are selected to yield a good empirical model. The p(E) values of each model are given in Table 1.

Application of p(E) to ³H and ¹⁴C Spectra

The p(E) models given of Table 1 were applied to the collected ³H and ¹⁴C spectra. In particular the decay rates for these spectra were calculated as

$$r_d = \Sigma \ S(E_i)/p(E_i) \tag{13}$$

where the sum is over all channels of the spectrum. The results for each model are given in Figure 6, where the ratio of predicted/known r_d values are plotted against the known efficiency, ϵ. For higher ϵ the ratio is close to 1, indicating that model predictions agree well with known values; however, for lower ϵ, the agreement falls off. This reflects the fact that the p(E) corrects a larger portion of the spectrum for lower-energy (and thus lower ϵ) cases, as implied by Table 1 and Equation 13. Model #1, which corresponds more closely to earlier effi-

Table 1. Models of p(E)

Channel #	E keV	p(E) Nidek #1	Model #2	Model #3	Model #4
0	0.25	0.01	0.01	0.01	0.0039
1	0.75	0.07	0.16	0.10	0.200
2	1.25	0.16	0.38	0.30	0.555
3	1.75	0.27	0.56	0.50	0.797
4	2.25	0.38	0.70	0.70	0.920
5	2.75	0.48	0.81	0.81	0.964
6	3.25	0.56	0.88	0.88	0.983
7	3.75	0.63	0.915	0.915	0.992
8	4.25	0.70	0.94	0.94	1.000
9	4.75	0.76	0.958	0.958	
10	5.25	0.81	0.971	0.971	
11	5.75	0.85	0.983	0.983	
12	6.25	0.88	0.992	0.992	
13	6.75	0.90	1.000	1.000	
14	7.25	0.915			
15	7.75	0.93			
16	8.25	0.94			
17	8.75	0.95			
18	9.25	0.958			
19	9.75	0.965			
20	10.25	0.971			
21	10.75	0.977			
22	11.25	0.983			
23	11.75	0.988			
24	12.25	0.992			
25	12.75	0.996			
≥ 26	≥ 13.25	1.000			

Notes: Model #1: $p_1 = p_2 = 0.042m$, $p_0 = 0.916$, $n = 1$, Model #2: $p_1 = p_2 = 0.084$, $p_0 = 0.832$, $n = 1$, Model #3: Model #2 modified at lower channels, Model #4: $p_1 = p_2 = 0.20$, $p_0 = 0.60$, $n = 2$.

ciency models,[4,9,12] yields worse agreement than the other models, which were developed from semiempirical considerations.

A bootstrap technique can extend the application to lower efficiencies. If the measured total spectral countrate, r_c, is divided by the predicted decay rate r_d, an experimental pseudo-efficiency, $\epsilon^|$, may be defined as r_c/r_d. If the ratios of predicted/known r_d are plotted against their $\epsilon^|$, the plots of Figure 7 result for each model, thus, any measurement of $\epsilon^|$ yields a predicted/known correction ratio k, from which the true decay rate may be calculated as r_d/k and $n = E/E^|$, so that Figure 7 yields $E = nE^|$ per measurement of $E^|$.

DISCUSSION

Each p(E) model is demonstrated to yield sample efficiencies to $\approx 10\%$ for efficiencies ranging from 0.1 to 1.0, provided that the bootstrap technique can be applied, as seen in Figure 7. No p(E) was found to yield efficiencies that agreed with the known ones over the entire efficiency range tested, as shown in Figure 6. The δ-function model, for which the p(E) models are based, appar-

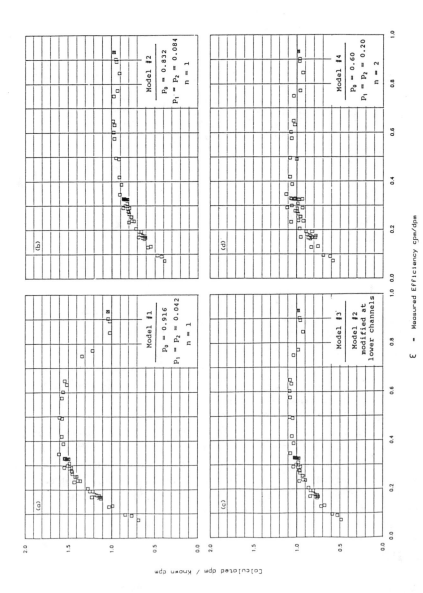

Figure 6. Comparison of predicted/known decay rates as function of efficiency ε, for the four models examined.

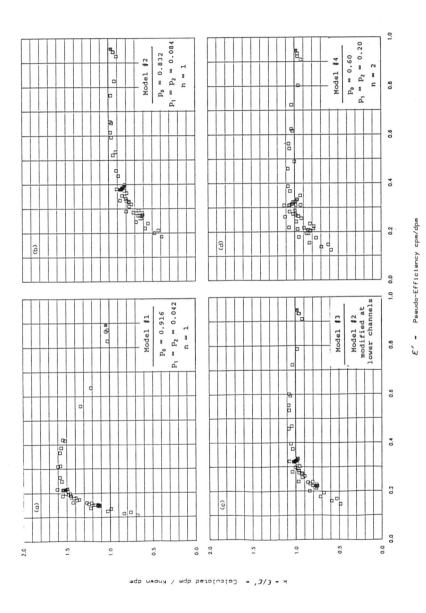

Figure 7. Comparison of predicted/known decay rates as function of pseudo-efficiency, $\epsilon^{|}$, for the four models examined.

ently breaks down for the lower efficiency cases. Improvement beyond the present treatment will require development of a more realistic inverse transformation kernal $f(E|E^|)$, to convolute the observed spectrum $S(E^|)$ to its ideal form $R(E)$.

Liquid waste samples complying with DOT regulations must be less than 2 nCi/g, or effectively 4440 dpm/mL. For a 3 mL sample and a 10 min counting time, a total of 133,200 decays would occur. For ϵ down to 0.1 (per measurement of $\epsilon^|$), at least 13,000 counts should be observed, and most of these would be at the lower end of the spectrum so that the p(E)-corrected counts would have reasonable counting errors. For higher ϵ, the counting errors will be even smaller; thus, the method should have application for certifying liquid waste for transport. Model #1, which yields overestimates as high as 60%, could be a useful conservative analysis method for this application.

Some liquid waste can fall outside the above range of application. For example, some liquids may require dilution so that a 3 mL scintillation sample would contain less effective activity than its original solution. Also, the sample may have quenching that renders an efficiency below 0.1. In these cases, the p(E) for the upper channels would still be 1; however, a low energy standard such as tritium can be used to spike an identical sample to appraise the low energy portion of the spectrum.

For the β-components of the spectrum, the calibration of Figure 2c could be used upon proper identification of the endpoints $E_\beta^|$. The endpoints on the $E^|$ scale are distorted as $aE_\beta + bE_\beta^{1/2}$ or $N + b^|N^{1/2}$, according to resolution broadening predicted by Equation 12 and illustrated in Figure 4. Here aE_β is the true spectral endpoint; which appears higher by $bE_\beta^{1/2}$ due to resolution broadening. The factor $a = F(E)$ is the scintillation efficiency, which has a broad maximum at ≈ 200 keV.[9] Using Figure 4, $b = 3.45$ keV$^{1/2}$ was estimated from the ce peak. Because the $E^|$ scale is normalized to yield a good fit with the E_β endpoints, it is not strictly proportional to E_β, although the linear approximation is reasonably good. The results of Figure 2c assume that the unquenched ^3H and ^{14}C cases correspond to their endpoints of $E_\beta^| = 18.6$ and 156.5 keV, but their unquenched cases are given by

$$E_\beta^| = (tSIE/1000) \, e_\beta + b \, [\, (tSIE/1000) \, e_\beta]^{1/2} \qquad (14)$$

where e_β is the effective aE_β pertaining to the normalized $E_\beta^|$ scale for the unquenched sample (tSIE = 1000). It is determined for ^3H and ^{14}C using their corresponding unquenched $E_\beta^|$. The $E_\beta^|$ of Figure 2c and Equation 14 are given in units of $t^|$keV. It should be recalled that this method assumes that the β-spectra have the same relative profiles, but this is not always true. For example, ^{241}Pu with $E_\beta = 20.82$ keV has a lower efficiency than ^3H with $E_\beta = 18.6$ keV,[12] this is contrary to the monotonic energy trend in Figure 2c. Here the profiles differ so much that the ^{241}Pu average β-energy of 5.23 keV is lower than that of 5.69 keV for ^3H.[17] Some care must be exercised in using this technique.

ACKNOWLEDGEMENTS

The author acknowledges useful discussions with K.J. Hofstetter, W.L. McDowell, and R.W. Taylor. The author also thanks C.D. Ouzts for conducting many of the laboratory measurements.

The information contained in this paper was developed under Contract No. DE-AC09-88SR18035 with the U.S. Department of Energy.

REFERENCES

1. Code of Federal Regulations, 49CFR173.403(g), June, 1986.
2. Ishikawa, H. and M. Takiue. *Nucl. Instr. and Methods*, 112:437–442 (1973).
3. Ishikawa, H., M. Takiue, and T. Aburai. *Int. J. Appl. Radiat. Isot.* 35:463–466 (1984).
4. Coursey, B.M., W.B. Mann, A. Grau Malonda, E. García-Toraño, J.M. Los Arcos, J.A.B. Gibson, and D. Reher. *Appl. Radiat. Isot.* 37:403–408 (1986).
5. Kobayashi, Y., M. Kessler, and S. van Cauter, *Amer. Lab.*, 20:160–167 (1988).
6. Houtermans, H., *Nucl. Instr. and Methods*, 112:121–130 (1973).
7. De Filippis, S.J. *Trans. Am. Nuc. Soc.* 50:15–16 (1985).
8. De Filippis, S.J. and S.C. van Cauter. *Trans. Am. Nuc. Soc.* 50: 22–23 (1985).
9. Broda, R., K. Pochwalski, and T. Radoszewski. *Appl. Radiat. Isot.*39:159–164 (1988).
10. Pochwalski, K., R. Broda, and T. Radoszewski. *Appl. Radiat. Isot.* 39:165–172 (1988).
11. Grau Malonda, A. and B.M. Coursey. *Appl. Radiat. Isot.* 39:1191–1196 (1988).
12. Coursey, B.M., A. Grau Malonda, E. García-Toraño, and J.M. Los Arcos, *Trans. Am. Nuc. Soc.* 50:13–15 (1985).
13. Grau Malonda, A. and E. García-Toraño. *Int. J. Appl. Radiat. Isot.* 33:249–253 (1982).
14. Jordan, P. *Nucl. Instr. and Methods*, 97:107–111 (1971).
15. Model 2000CA TRI-CARB Liquid Scintillation Analyzer/ Operation Manual, Publication No. 169-3029, Packard Instrument Co, 1987.
16. *Radionuclide Transformations—Energy and Intensity of Emissions*, Annals of ICRP—Publication 38, 11–13 (1983).
17. Kocher, D.C. Radioactive Decay Data Tables, DOE/TIC-11026 (1985).

CHAPTER 59

Absolute Activity Liquid Scintillation Counting: An Attractive Alternative to Quench-Corrected DPM for Higher Energy Isotopes

Michael J. Kessler, Ph.D.

The liquid scintillation efficiency tracing technique is a new and powerful method of quantitating radionuclides being analyzed in a liquid scintillation analyzer.[1-5] This technique has several advantages over conventional DPM analysis. First, no quench curve (quenched standard set) is required for each nuclide. Second, the technique can be used effectively for almost all pure beta and beta-gamma emitters. Third, only a single unquenched ^{14}C sample (same as that used to normalize the liquid scintillation analyzer) is required to calculate DPM results. Fourth, the efficiency tracing technique provides a simple method for DPM quantitation in the sample. Fifth, relatively small errors (105%) in the calculation of DPM can be achieved using this technique. Sixth, different radionuclides can be intermixed in the sample batch.

This efficiency tracing technique is based on a patented procedure which requires a liquid scintillation analyzer containing a multichannel analyzer, a sophisticated data reduction system, and a standard radionuclide (sealed calibration standard) of known absolute activity. The efficiency tracing procedure is accomplished in the following manner. First, the system is standardized (normalized) with an unquenched ^{14}C standard. The reference spectrum of this standard is analyzed, and the counting efficiency is determined in six separate regions simultaneously. Second, the % efficiency in each of the six regions is calculated and plotted against the actual number of counts in each region. Figure 1 shows a typical % efficiency vs CPM plot for the ^{14}C standard. Third, an extrapolated CPM value for the 100% efficiency is determined based on the plot of the resultant line analyzed by the method of least squares. Fourth, when an unknown sample is analyzed, its spectrum is analyzed in the same six regions, and the results are plotted using the same x-axis (% efficiency) generated by the ^{14}C standard. A line is then drawn through the resultant points. A curve is fitted (least square) through these six points and extrapolated to 100%

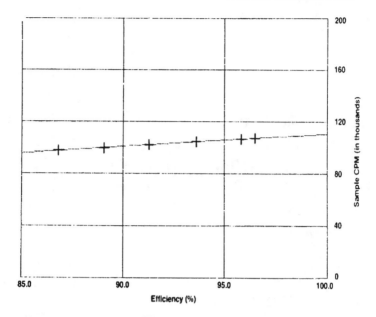

Figure 1. Typical efficiency tracing™ plot for ¹⁴C unquenched sample.

efficiency. The extrapolated CPM value at this point is equal to the number of DPM in the sample.

Over a dozen radionuclides (¹⁴C, ³²P, ³⁶Cl, ⁴⁶Sc, ⁵⁹Fe, ⁶⁰Co, ⁶³Ni, ⁸⁶Rb, ⁹⁰Sr-Y, ¹³¹I, ¹³⁴Cs, ¹⁴⁷Pm) have been assayed[6,7] using this technique with excellent statistical results (SD = 1.4%) between the DPM calculated by the efficiency tracing and the absolute activity (DPM) of the sample. Specific examples of the DPM results achieved using the efficiency tracing technique on the Packard LSA Model 2500TR for radionuclides ³⁶Cl, ⁵⁹Fe, ⁶³Ni, and ¹⁴C are shown in Table 1.

Specific examples of the actual efficiency tracing plots of the data show excellent statistical results; % recovery, equal 100% ± 2% for ³⁶Cl, ⁵⁹Fe, and ⁶³Ni are shown in Figures 2 to 4.

As can be seen from the previous table and results, excellent DPM values can be obtained for radionuclides ranging from ⁶³Ni to ³²P over the energy range of 60 to 1700 keV.

In order to further assess the reliability and accuracy for determining DPM values, a series of samples with various types of counting vials, scintillation cocktails, sample volume, microvolume samples, and color or chemical quenching degree were quantitated using the efficiency tracing technique (Table 2).

The mean % recovery is 99.91% with a very small standard deviation of only 1.15. These results conclusively demonstrate that using the efficiency tracing technique produces DPM results which are independent of:

Table 1. Efficiency Tracing Results of Various Nuclides at Various Quench Levels

Radionuclide	SIS	tSIE	DPM (ET)	DPM (Actual)	% Rec.
^{36}Cl	973	919	117,381	118,414	99.1
^{36}Cl	580	537	117,607	118,414	99.3
^{36}Cl	188	174	116,606	118,414	98.5
^{36}Cl	117	109	117,154	118,414	99.0
^{59}Fe	241	460	2,231,719	2,310,000	97.0
^{59}Fe	223	428	2,204,944	2,310,000	95.4
^{59}Fe	151	315	2,283,158	2,310,000	98.8
^{59}Fe	115	305	2,306,399	2,310,000	99.9
^{63}Ni	30	583	200,826	200,000	100.4
^{63}Ni	29	535	197,909	200,000	99.0
^{63}Ni	16	235	195,066	200,000	97.0
^{63}Ni	15	206	185,546	200,000	93.0
^{14}C	173	1,000	111,280	111,700	99.6
^{14}C	86	505	112,603	111,700	100.8
^{14}C	39	209	116,373	111,700	104.1
^{14}C	24	120	125,131	111,700	112.0

1. cocktail density variation
2. different vial sizes
3. varying sample volume
4. color quenching
5. chemical quenching
6. vial composition

The independence of the DPM values on the chemical quench level of the sample can be further demonstrated using ^{63}Ni at various quench levels (629,

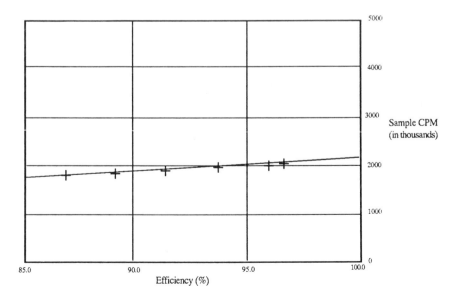

Figure 2. Efficiency Tracing Curve for ^{59}Fe at tSIE = 437 Quench Level.

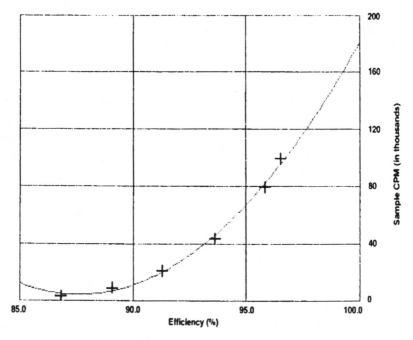

Figure 3.   Efficiency tracing of ^{63}Ni at tSIE = 204 (quenched).

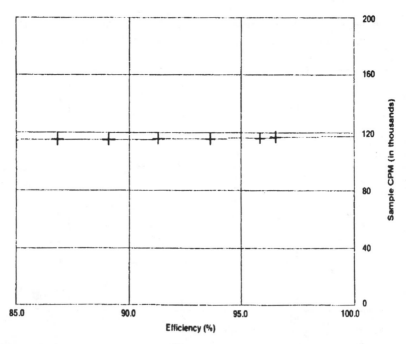

Figure 4.   Efficiency tracing curve for ^{36}Cl at tSIE = 953 (unquenched).

Table 2. Efficiency Tracing Results for ^{14}C Using Various Conditions

Sample	tSIE	DPM (ET)	DPM (Actual)	% Recovery
1. Std. vial Color quenched	625	143,457	144,000	99.6
2. Mini vial Color quenched	644	144,020	144,000	100.0
3. Std. vial (2.0 mL)	576	128,536	129,500	99.3
4. Std. vial (8.0 mL)	967	127,808	129.500	98.7
5. Std. vial (15.0 mL)	959	128,432	129,500	99.2
6. Mini vial (0.5 mL)	608	128,608	129,500	99.3
7. Mini vial (2.0 mL)	776	128,099	129,500	98.9
8. Mini vial (5.0 mL)	862	128,768	129,500	99.4
9. Std. vial Chemical quench	690	142,734	144,000	99.1
10. Mini vial Chemical quench	653	143,072	144,000	99.4
11. Std. vial Bray's cocktail	207	119,173	120,800	98.7
12. Micro volume 400 μL	739	10,992	10,760	102.1
13. Micro volume 200 μL	693	11,209	11,080	101.1
14. Micro volume 100 μL	643	11,043	10,860	101.7
15. Micro volume 50 μL	590	10,747	10,650	100.9
16. Micro volume 25 μL	548	10,715	10,460	102.4
17. Std. Vial - No Chemical quench	989	110,129	111,700	98.6
18. Std. vial Chemical quench	785	111,312	111,700	99.7
19. Std. vial Chemical quench	281	112,279	111,700	100.5
20. Mini Vial - No Chemical quench	862	132,904	133,500	99.6

501, 381, 221) plotted in Figure 5. The plot indicates that each efficiency tracing plot (different tSIE values) has a different slope, but all intersect the 100% activity line at approximately 197,000 DPM. This clearly indicates that the final DPM results are independent of the color or chemical quench level of the sample.

In addition, if efficiency tracing is used for low level DPM samples, results similar to those shown in Figure 6 can be expected. A set of samples containing 100 DPM was analyzed by this technique, and found efficiency tracing 98.813 DPM with a coefficient of variation of 0.714. Similar results were obtained from Dr. Ishikawa in Japan, DPM as low as 22.53 ± 1.11 for a set of low level samples. These results indicate that the efficiency tracing method can be used accurately to determine DPM in low level samples.

This technique can be used for most pure beta and beta-gamma emitters. The one exception is tritium, whose efficiency tracing plot can result in large errors (up to 25%) for highly quenched samples. In addition, the efficiency tracing technique is not applicable to radionuclides which decay by isometric transitions and electron capture (EC). The reason for this is that radionuclides that decay by electron capture are followed by the emission of an X-ray or

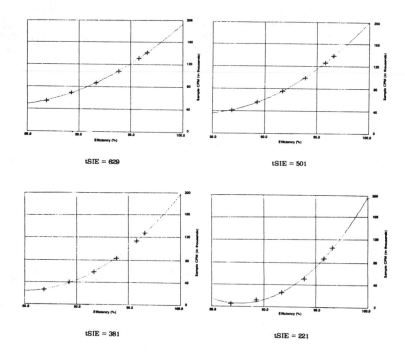

Figure 5.   Efficiency tracing curves for ^{63}Ni at four quench levels.

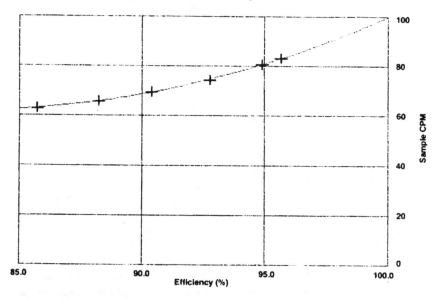

Mean (4) = 98.813 DPM   **S.D. = 0.705**   **C.V. = 0.714**

Figure 6.   Efficiency tracing of ^{14}C sample containing 100 DPM and statistics for sample counted four times.

Auger electron. This makes it difficult to find the true absolute activity of these radionuclides using this technique.[6]

In summary, the efficiency tracing technique is an accurate and reliable method of calculating DPM (absolute activity) for most pure beta and pure beta-gamma emitters (minimum energy = 60 keV). This method only requires counting an unquenched ^{14}C standard of known activity and does not require the preparation of a quenched curve set for each individual radionuclide analyzed. The final DPM for each sample is calculated from the respective spectrum compared to the standard unquenched sample using six separate spectral regions. The data (DPM) are calculated using a special curve fitting routine and extrapolation technique to determine the absolute activity (DPM) of the sample. This efficiency tracing method is available on all Packard liquid scintillation analyzers.

REFERENCES

1. Grau Malonda, A., E. Garcia-Toraño, and J.M. Los Arcos. *Int. J. Appl. Radiat. Isot.* 36(2).157 (1985).
2. Grau Malonda, A. and E. Garcia-Toraño. *Int J. Appl. Radiat. Isot.* 33:249 (1982).
3. Grau Malonda, A. *Int. J. Appl. Radiat. Isot.* 33:371 (1982).
4. Coursey, B.M., J.A.B. Gibson, M.W. Gertzmann, and J.C. Leak. "Standardization of Technecium-99 by Liquid Scintillation Counting," *Int. J. Appl. Radiat. Isot.* 35(12):1103 (1984).
5. Yura, O. *Radioisotopes* 20:493 (1971).
6. Ishikawa, H.M. Takiue, and T. Aburai. *Int. J. Appl. Radiat. Isot.* 35(6):463 (1984).
7. Ishikawa, H. and M. Takiue. *Nuc. Instrument Methods* 112:437 (1983).

CHAPTER 60

Chemical and Color Quench Curves Over Extended Quench Ranges

Charles Dodson

ABSTRACT

Chemical and color quench curves are presented for five nuclides over extended quench ranges. These curves may approximate the total domain of chemical/color quench because of the quenching agents used. The difference between chemical and color curves do not increase continuously as quench increases, rather they pass through a maximum. Study of that relationship suggested a heuristic approach for obtaining color efficiencies corrected from chemical quench curves.

INTRODUCTION

Chemical and color quench curves used in liquid scintillation counting do not superimpose over the complete range of counting efficiencies for a given radionuclide. A study was conducted to define these curves for a few nuclides over extended quench ranges. An approach was taken to define the entire quench domain for a given nuclide. That information led to the capability of providing color quench correction automatically from conventionally prepared chemical quench curves.

The radionuclides studied were ^3H, ^{63}Ni, ^{125}I, ^{14}C, ^{35}S, ^{36}Cl, ^{32}Na and ^{32}P. Results reported here are for ^3H, ^{125}I, ^{14}C, ^{35}S, and ^{32}P.

QUENCHING AGENTS

A variety of chemical quenching agents was tried, but nitromethane and nitrobenzene provided representative results over extended quench ranges. Representative color quench agents used in the study included:

- erythrosine: di-sodium salt of 9(o-carboxyphenyl)-6-hydroxy-2,4,5,7-tetraiodo-3H-xanthene-3-one

  FD&C red #3

- fluorescein: 3′,4′-dehydroxyfluoran
- tartrazine: tri-sodium salt of 4.5-dihydro-5-oxo-1-(4-sulfophenyl)-4-[(4-sulfophenyl)azo]-1H-pyrazole-3-carboxylic acid

  FD&C yellow #5
- methyl Red: 4′-dimethylaminoazobenzene-2-carboxylic acid
- ferrocene: dicyclopentadienyl iron
- allura Red AC: disodium salt of 6-hydroxy-5-[(2-methoxy-5-methyl-4-sulfophenyl)azo]-2-naphthalenesulfonic acid

  FD&C red #40

Mixtures of some of the above dyes were used, but none was as useful as Alberta PAQ,[1] (Photon Absorbing Quencher) which consists of:

- 262.8 mg 4-phenylazo-1-naphthol
- 578 mg 1-(2′-methyl-4′-(2″methylphenylazo)-phenylazo)-2-naphthol
- 137.3 mg 4-phenylazoaniline
- 176.7 mg 4-phenylazo-1-naphthylamine in 100 mL toulene

The importance of PAQ for this study is explained below.

Some problems with precipitation occurred. For example, ^{36}Cl was in the form of 1NHCl and produced a precipitate with FD&C red #3. Apparently exchange of ^{36}Cl with iodide in the dye occurred before precipitation, or ^{36}Cl was carried by the precipitate into a 2π counting condition, because the count rate of the sample decreased with time.

QUENCH CURVES

The general features of these quench curves are represented by Figures 1 to 5 for the nuclides ^{3}H, ^{125}I, ^{14}C, ^{35}S, and ^{32}P. Each plot presents chemical and color quench curves for the stated nuclide as measured by H#, the inflection point on the edge of the ^{37}Cs Compton spectrum. The upper curve is the chemical quench curve, the lower is the color curve.

These curves cover 600 H#'s, almost to 0% counting efficiency for each nuclide except ^{32}P, and are cubic polynomial fits to the original data. The color quench curve was generated using Alberta PAQ in standard 20 mL Pyrex vials. All other color quench curves generated in this study lie above the color curves shown here, whether composed of single or mixed dyes, whether the dyes were organic or inorganic soluble, whether mixtures of chemical and color quenching agents were used, and regardless of the chemical composition of the vial or volume of the sample up to 20 mL.

That the absorption spectrum[1] of the PAQ mixture absorbs essentially all photons between 375 and 525 nm was reconfirmed. Consequently each pair of chemical/color quench curves for each nuclide provides a reasonable approximation to the complete chemical/color quenching domain in liquid scintillation counting, if constrained in pathlength by the geometry of a standard 20 mL vial. This is supported by the fact that about 1690 μmol of nitromethane

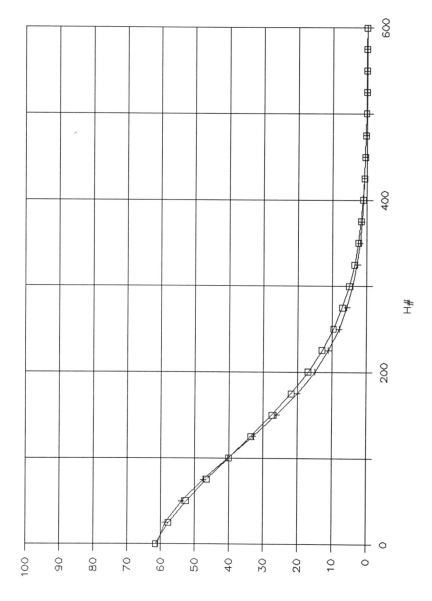

Figure 1. Chemical and color quench curves as a function of H# for ^{3}H.

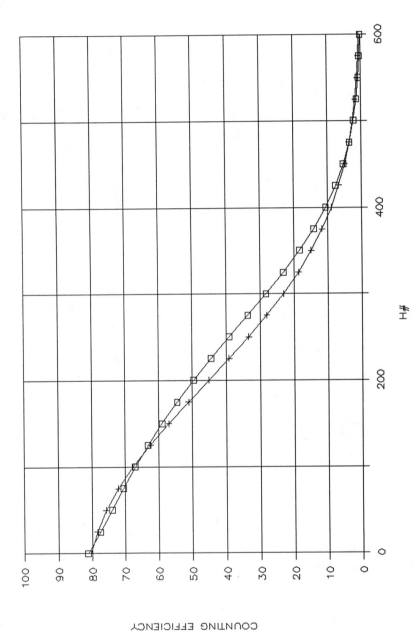

Figure 2. Chemical and color quench curves as a function of H# for ^{125}I.

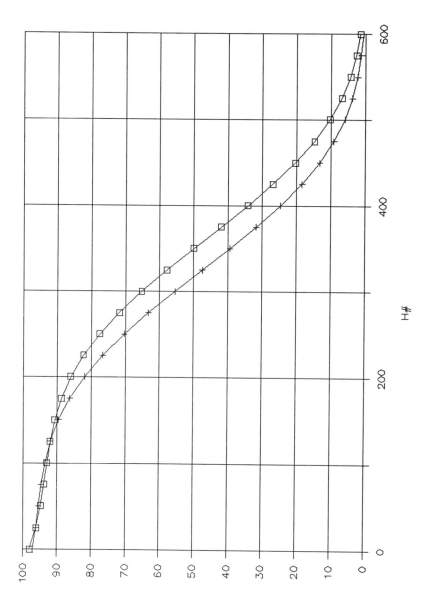

Figure 3. Chemical and color quench curves as a function of H# for ^{14}C.

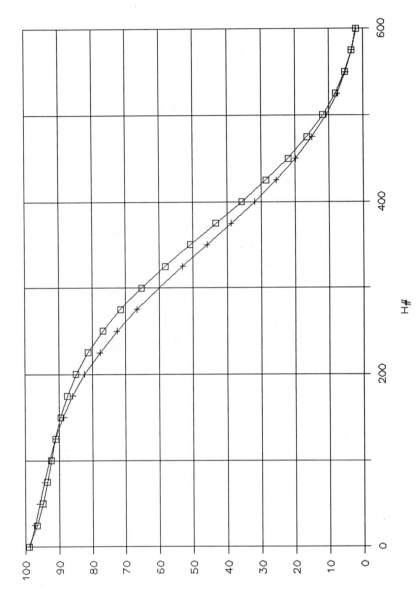

Figure 4. Chemical and color quench curves as a function of H# for ^{35}S.

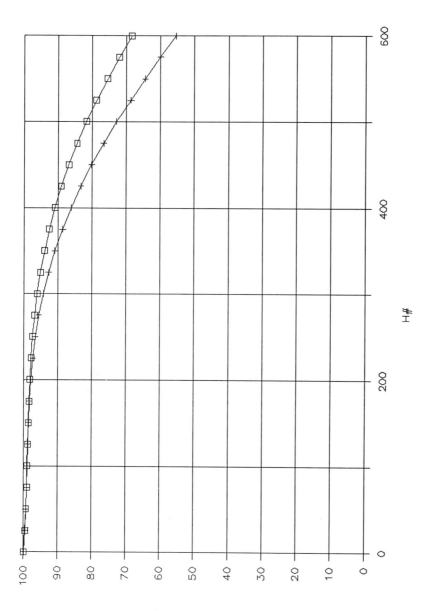

Figure 5. Chemical and color quench curves as a function of H# for ^{32}P.

Table 1. Chemical Quenched Unknowns vs Chemical Quench Curves

Nuclide	Minimum Error	Maximum Error	Mean Error
^{3}H	0.1	3.5	1.6
^{125}I	0.4	4.8	1.4
^{14}C	0.3	1.9	1.0
^{35}S	0.1	2.0	1.1
^{32}P	0.1	0.7	0.3

are required to produce the same counting efficiency as 1 μmol of the Alberta mixture.

These plots show that the difference between chemical and color quench curves does not increase continuously with quench but passes through a maximum. Although the detailed shape of quench curves prepared from the same standards varies if different photomultiplier tubes are used, the difference between chemical and color curves is constant.

DPM RECOVERY

If means existed to determine whether color correction were required and to what extent, the actual correction could be obtained from the quench curve differences presented here. Such a proprietary color correction monitor has been developed.[2] It depends upon an appropriate calibration of the liquid scintillation counter in terms of a generalization of the Beer-Lambert law.

Chemical quench curve results for ^{3}H, ^{125}I, ^{14}C, ^{35}S, and ^{32}P obtained from the color correction monitor follow. Table 1 gives results in terms of %error in DPM recovery for chemical standards run against chemical quench curves for each of the five nuclides. This provides a measure of accuracy for each nuclide quench curve. Each sample set consisted of 12 samples in standard 20 mL vials and 12 samples in 7 mL vials. The least, largest, and mean errors are given for each data set.

Table 2 provides exactly the same information for colored samples run against previously prepared colored quench curves. Table 3 provides the same information for a set of 24 unknowns for each nuclide. For the unknowns, volumes varied from 2 to 16 mL, and quench varied from pure chemical to pure color; a variety of dyes were used. Table 4 illustrates the %error in DPM recovery for pure colored samples run against chemical quench curves.

Table 2. Color Quenched Unknowns vs Color Quench Curves

Nuclide	Minimum Error	Maximum Error	Mean Error
^{3}H	0.1	2.9	1.4
^{125}I	0.7	2.5	1.6
^{14}C	0.4	4.5	1.8
^{35}S	0.1	3.5	1.3
^{32}P	0.1	0.3	0.2

Table 3. Chemical/Color Unknowns vs Chemical Quench Curves via Color Correction Monitor

Nuclide	Minimum Error	Maximum Error	Mean Error
^{3}H	1.7	10.9	7.7
^{125}I	0.3	5.0	3.6
^{14}C	0.1	2.1	1.2
^{35}S	0.1	5.6	3.2
^{32}P	0.1	3.5	0.8

Table 4. % Error, DPM Recovery No Color Correction Applied

Nuclide	Mean Error H# = 250	H# = 300	H# = 350	H# = 400
^{3}H	12.9	16.0	19.2	2.9
^{125}I	12.2	16.0	18.8	18.4
^{14}C	9.9	14.9	23.0	29.4
^{35}S	6.1	7.7	9.8	10.1
^{32}P	1.4	1.4	2.4	5.6

SUMMARY

A comparison of chemical and color quench curves for 5 nuclides over extended quench ranges shows that their differences pass through a maximum, and that this difference is independent from PMTs. Development of a color quench corrector combined with chemical quench curves and the fixed differences between chemical and color curves proved useful in providing efficiency correction for colored samples.

REFERENCES

1. Ediss, C., R.J. Flanagan, S.A. McQuarrie, and L.I. Wiebe. *Int. J. Appl. Radiat. Isot.* 33:296 (1982).
2. Dodson, C. To be published.

CHAPTER 61

The Characteristics of the CH Number Method in Liquid Scintillation Counting

Wu Xue Zhou

ABSTRACT

The main characteristics of the CH number method in liquid scintillation counting is described in the paper. The CH number method is compared to the methods of H number channel ratio, and spectral index widely used in liquid scintillation counting.

Key words: CH number, H number, spectral index, channel ratio.

INTRODUCTION

Liquid scintillation counting is widely used in many fields. It is an important method in radionuclide metrology. This chapter describes a new approach to measure zero probability and radionuclide activity.[1,2] Four (LS)-coincidence national standards have been established with which ^{3}H (hexadecane), ^{14}C (hexadecane), and tritium water national standards solution are caliblated.

A new quench moniter method — CH number method 3 was described for measurement of ^{3}H and ^{14}C activity. The relative measurement of radionuclide activity with LSC and calibration of a series of quenched standard are now being researched, particularly for quench monitor method.

The research of quench monitor method in LSC is very important. At present there is H number, spectral index, and channel ratio etc. in most popular quench monitor methods.

The characteristics of CH number method will be introduced here, and a comparison of CH number with other quench monitor methods will be made.

The combined uncertainty of radionuclide activity measurement with CH number method is small, an important parameter for radionuclide metrology.

The uncertainty in measurement results generally consists of several components that may be grouped into two categories according to the way in which their numerical value is estimated: 1. Those evaluated by applying statistical method to a series of repeated determinations and 2. those evaluated by other means.

Table 1.

Quench Monitor Method	Uncertainty Component of	
	A (%)	B (%)
CH number	0.20	0.10
Channel ratio	0.60	0.45
H number	0.50	1.06
Spectral Index (tSIE)	0.40	0.86

CH number method is a new quench monitor method in LSC; it is a relative measurement method. The components of B must include a component (about 1%) due to activity measurement of a series of quenched standards. It has to be excluded, when combined uncertainty of measured radionuclide activity is compared with different quench monitor methods to obtain an efficiency calibration curve, with which the activity of a quenched standard series is measured. The components of A and B for measured activity of a series of quenched standards would be obtained with CCl_4 as quench agent. The glass vial has two types made separatly in China and the U.S. A series of quenched standards generally consists of ten sources with different of quench agents.

The results of measured activity of a series of quenched standards are shown in Table 1.

Because the content of quench agents is different with different series of quenched standards, the uncertainty component of measured activity is not quite exact.

Independent of Color Quench with CH Number Method

Measured samples usually have color, so it is important to measure the color sample activity. The CH number method measures the activity for a series of color quenched standards which make up the efficiency calibration curve. The color quenched agent is methyl red, the content of which is from 0 to 300 L. The results of measured activity of a series of color quenched samples is shown in Table 2.

Table 2 shows that the difference between the mean result of measured

Table 2.

Number of Series	Standard Deviation of Mean (%)	Difference Between the Mean of Measurement of Activity of a Series Standards and Standard Value (%)
88092721 —29	0.20	—1.10
88092221 —31	0.20	—0.54
88092721 —29	0.37	—0.70

activity of a series of colour quenched sample with CH number method and standard value is about 1%. The other is quite large.

Independent of Volume Sample with CH Number Method

Different cocktail volumes may affect activity measurement, but not for the CH number method. The three ^{3}H cocktail sample volumes (10 mL, 6.6 mL, and 4.4 mL) are measured by CH number and tSIE. The maximum difference between activity of the three samples was 2% for CH number method, and 3.4% for tSIE.

Independent of Vertical Position of Sample Measured

The vertical position of sample measured is varied from 0 to 23 mm. Maximum variation of measured sample activity with CH number method is about 0.6%. Because the phototube is symetrical, the sample position measured could be varied about 23 mm, without varying the results of measured activity.

Independent of High Count Rate with CH Number Method

Measuring the activity of high sample count rate reduces statistical count and saves measurement time in radionuclide metrology. When counting a sample with high activity, the counting uncertainty is reduced, allowing for a more statistical measurement within a given measurement. The count time for high activity samples can be reduced and still yield good counting statistics; whereas, lower activity samples require longer count times to produce similar statistics. A series of quench standards, total activity 1700 Bq, were used to determine the activity of a sample containing 3800 Bq. The measurement results were unaffected.

The activity of short half-life radionuclides are also measured with CH number method. Many short half-life radionuclides are very useful, particularly many electrical capture radionuclides widely used in medicine. The correction of short half-life radionuclide decay is considered in calculation programs of the CH number method. The uncertainty component due to half-life measurement is the only need to consider.

Tritium water activity is measured with an efficiency calibration curve calibrated with ^{3}H n-hexadecane series of quenched standards. Also measured activity of ^{60}Co is used with a ^{14}C n-hexadecane efficiency calibration curve. The difference between measurement result of activity and standard value is less than 3% for tritium water and ^{60}Co.

CH number method is a new quench monitor method that has many advantages over other methods. It also has other advantages which would be researched further.

ACKNOWLEDGEMENT

The author wishes to thank Y.D. Yang, L. Song, and Z.E. Wang for preparing the liquid scintillation sources and Q.G. Li for his help.

REFERENCES

1. Song, L., X.Z. Wu, et al. *Nucl. Inst. and Meth.* 175:503 (1980).
2. Wu, X.Z., L. Song, et al. *Int. J. Radiat. Appl. Instrum.* Part A Vol 38, No 10:891 (1987).
3. Wu, X.Z., *Nucl. Elec. and Det.* (in press).

CHAPTER 62

Low-Level Scintillation Counting with a LKB Quantulus Counter Establishing Optimal Parameter Settings

Herbert Haas and Veronica Trigg

The use of spectrophotometric quartz vials with a square or rectangular cross section was reported in 1977.[1] The original application was in a counter with a single loading drawer (Intertechnique LS 20). Subsequently, counters from different manufacturers were used with the same type of vials. Today, all commercial counters have multiple sample loading mechanisms designed to handle standard glass vials. Our square vials are smaller than the standard vial; therefore, we need a holder with the outside dimensions of the standard vial to stabilize the square vials. These vial holders are lifted into the counting chamber by a piston. In some counters the lifting piston does not have a fixed rotational position. Since it is important that the vial face is always perpendicular to the photomultiplier (PM) tube axes, the loading mechanism had to be redesigned to insure that the vial always assumes the same position with no rotation. Some counters have a complicated system of shutters to protect the energized PM tubes from ambient light sources. The presence of shutters makes the design may be quite involved.

This modification of the piston mechanism was particularly simple for the Quantulus. The loading piston is part of a lever system and does not rotate. A V-shaped groove was cut in the flat surface of the piston. A ridge on the bottom of the black nylon holder fits in this groove and determines the vial position precisely.

The holder has the size of a standard counting vial (28 mm in diameter and 60 mm in length). A rectangular window is cut through the holder, along and centered around the PM tube axis. The quartz vial is positioned in this window and always exposes the same surface area to the PM tubes. The inner surfaces of the holder are made reflective by attaching a thin film of aluminized mylar. The benzene filling level in the vial is kept constant, and different sample volumes are accommodated by using vials with different volumes. We have vials available with volumes of 4.5, 3.0, 1.5, 0.6, and 0.3 cc.

The advantage of this design lies in the straight and direct optical link between vial and PM tubes. Optical conditions are precisely the same after each loading cycle. In contrast, most round standard vials are not precise cylinders. It is especially difficult to produce round quartz vials without distortions.

We prepare our sample benzene for scintillation counting by adding 0.91 wt% of Butyl-PBD primary scintillator. The precise amount of scintillator powder is added directly to the sample benzene. For our smallest vial (0.3 cc) the exact scintillator weight is 2.35 mg. This amount can be weighed routinely and introduced into the vial with an error of ±5%. Recently, new scintillation techniques have been introduced which allow for the photon pulse length modifications of beta decays. This is achieved by adding a small percentage of a secondary scintillation fluor to the primary scintillator. We do not believe that the extremely small amounts of required secondary fluor can be weighed with sufficient precision to accommodate our laboratory methods; therefore, we have not investigated secondary fluors. The advantages of adding the scintillation powder directly to the sample are: (1) the full volume of the vial is available for sample benzene and (2) a potential source of contamination is avoided. Our concern lies with impurities in the cocktail solution that may cause quenching or contain trace amounts of ^{14}C.

Recent design modifications of the Quantulus counter are making possible a wide range of discriminator adjustments for pulse height difference (PAC) and variation of pulse length (PSA). The range of the digitally controlled adjustment is between 1 and 256 for both discriminators. With the PAC discriminator, pulses from the left and the right PM tubes are compared. Dissimilar pulses are rejected and the rigor of comparison becomes stricter as the setting number is increased. The PSA discriminator passes longer pulses as the setting number is increased.

We have varied both settings over nearly their full range and also made some tests without the PSA discriminator. Four sets of measurements were made. A clear, fully transparent vial was used for measuring a background (anthracite) and a modern standard (NBS oxalic acid) solution. The series was repeated with a vial which had the outside of the quartz faces frosted with no. 400 grinding powder. These measurements were made without pulse energy restriction, i.e., the energy window was open to the full range of the observed beta energy.

Results are shown in Figures 1 to 3. Figure 1 presents the background benzene measurements. The benzene was synthesized in our laboratory from anthracite coal. Each measurement represents about 12 hours of counting time and has a standard error of about 0.02 cpm. Some scatter in the data was produced by the use of different vials and background solutions. One has to bear in mind that the very low background of the Quantulus counter will cause the above mentioned differences to appear as major fluctuations. We have picked from our tabulation of data two series of PSA values, one for PAC set at 5 (pulses with large amplitude difference are passed) and one for PAC set at

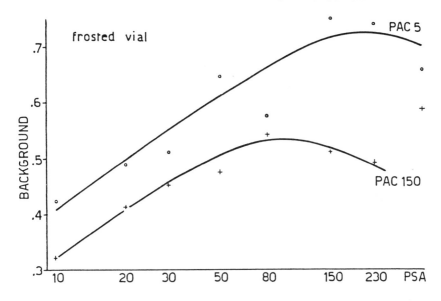

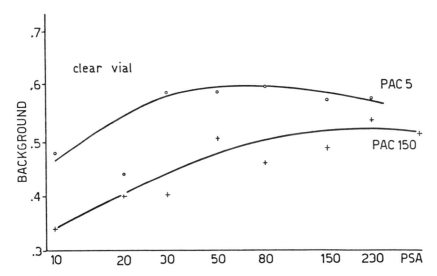

Figure 1. Background in cpm of a square quartz spectrophotometric vial with a light path of 10 mm, filled with 3 cc ^{14}C-free benzene and 0.9% butyl PBD scintillator. Pulse shape discriminator PSA restricts length of PM pulses strongly with setting 10 on scale and weakly with setting 230. Points on far right are measurements made without PSA discriminator. PAC curves show effect of discriminator for pulse amplitude differences. With PAC = 5 curve (open circles) only extreme differences are filtered out, with PAC = 150 (crosses) also smaller differences are filtered out.

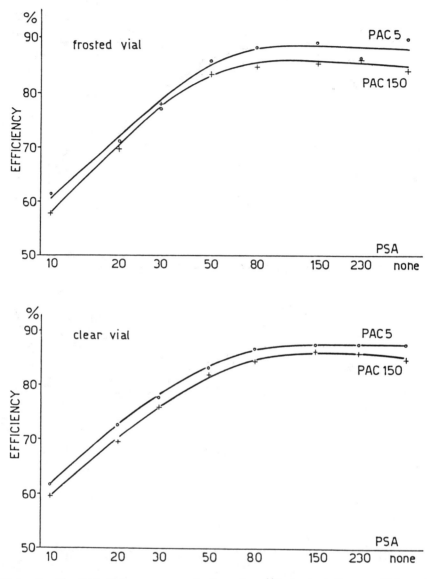

Figure 2. Counting efficiency measured with modern ^{14}C standard. Percent efficiency is based on the ratio of measured net count rate over theoretical rate of decay of the standard. Parameters shown are discussed in caption to Figure 1.

150 (moderately small pulse difference). We have also recorded two measurements with the PSA discriminator switched off. These results showed the largest deviation from the expected trend, causing some points to lie outside of the background graph. Pulse amplitude discrimination has a substantial influence on lowering the background. At PAC = 150 a larger number of pulses is

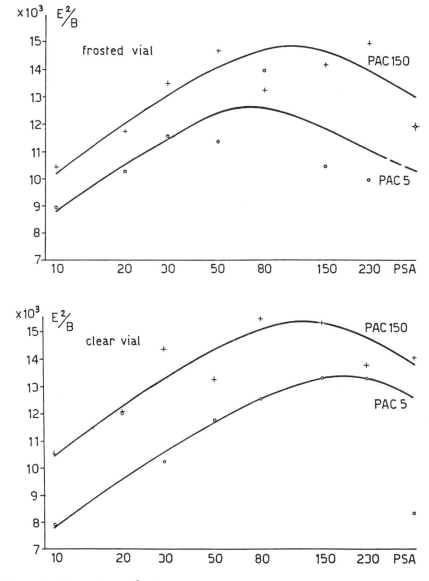

Figure 3. Figure of merit E^2/B. E is efficiency shown in Figure 2 and B is background shown in Figure 1.

rejected, which suggests that assymetrical pulses are abundant. These may originate from the glass envelope of either PM tube rather than from the vial. Frosted vials appear to increase the pulse height assymmetry. This is shown by the higher background level when the discrimination setting is at PAC = 5. Even at PAC = 150 a slightly higher background is observed for frosted vials.

Figure 2 shows the relation of counting efficiency to the PAC and PSA adjustments. Settings are the same as in the background study. The count rate of the modern standard used in this experiment is about fifty times higher than background; therefore, the scatter of the background data is only an insignificant contribution. The small separation of the two curves shows that pulse amplitude discrimination plays only a minor role. This implies that the vials, set in their holders, are radiating the photons equally from both windows of the holder, regardless of the point of origin of the photons within the vial. The large dependence of efficiency on the pulse length discrimination shows a substantial rate of ^{14}C pulse rejection up to a setting of 80.

The difference in efficiency between frosted and clear vials is insignificant. At PAC = 5 the frosted vials are more efficient by about 3%. At PAC = 150 the difference disappears. Clear vials have a lower background; thus, we have given preference to this vial type. Furthermore, clear vials have the added advantage of allowing sample filling and withdrawal process observation and residual matter detection in the vials. For example, wax-like coatings tend to build up over a 6-month period and can be seen in clear vials as a faint greyish layer on the inside up to the benzene filling level. This contaminant is a byproduct from the catalytic conversion of acetylene to benzene.[2]

The E^2/B, or figure of merit, is not controlled by the counter alone. It is strongly influenced by the vial and the reflectors installed around the vial. Figure 3 clearly shows the advantage of using pulse amplitude discrimination at a level of about 150. Additional data, which is not shown here, indicates that higher settings will not substantially decrease background but will negatively affect efficiency, causing a decline in E^2/B.

The ideal PSA setting is more difficult to assess. Efficiency and E^2/B graphs indicate an optimal setting of PSA = 80 or slightly higher; however background is substantially higher at the PSA = 80 setting than at the PSA = 30 setting, i.e., 0.5 compared to 0.43 cpm. This may be an important consideration if small or old age samples with net count rates of 0.1 cpm or less are to be dated.

CONCLUSION

Simple mechanical adaptations make it possible to use precisely manufactured square and rectangular quartz vials in most counters. Archaeological samples and geochemical probes (water and atmosphere) are usually small and seldom yield more than 4 cc of synthesized benzene; therefore, the reduced volume of these vials is not a disadvantage.

Our detailed study of this vial type in a Quantulus counter demonstrates the usefulness of electronic pulse filtering. Choosing clear vial faces, a sample volume of 3 cc benzene, a pulse amplitude (PAC) discrimination setting of 150 and pulse length discrimination setting of 80, and using no pulse energy limitations in the range of ^{14}C pulses, one can obtain a background of 0.48 cpm, an efficiency of 84% and a E^2/B factor of 15,000. For 0.6 cc vials we recently measured a background of 0.14 cpm, an efficiency of 75% and an E^2/B factor of 40,000.

Additional improvements on these numbers are possible by restricting the energy windows for the ^{14}C pulses to the main portion of the peak and cutting off the low energy tail of this peak. A brief, current study involving 3 cc vials indicates a lowering of background by 0.1 cpm, a decrease in efficiency of 4.5%, and a slight increase in E^2/B of about 2000.

The major application of the Quantulus installation will be in dating old archaeologic samples, as well as small carbon samples in the weight range of 200 to 600 mg.

ACKNOWLEDGEMENT

We acknowledge the support received from the NSF archeometry program with the grants BNS-8845200 and BNS-8801697.

REFERENCES

1. Haas, Herbert. *Specific Problems with Liquid Scintillation Counting of Small Benzene Volumes and Background Count Rate Estimation: in Radiocarbon Dating* R. Berger and H.E. Suess Eds., (1979), pp 246–255.
2. Coleman, D.D. et. al. "Improvement in Trimerization of Acetylene to Benzene for Radiocarbon Dating with a Commercially Available Vanadium Oxide Catalyst," Proceedings 8th International Conference on Radiocarbon Dating, 1972, p 158.

CHAPTER 63

Quench Correction of Colored Samples in LSC

Kenneth Rundt, Timo Oikari, and Heikki Kouru

ABSTRACT

Liquid scintillation counting of weak β-emitters like tritium and ^{14}C, usually involves quench correction with a stored quench curve, which traditionally expresses the counting efficiency as a function of a quench monitor value. Generally, the quench curve is produced by measuring chemically quenched (uncolored) calibration standards; however, if the unknown samples contain color, the computed activity will systematically be in error when a basic quench monitor, like the isotope mean pulse height or the external standard endpoint, is used. The errors will be even more pronounced in dual label counting, as the shapes of the spectra depend on the relative amounts of color and chemical quench. In this chapter, we outline an improved quench correction method based on the dynamic determination of both the overall quench level and the amount of color quench. The counting efficiency is computed from a quench curve which in fact should be regarded as a surface. This solution was first implemented in the Wallac 1219 Rackbeta liquid scintillation counter. In a more general solution, applicable in multi-label counting, the shape of the spectrum can be considered a function of two quench parameters. This extended version has been implemented in the new Wallac 1410 liquid scintillation counter.

INTRODUCTION

Color quenching, or more generally, photon quenching, involves the absorption of photons by colored compounds, solid particles, or macromolecules suspended in the liquid scintillation solution. As the emitted light is mainly in the blue region, around 400 nm, yellow compounds are the strongest color quenchers. Color quenching differs from chemical quenching in that it depends on the geometry of the vial and on the spatial coordinates of the disintegration.[1,2] This difference in the modes of the two quenching mechanisms leads to a difference in the pulse height distribution; it becomes reflected in the quench index used for efficiency determination. Traditionally, there have been two ways of handling color: bleaching with a peroxide so that chemical quench prevails, or dividing the samples into two groups, chemically quenched and color quenched, and using two different quench curves. Bleaching is an extra laborious step that may not even reduce the amount of color completely. Furthermore, bleaching also increases the amount of chemical quench in the sample. Segregating the samples may sound very simple, but for two reasons this mode of operation is not satisfactory. First, it demands an extra operation by the user, and second, as it will be shown later, samples that have high amounts of both chemical and color quench should be treated as chemically quenched samples and not as color quenched samples. The second reason is the most difficult to handle, as the human eye cannot judge the amount of chemical quench in the samples.

The presence of color in the samples imposes certain demands on the quench correction feature of the instrument — the counter should have an automatic method to detect the presence of color and compensate for the difference between color and chemical quench. Several attempts have been made to improve counters in this respect. One solution that has been applied commercially is based on using the spectrum produced by the lesser of the two pulses from the photomultiplier tubes.[3] According to tests made by McQuarrie et al.,[4] the lesser pulse height method does indeed perform slightly better than other conventional methods.

In current literature there are a number of other proposed methods. For example, Ross[5] has proposed a method wherein the sample absorbance is determined by immersing a small glass ampule containing an unquenched scintillation solution into a given volume of uncolored scintillation solution and the colored unknown sample. The count rates of the uncolored solution and colored solution are denoted by C_0 and C_1, respectively. The counting efficiency for the colored sample is calculated by multiplying the counting efficiency derived from the calibration curve by the ratio C_1/C_0. This method is not applicable in an automatic LS counter, as the sealed light source must be inserted and removed manually, and wiped carefully between measurements. The counting time per sample is also prolonged.

Lang[6] has proposed a method based on using both the count rate induced by the external standard and the external standard channels ratio, ESCR. In this method, four calibration curves are needed: the counting efficiency as a function of the external standard count rate and the ESCR for both colored and uncolored standards. Lang's method could in general be applied in an automatic counter, but as the external standard count rate is dependent on the volume of the liquid as well, this method has not been commercially applied. Moreover, both the count rate and the ESCR are very dependent on the so called plastic vial wall effect, which is a result of penetration of solvent into the plastic wall.

Takiue et al.,[7] have proposed a method quite similar to Lang's method. Their method, which is based on using sample channels ratio, SCR, together with ESCR, also requires two sets of quenched standards and will result in four quench equations. This method is an analogy of the homogeneity monitor proposed by Bush[8] and is thus dependent on sample homogeneity. Also this method is sensitive to the plastic vial wall effect by the use of ESCR. Furthermore, the method is applicable only when the activity of the isotope is high enough, so that SCR can be determined with high accuracy.

Ring et al.,[9] have suggested a method based on two parameters calculated from the external standard pulse amplitude spectrum. One parameter, called the quench index (QI), is proportional to the mean pulse height, the other parameter, called the color index (CI), is equal to the pulse height (channel number) below which the number of pulses is a constant fraction of the total number of pulses in the spectrum. An example of such a parameter is the median, which divides the spectrum into two parts having the same number of counts. In this method, two equations are needed: one expresses the counting efficiency of chemically quenched samples as a function of QI, and the other expresses the ratio between the counting efficiency of a colored sample and an uncolored sample, as a function of both QI and CI. By applying these two equations, the correct counting efficiency can be determined. Also this method depends on the plastic vial affect through the color index CI. The method is also affected by chemiluminescence in the solution as the CI is calculated using all pulses from channel one upwards.

The methods by Lang, Takiue et al., and Ring et al., require at least two sets of quenched standards to be prepared by the user. Generally, the correction can be performed two ways. Either the counting efficiency is corrected or the quench index is corrected. If P denotes the total quench index, equal to, e.g., the endpoint of the external standard spectrum, and R denotes a color index, then the counting efficiency E of any sample can be written as a product of two functions:

$$E = E(P) \cdot F(R,P)$$

E(P) is the function representing the overall quench curve, while F(R,P) is a function giving the correction to E(P) needed for a colored sample. More generally, it may be a better solution to try to combine R and P into a single parameter, Q, the actual quench index seen by the user. This means writing

$$E = E(Q) = E(Q(R,P))$$

which means that the quench curve is a function of Q, which is a function of both R and P. This is the solution taken by Wallac[10] in the Rackbeta series of instruments (Rackbeta 1219, 1214, and optionally 1209). Our external standard quench index, SQP(E) (Sample Quench Parameter of External standard), is based on two parameters: P which is the channel number below which 99% of the external standard spectrum is located, and R, which is described in more detail in this article. In the new 1410 LS counter, the actual quench curve is not a curve but a surface, i.e.,

$$E = E(R,P)$$

The SQP(E) value printed out by this instrument is equal to P, and the Color Index, that can be printed out as well, is equal to R. This solution was selected because the 1410 counter stores the shape of the spectra, as well as the counting efficiency, as functions of both P and R.

The parameter R is also computed from the external standard spectrum. The method implemented in Rackbeta instruments and in the new LS counter 1410 can best be described by using Figures 1 and 2. Figure 1a and 2a show three-dimensional plots of the [152]Eu external standard Compton spectrum. Similar plots for the beta-isotope dissolved in the scintillation liquid have been produced by Laney earlier.[11] The sample that produced Figure 1 was purely chemically quenched, and the sample that produced Figure 2 was purely color quenched. The counting efficiency of both samples was nearly the same (about 30% for tritium). Figure 1b shows a projection of the three-dimensional spectrum on a plane perpendicular to the pulse height plane in Figure 1a. The instrument does not actually record the spectra in these two figures, but the summed pulse height spectra of the two regions A and B are recorded. This is accomplished by using the left/right comparator and dual MCA technology. Dual MCA technology generally means that a pulse can be directed to one of two MCAs, depending on a certain criterion, which can be selected through software. When the external standard is measured, all pulses for which PHR/PHL > 1.5 or PHL/PHR > 1.5 (PHL and PHR are the linear pulse heights) are directed to the MCA-B while all other pulses are directed to the MCA-A. The resulting spectra are shown in Figure 3a and 3b. After the external standard has been measured, the P-value (equal to the 99% endpoint of the total spectrum A + B) and the color index R are calculated. R is equal to the ratio of counts (A + B)/A in a certain pulse height window.

EXPERIMENTAL

A set of 64 tritium samples was prepared for this work. The counting efficiency of the samples ranged from 3 to 62% and the color index R ranged from 1.0 to 6.5. The scintillation cocktail used was OptiScint HiSafe (Pharmacia-Wallac), a proprietary, high-flashpoint cocktail for organic samples. Standard 20 mL glass vials with 10 mL of cocktail were used. The quenchers used were nitromethane and Sudan 1 (yellow). The samples were prepared as one set of purely chemically quenched samples, one set of purely color quenched samples, and several sets of samples with a varying mixture of chemical and color quench. Figure 4a shows the amounts of quenchers and Figure 4b shows the recorded color index and total quench index for all the samples. The samples were measured in a prototype of the 1410 LS counter, with [152]Eu as external standard. The plastic vial effect was assessed by using a traditional toluene based cocktail in standard plastic vials.

Figure 1a

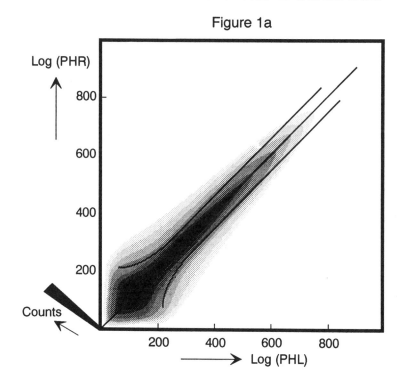

Figure 1b

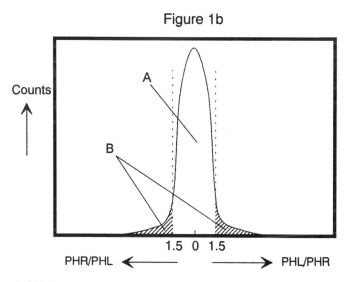

Figure 1. Left/right pulse height distribution for a chemically quenched sample (sample 'C' in Figure 4). a) three-dimensional plot showing the full distribution and the comparator limits (1:1.5). b) two-dimensional projection of the distribution in a). The color index R of the sample is equal to the ratio of counts (A + B)/A. For this sample R = 1.08.

Figure 2a

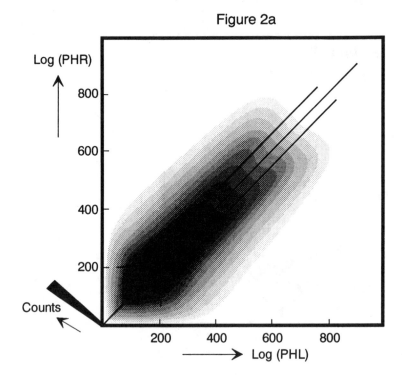

Figure 2b

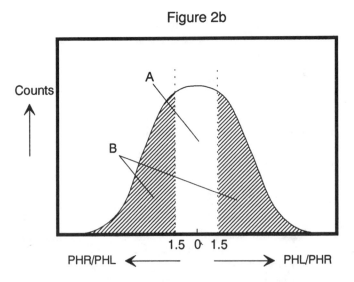

Figure 2. Left/right pulse height distribution for a chemically quenched sample (sample 'Y' in Figure 4). a) three-dimensional plot showing the full distribution and the comparator limits (1:1.5). b) two-dimensional projection of the distribution in a. The color index R of the sample is equal to the ratio of counts $(A + B)/A$. For this sample $R = 2.68$.

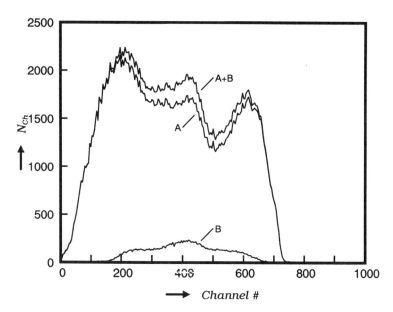

Figure 3a. Dual-MCA spectra of a chemically quenched sample "C". A is a spectrum comprising pulses within the two comparator limits and B is a spectrum comprising pulses outside the two comparator limits. A + B represents the normal spectrum.

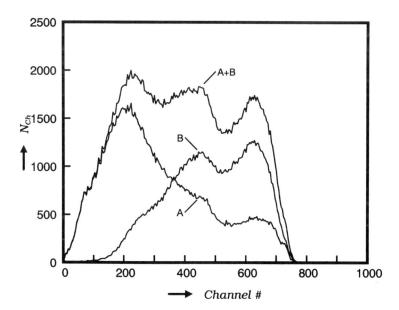

Figure 3b. Dual-MCA spectra of a color quenched sample "Y".

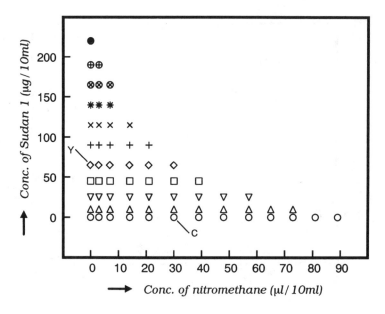

Figure 4a. The composition of the 64 tritium samples used in this work. The symbols are the same in each set of constant color quench but increasing amount of chemical quench.

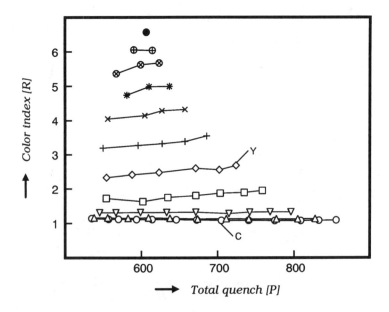

Figure 4b. The color index and the total quench of the 64 samples. In both figures, sample "C" is a purely chemically quenched sample "Y" is purely color quenched. Both have counting efficiency near 23%.

RESULTS

The need for color correction becomes obvious when looking at Figure 5. This figure shows the relative error in the calculated activity when calculating the activities of all the 64 samples with a traditional calibration curve, *without any color correction*. Especially when the counting efficiency goes below 10%, the error increases rapidly. Similar behavior when using the mean pulse height of the tritium spectrum as quench monitor has been documented by one of the authors elsewhere.[12] Figure 5 also shows that adding chemical quench to a color quenched sample makes the deviation from the chemical quench curve noticeably smaller (observe, e.g., the sample set marked by the symbol +). This means that it does not suffice to correct for pure color quench only, but that the instrument must be able to determine the relative amounts of color

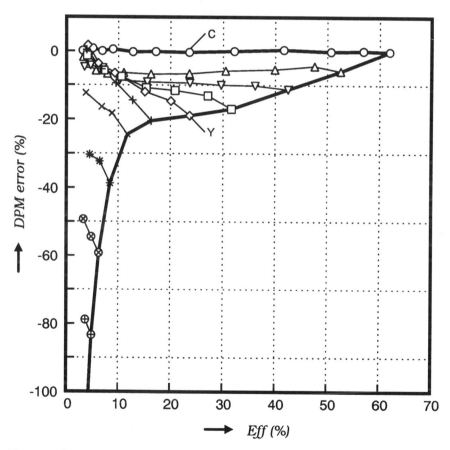

Figure 5. The relative error in tritium activity for all samples when a traditional quench curve involving no color correction is employed. The different symbols refer to the sample sets as shown in Figure 4.

and chemical quench. This is important as real samples mostly have both types of quench.

A natural question is whether the color index R and the total quench index P are good enough to describe the behavior of all kinds of samples. In order to give an indication of how well P and R work, the 64 samples were used to prepare a counting efficiency surface. The function selected to describe the surface was a linear combination of P and ln (R) having 13 parameters. This function was fitted to all the 64 points on the surface by using the method of least squares; thereafter, the function was used to predict counting efficiencies for all the samples. The result is shown in Figure 6. Notice the difference in ordinate scale between Figure 5 and Figure 6. Generally, for samples having a counting efficiency above 10%, the relative activity error is negligible, and for samples with counting efficiency between 3 and 10%, the error is below 5%.

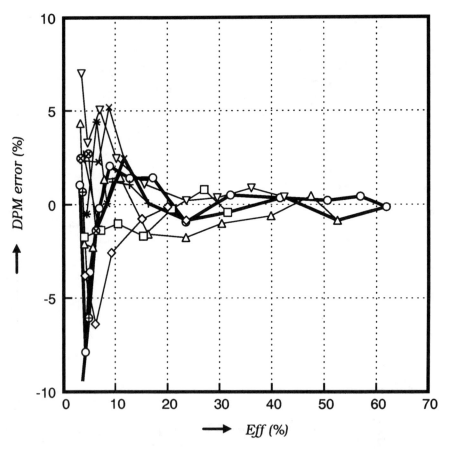

Figure 6. The relative error in tritium activity for all samples when a quench surface involving both total quench and color quench is employed. The different symbols refer to the sample sets as shown in Figure 4.

Even better results can be accomplished by selecting a more elaborate surface function.

Another important point of concern is the shape of the isotope spectrum. It has been noticed that the shape of the spectrum is different when comparing chemically quenched and color quenched samples having the same counting efficiency.[2] This is once again verified in Figure 7, which shows the tritium spectra of the two samples 'C' (pure chemical quench) and 'Y' (pure color quench). For single label counting, the spectrum shape is not very critical, but for dual label counting, the accuracy of, e.g., tritium is directly dependent on how well the spectra of the calibration standards describe the spectra of the samples. In Rackbeta counters, the "Three-Over-Two" method was equipped with an algorithm to correct the relative intensity curves for color. In the new 1410 counter, using Digital Overlay Technique (described elsewhere in this publication[13]), the single isotope spectra fitted to the unknown composite spectrum are retrieved from a spectrum library, wherein the spectra are stored

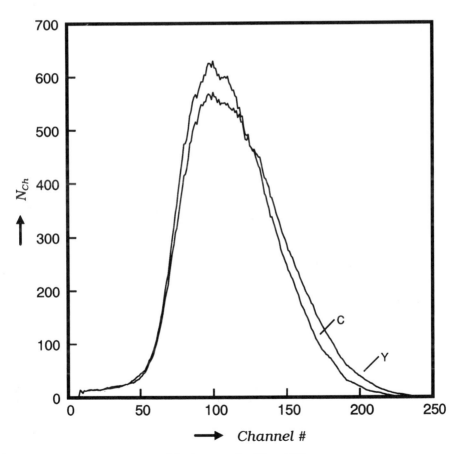

Figure 7.   The tritium spectra of the two samples 'C' and 'Y'.

Table 1. The Effect on Color Index R and Total Quench Index P of Diffusion of Toluene into the Plastic Vial Wall

Time/hour	R	P
0	1.30	699
5	1.32	698
28	1.37	698
46	1.38	699
75	1.38	698

as a function of the two quench parameters P and R. This guarantees that the correct spectrum shape is always fitted to the sample spectrum.

Volatile liquids in plastic vials give rise to the so called plastic vial effect that significantly affects external standard quench indices based on the channels ratio or mean pulse height. The influence of the plastic vial effect on the color index R was tested with one chemically quenched sample having 10 mL of a toluene based cocktail. The initial value of the color index is as high as 1.3, although no color has been added. The color index is normally slightly greater in diffuse plastic vials than in clear glass vials. The result of the measurement (Table 1) indicates that there is a small but negligible increase in the color index with time as a result of diffusion of toluene into the plastic material. As the quench level increases, the plastic vial effect becomes more pronounced, but it is always at quite a moderate level.

SUMMARY

An automatic quench correction method based on determining both the amount of color quench and chemical quench in liquid scintillation samples has been described. The method depends on using the external standard spectrum and pulse amplitude comparison for determining two quench parameters: the total quench level and a color index. Using two parameters instead of one means that quench correction must be based on quench surfaces instead of quench curves. In the simplest case, the two parameters are used to retrieve a counting efficiency value from such a quench calibration surface. In a more elaborate structure, the two quench parameters are used to retrieve model spectra for the isotopes of interest when using Digital Overlay Technique for resolving multi-labeled samples.

REFERENCES

1. ten Haaf, F.E.L. "Colour Quenching in Liquid Scintillation Coincidence Counters," in *Liquid Scintillation Counting, Vol 2*, M.A. Crook, P. Johnson, and B. Scales, Eds. (London: Heyden & Son Ltd, 1972), pp 39–48.
2. Neary, M.P. and A.L. Budd. "Color and Chemical Quench," in *The Current Status of Liquid Scintillation Counting*, E.D. Bransom Ed., (New York: Grune & Stratton, 1970), pp 273–282.

3. Laney, B.H. "Method and Apparatus for Determining Efficiency in a Liquid Scintillation Counting System," U.S. Pat. No. 3,725,657 (1973).

4. McQuarrie, S.A., C. Ediss, and L.I. Wiebe, L.I. "Some Performance Characteristics of Modern Liquid Scintillation Counters," *Int. J. Appl. Radiat. Isot.*, 34(7):1009–1012 (1983).

5. Ross, H.H. "Color Quench Correction in Liquid Scintillator Systems Using an Isolated Internal Standard," *Anal. Chem.*, 37(4):621–623 (1965).

6. Lang, J. F. "Chemical vs. Color Quenching in Automatic External Standard Calibration. Application of Empirical Observations in a Computer Program," in *Organic Scintillators and Liquid Scintillation Counting*, D.L. Horrocks and C.T. Peng, Eds. (New York: Academic Press, 1971), pp. 823–833.

7. Takiue, M., T. Natake, and M. Hayashi. "Double Ratio Technique for Determining the Type of Quenching in Liquid Scintillation Measurement," *Int. J. Appl. Radiat. Isot.*, 34(10):1483–1485 (1983).

8. Bush, E.T. "A Double Ratio Technique as an Aid to Selection of Sample Preparation Procedures in Liquid Scintillation Counting," *Int. J. Appl. Radiat. Isot.*, 19:447–452 (1968).

9. Ring, J.G., D.C. Nguyen, and L.J. Everett. "Liquid Scintillation Counting from Gross Counts to Spectral Analysis," in *Liquid Scintillation Counting, Recent Applications and Development, Vol.1*, C.T. Peng, D.L. Horrocks, and E.L. Alpen, Eds. (New York: Academic Press, 1980), pp. 89.

10. Oikari, T. and K. Rundt. "Method for Determining Counting Efficiency in a Liquid Scintillation Counting System," US Patent 4,700,072, 1987.

11. Laney, B.H. "Two-Parameter Pulse Height Analysis in Liquid Scintillation," in *Liquid Scintillation Counting, Vol 4*, M.A. Crook and P. Johnson, Eds. (London: Heyden & Son Ltd, 1977), pp 74–84.

12. Rundt, K. "The Effect on Quench Curve Shape of the Solvent and Quencher in a Liquid Scintillation Counter," in *New Trends in Liquid Scintillation Counting and Organic Scintillators*, 1989.

13. Kouru, H. and K. Rundt. "Multi-Label Counting Using Digital Overlay Technique," in *New Trends in Liquid Scintillation Counting and Organic Scintillators*, 1989.

CHAPTER 64

Low-Level Liquid Scintillation Counting Workshop

Gordon Cook and Henry Polach

Two workshop type discussion groups met during and after the International Conference on New Trends in Liquid Scintillation Counter and Organic Scintillators, 1989. Researchers discussed problems of common interest relating to low-level counting technology. The well attended discussions, attesting to the significance of the subject, were lead by Henry Polach. Selected for discussion were topics considered timely and pertinent:

Performance of the New Packard Vial Guard

It was resolved that the performance of the Packard Vial Guard, using vials made of different materials and various cocktails, should be tested in any LS spectrometer fitted with pulse shape analyses. Ed Robertson, Vice President Engineering, Packard, Downers Grove, IL 60515, U.S. would like to receive reprints of published research relating to this subject.

Background and Resolution of Low Count Rate Signals

It was recognized that there is a lack of publications addressing themselves to the: (1) significance of low background in applied research (2) definition of the theoretically attainable detection limit, and (3) testing of the practical resolution limits of weak signals close to background. Many radiocarbon daters, for example, observed contamination with ^{14}C during conversion of infinite age (no ^{14}C) sample carbon to synthetic benzene. Research in these areas is needed.

Low-Level Counting Quality Control and Assurance

It was generally recognized that low level counting, however it is defined and irrespective of type and origin of counters used, is both an art and science. Practitioners agreed that quality control within the laboratories is the only means of giving quality assurance outside of the laboratories. Exchange of information relating to equipment stability and long term reproducibility of very low count rates, participation in international calibrations and cross-checks, were seen as essential precursors to global validity of radiometric results in all branches of applied sciences.

Application of LS Low-Level Spectrometers to Environmental Studies, Food Quality, and Health Research

The initial usage of the LS technique was confined largely to biological and medical studies using beta particle emitters such as ^3H, ^{14}C, and ^{32}P; it was extended, only at a later stage, to environmental, chronological or isotope origin, abundance, and cycle type studies. Instrumental development included evolution from fixed window counters to multichannel, often multi-parameter spectrometry, pulse shape analysis, alpha and beta particle resolution, low-level spectrometers, software assisted DPM, spectral analyses, and quench corrections. One of the next most significant advances was the development of scintillation cocktails holding water and subsequently, gel scintillants. Applications of the method are now steadily growing with the development of detection techniques for ^{36}Cl, ^{90}Sr, ^{85}Kr, Th, Ra, and Rn isotopes. There is now much interest in the measurement of alpha particle emitting radionuclides, and it is envisaged that LS spectrometry will compete with Geiger-Muller and Gas Proportional counting techniques due to (1) simplified preparation procedures, (2) low inherent background, (3) counting efficiencies approaching 100% and (4) available selective extractants/ scintillants for actinide series elements. However, surface barrier counting techniques are unlikely to be superseded due to their high resolution.

The concensus was that liquid scintillation technology (instrumentation) is now generally ahead of application technology and know-how. Thus increased emphasis needs to be placed on research leading to improved sample separation and usage of modern cocktails leading to wider applications.

User Education

It was recognized by all that manufacturers and experienced researchers need to participate in "user education". The conclusions were as follows:

• Advances in instrumentation, techniques of sample preparation, and development of new applications have reached such a momentum that three years between LS conferences is far too long. Manufacturer supported workshops, lead by experienced researchers, were suggested to fill the gap and would fulfill the educational needs, especially if the workshop proceedings were to be published without the restrictions imposed by larger conferences and research journals.

• Many developments are too well protected by patents and we are now in a situation where the LS technique is rapidly becoming "black-box" technology. The manufacturer will have to accept much more responsibility for presenting information to help their customers.

• It was reported that sales backup leaves much to be desired. The policy of switching the equipment on and presenting the customer with an operations manual is now no longer sufficient.

• There are literature references that the "user friendly hardware and software" did not perform to specifications and/or did produce errors. Yet, in the hands of many (majority ?) these items perform well. "User friendliness" clearly is not a substitute for user experience. It was suggested that peer assistance, rather than sales assistance, would serve better to resolve the problems often caused by misunderstanding of functions and/or design aims.

• The expert customer (user) will need to assist the manufacturer in the production of application oriented manuals. In the past there has been a degree of apathy from the users (eg., Wallac publishes a Quantulus User Club newsletter for which no significant feedback from the user has been received). It is understood that Packard has taken positive steps in having an all embracing low-level manual produced. This is, from the user's point of view, a very encouraging move.

• Ideally the sales force should receive or have more scientific training. Instrumentation is now so complex (or so new), that during this conference for example, the sales engineers and senior personnel could not demonstrate, explain, or document all the features of the LS counters on their stands. Perhaps some expert users could have made themselves available for consultations, thus contributing to dissemination of knowledge and application of advanced technologies.

CONCLUSIONS

Research literature is, by definition of purpose, very restrictive in documenting "know-how". The liquid scintillation counting and spectrometry technique and technology covers a vast range of topics, all technically very complex until they are mastered. Increased interaction, communication and exchange of technical and procedural information between researchers is seen as the major key to success; rediscovering the wheel is an unnecessary process! We recommend and will support the organization of workshops (eg., low-level counting) of the kind that lead, through researcher/manufacturer interaction, to user education and production of detailed application oriented manuals.

Comment Made by Henry Polach as Chairman of the Waste Disposal Session.

I was fascinated to hear that as LS users we not only solve problems but also create them. The magnitude of the waste disposal problem that we have created is staggering. The diversity of regulations governing the disposal of the waste is unbelievable.

So, it seems to me that we are both: the scientist and the public, affected by our own activities.

The question that now must be asked, indeed the question that now goes begging is:

"How can we as scientists, as well as members of the concerned public, assist those who seek to assist us?"

I invite Jeane Kreiger from Du Pont to respond.

List of Attendees

Clement Aime, Beckman Instruments, Fullerton, CA.

Robert Anderson, Scottish Universities Reasearch and Reactor Centre, East Kilbride, Glasgow, G75 0QU, Scotland, UK.

Chris B. Ashford, E G & G Energy Measurements, Inc., Santa Barbara Operations, 130 Robin Hill Rd., Goleta, CA 93117.

Ed Bard, Wamamatsu, Bridgewater, NJ.

J. Matthew Barnett, Radiation Measurements Facility, Arizona State University, Tempe, Arizona.

Lindsey Bender, U.S. Army, Ft. Belvoir, VA.

I. B. Berlman, Racah Institute of Physics, Hebrew University of Jerusalem, Jerusalem, Israel.

Fritz Berthold, Laboratory Professor, Wildbad, Germany.

Fred Best, RJR Tobacco Co., BGTC Bldg. 611-13, Winston-Salem, NC, 27102.

Carl Borror, Westinghouse Electric, Idaho Falls, ID.

Joe Bradley, The Nucleus/Tennelec, Oak Ridge, TN. 37830.

Alen Brown, Environmental Safety Ser, University of Georgia, Athens, GA 30602.

Karl Buchtela, Atominstitute of the Austrian Universities, Vienna, Austria.

Rickie Byrd, Martin Marietta, Hobbs Rd., P. O. Box L410, Paducah, KY 42001.

Jackie M. Calhoun, Ionizing Radiation Division, Center for Radiation Research, National Institute of Standards and Technology, Gaithersburg, MD 20899.

Tom Callagham, 1436 Chelmsford Sq. W., Columbus, OH 43229.

A. Grau Carlés, Basic Research Institute, Ciemat, Spain.

M. T. Martin Casallo, Investigacion Basica, Ciemat, 28040 Madrid, Spain.

Staf van Cauter, R & D Department, Packard Instrument Company, 2200 Warrenville Road, Downers Grove, Il 60515.

Fred Cazer, Norwich Eaton Pharmac., P.O. Box 191, Norwich, NY 13815.

Robert Ceo, Oak Ridge National Lab, Oak Ridge, TN 37831.

Jeffrey Chandler, Lawrence Livermore, University of California, Livermore, CA 94550.

Lih Ching Chu, IL Dept. of Nuclear Safety, 1301 Knotts Street, Springfield, IL 62703.

Timmy Chio, American Cyamanid, Box 400, Princeton, NJ 08540.

Amon Clay, Brooks, AFB TX, San Antonio, TX.

Richard Cokefair, Watco Corp., King George Post Rd., Fords, NJ 08863.

Phillip Cole, Oak Ridge, TN.

Dale Condra, IT Corporation, 1550 Bear Creek Road, Oak Ridge, TN.

Gordon Cook, Scottish Universities Research and Reactor Centre, East Kilbride, Glascow 675 0QU, Scotland, UK.

Linda Coughenour, Parke-Davis Pharmac Res.

Bert M. Coursey, Ionizing Radiation Division, Center for Radiation Research, National Institute of Standards and Technology, Gaithersburg, MD 20899.

Anamaria Cox, The University of Texas School of Public Health, P O Box 20186, Houston, TX.

Thomas Curtis, Norwich Eaton Pharmac., Box 191, Norwich, NY 13815.

Thomas Davis, Procter & Gamble Co., Mvl., P. O. Box 398707, Cincinnati, OH 45239-87071.

Stan De Filippis, University of Georgia, Center for Applied Isotope Studies, Athens, GA 30602.

George Deffner, Radiometric Instruments, Tampa, FL.

James, M. Devine, Laboratory of Isotope Geochemistry, Department.
of Geosciences, University of Arizona, Tucson, AZ 85721.

Thomas, Devito, GE/KAPL, P. O. 1072, Schenectady, NY 12301.

Charles Dodson, Beckman Instruments Inc., Fullerton, CA.

Keith Doran, Ecotek Laboratory Serv., 219 Banner Hill Road, Erwin, TN 37650.

Leann Eastman, Hazleton Laboratories AM., P. O. Box 7545, Madison, WI 53707.

Albert Eccker, Research Products Int'l, 410 N. Business Center Dr., Mt. Prospect, IL 60056.

Nancy Eccker, Research Products Int'l. 410 N. Business Center Dr., Mt. Prospect, IL 60056.

Chris Ediss, University of Alberta, Alberta, Canada.

R. P. Effler, Oak Ridge National Labortory, P. O. Box 2008, MS 6027, Oak Ridge, TN 37831.

Sisurdur Einarsson, University of Iceland, Dunhaga 3, 107 Reykjavik, Iceland.

Jim Eldrige, Oak Ridge National Laboratory.

Aarno Eskola, Pharmacia Wallac Oy.

Jill Eveloff, Office of Environmental Health and Safety, University of California at San Francisco.

Roy Everett, Test-ER, Inc. 225 Mitchell Court, Addison, IL 60101.

Mark Farnsworth, Condoc Inc., P. O. Box 1267, Ponca, AZ 74603.

David Fatlonitt, Wamamatsu, Bridgewater, NJ.

Susan Feierberg, DuPont, 575 Albany St., Boston, MA 02118.

Yolanda Fields, Martin Marietta, Oak Ridge, TN 37831.

Olive Figuereda, University of Alberta, Room 350, Earthsciences B., Alberta, Canada.

David Filler, Westinghouse Srs.

James M. Floeckher, Packard Instrument Co., One State St., Meriden, CT 06450.

John M. Flournoy, E G & G Energy Measurements, Inc., Santa Barbara Operations, Goleta, CA 93117.

Bob Fortner, Nuclear Data Systems, Schaumburg, IL 60196.

Gary Frank, Florida State University, Tallahassee, FL 32306.

Charles Fredrick, TVA, Muscle Shoales, AL 35660.

Manfred Friedrich, Federal Institute for Food Control and Research, Vienna, Austria.

Haruo Fujii, Department of Radiology, School of Medicine, Tokyo Medical and Dental University, 1-5-45, Yushisma, Bukyo-ku, Tokyo, Japan.

E. Garcia-Toraño, Investigación Basica, Edificio 3 CIEMAT, Avda, Complut.22, Madrid, Spain.

Gerard Giblin, Fisons Sci. Bishop Mea., Loughborough, Leicestershire, England.

Jane Gilsinser, ICN Biomedicals, Inc., Costa Mesa, CA 92626.

Gregory Gowdy, SC Electric & Gas Co., P. O. Box 47, Jenkinsville, SC 29065.

D. Gray, Ionizing Radiation Division, Center for Radiation Research, National Institute of Standards and Technology, Gaithersburg, MD 20899.

Ronald Grebel, Nuclear Data Systems, Golf & Meacham Rds., Schaumburg, IL 60196.

Michael Greeno, Union Electric Nuclear, P. O. Box 620, Fulton, MD 65251.

Kay Griffith, Beckman Instruments, Inc., Fullerton, CA 92634.

S. Y. Guan, Beijing Nuclear.

Adele Gulfo, Fisher Scientific Co., 1 Reagent Lane, Fair Lawn, NC 07410.

Hans Güsten, Kernforschungszentrum Karlsruhe Institut fur Radiochemie, 7500 Karlsruhe 1, Postfach 3640, Karlsruhe, Germany.

Herbert Haas, Radiocarbon Laboratory, Institute for the Study of Earth and Man, Southern Methodist University, Dallas, TX 75238.

Jiange Han-ying, Institute of Biophysics, Academia Sinica, Beijing, China.

D. D. Harkness, NERC Radiocarbon Laboratory, East Kilbride, Glasgow, G75 0QU, Scotland, UK.

Wayne Harris, ICN Biomedicals, Inc., 67 Overlook Drive, Framingham, MA 01701.

Arnold Harrod, Oak Ridge National Laboratory.

Deborah Hartley, Springborn Laboratories, 790 Main St., Wareham, MA 02571.

Ray Harley, Packard Instruments, Little Rock, AR.

Joy Harvey, Wastec Inc., Oak Ridge, TN.

Scott Hay, SAIC, 3 Choke Cherry Road, Rockville, MD 20850.

Theo Hegge, Chemical Research, Packard Instrument Company, B. V., Ulgersmaweg 47, 9731 BK Groningen, The Netherlands.

Gerhard, Hoess, Pharmacia Lkb., Vienna, Austria.

Ken J. Hofstetter, Westinghouse Savannah River Company, Savannah River Laboratory, Aiken, SC 29808.

Alan Hogg, Packard Instrument Co., 2200 Warrenville Rd., Downers Grove, IL 60515.

A. Hogg, Radiocarbon Dating Laboratory, University of Waikato, Hamilton, New Zealand.

Theodore Houser, Beckman Instruments Inc., 3311 N. Kennicott Ave., Arlington Height, IL 60004.

John Hsu, DuPont Medical Products, NEN Research Products, Boston, MA 02118.

Ming Hu, Department of Isotope Research, Institute of Clinical Medical Sciences, China-Japan Friendship Hospital, Beijing 100013, China.

Mark Huntley, Idaho National Engineering Laboratory, P. O. Box 1655, Idaho Falls, ID 83415.

Hiroaki Ishikawa, 2-25-19, Wakamura-Minami, Nerima-Ku, Tokyo.

Stefan Jarnstorm, Wallac Oy, Turku 10, Finland.

Bill Johnson, Medical College of Penn., 3300 Henry Ave., Philadelphia, PA 19129.

Robert Jones, Pharmacia Lkb Nuclear, Gaithersburg, MD 20877.

Lauri Kaihola, Instrument Research Department, Wallac Oy, P. O. Box 10, SF-20101 Turku, Finland.

Robert M. Kalin, Laboratory of Isotope GGeochemistry, Department. of Geosciences, Universuty of Arizona, Tucson, AZ 85721.

L. Karam, Ionizing Radiation Division, Center for Radiation Research, National Institute of Standards and Technology, Gaithersburg, MD 20899.

Joel M. Kauffman, Chemistry Department, Philadelphia College of Pharmacy and Science, 43rd Sts. and Kingsessing Mall, Philadelphia, PA 19104.

Thomas Kellogg, Mississippi St. University, P. O. Drawer BB, Mississippi St.., MS 39762.

Michael Kessler, Senior Scientist, Packard Instrument Company, One State Street, Meriden, CT 06450.

John Killmar, Univ. of Tenn-Memphis, 956 Court, RM A 218, Memphis, TN 38163.

Yutaka Kobayashi, Ko-By Assoc., 60 Audubon Rd., Wellesley, MA 02181.

Michael Koets, The Upjohn Company, 301 Henrietta St., Kalamazoo, MI 49001.

Junji Konishi, Department of Radiology and Nuclear Medicine, Kyoto University School of Medicine, Khoto, Japan.

Daniel Kopp, Ordela, Inc., Oak Ridge, TN 37830.

Manfred Kopp, Ordela, Inc., Oak Ridge, TN.37830.

Heikki Kouru, Instrument Research Department, Wallac Oy, P. O. Box 10, SF-20101 Turku, Finland.

Jeanne K. Krieger, Medical Products, NEN Research Products, Boston, MA 02118.

Susumu Kuwashima, Packard - Japan K. K., Iwanoto-Cho, Chiyoda-Ku, Tokyo, Japan.

Mark Laxton, Pharmacie-Lkb Nuclear, Gaithersburg, MD.

Yun Ko Lee, Reynolds Elec. & Engin., P. O. Box 98521, Las Vegas, NV 89192.

Jie Li, Department of Isotope Research, Institute of Clinical Medical Sciences, China-Japan Friendship Hospital, Beijing 100013, China.

Zhong-hou Liu, Department of Isotope Research, Institute of Clinical Medical Sciences, China-Japan Friendship Hospital, Beijing 100013, China.

Jim Littlefield, IT Corporation, 1550 Bear Creek Road, Oak Ridge, TN. 37830.

Austin Long, Laboratory of Isotope Gerchemistry, Dept. of Geosciences, University of Arizona, Tucson, AZ 85721.

J. M. Los Arcos, Dpto de Mecanica, ETSII, Universidad Naciónal de Educación a Distancia, Ciudad Universitaria, 28040-Madrid, Spain.

S. S. Lutz, E G & G Energy Measurements, Inc., Santa Barbara Operations, Goleta, CA 93117.

A. B. Mackenzie, Scottish Universities Research and Reactor Centre, East Kilbride, Glasgow, G75 0Qu, Scotland, UK.

A. Grau Malonda, Investigacion Bagica, Edificio 2 Ciemat, Madrid, Spain.

Ed Maswood, The Nucleus/Tennelec, Oak Ridge, TN.

Rowena Maxwell, Atomic Energy Control B, Ottawa, Canada.

Dale McClendon, Beckman Instruments, Hendersonville, TN.

John McCormick, Quadrex HPS, Inc., Gainesville, FL.

Tom McCrea, Pharmacia, Atascocita, TX.

Betty McDowell, ETRAC, Inc., 10903 Melton View Lane, Knoxville, TN.

W. Jack McDowell, ETRAC, Inc. 10903 Melton View Lane, Knoxville, TN.

William McDowell, Westinghouse Savannah River Company, Savannah River Laboratory, Environmental Technology Section, Aiken, SC 29808.

John McKlveen, Radiation Measurements Facility, Arizona State University, Tempe, AZ.

Wayne McLemone, Beckman Instruments, Norcross, GA 30071.

Patricia McQueen, Desert Research Inst., P. O. Box 60220, Reno, NV 89506.

P. Naysmith, University of Glasgow Radiocarbon Dating Laboratories, Scottish Universities Research and Reactor Centre, East Kilbride G75 0QU, Scotland.

P. Mertens, Packard Instruments Co., 2200 Warrenville Rd, Downers Grove, IL 60515.

Yoshiaki Miayamoto, Aloka Co., LTD 6-22-1 Mure, Mitaka, Tokyo, Japan.

John Miglio, Los Alamos Nat'l Lab, Los Alamos, NM 87545.

Steven Miller, Beckman Instruments Inc., 8920 Rt 10B, Columbia, MD 21047.

Jeffrey Mirsky, National Diagnostics, 1013-1017 Kennedy Blvd., Manville, NJ 08835.

Dieter Molzamn, Kernchemie, Hans-meer Wein-Str, D-3550 Marburg/L, Germany.

Irene Morgan, Ordela, Inc., Oak Ridge, TN 37830.

Carolyn Morrell, University of Georgia, Athens, GA..

Wayne Moser, Bicron Corporation, 1234 Kinsman Rd., Newbury, OH 44065.

Lowell Muse, Environmental Safety Ser., Univ.ersity of Georgia, Athens, GA 30602.

Raymond Narushko, Environmental Air Cnt.

James Neel, Dept. of Biology, San Diego St. University, San Diego, CA 92182.

John Noakes, Center for Applied Isotope Studies, University of Georgia, Athens, GA.

Kenneth E. Neumann, Research and Development Department, Packard Instrument Company, 2200 Warrenville Road, Downers Grove, Illinois.

Timo Oikari, Instrument Research Departtment,Wallac Oy, P. O. Box 10, SF-20101 Turku 10, Finland.

Emmanuel Omoba, University College Hospital, P. O. Box 29757, Ibadak Oyo-St, Nigeria.

F. Ortiz, Dpto de Mecanica, ETSII, Universidad Naciónal de Educación a Distancia, Ciudad Universitaria, 28040-Madrid, Spain.

Donald Osten, Packard Instruments, Co. 2200 Warrenville Td., Downers Grove, IL 60515.

Jan Ostrup, Wallac Oy, Turku 10, Finland.

Zi-ang Pan, Department of Isotope Research, Institute of Clinical Medical Sciences, China-Japan Friendship Hospital, Beijing 100013, China.

Kevin Parnham, NE america, Monmouth Jct., NC, 08852.

Charles J. Passo, Jr., Packard Instrument Co., Meriden, Ct. 06450.

Vaughn Patania, Oak Ridge National Lab, Oak Ridge, TN 37830.

Kathryn Peacock, Wastec, Inc., 114 Tulsa Road, Oak Ridge, TN 37831.

Chin-Tzu Peng, University of California, Dept. of Pharmac, Chem., San Francisco, CA 94143.

Daniel Perlman, Brandeis University, Rosenstiel Research Center, Waltham, MA 02254-9110

Linda Plemons, Oak Ridge National Laboratory, P. O. Box 2008, Oak Ridge, TN 37831 (615) 574-4881.

Henry Polach, Radiocarbon Dating Research Unit, Research School of Pacific Studies, Australian National University, Canberra 2601.

Charles Pospisil, Costa Mesa, CA 92626.

Colin G. Potter, Nuffield Department of Clinical Medicine, John Radcliffe Hospital, Oxford, OX3, 9DU.

Howard Prichard, University of Texas School of Public Health, P. O. Box 20186, Houston, TX 77225.

Paul Quayle, N. W. America, Knoxville.

Kretner Raif, GSF, Germany.

Raymond Randolph, 1733 Whitnew, Idaho Falls, ID 83402.

Andrew Riech, Radiomatic Insts. & Chem., Tampa, FL 33611

Debbie Roberts, Oak Ridge National Laboratory, Oak Ridge, TN 37831.

Ed Robertson, Packard Inst., Downers Grove, IL

S. Robertson, Radiocarbon Dating Research Unit, Research School of Pacific Studies, Australian National University, Canberra 2601.

L. Rodriguez, Investigacion Basica, Avda. Complutense, 22, Madrid, Spain.

Norbert Roessler, Ph.D., R & D Department, Packard Instrument Company, 2200 Warrenville, Road, Downers Grove, IL 60515.

Brent Rognlie, Rockwell International, P. O. Box 464 Bldg 123, Golden, CO 80402.

Harley Ross, Oak Ridge NAtional Laboratory, Building 5505, MS 6375, Oak Ridge, TN.

Thomas Rucker, SAIC, P. O. Box 2501, Oak Ridge, TN 37830.

Kenneth Rundt, Instrument Research Department, Wallac Oy, P. O. Box 10, SF-20101 Turku, Finland.

Bill Rupert, Beckman.

Saleem Salaykeh, Savannah River Lab, Aiken , SC, 29808.

LainaSalonen, Finnish Centre for Rad., P.O. Box 268 00101 Helsinki, Finland.

Felix Sanchez, ETSII, Madrid 28040, Spain.

Franz Schönhofer, Federal Institute for Food, Kinderspitalg 15, A-1090 Vienna, Austria.

Bill Severance, Dupont, Wilmington, DE.

Lu Shau-wan, Institute of Biophysics, Academia Sinica, Beijing, China.

Peter Shaw, EG & G Idaho, P. O. Box 1625, Idah Falls, ID 83415.

Donald Sherman, Radiation Protection Ser, University of Buffalo, Buffalo, NE 14215.

Wang Shu-xian, Institute of Biophysics, Academia Sinica, Beijing, China.

Gene Sinclair, Nuclear Data Systems, Schaumburg, IL 60196.

Roger Sit, UC San Francisco, San Francisco, CA.

Paul Skierkowski, University of Oklahoma, 905 Asp, Rm 112, Norman, OK 73019.

David Sloviter, Curtiss Laboratories, 2510 State Rd., Bensalem, PA 19020.

John Smalling, Nuclear Data Systems, Johnson City, TN.

Jim D. Spaulding, Center for Applied Isotope Studies, University of Georgia, Athens, GA.

Jerry Stipp, 10005 S.W. 63rd Place, Miami, FL 33156.

John Sutton, Amershak Internationall Plc, Forest Farm Road, Whitc., Cardiff, Wales, UK.

Ara Tahmassian, Office of Environmental Health and Safety, University of California at San Francisco, San Francisco, CA 24143.

J. Thomson, Fisons Scientific Equipment PLC, Bishop Meadow Road, Loughborough, Leicesterhire, England, LE11 ORG.

Graham J. Threadgill, Beckman Instruments, Inc., Scientific Instruments Division 2500 Harbor Blvd., Fullerton, CA 92634.

Zhang Ting-kui, Institute of Biophysics, Academia Sinica, Beijing, China.

Howard Tisdale, Office of Environmental Health and Safety, University of California at San Francisco.

Kanju Torizuka, Department of Radiology and Nuclear Medicine, Kyoto University of Medicine, Khoto, Japan.

Giancarl Torri, ENEA-DISP, Rome, Italy.

Veronica Trigg, Radiocarbon Laboratory, Institute for the Study of Earth and Man, Southern Methodist University, Dallas, Texas, 75275.

Chr. Ursin, Danish Isotope Centre, Academy of Technical Sciences, 2 Skelbaekgade, DK-1717 Copenhagen V, Denmark.

Robert J. Valenta, R & D Department, Packard Instrument Company, 2200 Warrenville Road, Downers Grove, IL 60515.

Raine Vesaner, Univ.ersity of Gothenburg, Sweden.

Claus Vestergaard, Danish Isotope Centre, Academy of Technical Sciences, 2 Skelbaekgade, DK-1717 Copenhagen V, Denmark.

Jim Vondran, Packard Instruments, Downers Grove, IL 60515.

Wallace Walker, Jr., 40 Hill St., Randolph, NJ 07869.

Ivan Wallace, National Inst. of Health, Bldg. 21, Rm 134, Bethesda, MD 30305.

Hiu-ming Wang, Department of Isotope Research, Institute of Clinical Medical Sciences, China-Japan Friendship Hospital, Beijing 100013, China.

Donna Wardle, 549 Albany St., Boston, MA 02118.

Gerald T. Warner, University of Oxford UK, Oxford, UK.

Larry Webb, Tennelec//Nucleus, Oak Ridge, TN 37830.

Steve Weeks, Hamamatsu Corp., Bridgewater, NJ 08807.

Louis Weinrich, Idaho Nat'l Eng. Lab, P. O. Box 1655, Idaho Falls, ID 83415.

Lillian Weitzel, 7500 Silver Star Road, Orlando, FL 32818.

Zhang Wen-xin, Institute of Biophysics, Academia Sinica, Beijing, China.

Jan ter Wiel, Chemical Research, Packard Instrument Company, B. V., Ulgersmaweg 47, 9731 BK Groningen, The Netherlands.

Lisa Wilburn, CEP.

Barry Wilson, NE America, Monmouth Junction, NJ 08852

Heather Wilson, National Diagnostic, Inc. 1013-1017 Kennedy Blvd., Manville, NJ 08835.

Willard G. Winn, Westinghouse Savannah River Company, Savannah River Laboratory, Aiken, SC 29808.

Larry Wright, Department of Mathematics, University of Arizona, Tucson, AZ 85721.

Stephen W. Wunderly, Beckman Instruments, Inc., Scientific Instruments Division, 2500 Harbor Blvd., Fullerton, CA 92634.

Itsuo Yamamoto, Department of Radiology and Nuclear Medicine, Kyoto University School of Medicine, Kyoto, Japan.

Satoko Yamamura, Department of Isotope Research, Institute of Clinical Medical Sciences, China-Japan Friendship Hospital, Beijing 100013, China.

Shou-Li Yang, Department of Isotope Research, Institute of Clinical Medical Sciences, China-Japan Friendship Hospital, Beijing 100013, China.

Tapio Yrjonen, Pharmacia-Wallac Oy., Turku 10, Finland.

Jie-chang Yue, Department of Isotope Research, Institute of Clinical Medical Sciences, China-Japan Friendship Hospital, Beijing 100013, China.

Hu Yue, Department of Isotope Research, Institute of Clinical Medical Sciences, China-Japan Friendship Hospital, Beijing 100013, China.

Wu Xue Zhou, National Institute of Metrology.

Index